POLARIZED GAS TARGETS AND POLARIZED BEAMS

POLARIZED GAS TARGETS AND POLARIZED BEAMS

Seventh International Workshop

Urbana, IL August 1997

EDITORS
Roy J. Holt
Michael A. Miller
University of Illinois at Urbana-Champaign

American Institute of Physics

AIP CONFERENCE
PROCEEDINGS 421

Woodbury, New York

Editors:

Roy J. Holt and Michael A. Miller
University of Illinois
Loomis Laboratory of Physics
1110 W. Green Street
Urbana, IL 61801
Email: r-holt@uiuc.edu
          miller5@uiuc.edu

Authorization to photocopy items for internal or personal use, beyond the free copying permitted under the 1978 U.S. Copyright Law (see statement below), is granted by the American Institute of Physics for users registered with the Copyright Clearance Center (CCC) Transactional Reporting Service, provided that the base fee of $15.00 per copy is paid directly to CCC, 222 Rosewood Drive, Danvers, MA 01923. For those organizations that have been granted a photocopy license by CCC, a separate system of payment has been arranged. The fee code for users of the Transactional Reporting Service is: 1-56396-700-6/ 98 /$15.00.

© 1998 American Institute of Physics

Individual readers of this volume and nonprofit libraries, acting for them, are permitted to make fair use of the material in it, such as copying an article for use in teaching or research. Permission is granted to quote from this volume in scientific work with the customary acknowledgment of the source. To reprint a figure, table, or other excerpt requires the consent of one of the original authors and notification to AIP. Republication or systematic or multiple reproduction of any material in this volume is permitted only under license from AIP. Address inquiries to Office of Rights and Permissions, 500 Sunnyside Boulevard, Woodbury, NY 11797-2999; phone: 516-576-2268; fax: 516-576-2499; e-mail: rights@aip.org.

L.C. Catalog Card No. 97-78196
ISBN 1-56396-700-6
ISSN 0094-243X
DOE CONF- 970894

Printed in the United States of America

CONTENTS

Sponsors xiv
Preface xv

INVITED TALKS
- ^{3}He Targets -

New Results from Spin-Exchange Optical Pumping 3
  G.D. Cates

The HERMES $^3\vec{He}$ Target 16
  D. De Schepper for the HERMES Collaboration

Quasi-Elastic Electron Scattering from Polarized ^{3}He 26
  H.J. Bulten, R. Alarcon, Th. Bauer, D. Boersma, T. Botto, J.F.J. van den Brand, L. van Buuren, R. Ent, M. Ferro-Luzzi, D. Geurts, M. Harvey, P. Heimberg, D. Highinbotham, C.W. de Jager, B. Norum, I. Passchier, H.R. Poolman, M. van den Putte, E. Six, J. Steijger, D. Szczerba, and H. de Vries

Determination of the Neutron Electric Form Factor G_{En} in
Double Polarized Electron Scattering from ^{3}He 36
  A3 - collaboration at MAMI: H.G. Andresen, J. Annand, K. Aulenbacher, *J. Becker*, K. Beuchel, J. Blume-Werry, Th. Dombo, P. Drescher, M. Ebert, D. Eyl, H. Fischer, A. Frey, P. Grabmayr, T. Großmann, S. Hall, P. Hartmann, T. Hehl, W. Heil, C. Herberg, J. Hoffmann, D. Ireland, J. Kellie, F. Klein, M. Leduc, M. Meyerhoff, G. Miller, H. Möller, Ch. Nachtigall, A. Natter, M. Ostrick, E.W. Otten, R.O. Owens, S. Plützer, E. Reichert, D. Rohe, M. Schäfer, H. Schmieden, R. Sprengard, K.-H. Steffens, R. Surkau, Th. Walcher, R. Watson, and E. Wilms

Polarized High Pressure $^3\vec{He}$ Target at MAMI 41
  D. Rohe for the A1-Kollaboration: P. Bartsch, D. Baumann, J. Becker, J. Bermuth, L.J. de Bever, R. Böhm, S. Buttazoni, T. Caprano, N. Clawiter, A. Deninger, S. Derber, M. Ding, M. Distler, A. Ebbes, M. Ebert, I. Ewald, J. Friedrich, J.M. Friedrich, R. Geiges, T. Großmann, A. Klein, M. Hauger, W. Heil, D. Hofmann, A. Honegger, P. Jennewein, J. Jourdan, M. Kahrau, M. Korn, K.W. Krygier, G. Kubon, L. Lauer, A. Liesenfeld, H. Merkel, K. Merle, P. Merle, U. Müller, M. Mühlbauer, R. Neuhausen, E.W. Otten, Th. Petitjean, Th. Pospischil, D. Rohe, G. Rosner, H. Schmieden, I. Sick, R. Surkau, A. Wagner, Th. Walcher, G. Warren, M. Weis, S. Wolf, H. Woehrle, J. Zhao, and M. Zeier

- ³He Polarimeters & HELION '97 -

The SLAC E-154 ³He Polarimeter 46
    M.V. Romalis, P.L. Bogorad, G.D. Cates, T. E. Chupp, K.P. Coulter, E.W. Hughes, J.R. Johnson, K.S. Kumar, T.B. Smith, A.K. Thompson, and R. Welsh

³H Polarimetry in the HERMES Experiment 53
    A.P. Dvoredsky representing the HERMES Collaboration

Highlights from HELION 97 58
    M. Tanaka

- H, D Targets -

The HERMES Polarized Hydrogen Internal Gas Target 69
    J. Stewart for The HERMES Collaboration

Polarized Deuterium Internal Target at AmPS (NIKHEF) 79
    M. Ferro-Luzzi, Z.-L. Zhou, J.F.J. van den Brand, H.J. Bulten, R. Alarcon, N. van Bakel, T. Botto, M. Bouwhuis, L. van Buuren, J. Comfort, M. Doets, S. Dolfini, R. Ent, D. Geurts, P. Heimberg, D.W. Higinbotham, C.W. de Jager, J. Lang, D.J. de Lange, B. Norum, I. Passchier, H.R. Poolman, E. Six, J. Steijger, D. Szczerba, O. Unal, and H. de Vries

The Wisconsin-IUCF Polarized Gas Target 89
    F. Rathmann, W. Haeberli, B. Lorentz, P. Quin, B. Schwartz, T. Wise, H.O. Meyer, R.E. Pollock, J. Doskow, M. Dzemdizic, J.H. Hardie, B.v. Przewoski, T. Rinckel, F. Sperisen, M. Wolanski, P.V. Pancella, P.B. Ugorowski, W. Daehnick, R. Flammang, and D. Tedeschi

The EDDA Experiment at COSY 99
    H. Rohdjess for the EDDA Collaboration

Cryogenic Atomic Beam Source at VEPP-3 109
    L.G. Isaeva, B.A. Lazarenko, S.I. Mishnev, D.M. Nikolenko, A.N. Osipov, S.G. Popov, I.A. Rachek, Yu.V. Shestakov, A.A. Sidorov, V.N. Stibunov, *D.K. Toporkov*, D.K. Vesnovsky, and S.A. Zevakov

Ultra-Cold Methods for Polarized Atomic Hydrogen 119
    V.G. Luppov, J.D. Arnold, B.B. Blinov, M.A. Bychkov, S.E. Gladycheva, A.D. Krisch, A.M.T. Lin, R.S. Raymond, V.V. Fimushkin, V.V. Mochalov, and P.A. Semenov

- Laser-Driven H/D Targets -

The Argonne Laser-Driven D Target: Recent Developments and Progress 129
    J.A. Fedchak, K. Bailey, W.J. Cummings, H. Gao, C.E. Jones, R.S. Kowalczyk, J. Magnes, and M. Pipes

Nuclear Spin Polarized H and D by Means of Spin-Exchange Optical Pumping 139
   J. Stenger, C. Grosshauser, W. Kilian, W. Nagengast, B. Ranzenberger, K. Rith, and F. Schmidt

First Use of a Laser-Driven Polarized H/D Target at the IUCF Cooler 148
   M.A. Miller, K. Bailey, J. Brack, R.V. Cadman, W.J. Cummings, J. Fedchak, B. Fox, H. Gao, C. Grosshauser, R.J. Holt, C. Jones, E. Kinney, R. Kowalczyk, Z.-T. Lu, W. Nagengast, B. Owen, K. Rith, F. Schmidt, E. Schulte, J. Sowinski, F. Sperisen, J. Stenger, E. Thorsland, and S. Williamson

- Target Polarimetry and Beam Interaction -

The HERMES Polarimeter 156
   B. Braun for the HERMES Collaboration

Beam-Induced Resonant Depolarization Effects in a Polarized Atomic Hydrogen Gas Target at HERA 162
   H. Kolster for the HERMES Collaboration

- Beam Polarimetry -

Novel Polarization Measurements at IUCF: Recent Work and Future Plans of the PINTEX Collaboration 172
   P.V. Pancella for the PINTEX collaboration

Beam Polarimetry at HERA 181
   W. Lorenzon on behalf of the HERMES collaboration and the HERA polarimeter group

Radio-Frequency Polarimetry 191
   Ya.S. Derbenev

- NMR Tomography with Polarized Noble Gases -

Magnetic Resonance Imaging with Laser Polarized ^{129}Xe 200
   S.D. Swanson, M.S. Rosen, B.W. Agranoff, K.P. Coulter, R.C. Welsh, and *T.E. Chupp*

Provision of Hyperpolarized $^3\vec{\text{He}}$ and its Application in MRI 208
   P. Bachert, J. Becker, J. Bermuth, M. Bock, A. Deninger, M. Ebert, *T. Großmann*, W. Heil, D. Hofmann, H.U. Kauczor, M.W. Knopp, K.F. Kreitner, L. Lauer, M. Leduc, H. Nilgens, E.W. Otten, L.R. Schad, R. Surkau, T. Roberts, and M. Thelen

Polarized Noble Gas MRI 213
   J.R. Brookeman, J.P. Mugler III, P. Bogorad, T.M. Daniel, E.E. de Lange, B. Driehuys, J. Knight-Scott, T. Maier, J.D. Truwit, G. Cates, and W. Happer

- Polarized Electron Sources -

Polarized Electrons at Jefferson Laboratory 218
  C.K. Sinclair

Status of the MAMI-Source of Polarized Electrons 229
  B2-Collaboration at MAMI: K. Aulenbacher, H. Euteneuer, D.v. Harrach, P. Hartmann, J. Hoffmann, P. Jennewein, K.-H. Kaiser, H.J. Kreidel, M. Leberig, E. Reichert, M. Schemies, J. Schuler, M. Steigerwald, and M. Zalto

Polarized Electrons at MIT-Bates 240
  M. Farkhondeh, D. Barkhuff, G. Dodson, E. Tsentalovich, B. Yang, T. Zwart, E. Ihloff, and C. Tschalaer

The SLAC Polarized Electron Source 250
  J.E. Clendenin, R. Alley, J. Frisch, T. Kotseroglou, G. Mulhollan, D. Schultz, H. Tang, J. Turner, and A.D. Yeremian

The Polarized Electron Source at NIKHEF 260
  M.J.J. van den Putte, C.W. de Jager, S.H. Konstantinov, V.Ya. Korchagin, F.B. Kroes, E.P. van Leeuwen, B.L. Militsyn, N.H. Papadakis, S.G. Popov, G.V. Serdobintsev, Yu.M. Shatunov, S.V. Shevelev, T.G.B.W. Sluijk, A.S. Terekhov, and Yu.F. Tokarev

A Diode Laser System for Synchronous Photoinjection 270
  M. Poelker and J. Hansknecht

Strained Semiconductor Structures for Polarized Electrons 276
  C.W. Tu

Spin Polarization of Photoelectrons from Ordered Semiconductor Alloys 284
  S.-H. Wei

Emission of Polarized ps-Electron Bunches from III-V Semiconductor Cathodes 296
  P. Hartmann, J. Bermuth, J. Hoffmann, S. Köbis, H.G. Andresen, K. Aulenbacher, P. Drescher, H. Euteneuer, H. Fischer, K. Grimm, Th. Hammel, D. v. Harrach, H. Hofmann, K.-H. Kaiser, E.-M. Kabuß, H.J. Kreidel, A. Lopes-Ginja, F.E. Maas, Ch. Nachtigall, S. Plützer, E. Reichert, M. Schemies, E. Schilling, J. Schuler, K.-H. Steffens, M. Steigerwald, and H. Trautner

Highly Polarized Electrons from Superlattice Photocathodes 300
  T. Nakanishi, S. Okumi, K. Togawa, C. Takahashi, C. Suzuki, F. Furuta, T. Ida, K. Wada, T. Omori, Y. Kurihara, M. Tawada, M. Yoshioka, H. Horinaka, T. Matsuyama, T. Baba, and M. Mizuta

Laser Sources for MIT-Bates and IASA 311
  D. Fraser, A. Hatziefremidis, M. Ciarrocca, V. Markou, T. Papakyriakopoulos, T. Houbavlis, and H. Avramopoulos

- Electron Polarimeters -

A Compton Backscattering Polarimeter for Measuring Longitudinal Electron Polarization 316
  I. Passchier, D.W. Higinbotham, N. Vodinas, N. Papadakis, C.W. de Jager, R. Alarcon, T. Bauer, J.F.J. van den Brand, D. Boersma, T. Botto, M. Bouwhuis, H.J. Bulten, L. van Buuren, R. Ent, D. Geurts, M. Ferro-Luzzi, M. Harvey, P. Heimberg, B. Norum, H.R. Poolman, M. van den Putte, E. Six, J.J.M. Steijger, D. Szczerba, and H. de Vries

The TJNAF Hall A Möller Polarimeter 321
  D.S. Dale, A. Gasparian, B. Doyle, T. Gorringe, W. Korsch, V. Zeps, A. Glamazdin, V. Gorbenko, R. Pomatsalyuk, J.P. Chen, S. Nanda, and A. Saha

Mott-Scattering of Multi-MeV Electrons from Heavy Nuclei 326
  D. Conti, S. Navert, K. Bodek, W. Haeberli, S. Kistryn, J. Lang, O. Naviliat, E. Reichert, *J. Sromicki*, M. Steigerwald, E. Stephan, and J. Zejma

- Polarized Ion Sources -

Review of High Intensity Polarized H and D Ion Sources: Recent Progress and Future Projections 336
  T.B. Clegg

Status of the Polarized D⁻ Ion Source at KEK 347
  A. Takagi, M. Kinsho, K. Ikegami, Z. Igarashi, and Y. Mori

The RCNP Ion Source 352
  K. Hatanaka, K. Takahisa, and H. Tamura

Polarized Ions from a Storage Cell 362
  A.S. Belov, S.K. Esin, L.P. Netchaeva, V.S. Klenov, A.V. Turbabin, and G.A. Vasil'ev

OPPIS Development for Precision Experiments and High Energy Colliders 372
  A.N. Zelenski, V.I. Davydenko, G. Dutto, A.A. Hamian, V. Klenov, C.D.P. Levy, I.I. Morozov, P.W. Schmor, W.T.H. van Oers, and G.W. Wight

- Developments in Atomic Beams -

A New Method to Produce Cold Atomic Hydrogen and Deuterium Beams 381
  V.L. Varentsov, A.A. Ignatiev, E. Steffens, and N. Koch

Polarization of Deuterium Molecules 389
  J.F.J. van den Brand, H.J. Bulten, M. Ferro-Luzzi, Z.-L. Zhou, R. Alarcon, T. Botto, M. Bouwhuis, R. Ent, P. Heimberg, D. Higinbotham, C.W. de Jager, J. Lang, D.J. de Lange, I. Passchier, H.R. Poolman, J.J.M. Steijger, O. Unal, and H. de Vries

Has the Atomic Beam Source Reached a Hard Intensity Limit? 399
  E. Steffens

CONTRIBUTED TALKS
- Atomic Beam and Ion Sources -

The Polarization Technology at the High Energy Laboratory of The Joint Institute for Nuclear Research 411
    V.P. Ershov, V.V. Fimushkin, G.I. Gai, M.V. Kulikov, L.V. Kutuzova, M. Yu. Liburg, *Yu. K. Pilipenko*, A.I. Valevich, and A.S. Belov

A Polarized Atomic-Beam Target for COSY-Jülich 419
    P.D. Eversheim, M. Altmeier, O. Felden, M. Glende, M. Walker, A. Hiemer, and R. Gebel

Polarized Beam for the IUCF Cooler Injector Synchrotron 422
    V.P. Derenchuk and A.S. Belov

A Microwave based Dissociator for the HERMES-ABS 427
    N. Koch

- Optical Pumping -

^{3}He Neutron Spin Filter at ILL 430
    H. Humblot, W. Heil, F. Tasset, and D. Hoffmann

A Novel Approach for a Spin-Exchange High Density ^{3}He Target 431
    R.C. Welsh, J.N. Zerger, *K.P. Coulter*, T.B. Smith, and T.E. Chupp

Development of Optical-Pumping Polarized Deuteron Target 434
    T. Tamae, T. Yokokawa, I. Nishikawa, K. Abe, O. Konno, H. Miyase, I. Nakagawa, M. Sugawara, E. Tanaka, H. Tsubota, N. Yamaguchi, and H. Yamazaki

Performance of the Laser Driven Polarized Hydrogen Source at IUCF .. 437
    R.V. Cadman, K. Bailey, J. Brack, W.J. Cummings, J. Fedchak, B. Fox, H. Gao, C. Grosshauser, R.J. Holt, C.E. Jones, E. Kinney, W. Kirwin, R.S. Kowalczyk, Z.-T. Lu, M.A. Miller, W. Nagengast, B.R. Owen, K. Rith, F. Schmidt, E.C. Schulte, J. Sowinski, F. Sperisen, J. Stenger, E. Thorsland, and D.K. Toporkov

- Polarized Electron Sources -

Status of the Helium Afterglow Injector for MAMI 440
    J. Arianer, S. Cohen, *S. Essabaa*, R. Frascaria, O. Zerhouni, F. Szeremeta, R. Wurzinger, and K. Aulenbacher

Photocathode Research at SLAC 443
    G. Mulhollan, J. Clendenin, E. Garwin, R. Kirby, T. Maruyama, R. Prepost, and H. Tang

5 MeV Mott Polarimeter Development at Jefferson Lab 446
    J.S. Price, B.M. Poelker, C.K. Sinclair, K.A. Assamagan, L.S. Cardman, J. Grames, J. Hansknecht, D.J. Mack, and P. Piot

Polarization Transport at TJNAF: Simulations and Measurements 451
    J.M. Grames, D.H. Beck, L.S. Cardman, J.H. Mitchell, B.M. Poelker, J.S. Price, C.K. Sinclair, and B. Zihlmann

POSTER SESSION

Fiber Optic Applications for Laser Polarized Targets 457
    W.J. Cummings and R.S. Kowalczyk

Polarisation and Compression of ^{3}He for Magnetic Resonance Imaging Purposes 459
    D.G. Geurts, J.F.J. van den Brand, H.J. Bulten, M. Ferro-Luzzi, K. Nicolay, and H.R. Poolman

The UNH Polarized ^{3}He Program 461
    F.W. Hersman, R.H. Carrier, and V.R. Pomeroy

Polarized ^{3}He Spin Filters for Polarized Neutron Experiments at LANSCE 463
    V.R. Pomeroy, F.W. Hersman, and M.B. Leuschner

Optimization of Spin-Exchange Cell Geometry for a Laser Driven Target 465
    E.C. Schulte, R.J. Holt, B.R. Owen, and E.L. Thorsland

Mechanical Filter for Alkali Atoms 467
    D. Toporkov and B. Wojtsekhowski

Beam Saturation in the Munich Atomic-Beam-Source 469
    R. Hertenberger, Y. Eisermann, A. Hofmann, A. Metz, P. Schiemenz, S. Trieb, and G. Graw

Optimized Atomic Beam Sources 471
    S. Lemaitre, R. Brüggemann, V. Nelyubin, and H. Paetz gen. Schieck

Design Studies for the ABS of ANKE in COSY Jülich by 3-Dimensional Magnetic Field Calculations 473
    V. Nelyubin and H. Seyfarth

Tests of a Prototype Large-Bore, Low-Power 2×4 RF Transition Unit 475
    R.S. Raymond

The Polarized Storage Cell Gas Target for the ANKE Spectrometer at COSY - Status and Design Considerations 477
    H. Seyfarth for the ANKE-ABS working group

Cryogenic Ne Heat Pipe Nozzle Cooling System for an Atomic Beam Source 479
    A. Vassiliev, V. Koptev, S. Kotov, and H. Seyfarth

Ion-Extraction Polarimetry for a Polarized ^{1}H/^{2}H Internal Target 481
    Z.-L. Zhou, M. Ferro-Luzzi, J.F.J. van den Brand, H.J. Bulten, and J. Lang

The Photocathode Gun of the Polarized Electron Source at NIKHEF 483
    C.W. de Jager, V.Ya. Korchagin, B.L. Militsyn, V.N. Osipov, N.H. Papadakis, S.G. Popov, M.J.J. van den Putte, Yu.M. Shatunov, and Yu.F. Tokarev

Photocurrent Saturation at GaAs (Cs,O) 485
  A.S. Jaroshevich, M.A. Kirillov, D.A. Orlov, A.G. Paulish,
  H.E. Scheibler, and A.S. Terekhov
Transverse Energy Measurements on NEA-GaAs Photocathodes 487
  S. Pastuszka, D. Kratzmann, D. Schwalm, A.S. Terekhov, and
  A. Wolf
**Polarization Anomalies in Polarized Electron Emission and
Luminescence from Highly p-doped Semiconductor Structures** 489
  A.V. Subashiev and E.P. German
**Surface Potential Fluctuations in Negative Electron Affinity
State Formation** 491
  B.D. Oskotskij, A.V. Subashiev, and L.G. Gerchikov
**Elucidation of Activation Layer Model by Means of Measurements
of Photoelectron Energy Distribution Curves** 493
  S. Pastuszka, D. Kratzmann, D. Schwalm, A. Wolf, D.A. Orlov,
  A.G. Paulish, H.E. Scheibler, A.S. Terekhov, and O.E. Tereshchenko
**Surface Charge Limit Observed in an NEA Photocathode
of a 100 keV Polarized Electron Gun** 495
  K. Togawa, T. Nakanishi, S. Okumi, C. Takahashi, C. Suzuki,
  F. Furuta, T. Ida, K. Wada, Y. Kurihara, H. Matsumoto, T. Omori,
  Y. Takeuchi, M. Yoshioka, T. Baba, and M. Mizuta
Polarized Electrons in ELSA (Preliminary Results) 497
  S. Nakamura, W. von Drachenfels, D. Durek, F. Frommberger,
  M. Hoffmann, D. Husmann, B. Kiel, F.J. Klein, D. Menze, T. Michel,
  T. Nakanishi, J. Naumann, T. Reichelt, C. Steier, T. Toyama,
  S. Voigt, and M. Westermann
**APOLLON at DESY: Spin-Dependent Photoproduction of
Charm** 499
  C. A. Miller, on behalf of the APOLLON Working Group
**A P-even Test of Time-Reversal Invariance in $\vec{p}-\vec{d}$ Scattering
at COSY-Jülich** 501
  P.D. Eversheim, F. Hinterberger, J. Bisplinghoff, R. Jahn, J. Ernst,
  H. Paetz gen. Schieck, W. Kretschmer, and H.E. Conzett
**The Polarized Ion Source at the Cooler Synchrotron COSY
in Jülich** 503
  R. Gebel, P.D. Eversheim, M. Altmeier, O. Felden, and M. Glende
**Performance of the Polarized Ion Source POLIS used at the
AGOR Accelerator Facility** 507
  H.R. Kremers and A.G. Drentje

Polarized Negative Ion Source at the Kyoto University Tandem Accelerator 509
    M. Nakamura, S. Kuwamoto, S. Takahashi, M. Hirose, K. Imai, T. Murakami, M. Yosoi, M. Yoshimura, Y. Mori, A. Takagi, K. Ikegami, and M. Kinsho

Polarized Beams at the 10 GeV Machine of JINR (Dubna) 511
    Yu. K. Pilipenko, P.A. Rukoyatkin, and L.S. Zolin

APPENDICES

A. Program of The Seventh International Workshop on Polarized Gas Targets and Polarized Beams 517
B. List of Participants 525
C. Author Index 543

SPONSORS

U. S. National Science Foundation

International Committee for High Energy Spin Physics Symposia

Thomas Jefferson National Accelerator Facility
(Jefferson Laboratory)

Department of Physics at the University of Illinois at Urbana-Champaign

PREFACE

The Seventh International Workshop on Polarized Gas Targets and Polarized Beams was held at the University of Illinois in Urbana-Champaign, Illinois from 18 August to 22 August, 1997. The Workshop was attended by more than 100 scientists. There were 47 invited talks, 12 contributed talks and 27 posters. The Workshop program included polarized hydrogen and deuterium targets based on the atomic beam method as well as spin-exchange optical pumping, optically pumped ^{3}He targets, target polarimetry, polarized ion beam sources based on the atomic beam method and optical pumping, polarized electron sources and beams, beam polarimetry, and NMR tomography with polarized noble gases. The tremendous advances made in these fields during the past two years is nothing short of breathtaking as this Proceedings will disclose.

The Workshop participants enjoyed a reception, a banquet, and an excursion to the Krannert Art Museum. We thank Maarten van de Guchte, the Director, and Dr. Linda Duke, the Director of Education, for conducting private tours of the museum.

We are especially grateful to the Workshop Secretary, Penny Sigler, who had an indispensible role in dealing with every aspect of the organization of this event. In addition, thanks are due to Gaylon Reeves for his assistance during the Workshop. Also, we are grateful to Marion Evans for his assistance with the audio/visual equipment necessary for the presentations.

The organizing committee of J. Cameron (IUCF), T. Clegg (Duke), C. W. deJager (NIKHEF), W. Haeberli (Wisconsin), R. Holt, Chair (Illinois), S. Kowalski (MIT), A. Krisch (Michigan), M. Leduc (Paris), P. Levy (TRIUMF), R. McKeown (Caltech), Y. Mori (KEK), E. Reichert (Mainz), C. Sinclair (Jefferson), E. Steffens (Erlangen), and D. Toporkov (Novosibirsk) developed the overall program, including suggestions for topics and speakers.

Finally, thanks are due to the speakers, session chairs and the participants, who made this Workshop a tremendous success.

The Workshop was supported by grants from the NSF (PHY-97-13328), the International Committee on High Energy Spin Physics, Thomas Jefferson National Accelerator Laboratory and the Department of Physics at the University of Illinois.

Roy J. Holt
Michael A. Miller

INVITED TALKS

New Results from Spin-Exchange Optical Pumping

Gordon D. Cates

Department of Physics
Princeton University
Princeton, NJ 08544

Abstract. Recent results involving spin-exchange optical pumping are reviewed, particularly as they pertain to polarized ^{3}He targets. Included are recent measurements of relevant spin-exchange parameters, operational experience running the polarized ^{3}He target that was used for SLAC E154, and a discussion of the polarized ^{3}He target currently being constructed for TJNAF.

INTRODUCTION

Spin-exchange optical pumping has proven to be an effective technique for producing high nuclear polarizations in dense ^{3}He targets. Such targets can also withstand high beam currents, so despite the fact that they are much less dense than polarized solid targets, the luminosities that can be achieved are excellent. The value of spin-exchange polarized ^{3}He targets has been particularly apparent in two recent polarized deep inelastic scattering studies at SLAC: E142 and E154. The recent E154 results stand as the most precise determination of the neutron spin structure function g_1^n to date [1]. At the time they were published, the same was true for the E142 results [2,3]. The impact that the SLAC work has had on the field is illustrated in Figure 1, in which the E154 results are shown together with the previous world results on the neutron.

Based solely on Figure 1, it might be tempting to assume that spin-exchange polarized ^{3}He targets have already reached a state of maturity such that further large improvements are unlikely. Fortunately or unfortunately, that is not the case. A great deal remains to be learned before these targets reach their full potential. Important issues are still being resolved, both in the basic underlying physics upon which these targets are based, as well as in matters that might be considered engineering. In fact, the operation of the target for E154 was far more of an adventure than would have been desirable, as is illustrated by Figure 2, in which the target polarization is shown as a function of time for E154. Over the course of the experiment, nine separate targets were used, six of which blew up in the beam. Some targets were removed not because they exploded, but rather because their performance was so bad. The factors that contributed to the limits in the E154 target performance are worthy of some examination, and will be dealt with below. It is important to note, however, that the E154 target, despite the unpleasant surprises, yielded performance that significantly surpassed the target performance of E142, and set new standards in the field. As will also be discussed, tests currently being conducted as part of the development of a polarized ^{3}He target for TJNAF show promise that the trend of improvement will continue.

SPIN EXCHANGE

A thorough understanding of the basic physics of spin exchange and optical pumping is critical to the development of optimized polarized ^{3}He targets. While a great deal is already known, there are still many issues of great importance. I will review here some recent results by Baranga *et al.* that help to resolve some outstanding questions.

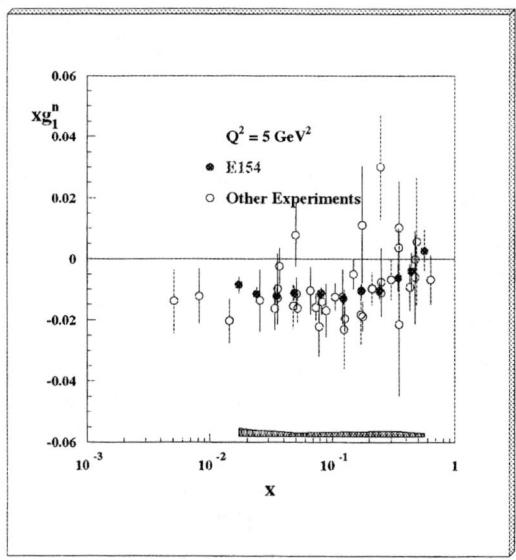

FIGURE 1. Shown is the product of the spin structure function of the neutron, $g_1^n(x)$ and the Björken scaling parameter x as a function of x. The E154 data are shown with closed points, and all other world data (including the earlier E142 results) are shown with open points. The error bars are statistical only.

Spin-exchange optical pumping refers to a two step process in which, (1) alkali-metal atoms are polarized by optical pumping, and (2) the electronic polarization of the alkali-metal atoms is transferred to the nuclei of the ^{3}He atoms by spin-exchange collisions. In what follows, I will assume that the alkali-metal is rubidium (Rb) unless otherwise specified. The optical pumping is accomplished by driving transitions from the Rb $5^2S_{1/2}$ ground state to the $5^2P_{1/2}$ first excited state using circularly polarized light from lasers. The wavelength of this transition is 795 nm. Within a timescale of milliseconds, one of the two substates of the ground state is selectively depopulated, resulting in very high atomic polarization. Spin exchange takes place when the polarized Rb atoms undergo binary collisions with the ^{3}He atoms. During the collisions, the Rb valence electron penetrates to the core of the ^{3}He atom, and through a hyperfine interaction, transfers angular momentum to the spin-1/2 ^{3}He nucleus. The ^{3}He electrons, which are paired in a $^1S_0$ ground state, do not participate in the collision from a "spin" point of view.

At saturation, the Rb polarization is given by $P_{\text{Rb}} = \gamma_{\text{op}}/(\gamma_{\text{op}} + \gamma)$, where γ_{op} is the optical pumping rate, and γ accounts for all the relaxation mechanisms affecting the Rb spin, including spin exchange. As long as $\gamma_{\text{op}} \gg \gamma$, the Rb polarization will be quite high. With modern lasers, this is quite easy to arrange for a small sample of Rb. The relaxation processes included in γ however, represent loss mechanisms for the angular momentum being infused by the laser. Thus, the size of γ influences the the laser power required for a sample of a given size. Specifically, the minimum amount of laser power required to maintain a Rb polarization P_{Rb} is given by

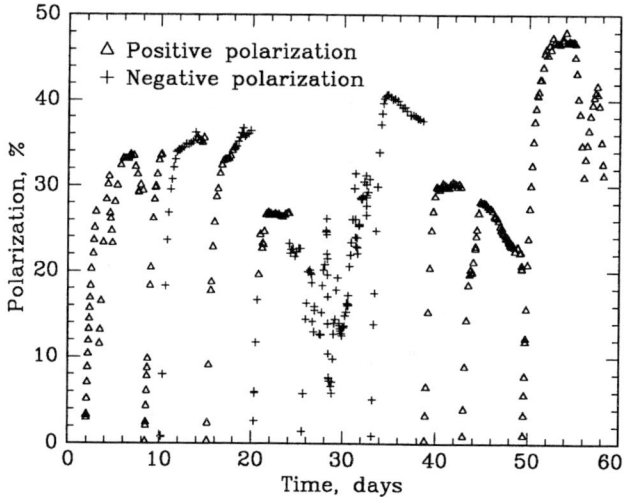

FIGURE 2. Shown is the E154 target polarization as a function of time during the run.

$$\text{Absorbed laser power} = \gamma[\text{Rb}]VP_{\text{Rb}} \qquad (1)$$

where V is the volume of Rb polarized and [Rb] is the Rb number density. We can write γ as

$$\gamma = \kappa'_{\text{sd}}[\text{Rb}] + \kappa''_{\text{sd}}[\text{N}_2] + (\kappa_{\text{sd}} + \kappa_{\text{se}})[\text{He}], \qquad (2)$$

where κ_{sd}, κ'_{sd}, and κ''_{sd} represent rate coefficients characterizing spin destruction due to He, Rb, and N_2 respectively. Spin exchange with ^{3}He is characterized by κ_{se}. Knize measured the Rb-Rb relaxation rate $\kappa'_{\text{sd}} = (7.8 \pm 0.8) \times 10^{-13}\,\text{cm}^3/\text{s}$ [4]. This rate was also measured by Wagshul and Chupp to be $\kappa'_{\text{sd}} = (8.11 \pm 0.33) \times 10^{-13}\,\text{cm}^3/\text{s}$ [5]. This rate is important because it imposes practical limits on the Rb number densities that can be employed. Wagshul and Chupp measured the Rb-N_2 $\kappa''_{\text{sd}} = (9.38 \pm 0.22) \times 10^{-18}\,\text{cm}^3/\text{s}$, and put limits on the Rb-^{3}He relaxation rate $(\kappa_{\text{sd}} + \kappa_{\text{se}}) \leq (2.29 \pm 0.23) \times 10^{-18}\,\text{cm}^3/\text{s}$ [5].

Most of the coefficients listed in eq. (2) have been remeasured in a recent set of experiments at Princeton [6]. The experiments were of two distinct types: one aimed at making a direct measurement of the efficiency of spin exchange, and the other aimed at measuring Rb spin relaxation rates. Taken together, these experiments allow the extraction of κ_{sd}, κ'_{sd}, and κ_{se}.

The spin-exchange photon efficiency η is defined such that $1/\eta$ is the number of photons needed to provide $\hbar/2$ units of spin to an initially unpolarized ^{3}He nucleus, thus leaving it fully polarized. Measurements of η were performed using small cells containing gas mixtures much like those found in target cells. First, the cells were polarized until the ^{3}He polarization was 20–50%. Next, the laser was turned off, and measurements were made of both the ^{3}He polarization as well as the alkali-metal polarization. Even in the absence of the laser light, the alkali-vapor remains slightly polarized because of spin exchange with the ^{3}He. One can show that under these conditions:

$$P_{\text{Rb}} = \frac{\kappa_{\text{se}}[\text{He}]}{\gamma} = \eta P_{\text{He}}. \qquad (3)$$

Measurements were performed for both the Rb-^{3}He and the K-^{3}He systems, and the results are shown in Figure 3a. At the temperatures at which ^{3}He targets are typically run, the efficiency for the Rb-^{3}He system is seen to be about 2–3%. Remarkably, for the K-^{3}He system the efficiency is around 20%. Unfortunately, high power diode laser arrays are not reliable at the wavelength needed to pump K (780 nm). This result, however, should be kept in mind as the laser technology continues to develop.

In a separate set of experiments at Princeton, γ was measured over a wide range of temperatures and He number densities. These measurements yielded values for κ'_{sd}, and the sum $(\kappa_{\text{sd}} + \kappa_{\text{se}})$. When combined with the measurements of η described above, one can also disentangle κ_{sd} from κ_{se}. The results from these measurements are shown in Figure 3b. One striking feature of the data is the strong temperature dependence of κ_{sd}, the Rb-^{3}He spin destruction rate, an effect that was predicted theoretically by Walker *et al.* [7]. Baranga *et al.* at Princeton find that their results on κ_{sd} can be well represented by

$$\kappa_{sd} + \kappa_{se} = 1.0 \times 10^{-29} T^{4.259} \text{cm}^3\text{s}^{-1} \quad (4)$$

For the Rb-Rb rate, they find that

$$\kappa'_{sd} = (4.2 \pm 0.4) \times 10^{-13} \text{cm}^3\text{s}^{-1} \quad (5)$$

This value for κ'_{sd} is inconsistent with the earlier values measured by Knize and

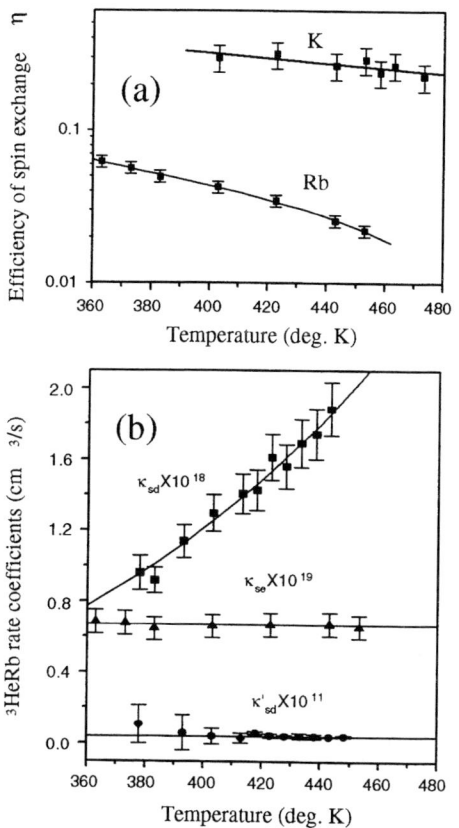

FIGURE 3. Shown are data from reference [6] showing (a) the efficiency of spin-exchange for the Rb-^{3}He and K-^{3}He systems as a function of temperature and (b) some of the key rates listed in equation (2), also as a function of temperature.

Wagshul and Chupp. It seems likely, however, that the strong temperature dependence in eq. (4) caused some confusion in the earlier measurements.

The numbers given above indicate that rather little laser power should be needed to operate the SLAC target. Assuming a pumping volume of 100 cm^3, a pumping cell temperature of 175°C, and a He density of 9 amagats, about 5.5 Watts of light is needed according to eq. (1). This is much less light than was used, and it appeared that the target benefited from maintaining maximal laser power. One resolution to this possible discrepancy comes from Wagshul and Chupp, who point out that Rb can also be depolarized by collisions with the glass cell walls. This effect results in a depolarized layer of Rb near the wall, and an additional sink on laser power [5]. Any careful accounting of laser power losses must take this into account as well as bulk effects.

The time evolution of the ^{3}He polarization, assuming the ^{3}He polarization $P_{He} = 0$ at $t = 0$, is given by

$$P_{He}(t) = P_{Rb} \left(\frac{\gamma_{se}}{\gamma_{se} + \Gamma_R} \right) \left(1 - e^{-(\gamma_{se} + \Gamma_R)t} \right) \qquad (6)$$

where $\gamma_{se} = \kappa_{se}[\text{He}]$ is the Rb $-$ ^{3}He spin-exchange rate per ^{3}He atom, Γ_R is the relaxation rate of the ^{3}He nuclear polarization through all channels other than spin exchange with Rb, and P_{Rb} is the average polarization of a Rb atom. From eq. (6), the polarization of the ^{3}He will saturate at a value determined by γ_{SE}, Γ_R, and P_{Rb}.

At the time of this writing, the literature contains two rather disparate values for the spin-exchange rate coefficient κ_{se}. Chupp et al. find that $\kappa_{se} = 1.2 \times 10^{-19}$ cm^{3}s^{-1} [8], whereas Larson et al. find that $\kappa_{se} = 6.1 \times 10^{-20}$ cm^{3}s^{-1} [9], about a factor of two smaller. As can be seen in Figure 3, Baranga et al. are consistent with Larson et al., finding

$$\kappa_{se} = (6.7 \pm 0.6) \times 10^{-20} \text{cm}^3\text{s}^{-1}. \qquad (7)$$

If we assume that the correct value of κ_{se} is that given by Larson et al. or equivalently Baranga et al., the polarization predicted by eq. (6) would be lower for any given value of [Rb]. It is thus easier to understand why the polarizations shown in, for instance, Figure 2 are not higher.

Lastly I discuss the non-spin-exchange ^{3}He relaxation rate Γ_R, to which several processes contribute. One contribution is relaxation that occurs during ^{3}He–^{3}He collisions due to the dipole interaction between the two ^{3}He nuclei [10]. Dipole induced relaxation provides a lower bound to Γ_R, and has been calculated to be $\Gamma_{dipolar} = (744 \text{ hours})^{-1}[^3\text{He}]$ at 23°C where [^{3}He] is the number density of ^{3}He in amagats (an amagat is a unit of density corresponding to 1 atm at 0°C). Another important contribution to Γ_R is relaxation that occurs during wall collisions, a relaxation rate we will designate Γ_{wall}. Both $\Gamma_{dipolar}$ and Γ_{wall} are intrinsic to a given target cell, making it useful to define the quantity $\Gamma_{cell} = \Gamma_{dipolar} + \Gamma_{wall}$ that accounts for all relaxation mechanisms that are associated with a specific cell. For properly made cells, it is generally possible to produce cells in which Γ_{cell}^{-1} is in the range of 30–70 hours. When $\Gamma_{cell}^{-1} \approx 70$ hours, relaxation is dominated by ^{3}He-^{3}He collisions.

In addition to Γ_{cell} there are interactions not inherent to the target cell which further increase the nuclear relaxation rate. Magnetic field inhomogeneities, for instance, can cause nuclear spin relaxation [11]. In high pressure cells, however, it is quite easy to keep this effect small. Another more troublesome effect is a relaxation rate Γ_{beam} due to the ionization caused by the electron beam, a phenomenon that is well understood both theoretically [12] and experimentally [13]. At SLAC, where beam currents were limited to a few microamps, Γ_{beam}^{-1} was 200 hours or more. At TJNAF, however, where approved experiments are scheduled to run with as much as 15 μA, Γ_{beam}^{-1} could be as short as 30–40 hours.

THE E154 TARGET

The E154 target is shown schematically in Fig. 4. The ^{3}He was contained in glass target cells, constructed of "resized" aluminosilicate glass [14]. Typically, the cells were filled to a density of 8.8–9.0 amagats. Each cell was also filled with a few milligrams of Rb metal, and about 60 Torr of N_2. The target cells were based on a double chamber design, [15] comprising an upper "pumping chamber" in which the optical pumping and spin exchange took place, and a lower "target chamber" through which the electron beam passed. The target chamber was about 30 cm in length, resulting in a ^{3}He target thickness of about 8×10^{21} atoms/cm^2. The pumping chamber was enclosed by flowing-hot-air oven, the temperature of which determined the Rb number density. The target cell and the oven are both illustrated in Figure 4. Polarimetry was accomplished using two methods. The NMR technique of adiabatic fast passage (which was used for on-line monitoring), and a technique in which the Rb electron paramagnetic resonance (EPR) frequency was monitored. The Rb EPR frequency becomes shifted as the ^{3}He becomes polarized because of an effective magnetic field created by the polarized nuclei. The polarimetry for E154 is discussed extensively in a separate paper by M.V. Romalis in this proceedings, and will not be discussed further here. The laser system included four argon-ion laser/Ti-Sapphire laser pairs, which collectively produced about 12–16 Watts. We also used 3 fiber coupled high power diode laser arrays. Each diode system included a 20 Watt bar, which after being coupled into the fibers, provided about 15 Watts of output power. We note that the linewidth of the diode systems was roughly 2 nm, and that only about 50% of the spectral width was absorbed by the Rb vapor. Additional losses in the laser transport line also limited the laser power actually delivered to the cell. Still, significantly more laser power was available than was the case in E142.

The largest improvement in the performance of the E154 target compared to the E142 target came from the dilution factor, the fraction of scatters that came from ^{3}He as opposed to glass. The dilution factor was about 0.55 for E154, in contrast to 0.33 for E142. This improvement alone made running nearly three times more efficient. At each end of the target chamber was a "window" for the electron beam, comprising a region of extremely thin glass. The dilution factor was brought closer to unity by making the glass windows thinner while making the ^{3}He pressure higher. This required a large increase in the strength of the windows which was accomplished by using an inverted window design, in which the window had a concave shape (see Figure 4). Since glass is much stronger under compression than it is under tension, the inverted window design enabled us to restrict the glass thickness to about 50μ, more than a factor of two thinner than in E142.

As mentioned earlier, a serious problem with the E154 target was the fact that cells exploded in the beam. To better understand this effect, we show a plot in Figure 5 of the total charge accumulated prior to exploding as a function of the window thickness [16]. A general trend seems to be evident. We note, however, that Minihaha, a cell used during E142 that had an average window thickness of $117\,\mu m$, accumulated 20×10^{18} electrons without exploding, which is perhaps four times more than what one might expect from the trend in Figure 5. The peak currents during the two experiments were about 20–60 mA for E154 and 8–32 mA for E142. As it turned out, most of the running during both experiments was done with a peak current around 30 mA, so it would be inappropriate to blame the E154 cell failures on higher

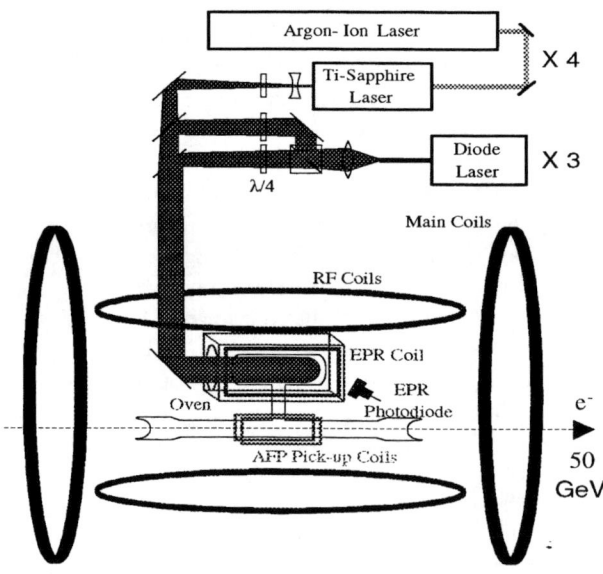

FIGURE 4. Shown is a schematic representation of the E154 polarized ^{3}He target.

peak current. I would be tempted to conclude that the inverted window design itself may be more prone to failure than a non-inverted design, perhaps because of a buckling instability.

One of the important lessons learned during E154 had to do with masing. Looking at Figure 2, the polarization between between days 23–31 is erratic and generally substantially less than was achieved during the rest of the run. The difficulty turned out to be due to masing. This was something of a surprise, because we had already taken measures to avoid such an occurrence. High density ^{3}He targets, however, operate in a regime in which the time required for precessing spins to dephase, due to magnetic field gradients, is much less than the time required for atoms to diffuse a distance characteristic of the cell's dimensions. This is a regime that is not yet well explored theoretically, and preliminary indications are that masing in this regime can under certain conditions occur at times that one would not expect naively [16].

A series of tests were conducted after the conclusion of E154 to confirm that masing was a problem during our run. Masing only occurs when the ^{3}He polarization reaches a certain threshold. The threshold, in turn, is inversely proportional to T_2, the transverse relaxation time for the nuclear spins. One of the most unambiguous signatures of masing is that making T_2 *shorter* (by increasing magnetic field inhomogeneities) increases the polarization threshold at which masing occurs. In Figure 6 tests are shown in which the cell "SMC" was polarized and allowed to saturate. A magnetic field gradient was then turned on, which had the effect of causing the polarization to begin rising

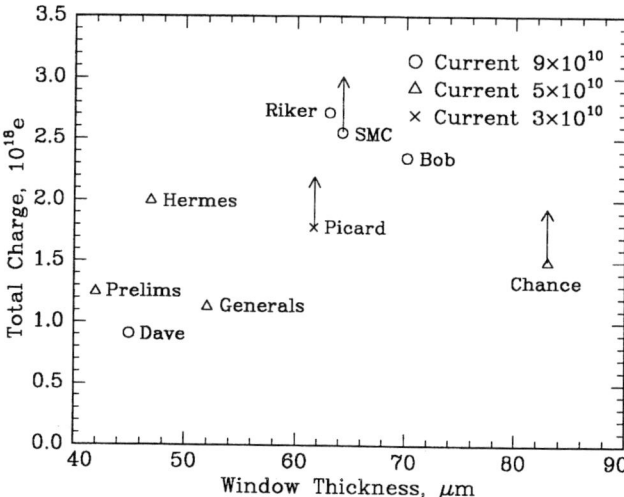

FIGURE 5. Shown is the charge accumulated on the various E154 target cells before they exploded as a function of window thickness.

sharply. Turning the gradient off had the effect of causing the polarization to quickly drop. Lowering the holding field, which has the effect of decoupling the spins from the NMR pick-up coils, also caused the polarization to rise. The tests illustrated in Figure 6 all point to masing.

TJNAF AND BEYOND

A collaboration has formed to build a spin-exchange polarized ^{3}He target for use at TJNAF. The collaboration includes researchers from Argonne, Caltech, Kentucky, MIT, Temple, TJNAF, and William and Mary. Planned experiments range from studies of spin structure at low Q^2 to measurements of the electric and magnetic form factors of the neutron. The first experiments will be in Hall A, and are expected to begin in August of 1998.

There are several important differences between the target that is being built for TJNAF and the targets that were used for E142 and E154 at SLAC. Firstly, the acceptance of the spectrometers in Hall A is such that, with an appropriately designed target cell, electrons scattering from the glass end windows will not be seen. This permits the use of thicker end windows, eliminating the need for the thin inverted windows that were used during E154. Furthermore, again because of the acceptance of the spectrometers, it is not necessary for the target to be in a vacuum. These factors simplify the design.

One feature being adopted for the TJNAF target involves the shape of

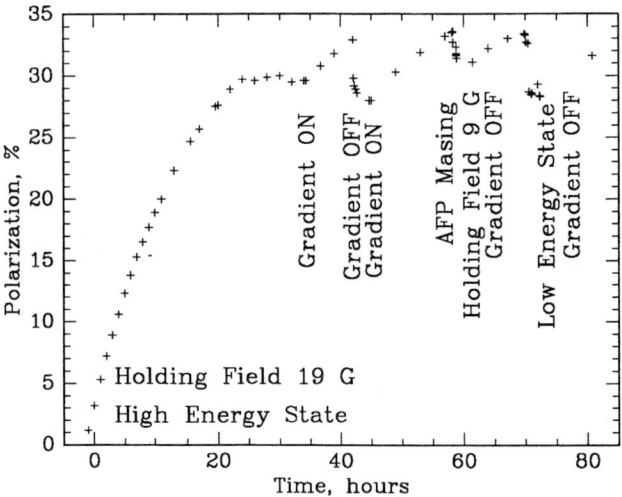

FIGURE 6. Tests are shown in which the cell SMC was polarized and allowed to reach saturation while an applied magnetic field gradient was turned on and off. The fact that polarization went up when a gradient was applied is a signature of masing.

the pumping chamber of the target cell. At SLAC a cylindrical shape was used. This somewhat impedes the direction with which the target spin can be oriented, however. Several of the planned experiments at TJNAF call for the target spin to be oriented either parallel or normal to the direction of the momentum transfer. It is thus desirable to have the ability to pump along an arbitrary direction, a constraint that makes a spherical pumping cell attractive.

A series of studies are currently underway at Princeton to test the TJNAF cell geometry, and explore means of optimizing target performance. These tests are being conducted with a 100 Watt fiber coupled diode laser system manufactured by Optopower. The studies are the first of their kind, in that it has never before been practical to conduct full scale target tests at a university because of limitations in the available laser power. Several diagnostics are being employed during the tests.

- The Rb EPR frequency is being monitored to provide an absolute polarization calibration.

- An "Optical Multichannel Analyzer" (OMA) is being used to monitor the absorption profile of the laser light by the Rb. The OMA is useful for monitoring the laser frequency, and provides a direct indication of the Rb density (since high Rb density results in excessive absorption).

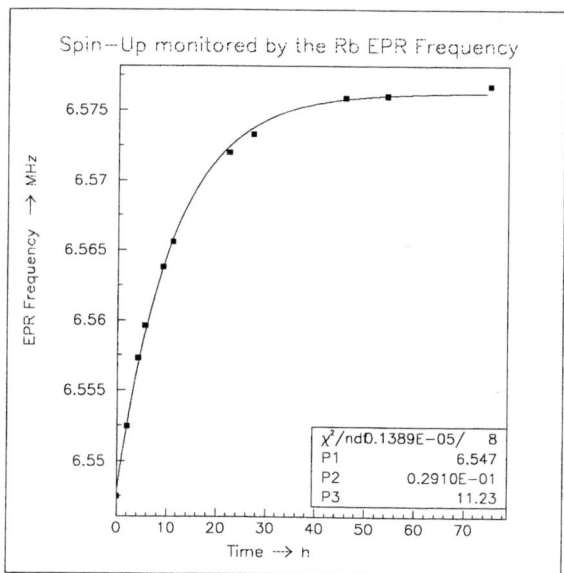

FIGURE 7. Shown is the Rb EPR frequency in Athena, the first of the TJNAF target cells, as a function of time during a pump-up. The maximum shift of the EPR frequency corresponds to nearly 70% polarization.

- The Rb polarization is being periodically mapped using an EPR-based MRI system [17]

The results of these tests have been quite encouraging, as well as illuminating. Tests involving the first cell of the TJNAF series, "Athena", have yielded polarizations from just under 40% to about 70%. An example of a polarization test is shown in Figure 7, in which the Rb EPR frequency is plotted as a function of time. The data are fit to a function of the form

$$\nu_{\text{EPR}}(t) = P_1 + P_2 \left(1 - e^{t/P_3}\right). \tag{8}$$

The fit indicates that the Rb EPR frequency is shifted by about 29 kHz at saturation. A rule-of-thumb for EPR polarimetry is that 0.5 kHz of shift corresponds to 1% polarization for a typical 9 amagat target cell. A more accurate number for Athena is 0.43 kHz/percent, indicating 68% for the pump-up shown. This result should be considered quite preliminary, but indications are that the polarizations are in excess of what was achieved at SLAC.

It is desirable to understand run-to-run differences, and differences between the SLAC target and the TJNAF target. For the Athena tests, the average Rb polarization was roughly 85%. We suspect, but have not yet shown, that this is higher than was achieved in the SLAC target. This could certainly contribute to the higher observed polarizations. Understanding run-to-run variations is tricky, but we have observed a curious phenomena in which the time constant that governs the saturation of the polarization, $(\gamma_{\text{se}} + \Gamma_R)^{-1}$ (see eq. (6)), varies from run-to-run. One could imagine several reasons for this, several of which we are exploring. One thing is clear, we are learning more about the operation of spin-exchange polarized targets now than was possible in previous years, and we are confident that at least some optimization will be possible.

CONCLUSIONS

The basic physics behind spin-exchange polarized ^{3}He targets continues to be better understood, and the accumulated experience working with these targets continues to grow. This, together with the rapid evolution of diode laser technology, will promote improvements in performance over the coming years. Larger targets and higher polarizations both appear likely. Spin-exchange polarized ^{3}He targets have had a large impact on nuclear physics, and it is likely that this impact will continue to grow over the coming years.

ACKNOWLEDGEMENTS

The work presented in this paper was drawn from a number of sources. I would particularly like to thank M.V. Romalis whose thesis work is extensively represented here, including four of his thesis figures. I would also like to thank A. Ben-Amar Baranga, S. Appelt, M.V. Romalis, C.J. Erickson, A.R. Young, and W. Happer at Princeton for permitting me to include results from an upcoming PRL. I would also like to thank the E-154 target group including E.W. Hughes of Caltech, T.E. Chupp, K.P. Coulter, T.B. Smith, and R. Welsh of U. of Mich., A.K. Thompson of NIST, J.R. Johnson of U. of Wisc., and P.L. Bogorad, M.V. Romalis and K.S. Kumar of Princeton. On the TJNAF work I would like to thank I.K. Kominis, G. Corrado, B. Humensky, and T. Pavlin of Princeton, Z.-E. Meziani of Temple, W.J. Cummings of Argonne, J.P. Chen of TJNAF, W. Korsch of Kentucky, H. Gao of MIT, and E.W. Hughes and R. McKeown of Caltech.

REFERENCES

1. K. Abe et al. (SLAC E154 Collaboration), Phys. Rev. Lett. **79**, 26 (1997).
2. P.L. Anthony et al. (SLAC E142 Collaboration), Phys. Rev. Lett. **71**, 959 (1993).
3. P.L. Anthony et al. (SLAC E142 Collaboration), Phys. Rev. D **54**, 6620 (1996).
4. R.J. Knize, Phys. Rev. A **40**, 6219 (1989).
5. M.E. Wagshul and T.E. Chupp, Phys. Rev. A **49**, 3854 (1994).
6. A. Ben-Amar Baranga, S. Appelt, M.V. Romalis, C.J. Erickson, A.R. Young, G.D. Cates and W. Happer, submitted to Phys. Rev. Lett. (1997).
7. T.G. Walker, J.H. Thywissen, and W. Happer, submitted to Phys. Rev. A (1997)
8. T.E. Chupp, M.E. Wagshul, K.P. Coulter, A.B. McDonald, and W. Happer, Phys. Rev. C **36**, 2244 (1987).
9. B. Larson, O. Häusser, P.P.J. Delheij, D.M. Whittal, and D. Thiessen, Phys. Rev. A **44**, 3108 (1991).
10. N. R. Newbury, A. S. Barton, G. D. Cates, W. Happer, and H. Middleton, Phys. Rev. A **48**, 4411 (1993).
11. R. L. Gamblin and T. R. Carver, Phys. Rev. A **138**, 946 (1965); L. D. Schearer and G. K. Walters, Phys. Rev. A **139**, 1398 (1965); G.D. Cates, S.R. Schaefer and W. Happer, Phys. Rev. A **37**, 2877 (1988).
12. K.D. Bonin, T.G. Walker, and W. Happer, Phys. Rev. A **37**, 3270 (1988).
13. K.P. Coulter, A.B. McDonald, G.D. Cates, W. Happer, T.E. Chupp, Nuc. Inst. Meth. in Phys. Res. **A276**, 29 (1989).
14. H. Middleton et al., in **Polarized Ion Sources and Polarized Gas Targets** (AIP Conf. Proc. 293), eds. L. W. Anderson and Willy Haeberli, American Institute of Physics, p. 22 (New York, 1994).
15. T.E. Chupp, R.A. Loveman, A.K. Thompson, A.M. Bernstein, and D.R. Tieger, Phys. Rev. C **45** 915 (1992).
16. M.V. Romalis, Princeton University Ph.D. thesis, unpublished (1997).
17. A.R. Young, S. Appelt, A. Ben-Amar Baranga, C. Erickson, and W. Happer, Appl. Phys. Lett. **70**, 3081 (1997).

The HERMES $^3\vec{\mathrm{He}}$ target

D. De Schepper for the HERMES collaboration

Massachusetts Institute of Technology[1]

Abstract. The HERMES experiment uses an internal target in the HERA positron ring at DESY to study polarized inclusive and semi-inclusive deep inelastic scattering processes. Several light nuclei will be used to disentangle the contributions of protons and neutrons. In 1995 HERMES used an optically pumped metastability exchange ^{3}He target. This target is described in this paper, with emphasis on the specific adaptations to the HERMES experiment.

Introduction

The metastability exchange technique for optically polarizing helium was developed in the early sixties (references [2], [3], [4] and [5]) culminating in the first use of a polarized ^{3}He gas target in the nuclear experiment described in [6]. Large progress was made after the LNA laser became commercially available. The high power and polarizability of the laser light allowed much larger polarizations and much faster pump-up times. Descriptions of this technique can be found in references [7], [8], [9] and [10]. An excellent overview of the different types of optically pumped targets is given in an article by Chupp, Holt and Milner [11]. Targets similar to the HERMES target have been operated by the MIT group for the IUCF ring in the CE-25 experiment [10,12] and as an external target for the BATES accelerator in the 88-25 experiment [13] and for the PELLETRON accelerator in the HERMES target test experiment [14].

The laser system and the holding field

An infrared laser system was used as the source for the 1.083 μm photons. A cylindrical crystal with dimensions 4 mm × 79 mm of La$_{.85}$Nd$_{.15}$MgAl$_{11}$O$_{19}$ (Nd:LNA) was used as the lasing rod in a standard Lasermetrics 9550 Nd:YAG cavity. Two uncoated etalons, one 0.3 mm and the other 1.0 mm thick, were placed inside the cavity to select the desired frequency. Both etalons were mounted in rotatable, temperature-controlled *ovens* to stabilize the tune. The power available at the optical transition was between 2 and 3 Watts at moderate currents (16-17 Amps) during the 1995 running period. The laser ran smoothly and retuning was only necessary about once a week.

Figure 1 shows a schematic view of the laser transport optics. To increase the optical pumping efficiency, a mirror attached to the opposite side of the pumping cell reflected the laser light back through it. All optical components used had commercially dielectric coatings set for the Nd:YAG wavelength of 1064 nm. All lenses used in the region near the target were made from fused silica, in order to have the highest radiation hardness. Circular polarization of the laser light was established using a linearly polarizing cube followed by a zero-order quarter-wave plate.

As the target system was designed to be able to deliver either longitudinal or transverse target polarization, a system for directing the laser beam along one of two final optics paths was needed. A half-wave plate near the laser and a polarizing cube near the pumping cell were employed for this purpose. When the half-wave plate *was not* in the laser beam path, the linear polarization of the light was such that it was *transmitted* through the first polarizing cube and the light then traveled through the optics for longitudinal target polarization. When the half-wave plate *was* in the laser beam path, the axis of linear polarization of the light was rotated by 90°, so that the light was *reflected* at the first polarizing cube and the light traveled through the optics for transverse polarization.

A holding field is required to determine the quantization axis of polarization, and the strength and uniformity of that field play a role in obtaining good polarization. Gradients in the transverse components of the holding field (*i.e.* transverse to the quantization axis) contribute a relaxation time τ_{GRAD} to the total relaxation time of the polarization. Taking into account the velocity, mean free path, and diffusion time in the pumping cell, one can show [15] that $\frac{1}{\tau_{GRAD}} \simeq \frac{1}{3} \frac{|\vec{\nabla} B_\perp|^2}{|\vec{B}|^2}$ for ^{3}He at room temperature and at 0.5 Torr pressure. The residence time in the pumping cell is about 125 seconds, so in the design we aimed for $\tau_{GRAD} \geq 300$ sec so that the gradients would not be the determining factor in the ultimate polarization. This implies that the gradient-to-field ratio above should be less than 0.1 m^{-1}. A design satisfying this criterium as well as the severe spatial constraints of the HERMES target region was developed using the MAFIA code [16]. As shown in figure 2 the longitudinal coils were a pair of rectangular coils whereas the transverse coils consisted of a set of four coils: two rectangular outer coilsand two inner coils of a more complicated design. The curved sections of the inner transverse coils are necessary to avoid the vacuum flanges of the scattering chamber. Despite this odd design, the field calculated at the pumping and target cells is uniform enough for our purposes. The stray field of the HERMES spectrometer magnet in the region of the target, while only a few Gauss in magnitude, has a significant gradient

$dB_y/dz \simeq 200$ mGauss/cm. In order to cancel this gradient, a rectangular correction coil (of the same construction as one of the outer transverse holding coils) was mounted above the target cell, with its principal field axis pointed parallel to the beam axis. The field produced by this coil at the pumping cell exactly cancels the stray field gradient. In tests with sealed cells of ^{3}He, the maximum polarization achieved with the HERMES magnet and this correction coil on was 94% of that seen with both the magnet and this coil off, compared to about 50% with the correction coil unused.

The pumping and storage cells

The quartz pumping cell assembly was developed in close collaboration with the scientific glassware company Finkenbeiner Inc.

The entrance capillary C1 (see figure 3) is necessary to bring the pressure before the pumping cell up into a more easily regulated range and to provide a low-conductance connection from inside the cell to the metal parts of the gasfeedsystem. The latter is necessary to avoid wall depolarization on the metal. A bypass valve was added to increase the conductance from the inside of the cell during cleaning procedures. This valve is made out of quartz, with sealing provided by a set of O-rings (KONTES 826450-0004). The C1 capillary is 2.96 inch long and has an ID of 0.017 inch. The exit capillary C2 to the target cell determines the actual flow rate of the system (given that the optimal pressure for polarization is around .5 Torr). Its length was chosen to be 4.1 inch for an ID of 0.045 inch. Both capillaries were obtained from Wilmad.

The attachment of the glass pumping cell to the storage cell was accomplished using a very thin (25 μm) aluminum tube that just fit over the C2 capillary extending from the pumping cell, with a 5 μm clearance (see figure 3). The tube slid onto a mating aluminum tube that was clamped into the storage cell feed tube and held rigid by the support rails. The sealing was provided by the large overlap between the tubes which was more than 2 inches. To make this possible C2 was laser-aligned to be normal to the quartz baseplate to less than 0.5°. The outer diameter was also specified to 6.5 ± 1/1000 mm to allow a tight fit to the sliding thin aluminum tube that provides the final connection to the target cell.

The pumping cell was constructed as a cube, to allow for two different polarization directions. Its inner side length was 3.5 inch, the largest that was achievable given the strength of quartz and the difficulty of working with it. The thickness of the glass windows was 0.25 inch. The windows were carefully annealed to minimize birefringent effects. The actual assembly is shown in figure 3. The pumping cell and the C2 capillary are mounted on opposite sides of a quartz baseplate. This baseplate is a 0.5 inch thick, circular plate with a 6 inch diameter and is mounted on the side of the vacuum chamber. The helicoflex seal sandwiched between the large, optically flat quartz baseplate and the side of the target chamber provides a UHV compatible seal. The entrance tube was bent at 90° because of the tight space constraints in the target region. The graded quartz-to-pyrex transition starts shortly after the bend. The transition is necessary to connect to the metal bellows.

The storage cell was a long elliptical cylinder with its central axis positioned along the beam axis. The axes were 29.0 and 9.8 mm long and the length of the cell was 550 mm. The effective target length was only the first 400 mm of the storage cell. The remaining 150 mm downstream of this was included to insure that particles coming from the end of the storage cell passed through the same thickness of storage cell. This avoids the complication of vertex-dependent corrections in the analysis due to differences in the material seen by the exiting particles. The elliptical shape of the cell mirrors the beam shape, and the size corresponds to a 20 σ clearance for the beam during injection.

The influence of HERMES on the operation of the HERA storage ring is mainly through the reduction of the stored beam lifetime through atomic bremsstrahlung from the target gas atoms. To minimize the influence of the target on the ring it is necessary to remove the target atoms as quickly as possible after they leave the storage cell. To this end HERMES built a differential pumping system consisting of 5 turbomolecular pumps. This system effectively removed the target gas before it could leave the HERMES experimental area. The target density was then determined by the requirement that the lifetime T_{target} due to the target was above 45 hours. This corresponds with a ^{3}He target thickness of 1.0×10^{15} nucleons/cm^2.

The storage cell was also cryogenically cooled to increase the target density without increasing the gas load to the storage ring. This was accomplished by flowing cryogenic helium gas through the storage cell support rails. The cooling system was designed to remove up to 10 Watt at 15 K. By operating at 15 K, the target density could be increased by a factor of 4.5 over a room temperature storage cell. Below this temperature, ^{3}He wall-depolarization effects become non-negligible as reported in reference [14]. To make the temperature of the target cell more uniform, the storage cell was fabricated from ultra-pure (99.999%) aluminum stock, which is over thirty times more thermally conducting than commercial pure (99%) aluminum at our operating temperatures. To allow the removal of the gas outside the active target region, holes were made in the sides of the 150 mm extension at the downstream end of the cell. These holes were outside the acceptance of the detector.

The storage cell was clamped to the mounting assembly that incorporated the cryogenics and the temperature sensors and allows for a three-point alignment. The mounting assembly is shown in figure 4. The mounts are thermally isolating and include a heat shield to minimize the radiative thermal load on the cell. The target cell support system was positioned outside the limits of the detector acceptance. The downstream face of the radiation shield was therefore missing. An earlier design for this system is described in reference [17].

Target polarimetry

The polarization of the target needs to be measured accurately to minimize the error on the extracted asymmetry. To this end there are two polarimeters that continuously measure the polarization in the pumping cell *and* in the storage cell. Both measurements use the circular polarization of the light emitted during deexcitation of electronic states created in the discharge. This technique was developed by Pavlovic and Laloë [18]. The principle and operation of the optical polarimeter has been described in references [10] and [19]. The relation of the circular polarization of this photon to the nuclear polarization of the gas has been determined experimentally using NMR ([19]). The pressure dependence of this relation has been studied in detail by Pinard and Van Der Linde ([20] and [21]). Other corrections to the relation between the circular polarization of the emitted light and the nuclear polarization follow from the angle of observation and the magnetic field. The angular dependence follows from the dipolar nature of the emission (see for example [19]): $P = M(B)\frac{P_m}{\cos\theta}$ The magnetic field dependence has been calibrated and can be parametrized for fields below 50 Gauss as follows: $M(B) = 1 + 4.08710^{-4} \cdot B + 5.58610^{-6} \cdot B^2$

The first polarimeter, the pumping cell polarimeter (PCP), was mounted 350 mm from the pumping cell, at an angle of 16.6° with respect to the laser beam. The polarimeter was found to be sensitive to the 15 MHz discharge frequency. This appeared as *a dc offset* depending on the discharge intensity. This offset was measured at every spin reversal and subtracted from the dc signal. A careful study of the systematic error in the polarization measurement gave an error of 2.2% in the correction factors, 1.1 of the ac signal and 1.0% in the measurement of the dc signal. This gives a total error of 2.7% in the polarization measurement.

However, apart from these well-known systematic errors, there were two other sources of measurement errors. The first is caused by the strong RF pulse starting the discharge. This pulse causes a strong light flux into the PMT, temporarily increasing its dark current. This in turn increases the dc signal. The dc signals for a typical sequence of spin flips is shown in figure 5. The dark current increase after the spin slip is shown clearly. This effect causes an error of about 1% right after the discharge is lit. By the time the polarization has built up the error is reduced considerably. The average polarization measurement suffers even less, since this effect only persists for the first few bursts after the spin flip (there are 60 bursts between spin flips). An upper estimate of the error is .1%. A second source of error is caused by an unexplained correlation between the dc level of the discharge and the target polarization state. This is obvious from figure 5. This difference was shown to be real, i.e. the dc level of the discharge light is correlated to the polarization of the gas. We have not been able to explain the effect. It seems to be caused by a higher multipole component of the emitted light and it has been treated as a systematic error for the current analysis of the data. The difference between the dc levels is about 2.5%. To determine the total error in the polarization it seems prudent to add this error linearly to the error coming from other sources, giving an combined error of 5.2%. The error introduced by the increased dark current after a spin flip should probably also be added linearly to this error, since its sign is constant. The final estimate for the error in the polarization measurement is therefore 5.3%.

A second polarimeter, the target optical monitor (TOM), was used to measure the polarization inside the target cell directly. The TOM uses the light coming from the recombination of excited states created by Coulomb excitation from the positron beam in the storage cell. The photon flux is of course much smaller, necessitating the use of photon counting techniques. An example of the simultaneous polarization measurements with the PCP and TOM is shown in figure 6. In this plot each datapoint represents the average polarization over 100 seconds of data. The statistical error on the TOM polarization is about 30% in this case. A full description of the technical details of the TOM can be found in reference [22].

The main use of the TOM is not in the absolute measurement of the polarization, but as a monitor of the relative polarization. As such it is extremely useful to study beam depolarization as well as temperature dependent wall depolarization effects. Figure 7 shows the ratio of TOM and PCP measurements of the polarization for various internal target temperatures. Over the entire measured range (18-60 K) the ratio changes less than 5%, indicating that no wall depolarization occurs. Although helium has a closed electron shell configuration, the presence of ions in the target allows for the possibility of beam-induced depolarization. This has been shown experimentally for the high-density external targets (see [23] and [24]). In the HERMES storage cell there is little time for the atom to become ionized, depolarized and DIS-scattered from. Hence we do not expect a significant beam-induced depolarization. This was tested by measuring the TOM to PCP ratio as a function of positron beam current. The result (shown in figure 7) does not indicate any decrease of the polarization in the target cell with increasing beam current. The relative change in polarization from 12 to 29 mA is less than 7%.

Target performance

The maximum polarization for a sealed cubic cell (volume .44 liter) was found to be 76%, for a pump-up time of 35 seconds. This was achieved with a 3 Watt, double etalon laser. The polarization that this cell should reach in a flow-through assembly at a nominal flow rate of 10^{17} was estimated from the formula $P_o^{flow} = P_o^{sealed} \times \frac{\tau_{res}}{\tau_p + \tau_{res}}$ where τ_{res} is the residence time of the helium atoms in the pumping cell and τ_p is the pump-up time in the sealed cell. The result gave an estimated polarization of about 61%. The maximum polarization observed in the target was 54%, but polarizations in excess of 50%

were normal. The reason for the somewhat lower than expected polarization is the reduced laser power available on the target platform (typically 1-2 Watts) and somewhat higher magnetic field gradients. The averaged polarization over the datataking period was 47%. The polarization was reversed every 10 minutes. Because the polarimeter DC offset was measured during the spin reversal each reversal took about 30 seconds.

REFERENCES

1. currently at the Argonne National Laboratory.
2. F. Colegrove and P. Franken. "Optical pumping of helium in the $^3S_1$ metastable state," *Physical Review*, **119**(2), 680–690 (1960).
3. G. Walters et al. "Nuclear polarization of He^3 gas by metastability exchange with optically pumped metastable He^3 atoms," *Phys. Rev. Lett.*, **8**(11), 439–442 june 1962.
4. L. Schearer et al. "Large ^{3} nuclear polarization," *Phys. Rev. Lett.*, **10**(3), 108–110 february 1963.
5. F. Colegrove et al. "Polarization of He^3 gas by optical pumping," *Physical Review*, **132**(6), 2561–2572 december 1963.
6. G. Phillips et al. "Demonstration of a polarized He^3 target for nuclear reactions," *Phys. Rev. Lett.*, **9**(12), 502–504 december 1962.
7. P. Nacher and M. Leduc. "Optical pumping in ^{3}He with a laser," *Journal de Physique*, **46**(12), 2057–2073 (1985).
8. J. Daniels et al. "Polarizing ^{3}He nuclei with neodymium La$_{1-x}$Nd$_x$MgAl$_{11}$O$_{19}$," *Journal of the Optical Society of America*, **4**, 1133–1135 july 1987.
9. R. Milner et al. "A polarized ^{3}He target for nuclear physics," *Nucl. Instr. and Meth. A*, **274**, 56–63 (1989).
10. K. Lee et al. "A laser optically pumped polarized ^{3}He target for storage rings," *Nucl. Instr. and Meth. A*, **333**, 294 (1993).
11. T. Chupp et al. "Optically pumped polarized gaseous H,D and ^{3}He gas targets," *Annual Reviews of Nuclear and Particle Science*, **44**, 373–411 (1994).
12. M. Miller et al. "Measurement of quasielastic ^{3}He(p,pN) scattering from polarized ^{3}He and the three body ground state spin structure," *Phys. Rev. Lett.*, **70**, 738 (1993).
13. H. Gao et al. "Measurement of the neutron magnetic form factor from inclusive quasielasitic scattering of polarized electrons from polarized He-3," *Phys. Rev. C*, **50**, 546–549 (1994).
14. W. Korsch et al. *Nucl Instr. and Meth. A*, **354**, 437 (1995).
15. C. E. Jones. *A Measurement of the Spin-Dependent Asymmetry in Quasielastic Scattering of Polarized Electrons from Polarized ^{3}He*. PhD thesis, 1992.
16. MAFIA computer code, ©Prof. Dr.-Ing. T. Weiland, Ohlystrasse 69, D-64285 Darmstadt, Germany.
17. L. Kramer et al. *Nucl. Instr. and Meth. A*, **365**, 49 (1995).
18. M. Pavlovic and F. Laloe *J. Phys.*, **31**, 173 (1970).

19. W. Lorenzon et al. "NMR calibration of optical measurement of nuclear polarization in ^{3}He," *Phys. Rev. A*, **47**, 468 (1993).
20. M. Pinard and J. V. D. Linde *Can. J. Phys.*, **52**, 1615 (1974).
21. M. Pinard. PhD thesis, Universite de Paris, France, 1973.
22. M. L. Pitt. In H. Paetz, editor, *Int. Workshop on Polarized Beams and Polarized Gas Targets*, pages 413–417. World Scientific, 1996.
23. R. Milner et al. "Study of spin relaxation by a charged particle beam in a polarized ^{3}He gas target," *Nucl. Instr. and Meth. A*, **257**, 286–290 (1987).
24. K. Bonin et al. *Phys. Rev. A*, **37**, 3270 (1988).

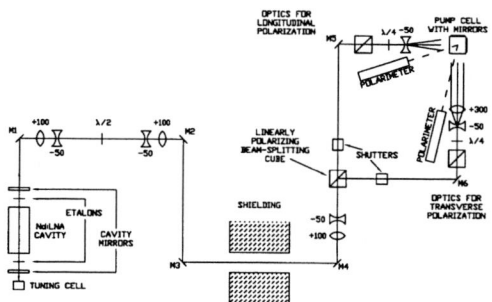

FIGURE 1. Schematic diagram of the laser transport optics system. Focal lengths of lenses are in mm. Retardation plates are labelled as "$\lambda/2$" (half-wave) or as "$\lambda/4$" (quarter-wave). The total distance from laser to pumping cell is about 10 meters.

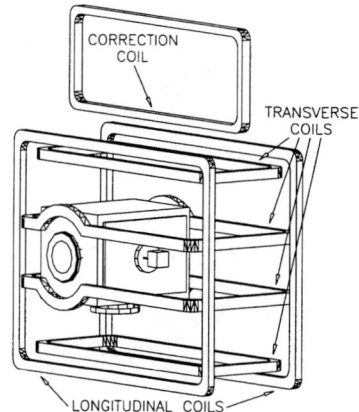

FIGURE 2. The transverse and longitudinal holding field coils. Also shown is the correction coil. The scattering chamber and pumping cell are shown inside the coils.

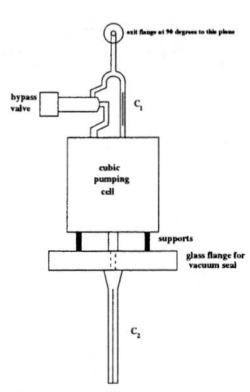
(a) The quartz cell

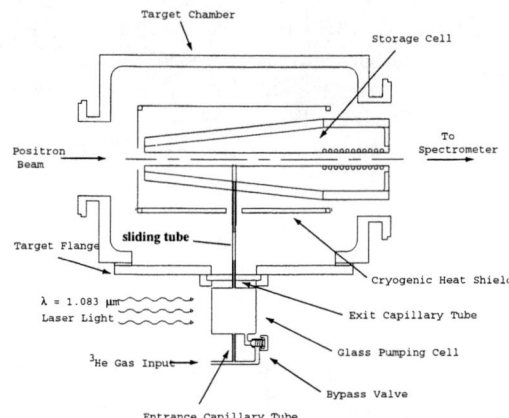
(b) Connection to the storage cell using a sliding tube

FIGURE 3. The pumping cell

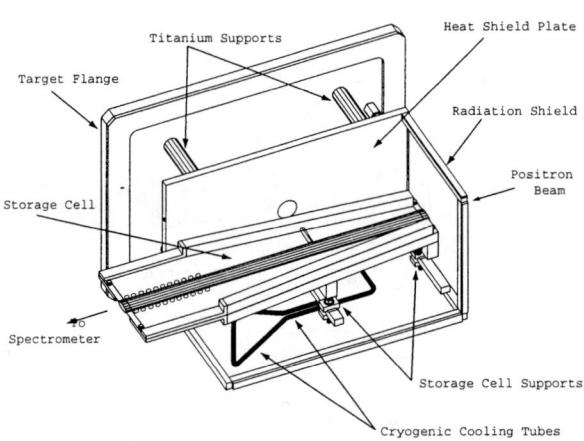

FIGURE 4. The HERMES cryogenic storage cell

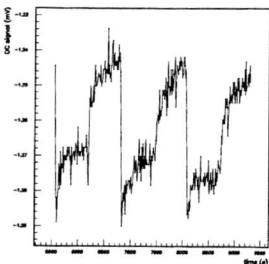

FIGURE 5. The dc component of the polarimeter signal

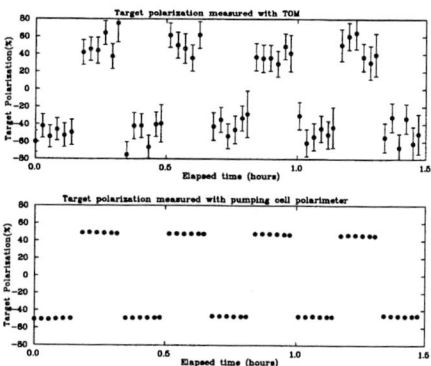

FIGURE 6. TOM and PCP polarizations measurements for 100 second intervals

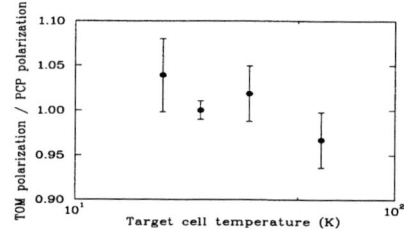
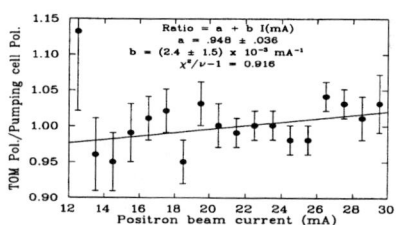

(a) Ratio of TOM to PCP polarizations for different temperatures (18-60 K)

(b) Ratio of TOM to PCP polarizations for different currents (12.5-29.5 mA)

FIGURE 7. Target depolarization studies

Quasi-elastic Electron Scattering from Polarized ^{3}He

H.J. Bulten[1], R. Alarcon[2], Th. Bauer[3], D. Boersma[3], T. Botto[4], J. F. J. van den Brand[1,4], L. van Buuren[1], R. Ent[5,6], M. Ferro-Luzzi[4], D. Geurts[1], M. Harvey[5,6], P. Heimberg[1], D. Highinbotham[7], C. W. de Jager[5], B. Norum[7], I. Passchier[4], H.R. Poolman[4], M. van den Putte[4], E. Six[2], J. Steijger[4], D. Szczerba[1], and H. de Vries[4].

[1] *Department of Physics and Astronomy, Vrije Universiteit, 1081 HV Amsterdam, The Netherlands*
[2] *Department of Physics, Arizona State University, Tempe, AZ 85287, USA*
[3] *Department of Physics, Rijks Universiteit Utrecht, The Netherlands*
[4] *NIKHEF, 1009 DB Amsterdam, The Netherlands*
[5] *TJNAF, Newport News, VA 23606*
[6] *Department of Physics, Hampton University, Hampton, VA 23668, USA*
[7] *Department of Physics, University of Virginia, Charlottesville, VA 22901, USA*

Abstract

Quasi-elastic electron scattering may provide precise information on the S' and the D-wave parts of the ^{3}He ground-state wave function, the neutron form factors, and the rôle of spin-dependent reaction mechanism effects. An experiment is being performed at the AmPS storage ring at NIKHEF (Amsterdam, the Netherlands), where polarized electrons (up to 900 MeV) are used in combination with large acceptance electron and hadron detectors. Preliminary results from data at four-momentum transfer squared $Q^2 = 0.15$ GeV^2 are presented.

1 Physics motivation

The ^{3}He nucleus plays a special rôle in subatomic physics. For the three-body nuclear system exact (Faddeev) calculations are available, hence our understanding of nuclear structure can be precisely compared with data. In addition, it is generally thought that polarized ^{3}He can serve as an effective polarized neutron target for experiments in nuclear [1] and particle [2] physics. Fundamental properties of the neutron such as its charge and spin distributions remain largely unconstrained experimentally. Thus, many experiments using polarized ^{3}He targets are underway worldwide in large part motivated by measurement of neutron elastic form factors[3, 4, 5] and deep inelastic structure functions [6, 7, 8]. It is imperative to understand the ground state spin structure of the ^{3}He nucleus to extract information on neutron structure from these measurements.

The nuclear structure information of the ^{3}He ground state is contained in the spin-dependent spectral function[9] $S_{\hat{\sigma}}^N(E, \mathbf{p})$ defined as the probability density of finding a nucleon N with separation energy E, momentum $\mathbf{p}$ and spin along(opposite) the $\overrightarrow{^3\text{He}}$ spin indicated by $\hat{\sigma}=+(-)$. Quasi-elastic spin-dependent electron-induced knockout of the constituent nucleons of ^{3}He offers the most direct experimental approach to constrain the spectral function.

Exclusive and semi-inclusive scattering of polarized electrons from polarized ^{3}He using the reactions $(\vec{e}, e'p)$ and $(\vec{e}, e'd)$ enables one to disentangle the S, S' and D-state components of the ground-state wave function[10, 11] In the non-relativistic approximation using Faddeev techniques, the S-state contribution to the spectral function is found to be dominant at low initial nucleon momenta [1, 12]. Measurements of the sideways target asymmetry, $\mathbf{A}'_\mathbf{x}$, in the reactions $(\vec{e}, e')$, and $(\vec{e}, e'n)$ will constrain the neutron charge form factor. Our method of simultaneously surveying several reactions over a wide range of kinematics is central in our approach to disentangle nucleon and nuclear structure from reaction mechanism effects such as final-state interactions (FSI) and meson-exchange currents (MEC). In addition, we measure induced asymmetries as an additional means to isolate the FSI and MEC contributions. As an example, we show in figure 1 calculations of spin-correlation parameters $\mathbf{A}'_\mathbf{x}$ and $\mathbf{A}'_\mathbf{z}$ and the transverse analyzing power $\mathbf{A}^0_\mathbf{y}$ for quasi-elastic neutron knock-out in the model of Laget [10]. The calculations are performed for kinematics with missing momentum equal to 0. Furthermore, we show the world data (figure from Platchkov et al., Ref. [13]) of G_E^n obtained from elastic scattering off the deuteron. The curves indicate the uncertainty on G_E^n, due to the choice of NN-potential in the analysis. Exclusive data yield an absolute normalization of G_E^n, provided that the description of the reaction mechanism is under control.

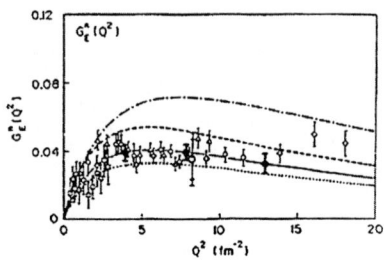
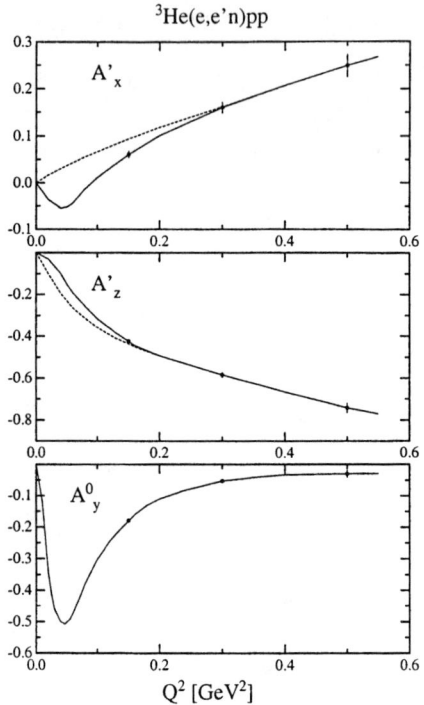

Figure 1: Top: World data[13] on G_E^n as a function of Q^2. The solid markers represent the expected accuracy of the present experiment. The open circle at $Q^2 = 8$ fm^{-2} represents the datum from Ref.[4], which is obtained with the ^{3}He(e,e'n) reaction. Right: The asymmetry $\mathbf{A}_y^0$ and the spin-correlation parameters $\mathbf{A}_x'$ and $\mathbf{A}_z'$ as a function of Q^2. The dashed and solid curves represent the calculations of Laget[10] for PWIA and including FSI and MEC effects, respectively.

2 The Experiment

We intend to measure the transverse asymmetry $\mathbf{A}_y^0$ and the sideways and longitudinal spin-correlation parameters $\mathbf{A}_x'$ and $\mathbf{A}_z'$ for quasi-elastic electron scattering. Large-acceptance detectors are applied to simultaneously obtain data for the reaction channels (e,e'), (e,e'p), (e,e'n), and (e,e'd) for missing momenta up to about 400 MeV/c. Furthermore, a silicon detector is used to measure low-energy recoiling particles, giving access to the coherent π^0 and π^+ production from ^{3}He and enabling us to detect triple coincidences (^{3}He(e,e'pp)n and ^{3}He(e,e'pd)). First data at Q^2 of 0.15 GeV^2 have been obtained, whereas further measurements will be performed for Q^2 of 0.15, 0.3, and 0.5 GeV^2. A consistent set of measurements of these different asymmetries will provide the means to disentangle different aspects of the reaction mechanism. The asymmetry $\mathbf{A}_y^0$ is identical to 0 in plane-wave impulse approximation; its measurement will isolate the effects of final-state interactions and meson-exchange currents. The measurements of $\mathbf{A}_x'$ and $\mathbf{A}_z'$ in the (e,e') and (e,e'n) reaction channels will constrain G_E^n and G_M^n. The spin-correlation parameters in the (e,e'p) channel are sensitive to the S'-state at low - and D-state at high missing momentum.

The experiment is performed in the internal target hall at NIKHEF. Polarized electrons of energies up to 900 MeV will be scattered from a ^{3}He target[14]. Recently, we obtained data with polarized electrons at 442 MeV. The polarization at the source amounted to 81 %. The polarized electron source is discussed elsewhere in these proceedings. The polarization of the electron beam in the storage ring could be accurately measured by means of Compton backscattering (see the contribution of I. Passchier et al., elsewhere in these proceedings).

We apply the metastability optical pumping technique. ^{3}He gas is flown into a pyrex pumping cell. A discharge is created by two RF-coils, driven by a 20 MHz generator. A second set of coils, operated at 170 kHz, was used to ignite the discharge. The discharge populates the 2^3S_1 triplet state. Circularly-polarized laser light with a wavelength of 1.083 μm impinges on the pumping cell, and induces transitions between the 2^3S_1 and 2^3P_0 states. The resulting polarization of the ^{3}He nuclei can be reversed by inverting the helicity of the laser light. We use a 0.3 mm thick etalon, optionally in combination with a 1 mm thick etalon, to tune the laser. The angles as well as the temperatures of the etalons are remotely controlled. The target polarization can be determined by analyzing the circular polarization of the 667 nm balmer line in the fluorescence spectrum[15]. The external magnetic field that is used to direct the spin of the ^{3}He nucleus is supplied by three mutually perpendicular sets of Helmholz coils, that allow for rotation of the spin to any given direction. As demonstrated by Figure 2., we obtained 68 % polarization on a sealed cell in the laboratory. During the experiment we obtained initial polarizations of about 43 (47) % at a flow of 1×10^{17} atoms/s (0.5×10^{17} atoms/s), which decreased with about 1 % per day due to dirt collection on the etalons. About once every five days we obtained hall access and cleaned the optical system. The polarization was limited by a number of effects, among which the magnetic stray fields of the electron spectrometer, which were not sufficiently compensated for.

The T-shaped storage cell (fabricated from 50 μm thick ultrapure aluminum) is cooled to 17 K in order to further increase the atomic density of the target. The storage cell contains a slot, covered mith a thin mylar foil, to allow very low energy particles to reach the recoil detector. The cryogenic target is shown in Fig. 3. The coldhead contains ferromagnetic materials, therefore it is located outside the Helmholtz coils, far away from the pumping cell. In order to suppress heating of the cell due to the RF field of the beam, wake field suppressors were installed between the storage cell and the beam pipe.

Fig. 4 shows part of the experimental setup. Scattered electrons are detected by a magnetic spectrometer with a solid angle of 100 msr and a momentum acceptance of 300-1000 MeV/c. The angular resolution of this detector is dominated by multiple scattering (a few mrad), the momentum resolution is

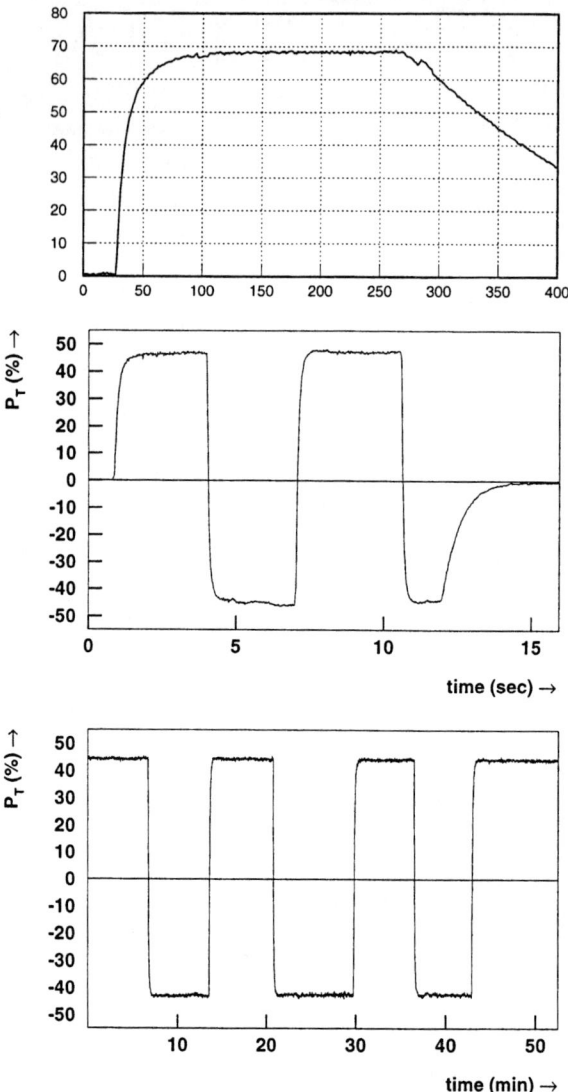

Figure 2: The top panel shows the polarization for the ^{3}He target as a function of time for a sealed cell. At $t \sim 270$ s the laser light is blocked. The middle and bottom panels show the measurements of the polarization performed during the present experiment at flows of 0.5 and 1×10^{17} atoms/s, respectively.

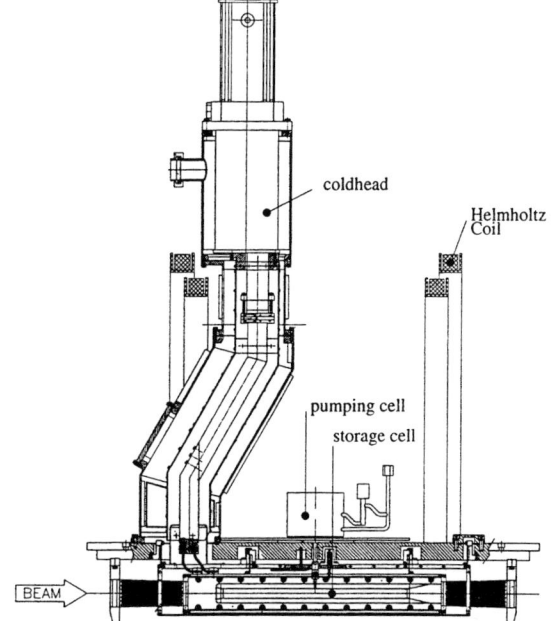

Figure 3: Overview of the cryogenic target. The coldhead is located outside the Helmholtz coils and is connected via a heat pipe with two bends to the storage cell.

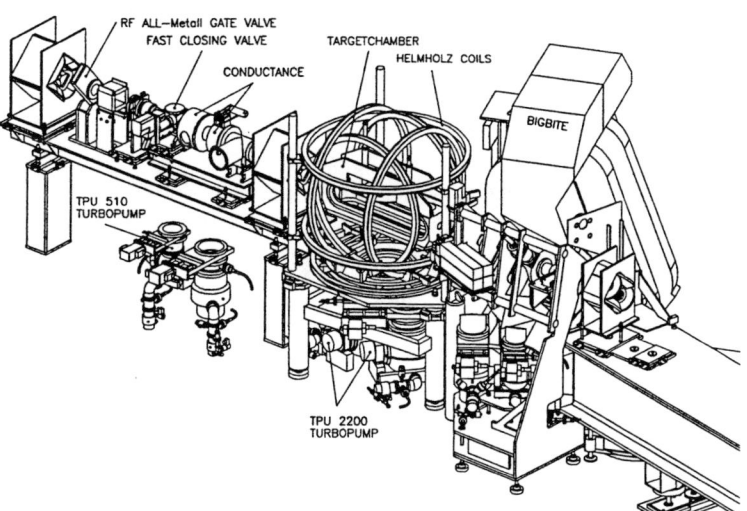

Figure 4: Overview of the experimental area. Part of the beam line is shown, as well as the target chamber, Helmholtz coils and electron detector.

expected to be $\Delta p/p \sim 5 \times 10^{-3}$ when all calibrations have been completed. Protons are detected in a range telescope, which contains two wire chambers

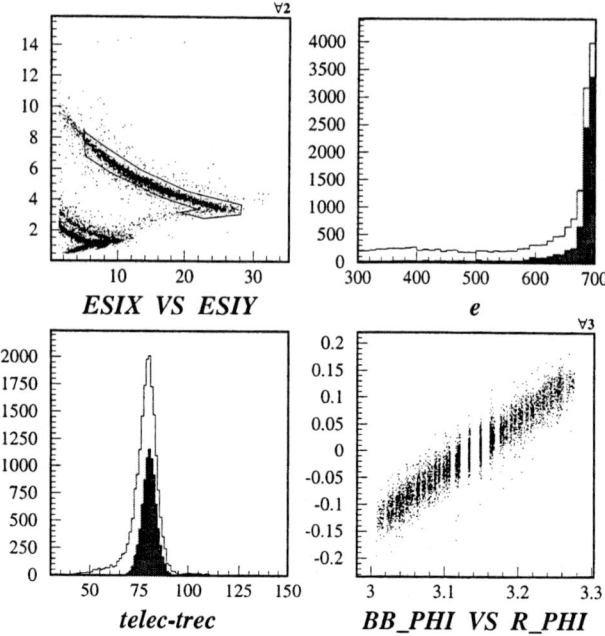

Figure 5: Coincident events between the electron detector and the silicon detector. Top Left Panel: correlation of the deposited energies in the first and second silicon layers. One can clearly distinguish bands for recoiling protons, deuterons, tritons, ^{3}He and ^{4}He nuclei. The area that yields recoiling ^{3}He nuclei is indicated. Top Right: Energy spectrum of the electron detector. The shaded area corresponds to the ^{3}He nuclei in the top left panel. Bottom Left: timing difference between the events in the electron and recoil detector. The shaded area corresponds to the ^{3}He nuclei in the top left panel. Bottom Right: Correlation in out-of-plane angles for the ^{3}He events. For elastic scattering, these events should be coplanar.

(with planes at 0°, 45°, and 90 °) for tracking, followed by a 30 strips wide hodoscope and 15 layers of 10 mm thick plastic scintillators. The range telescope has a solid angle of $\sim$ 200 msr and an energy resolution of $\sim$ 3 MeV. Behind the range telescope, at 3 m from the interaction point, a neutron detector is positioned, covering about the same solid angle. This neutron detector contains 8 scintillator bars of 0.2×0.2 × 1.6 m. Each bar is preceded by two veto scintillators of 4 mm and 10 mm thickness. Silicon detectors are located inside the vacuum to enable detection of recoiling protons and deuterons. Since the energy resolution of the electron and proton detector are not good enough to unambiguously separate the two-body and three-body break-up channel in the (e,e'p) reaction, this separation has to be obtained in another manner.

We will follow three strategies to assess the physics related to these different break-up channels: a) We simultaneously perform measurements for the (e,e'd) reaction channel; b) We will explore the (e,e'p) asymmetries as a function of missing energy by applying cuts to the low - and high energy side of the missing-energy peak corresponding to two-body breakup; c) we will analyze triple coincidences, in which a recoiling proton or deuteron is detected in the silicon detectors. Obviously, the latter approach will only allow separation of two- and three-body breakup for a limited part of the (e,e'p) phase space.

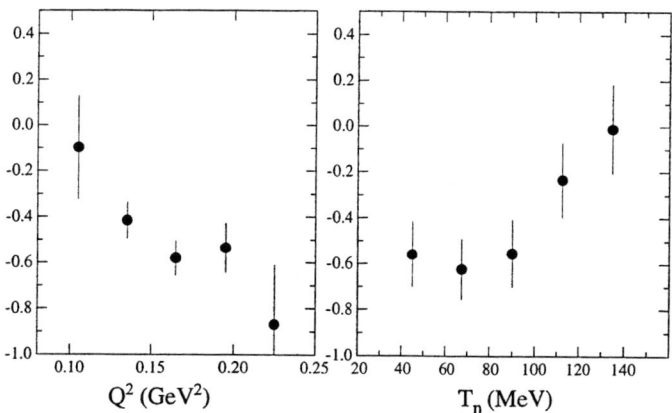

Figure 6: The analyzing power A_y^0 for the reaction ^{3}He(e,e'n) as a function of Q^2 (left) and T_n (right).

Fig. 5 demonstrates, that with the silicon detectors we can cleanly separate elastically scattered ^{3}He particles from background. Since the spin-spin correlation for elastic scattering from ^{3}He can be calculated from the electromagnetic form factors for ^{3}He, which are well-known in the range of Q^2, this coincidence reaction provides us with an alternative way to determine the product of target and beam polarization.

In the spring of this year, we obtained data for $Q^2 = 0.15$ GeV^2 with unpolarized electrons of 720 MeV and longitudinally polarized electrons of 442 MeV. The analysis is ongoing, especially the calibration of the detector response is not completed yet. However, we would like to present a few preliminary results on the (e,e'N) channel. In Fig. 6 we show $\mathbf{A}_y^0$ as a function of Q^2 and the neutron kinetic energy T_n. Due to the Watson-Fermi theorem [16] this asymmetry is strictly equal to 0 in PWIA, hence it is sensitive to the ingredients of the reaction mechanism (notably final-state interaction). It appears, that $\mathbf{A}_y^0$ as a function of Q^2 does not resemble the calculation of Laget (see Fig. 1). This is due to the fact that in our data, we cover the full quasi-elastic cone, whereas the result of Laget is valid for the condition $p_m = 0$, and hence Q^2 is strongly correlated with T_n. Note, that for $E_m = p_m = 0$ one obtains $T_n = Q^2/2M_{nucleon}$. Indeed, in the calculation of Laget $\mathbf{A}_y^0$ shows a minimum of -0.5 for kinematics with $T_n \sim 25 MeV$ ($Q^2 = 0.05$ GeV^2), reducing to about -0.1 at $T_n \sim 130$ MeV ($Q^2 = 0.25$ GeV^2), which is in qualitative agreement with the data. Obviously, one should take care to compare the data to theoretical calculations that cover the full large kinematical acceptance of the experimental setup.

3 Summary.

In summary, with the advent of CW polarized electron beams, polarized ^{3}He internal targets, and a large acceptance detector, a thorough study of spin observables in the ^{3}He system can be performed. Preliminary data for neutron knockout indicate, that final-state interaction strongly modifies the asymmetries at small kinetic energies. Simultaneous measurements of many reaction channels with the target spin oriented sideways, normal and parallel to the momentum transfer may lead to an unambiguous determination of the small components in the ^{3}He ground-state wave function and may enable to extract the neutron electromagnetic form factors with little uncertainties.

Acknowledgements

This work was supported in part by the Stichting voor Fundamenteel Onderzoek der Materie (FOM).

References

[1] B. Blankleider and R. W. Woloshyn, Phys. Rev. **C29**, 538 (1984).

[2] R.G. Milner, *Proceedings of the Workshop on Polarized ^{3}He Beams and Targets*, ed. R. W. Dunford and F. P. Calaprice (AIP 131, New York, 1984) p. 186.

[3] H. Gao et al., Phys. Rev. **C50**, R546 (1994).

[4] M. Meyerhoff et al., Phys. Lett. **B327**, 201 (1994).

[5] Bates Large Acceptance Spectrometer Toroid (BLAST) proposal, MIT-BATES, 1992.

[6] P. L. Anthony et al. Phys. Rev. Lett. **71**, 959 (1993).

[7] HERMES experiment at DESY, DESY rep. no. DESY-PRC-93-06, 1993, unpublished.

[8] Experiment E-154; spokesperson R. Hughes.

[9] S. Frullani and J. Mougey, Adv. Nucl. Phys. **14** (1984).

[10] J. M. Laget, Phys. Lett. **B 276**, 398 (1992).

[11] S. Nagorny and W. Turchinetz, Phys. Lett. **B 389**, 429 (1996).

[12] R. W. Schulze and P. U. Sauer, Phys. Rev. **C48**, 38 (1993).

[13] S. Platchkov et al., Nucl. Phys. **A510**, 740 (1990).

[14] H.R. Poolman, in *proceedings of the International Workshop on Polarized Beams and Polarized Gas Targets*, Cologne, June 6-9, 1995, eds. H. Paetz gen. Schiek and L. Sydow, World Scientific.

[15] W. Lorenzon et al., Phys. Rev. **A47**, 468 (1993).

[16] K. M. Watson, Phys. Rev. **95**, 278 (1954).; E. Fermi, Suppl. Nuovo Cimento **2**, 17 (1955).

Determination of the Neutron Electric Form Factor G_{En} in Double Polarized Electron Scattering from ^{3}He

A3 - collaboration at MAMI:

H. G. Andresen[1], J. Annand[2], K. Aulenbacher[3], J. Becker[3],
K. Beuchel[3], J. Blume–Werry[1], Th. Dombo[1], P. Drescher[3],
M. Ebert[3], D. Eyl[1], H. Fischer[3], A. Frey[1], P. Grabmayr[4],
T. Großmann[3], S. Hall[2], P. Hartmann[1], T. Hehl[4], W. Heil[5],
C. Herberg[1], J. Hoffmann[3], D. Ireland[2], J. Kellie[2], F. Klein[6],
M. Leduc[7], M. Meyerhoff[3], G. Miller[2], H. Möller[1],
Ch. Nachtigall[3], A. Natter[4], M. Ostrick[1], E. W. Otten[3],
R. O. Owens[2], S. Plützer[3], E. Reichert[3], D. Rohe[3], M. Schäfer[3],
H. Schmieden[1], R. Sprengard[1], K.-H. Steffens[1], R. Surkau[3],
Th. Walcher[1], R. Watson[2], E. Wilms[3]

presented by J. Becker

[1] *Institut für Kernphysik, Universität Mainz;* [2] *Department of Physics and Astronomy, University of Glasgow;* [3] *Institut für Physik, Universität Mainz;* [4] *Physikalisches Institut, Universität Tübingen;* [5] *Institut Laue–Langevin, Grenoble;* [6] *Physikalisches Institut, Universität Bonn;* [7] *École Normale Superiéure, Paris*

Abstract. The neutron electric form factor G_{En} has been measured at the cw electron accelerator MAMI in the double polarized exclusive reactions $^3\vec{H}e(\vec{e}, e'n)$ [1–3] and $D(\vec{e}, e'\vec{n})$ [4–7] in quasi elastic kinematics with one common detector system. Experimental set–up and analysis of a $^3\vec{H}e(\vec{e}, e'n)$-measurement in the range of momentum transfer Q^2=0.27-0.5 (GeV/c)^2 are discussed in this article. For events with low missing momentum ($|\vec{P}_m|\leq$100 MeV/c) the PWIA analysis yields G_{En}=0.0352±0.0033±0.0024 with statistical and systematical error, corresponding to a quadratically added total error of 11.6% relativ. This result will be compared with preliminary results obtained from the $D(\vec{e}, e'\vec{n})$–reaction and with data from elastic $D(e, e')$ scattering [8].

INTRODUCTION

The precise knowledge of the electromagnetic structure of the nucleon is of fundamental importance both in nuclear and particle physics in order to make significant statements on quark wave functions or models of the nucleon. While the data on the proton over many years gradually have been extended both in momentum transfer and precision, comparatively little progress has been made on the neutron electric and magnetic form factors G_{En} and G_{Mn}. In particular the electric form factor G_{En} is very difficult to determine using the conventional method of Rosenbluth separation. A much better technique determines the asymmetry in the scattering of high energy polarized electrons from polarized targets or the spin transfer to the recoiling neutron from unpolarized targets. In both cases G_{En} is extracted from an interference between the small electric and large magnetic scattering amplitude. As no dense neutron target exists the A3-collaboration at MAMI investigates G_{En} in the quasi elastic electron scattering from neutrons bound in the light nuclei D and ^{3}He through the exclusive reactions $D(\vec{e}, e'\vec{n})$ and $^3\vec{He}(\vec{e}, e'n)$ using one common detector set-up. Goal is to extrapolate from the two results to the form factor of the free neutron.

EXPERIMENTAL SET–UP

Longitudinally polarized electrons with energy E=855 MeV pass the polarized ^{3}He- or unpolarized D_2-target cell. The scattered electrons are detected in coincidence in a large solid angle ($\Delta\Omega$=100 msr) gas Čerenkov-detector and a lead glass calorimeter. The focussing Čerenkov-detector suppresses detection of electrons scattered in the entrance windows of the ^{3}He-target cell as well as high energy photons from π°-decay. The segmented lead glass calorimeter provides a good angular resolution of $\delta\varphi*\delta\vartheta$=0.1 msr and a modest energy resolution of $\Delta E/E$=20% FWHM that serves for suppression of π-production. In coincidence with the scattered electron the neutron is detected in a TOF detector array of solid angle $\Delta\Omega$=250 msr consisting of two walls of plastic scintillators, each of them covered with an additional layer of veto counters for charged particles. The neutron detector is shielded in target direction by a 5 cm lead brick wall and by 1 m of concrete at both sides. The scattering angles of electron and neutron and the neutron energy are used to reconstruct the kinematics of each scattering event. Polarization of the incident electron beam is monitored by a compton polarimeter integrated in the beam dump.

In the $^3\vec{He}(\vec{e}, e'n)$-experiment an electron beam of 7 μA at P_e=(50±5)% was accelerated from a GaAsP photocathode, irradiated with circularly polarized focussed laser light. Fatigue of electron emission was circumvented by shifting the laser spot at the cathode surface from time to time and eventually replacing and refreshing the cathode by on–line procedures [9,10]. A total of 2483 μAh of beam charge was taken.

The polarized target (Fig. 1) is a system of circulating ^{3}He gas. A LNA laser

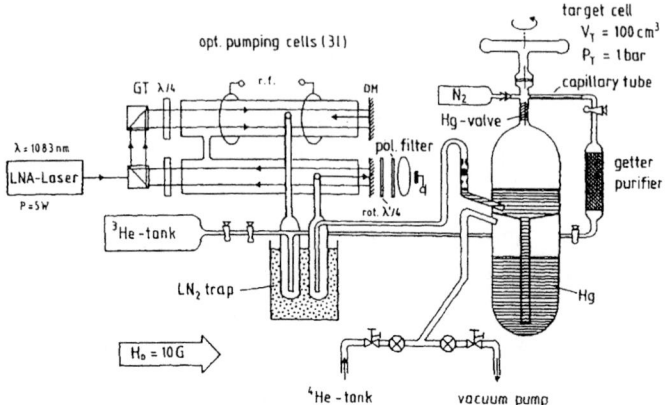

FIGURE 1. Schematic of polarized target (Toepler compressor).

with cw power of 5 W spin polarizes ^{3}He by optical pumping of metastable ^{3}He* atoms [11,12] at about 1 mbar in two optical pumping cells (OPC) (V_{OPC} = 3 l). The $^3\vec{He}$ gas is compressed by means of a non magnetic mercury compressor (Toepler compressor) into a Cs coated target cell (P_t=1 bar, V_t=100 cm^3). In order to compensate relaxation in the target cell a constant flow back of 2 mbl/min ($\approx 10^{18} \frac{^3He\ atoms}{sec}$) is circulated from the target cell via a capillary tube and, after passing a getter purifier, this gas reenters the optical pumping process. Target polarization is monitored by NMR and calibrated by optical measurement. Over 600 h of data taking a constant target polarization of P_{He}=(50±5)% was reached.

ANALYSIS

In the analysis of the $^3\vec{He}(\vec{e},e'n)$-experiment the polarized $^3\vec{He}$ nucleus serves as an effective polarized neutron target. The PWIA asymmetry A of the cross section with respect to a reversal of the electron helicity is for $\vec{e} - \vec{n}$ scattering:

$$A = P_e * P_n * \frac{a * sin\Theta * G_{En} * G_{Mn} + b * cos\Theta * G_{Mn}^2}{c * G_{En}^2 + d * G_{Mn}^2} \quad (1)$$

A depends on the polarization degrees of beam and target P_e and P_n, the form factors G_{En} and G_{Mn}, kinematical factors a,..,d (see [1]) and the angle ϑ

between neutron polarization $\vec{P}_n$ and momentum transfer $\vec{q}$. $\vec{P}_n$ can be orientated either perpendicular or parallel to $\vec{q}$ and was switched twice every hour by $90°$. Hence two asymmetries $A_\perp$ ($\vartheta=90°$) and $A_\parallel$ ($\vartheta=0°$) were measured in parallel. The asymmetry ratio

$$\frac{A_\perp}{A_\parallel} = \frac{a}{b}\frac{G_{En}}{G_{Mn}} \qquad (2)$$

is linear in the form factor ratio and independent of many sources of systematic errors like the absolute value of the polarization product $P_e \cdot P_n$. In order to consider the large detector acceptance and to select quasi elastic scattering, a kinematical reconstruction was performed on an event to event basis. Comparison of various reconstructed kinematical distributions to the respective Monte Carlo simulations provides a strong tool to select quasi elastic events. To consider the detector acceptance an analysis in two steps was necessary: In the first step the kinematical factors in eq. 1 were calculated for each event and averaged over all these events (for details see [1]). The second step is a Monte Carlo adaptation of this reconstruction which considers effects that are not determined for each event but are known in their statistical distribution: missing energy, radiative losses and energy loss of the neutrons in the lead shielding of the neutron detector. The missing momentum was restricted to $|\vec{P}_m| \leq 100$ MeV/c in order to reduce systematical uncertainties in both the kinematical reconstruction and the interpretation of the experimental asymmetries in the PWIA analysis as well as to reduce the background to a fraction of 8%. With a value of $G_{Mn}=(1.05\pm0.05)$ G_D slightly above the dipole value $G_D=\mu_n(1+Q^2/0.71(\text{GeV/c})^2)^{-2}$ the Q^2-averaged result $G_{En}=0.0352\pm0.0033\pm0.0024$ was obtained at $\bar{Q}^2=0.35$ (GeV/c)^2.

This result is in good agreement with the G_{En} value extracted from an exploratory experiment at a somewhat lower momentum transfer $\bar{Q}^2=0.31$ (GeV/c)^2 where a subset-up of the existing detector system was used [3]. Fig. 2 shows the G_{En} results of our full and exploratory ^{3}He-experiment, the preliminary results of our D-experiments and data extracted from elastic electron scattering from deuterium [8]. The four lines in the plot are two-parameter-fits to the same data set of [8] using different nucleon-nucleon potentials in the calculation of the deuterium wave function. Data points are plotted for the case of the analysis with the Paris potential only (fit: solid line).

The significant deviation of the A3-results with different targets at equal momentum transfer ($\bar{Q}^2=0.35$ (GeV/c)^2) is still an open question. Therefore the ^{3}He-measurement will be investigated in a further theoretical analysis in order to determine possible influences on the PWIA result due to final state interactions (FSI) and modifications of the neutron in the ^{3}He nucleus [13]. We will continue the ^{3}He-experiment at the higher momentum transfer $\bar{Q}^2=0.67$ (GeV/c)^2 at the MAMI spectrometer facility (A1-collaboration) [14].

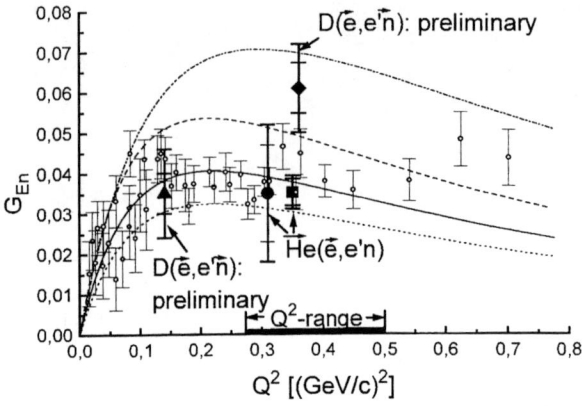

FIGURE 2. Results on G_{En}: full circle: exploratory ^{3}He–experiment [3]; full square: full ^{3}He–experiment (stat. and quadr. added total error); full triangle and full rhombus: D–experiment [4–7], preliminary; lines and open points: elastic D(e,e') [8], see text.

As a prerequisite to increase the luminosity, transportable target cells filled with polarized ^{3}He-gas up to 10 bar ($P_{He} \approx 50\%$) have been developed [15,16]. A successfull exploratory measurement at this momentum transfer has been carried out recently [17].

REFERENCES

1. J. Becker, Dissertation University of Mainz, 1997.
2. P. Grabmayr et al., Proc. of Conf. on Persp. in Hadr. Phys., (Trieste, Italy 1997).
3. M. Meyerhoff, D. Eyl, A. Frey et al., Phys. Lett. B 327 (1994) 201.
4. F. Klein, Proc. of PANIC96 (Williamsburg, USA 1996).
5. H. Schmieden, Proc. of SPIN96 (Amsterdam, The Netherlands 1996).
6. M. Ostrick, Dissertation in preparation, University of Mainz.
7. C. Herberg, Dissertation in preparation, University of Mainz.
8. S. Platchkov et al., Nucl. Phys. A 510 (1990) 740.
9. K. Aulenbacher et al., accepted for Nucl. Instr. and Meth.
10. M. Steigerwald et al., Proc. of SPIN96 (Amsterdam, The Netherlands 1996).
11. F. D. Colegrove, L. D. Schearer and G. K. Walters, Phys. Rev. 132 (1963) 2561.
12. G. Eckert et al., Nucl. Instr. and Meth. A 320 (1992) 53.
13. W. Glöckle and G. Ziemer, University of Bochum, private communication.
14. W. Heil (contact person), MAMI Proposal A1/4-95.
15. J. Becker et al., Nucl. Instr. and Meth. A 346 (1994) 45.
16. W. Heil et al., Phys. Lett. A 201 (1995) 337.
17. D. Rohe, Dissertation in preparation, University of Mainz.

Polarized high pressure $^3\vec{H}e$ Target at MAMI

Daniela Rohe[1]

for the

A1 - Kollaboration

P. Bartsch[2], D. Baumann[2], J. Becker[1], J. Bermuth[1], L.J. de Bever[4], R. Böhm[2], S. Buttazoni[4], T. Caprano[2], N. Clawiter[2], A. Deninger[1], S. Derber[2], M. Ding[2], M. Distler[2], A. Ebbes[2], M. Ebert[1], I. Ewald[2], J. Friedrich[2], J.M. Friedrich[2], R. Geiges[2], T. Grossmann[1], A. Klein[3], M. Hauger[4], W. Heil[5], D. Hofmann[2], A. Honegger[4], P. Jennewein[2], J. Jourdan[4], M. Kahrau[2], M. Korn[2], K.W. Krygier[2], G. Kubon[4], L. Lauer[1], A. Liesenfeld[2], H. Merkel[2], K. Merle[2], P. Merle[2], U. Müller[2], M. Mühlbauer[4], R. Neuhausen[2], E.W. Otten[1], Th. Petitjean[4], Th. Pospischil[2], D. Rohe[1], G. Rosner[2], H. Schmieden[2], I. Sick[4], R. Surkau[1], A. Wagner[2], Th. Walcher[2], G. Warren[4], M. Weis[2], S. Wolf[2], H. Woehrle[4], J. Zhao[4] and M. Zeier[4]

[1]Institut für Physik, Universität Mainz, D-55099 Mainz
[2]Institut für Kernphysik, Universität Mainz, D-55099 Mainz
[3]Dept. of Physics, Old Dominion University, Virginia USA
[4]Institut für Physik, Universität Basel, CH-4056 Basel
[5]Institut Laue - Langevin, F-38042 Grenoble

Abstract. In the frame of the A1-Collaboration at the Mainz Microtron a test measurement of doubly polarized $^3\vec{H}e(\vec{e},e'n)$ scattering from a high pressure target was performed in July aiming for the determination of the neutron electric form factor G_{en} at high momentum transfer ($Q^2 = 0.7(\text{GeV}/c)^2$) [1]. Due to the small value of G_{en} compared to G_{mn} a preferred procedure is to determine the asymmetry in the exclusive quasi elastic scattering of polarized electrons ($P \approx 70\%, I \geq 2\mu A$) from polarized $^3\vec{H}e$. The scattered electrons are detected in a high resolution magnetic spectrometer while the scattering angles of the outgoing neutrons are measured in a plastic scintillator. In this reaction the polarized $^3\vec{H}e$ nucleus serves as an effective polarized neutron target.

Because of the large magnetic field gradients caused by the spectrometer and limited space at the target place, the $^3\vec{H}e$ gas is polarized elsewhere and transported to the target place in specially prepared glass cells [2]. The glass cells are designed for high pressure (up to 10 bar) and with thin windows to prevent background. To reduce the relaxation due to the magnetic field gradients a μ-metal shielded guiding field of 4 Gauss is used. The guiding field is generated by three independent coils which permit the rotation of the target spin in any direction desired, especially perpendicular and parallel with respect to $\vec{q}$. So we are able to take the ratio of the asymmetries $A_\perp/A_\parallel$, which in first order depends only on kinematical factors and the ratio G_{en}/G_{mn}.

The method of metastable optical pumping is used to polarize the $^3\vec{H}e$ at about 1 mbar with a LNA laser ($\approx 6W$). Subsequently the gas is compressed by means of a two-stage-piston compressor up to 6 bar with a polarization of roughly 45% [3].

INTRODUCTION

After the sucessfull determination of the neutron electric form factor G_{en} at $Q^2 = 0.35$ (GeV/c)^2 in the A3-collaboration [4] we want to extend this measurement to higher $Q^2 = 0.67$ (GeV/c)^2. As before we use the double polarized exclusive reaction $^3\vec{H}e(\vec{e},e'n)$ in quasi elastic kinematics. At higher Q^2 some experimental modifications have to be performed. Because of the overlapping of the quasi elastic and inelastic peaks (Δ-resonance and π^o-decay) at higher Q^2 we now use a high resolution magnetic spectrometer ($\Delta E/E = 10^{-4}$) instead of a lead glass detector. The neutrons are detected by a segmented plastic szintillator specially adapted to stand the high background under forward scattering angles; its neutron efficiency is 30%. To compensate the smaller cross section at high Q a high pressure target cell of 6 bar was developed, which is made out of glass with thin glass windows and Havar foils to reduce background.

DETECTOR SETUP

The entire setup is shown in Figure 1. The nucleon detector covers a solid angle of 100 msr. It consists of 9 scintillator bars (50 x 10 x 1 cm^3) used as ΔE-detectors with 20 E-detectors (50 x 10 x 10 cm^3) behind it. So we are able to detect protons and neutrons simultaneously. Each detector has got two photomultipliers to reconstruct the vertical coordinate of the hit by TOF-difference. In direction of the target the neutron detector is shielded by 2 cm lead. Consequently, and due to a very efficient lead shielding around the target region and the neutron detector itself the background rates in the ΔE-detectors are about 50 - 100 kHz and in the E-detectors less than 5 kHz at 10 μA. The scattered electrons are detected in coincidence with the nucleon by spectrometer A with a solid angle of 28 msr, which fixes the momentum and the azimuthal and polar angles of the outgoing electrons. A gas Cerenkov-detector inside the spectrometer suppresses pions. Spectrometer B with a solid angle of 5.6 msr serves to determine the polarization product $P_e * P_{He}$ via the elastic scattering $^3\vec{H}e(\vec{e},e')$. Similar to the quasi elastic asymmetry we

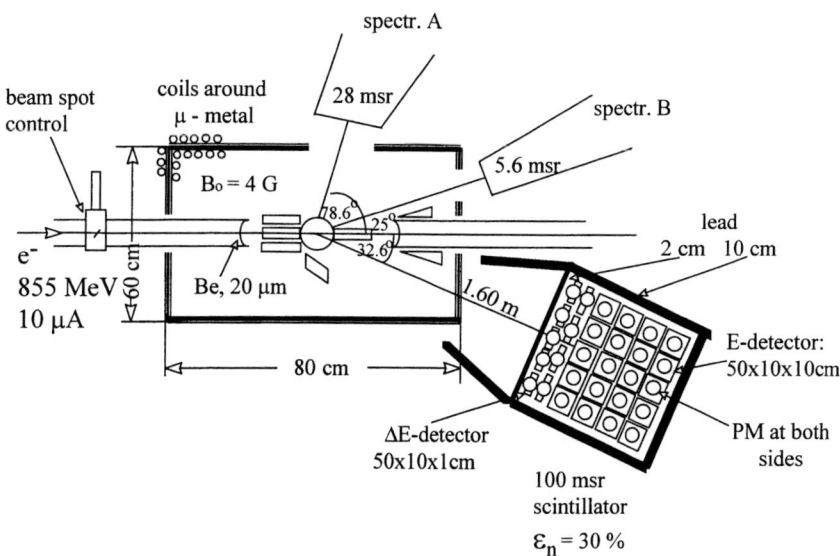

FIGURE 1. Setup in the spectrometer hall at MAMI for the G_{en}-measurement at $Q^2 = 0.67 (\text{GeV}/c)^2$

measure an elastic asymmetry which contains the elastic form factors of the ^3He-nucleus. The latter are precisely known (better than 2% [5]). Because of the high elastic cross section a statistical accuracy of 5% can be reached in 1 h for I = 10 μA, $P_e = 0.7$, $P_{He} = 0.4$ and a target pressure of 6 bar.

TARGET SETUP

The relative gradients of the magnetic guiding field for the polarized target should be less than $1 * 10^{-3}$ 1/cm in order to prevent relaxation of the polarization and to enable its measurement by NMR and AFP. Due to the stray field of the magnetic spectrometer with a gradient of $1.5 * 10^{-2}$ 1/cm and at a field value of $B_o = 2$ G, the guiding field (4 G) is produced within a μ-metal shielded box by 3 independent pairs of coils . With this construction we are able to rotate the target spin in any direction, so it permits the measurement of $A_\perp$ and $A_\parallel$, where the spin direction is perpendicular and parallel with respect to $\vec{Q}$.

During beam time the polarization and relaxation of the target is controlled by NMR and AFP. In the test run the target cells were filled up with 6 bar polarized ^3He (see below) and 6 mbar N_2 to suppress the formation of $^3He_2^+$ ions. After the filling procedure the target cells are transported in a

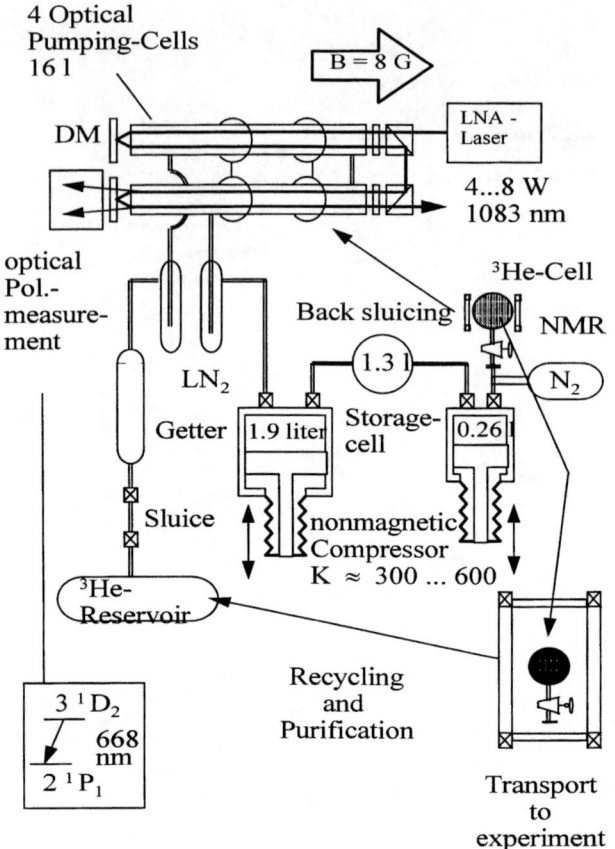

FIGURE 2. Polarizing of ^3He-gas with the Piston compressor

Helmholtz field to the target place. The relaxation rate of a Cs-coated target cell of Corning 7744 glass during a e^--current of 10 μA was $1/(32.5 \pm 1.8$ h); without beam $1/(40.5 \pm 0.5$ h). From this the relaxation due to beam is calculated to $1/(164.5 \pm 47$ h), which must be compared to the expected value of $1/(210$ h) [6]. The target cell itself was changed every 12 h, causing a break of 1 h.

POLARIZING THE ^3HE

The ^3He-gas is polarized up by metastable optical pumping [7]. First a fraction of 10^{-6} of 1 mbar ^3He is excited by a weak discharge into the metastable 2^3S_1 niveau. This state can be efficiently optical pumped by the circularly polarized light of a LNA-laser ($\lambda = 1083$ nm, 4-8 W). A magnetic field of 8 G serves as a guiding field for the spins. The nuclear polarization built up in the $^3S_1$-state by hyperfine coupling is then transferred to the ground state atoms by metastable exchange collisions which occur with a very large cross section of the order of 10^{-15} cm^2. In a sealed cell filled with 1.3 mbar ^3He we reached a polarization of nearly 80 %. To compress the gas up to 6 bar we use a two-stage-piston compressor made of titanium (see Figure 2) [3]. In the optical pumping region which consists of four 1 meter long cells (16 l) we reach a polarization of 65 % at a flux of 9 mbar l/min. In the target itself a polarization of roughly 45 % at 6 bar is obtained. After usage the cell is reconnected to the system, the depolarized ^3He is sliuced back through a two-stage-getter into the reservoir to remove the additional N_2 and other impurities and finally repolarized.

PRELIMINARY RESULTS

During a run time period of 100 h we tested the total setup for the G_{en}-measurement via the reaction $^3\vec{H}e(\vec{e},e'n)$. The electron source consisted of a strained layer GaAsP crystal and produced a polarization of 70 % and a current of at least 10 μA. The electron spin was flipped statistically with 1 Hz using a Pockels - cell. Altogether the collected charge on target was 384 μAh.

The starting polarization of the ^3He-target was between 42 and 47 % (measured by AFP) with relaxation times of 20 - 32 h. So we obtained a polarization of 30 % averaged over the run time. Alternately the spin of the ^3He was rotated parallel and perpendicular to $\vec{q}$. A first online analysis yielded for the parallel asymmetry, which is independent of the form factors a value of 18.3 % with a statistical error of 0.95 % in good agreement with the expected value of 19.3 %.

REFERENCES

1. W. Heil (contact person), MAMI Proposal A1/4-95.
2. W. Heil et al., *Phys. Lett. A* **201**, 337 (1995)
3. Becker et al., *Nucl. Instr. Meth. A* **346** 45 (1994)
4. thesis of J. Becker.
5. Amroun et al., *Nucl. Phys. A* **579**, 596 (1994).
6. Bonin et al., *Phys. Rev. A* **38**, 4481 (1988)
   Chupp, et al., *Phys. Rev. C* **45** 915 (1992)
7. Colgrove, Schearer and Walters, *Phys. Rev.* **132** 2561 (1963)

The SLAC E-154 ^{3}He Polarimeter

M.V. Romalis[4], P.L. Bogorad[4], G.D. Cates[4], and T.E. Chupp[2],
K.P. Coulter[2], E.W. Hughes[1], and J.R. Johnson[5], K.S. Kumar[4],
T.B. Smith[2], and A.K. Thompson[3], and R. Welsh[2]

[1] *California Institute of Technology, Pasadena, CA 91125*
[2] *University of Michigan, Ann Arbor, MI 48109*
[3] *National Institute of Standards and Technology, Gainesville, MD 20899*
[4] *Princeton University, Princeton, NJ 08544*
[5] *University of Wisconsin, Madison, WI 53706*

Abstract. We describe the NMR and Rb Zeeman frequency shift polarimeters used for determining the ^{3}He polarization in a recent precision measurement of the neutron spin structure function g_1 at SLAC (E-154). We performed a detailed study of the systematic errors associated with the calibration of the NMR polarimeter. A new technique was used for determining the ^{3}He polarization from the frequency shift of the Rb Zeeman resonance.

I INTRODUCTION

The ^{3}He polarimetry system for E-154 was designed to meet the challenges of a precision measurement with a small statistical error. The goal was to measure the ^{3}He polarization with a relative error of less than 5% and a good control over systematic errors. The polarization was measured by two independent methods. The first method used a traditional technique of Adiabatic Fast Passage NMR. The NMR signal was calibrated by detecting the Boltzmann polarization from a sample of water. The second method, used for calibration of the ^{3}He AFP signal, utilized the shift of the Rb Zeeman resonance frequency due to the Rb-^{3}He spin exchange. In each case we investigated several effects which can lead to systematic errors. The two method of polarimetry have comparable errors and are in good agreement with each other.

II NMR POLARIMETRY

Adiabatic Fast Passage (AFP) NMR was used to measure the ^{3}He polarization at regular intervals throughout the run [1,2]. The polarized target was placed inside AFP coils, which created an RF field H_1 orthogonal to the holding magnetic field H. The holding field H was swept linearly in time, $H = H_0 + \alpha t$, through the ^{3}He NMR resonance at $H_0 = 28$ G. The sweep rate was optimized to satisfy the AFP conditions. ^{3}He polarization losses per sweep were on the order of 0.1%. The NMR signal, induced in a set of pick-up coils orthogonal to the AFP coils, was measured by a lock-in amplifier as the field was being swept. Under ideal conditions the AFP signal is given by:

$$V(t) = \frac{GH_1\mu_{He}\,[^3\text{He}]\,P_{He}}{\sqrt{(\alpha t)^2 + H_1^2}} \quad (1)$$

where μ_{He}, $[^3\text{He}]$ and P_{He} are the magnetic moment, density, and the polarization of ^{3}He, and G is the gain of the detection system, which depends, among other things, on the dimensions of the ^{3}He cell and the pick-up coils. The broadening of the signal due to the field inhomogeneity and the lock-in time constant were negligible, and the data were well fit by Eq. (1), with residuals of less than 0.5%.

The NMR system was calibrated by detecting a signal from the Boltzmann polarization of protons in water. This calibration procedure is complicated by several factors. The Boltzmann polarization of protons in our conditions is $P_p = \mu_p H/k_B T = 7.5 \times 10^{-9}$, and their NMR signal is very small. For each set of calibration data we averaged 50 sweeps to reduce the random error to 1%. The shape of the proton AFP signal is different from Eq. (1) because the relaxation time in water is only about 2 sec., and the polarization of protons changes as the magnetic field is being swept. The relaxation process is parametrized by a longitudinal relaxation time T_1 and a transverse relaxation time T_2 [1]. Relaxation during the sweep affects both the height and the shape of the AFP signal. In the past [2] this has been the dominant source of the systematic error. It also makes the signal dependent on the speed and direction of the magnetic field sweep. To properly fit the water signals it is important to know the functional form of the signal.

Naively, one would expect that $T_2 = T_1$ in water, since the correlation time associated with the translation and rotation of the molecules is much shorter than the Larmor period [1]. However, measurements [3,4,7] show that $1/T_2 = 1/T_1 + 0.125\,\text{sec}^{-1}$ for neutral (i.e. pH=7.0) water. The reason for this turns out to be the presence of 0.037% of ^{17}O isotope in natural water [4]. ^{17}O has a nuclear spin of 5/2 and an effective scalar coupling to proton spins. It also has a relatively long correlation time of 10^{-3} sec, so that motional narrowing does not apply. The transverse relaxation time is also affected by the presence of the RF field H_1 [4]. For our conditions $1/T_2(H_1) = 1/T_1 + 0.033$ sec^{-1} and the height of the water signal is reduced by 0.4% compared to the case of $T_2 = T_1$. The longitudinal relaxation time T_1 is very sensitive to the temperature and the chemical impurities in water, such as dissolved oxygen [5,6]. We measured $T_1 = 2.4 \pm 0.3$ sec by comparing the size of the water signals obtained while the field was swept up and down through the resonance.

To fit the water signals it is convenient to have an analytic expression for their shape. If we set[1] $T_1 = T_2$ the polarization remains parallel to the effective field in the rotating frame, $\vec{H}_{eff} = (H - H_0)\hat{z} + H_1\hat{x}$, and its magnitude is governed by a single differential equation:

$$\frac{dP_{eff}}{dt} = \frac{1}{T_1}(P_{eq}(t) - P_{eff})$$

$$P_{eq}(t) = \frac{\mu_p}{kT}\left(\frac{H(H-H_0) + H_1^2}{\sqrt{H_1^2 + (H-H_0)^2}}\right) \quad (2)$$

It cannot be solved analytically. However, if $H_1/\alpha \ll T_1$ one can expand the resulting integral in powers of t/T_1 near the resonance and powers of $\alpha t/H_1$ away from the resonance, and get an analytic, though cumbersome, function for the signal. This function was used in our analysis. Figure 1 shows an averaged water signal with a fit based on equation (2). A simple fit to equation (1) is also shown for comparison. In each case we vary 5 parameters: the height, width, and center of the peak, as well as a linear background. By using a functional form that closely approximates the shape of the signal, we are reducing the sensitivity of the final result to small distortions caused by background fluctuations, etc. The residuals of the fit are consistent with random noise. The signal heights extracted from the sweeps up and down through the resonance are consistent within errors. In contrast, if one uses a simple fit in the form (1), the two heights are different by 20%. Our final result for the water signal height has a 1% statistical uncertainty, and a conservative 1.5% systematic error.

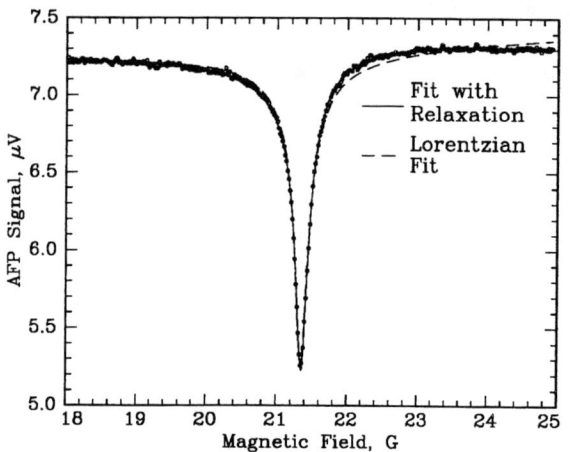

Figure 1: Average of 50 water signals. Solid line fit is based on Eq. (2), broken line fit on Eq. (1)

In using the signal from proton NMR for calibration of the ^{3}He signal, one also has to apply corrections due to slightly different dimensions of the ^{3}He and water cells, which are as large as 20%. To calculate these corrections we used a

---

[1] The correction due to $T_2 < T_1$ is applied separately.

complete geometrical model of the pick-up coils and ^{3}He cells. The model also allowed us to predict the absolute size of the water signal with an accuracy of 5%, which was in excellent agreement with the data. The final error of our AFP polarimetry is 3.4%, which comes in roughly equal proportions from uncertainties in the height of the water signal, ^{3}He density, dimensions and position of the cell, temperature of the water cell, and several other sources.

III ZEEMAN FREQUENCY SHIFT POLARIMETRY

The second method of polarimetry uses a ^{3}He polarization-induced shift of the Rb Zeeman resonance (also called Electron Paramagnetic Resonance). The resonance frequency is mainly shifted due to the Rb-^{3}He spin exchange interaction, the same interaction which is responsible for transferring the angular momentum to ^{3}He in spin exchange optical pumping. There is also a small additional shift due to the classical magnetic field created by ^{3}He, which depends on the geometry of the polarized sample. The total shift can be attributed to an additional magnetic field which is given for a spherical sample by the following equation [8–10]:

$$B_{He} = (8\pi/3)\,\kappa_0 \mu_{He} \left[^3\text{He}\right] P \qquad (3)$$

This field causes a shift in the frequency of the Zeeman resonance by $\Delta\nu = (d\nu\,(F,M)\,/dB)\,B_{He}$, where $d\nu\,(F,M)\,/dB$ is the derivative of the frequency of the Rb Zeeman transition ($F, M \to F, M-1$) with respect to the magnetic field, given by the well-known Breit-Rabi equation [11]. Here F and M are the quantum numbers of the Rb hyperfine manifold. The size of the shift is large (in our case about 20 kHz), and can be easily detected in a typical field of 20 G, where the Rb Zeeman frequency is 9.3 MHz. The constant κ_0, parameterizing the imaginary part of the Rb-^{3}He spin exchange cross-section, has been measured with an accuracy of 1.5% in [9,10]. Its value in the temperature range 100-180°C can be parameterized as follows: $\kappa_0 = 4.52 + 0.00934\,T$ (°C).

For this experiment we implemented a new method of measuring the frequency shift, which is ideally suited for monitoring the ^{3}He polarization during optical pumping. The system allowed measurements of the polarization without access to the target and proved robust under accelerator conditions. The EPR frequency was detected optically, by monitoring the fluorescence while optically pumping the cell. By applying an RF field at the frequency of the $F = 3$, $M = 3 \to 2$ transition we partially depolarized the Rb atoms. This, in turn, caused an increase in the intensity of the fluorescence emitted from the cell, which was detected by a photodiode.

The equipment setup for EPR measurements is shown in Figure 2. The RF field was created by a coil mounted on the side of the oven. The fluorescence

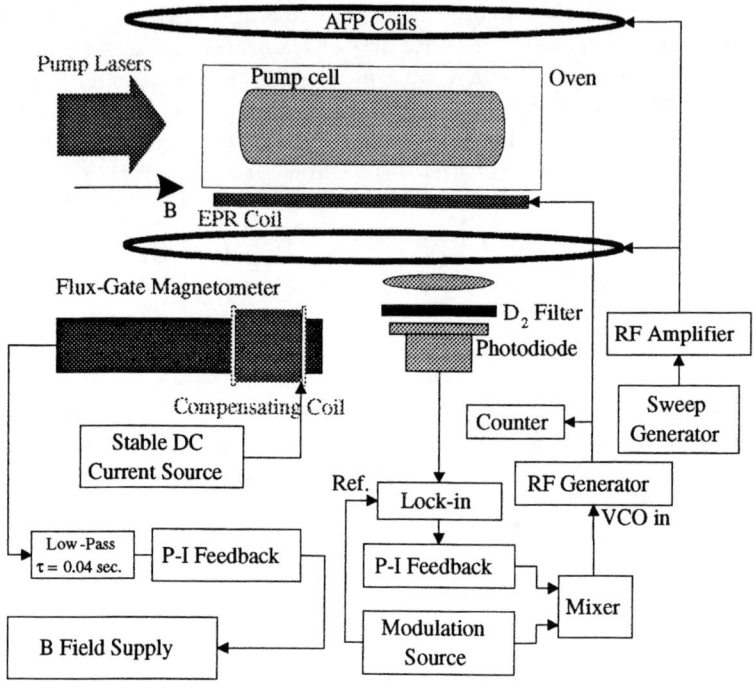

Figure 2: Equipment setup for the frequency shift polarimetry.

from the cell was detected by a photodiode with a D_2 filter to block the scatter from the pumping lasers. The frequency of the RF field was modulated using a Voltage Controlled Oscillator (VCO). The signal measured by the lock-in amplifier referenced to the modulation frequency was proportional to the derivative of the resonance line shape. The feedback circuit adjusted the DC level at the input of the VCO to keep the lock-in signal zero, i.e. locked to the center of the line. To reduce the noise in the frequency measurement the holding magnetic field was stabilized to one part in 10^5 by a feedback system based on a Bartington flux-gate magnetometer.

To isolate the frequency shift due to the ^{3}He polarization we periodically reversed the direction of the polarization. The reversal was done by AFP, only instead of sweeping the magnetic field through the resonance we swept the RF frequency. We utilized the same coils, RF amplifier and generator as used for NMR polarimetry. The measurement cycle consisted of recording the EPR frequency for about 1 min., flipping ^{3}He spins by AFP and recording the frequency for another minute. This procedure was repeated several times. A typical data set is shown in Figure 3. The data are fit allowing a small amount of polarization loss per cycle, which is due to the AFP losses and the decay of the polarization during one half of the cycle, when the lasers are pumping in the direction opposite to the ^{3}He polarization. The quality of the data is very good and the size of the frequency shift can be extracted with a error of less than 0.5%.

Two corrections have to be applied to the frequency shift data. The target pumping cell, where the measurements were performed, was not spherical, so a small additional shift due to the classical magnetic field should be added to equation (3). It resulted in a 4.6% correction and a 1.3% error. The error is due to our limited knowledge of the region in the pumping cell which was sampled by the photo-diode. Also, the frequency shift measures the polarization of ^{3}He in the pumping chamber of the cell. There is a small polarization gradient between the pumping and target cells due to a finite diffusion time. It results in a 3.8% correction and a 1.5% error. The correction was calculated by using a model of diffusion between the target and the pumping cells. The model was checked by measuring the polarization build-up in the target cell in the first hour of a spin-up. The error is due to the uncertainty in the target spin relaxation rates. The total error of the frequency shift polarimetry method is 3%, coming from the uncertainty in the value of κ_0, the density of ^{3}He, and the two corrections described above.

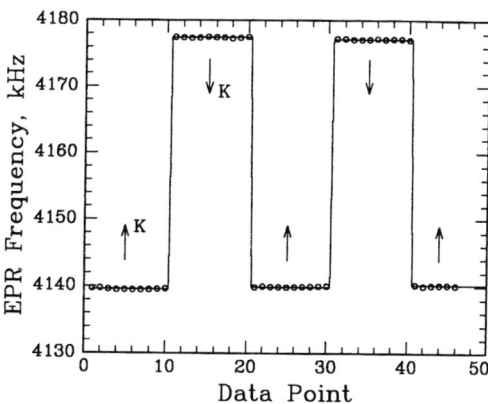

Figure 3: Typical set of EPR data.

IV POLARIMETRY RESULTS

Since the density of the ^{3}He is the only common parameter for both polarimetry techniques, it was also measured in two independent ways. We calculated the density with an accuracy of 1% using pressure measurements during the filling of the cells. In addition, the density was determined by measuring the width and shift of the Rb absorption lines D_1 and D_2, which are pressure-broadened by ^{3}He. Using the data from [12] as a calibration, the density of ^{3}He in the cells was determined with a error of 1%. The two methods are in excellent agreement.

The polarization of ^{3}He was measured by two independent methods, using Adiabatic Fast Passage and the Rb Zeeman frequency shift. The uncertainties are 3.4% for the AFP method and 3.0% for the frequency shift method. The results of the two methods differ by 5.7%. Since the errors in each case come from many independent sources, in comparing the two methods we combine their errors in quadrature, which gives a total error of 4.6%. Thus, their difference is 1.2 times larger than their combined error, and the two methods are in good agreement. For the final result we averaged the two results and used a conservative 4.8% error.

In conclusion, we described the two methods of ^{3}He polarimetry used in E-154. For the AFP method we considered in detail the effect of thermal relaxation of protons in water. We also described a novel implementation of the Zeeman frequency shift polarimetry suitable for a nuclear physics experiments. Because this technique was used for the first time, several refinements are still possible. The uncertainty due to the classical magnetic field shift can be reduced by restricting the region of the cell sampled by the photo-diode. The error due to the polarization gradient between the target and pumping cells can be reduced by making dedicated measurements designed to study the effect. Because the technique relies on a frequency shift, which can be measured with high accuracy, we believe that with these refinements the error can be reduced below our value of 3%.

This work was supported by the Department of Energy contracts: DE-AC03-76SF00515 (SLAC), DE-FG02-90ER40557 (Princeton) and by an NSF Grant 9217979 (Michigan).

REFERENCES

1. A. Abragam, *Principles of Nuclear Magnetism*, (Oxford University Press 1961).
2. P.L. Anthony et al., (the E-142 Collaboration), Phys. Rev. D **54**, 6620 (1996).
3. S. Meiboom, Z. Luz, D. Gill, J. Chem. Phys. **27**, 1411 (1957).
4. S. Meiboom, J. Chem. Phys. **34**, 375 (1961).
5. G. Chiarotti, G. Cristiani, L. Giulotto, Nuovo Cimento **1**, 863 (1955).
6. J.H. Simpson, H.Y. Carr, Phys. Rev. **111**, 1201 (1956).
7. R.E. Glick, K.C. Tewari, J. Chem. Phys. **44**, 546 (1966).
8. S.R. Schaefer, G.D. Cates, T.R. Chien, D. Gonatas, W. Happer, T.G. Walker, Phys. Rev. A **39**, 5613 (1989).
9. M.V. Romalis and G.D. Cates, (in preparation).
10. M.V Romalis, Ph.D. Thesis, Princeton University, 1997.
11. G.K. Woodgate, *Elementary Atomic Structure*, (Oxford University Press, Oxford, 1989).
12. M.V. Romalis, E. Miron, G. D. Cates, (submitted to Phys. Rev. A).

³He POLARIMETRY IN THE HERMES EXPERIMENT

A. P. Dvoredsky

W. K. Kellogg Laboratory, California Institute of Technology, Pasadena, CA 91125

Representing the HERMES Collaboration

Abstract

We describe two polarimetry techniques used in the HERMES experiment. They are both based on the principle of measuring the rate and circular polarization of photons emitted from excited states of target atoms and can be used together to directly access information regarding the target atoms which interact with the beam.

Polarimetry Measurements.

In 1995, the HERMES experiment began measurements of spin-dependent structure functions via deep inelastic scattering of positrons on nucleons in an internal ³He target used at the HERA ring at DESY. The target is described in detail in D. De Schepper's contribution to these proceedings and hence here we summarize briefly the method of polarization via optical pumping. A discharge is used to produce the metastable 2^3S_1 atomic state, and then circularly polarized 1083 nm laser light is used to produce electron polarization in the metastable state; this polarization is transferred to the neutrons via the hyperfine coupling. The atoms then undergo metastability exchange collisions where they retain their nuclear polarization in the ground state.

The polarization process takes place in a pumping cell external to the storage ring. In this pumping cell, the discharge also excites the 3^1D_2 level, where the hyperfine interaction transfers the nuclear polarization back to the electronic system. The resulting 667 nm photons from the decay to the 2^1P_1 level have circular polarization which can be measured and related to the nuclear polarization [1]. The device which performs this analysis is the pumping cell polarimeter (PCP), where the circularly polarized photons pass through a a rotating quarter-wave plate, a linear polarizer, and an interference filter centered at 667 nm. The light then enters a PMT tube, and the signal is split and directed into a lock-in amplifier to measure the AC component as well as a DC amplifier. The nuclear polarization P is simply given by

$$P = \alpha \frac{AC}{DC}. \quad (1)$$

The proportionality constant α is given by several factors [2], including the dependence of the nuclear polarization on the pressure in the pumping cell, the magnetic field, and the observation angle; these factors are all determined off-line.

Because of this signal integration, the only uncertainty is systematic rather than statistical. Various effects are discussed in detail in a general description of the HERMES target by De Schepper, *et al.*, [3], and so here we note only that the total relative error on this measurement is 3.4%. The greatest measurement error results from determination of the DC component, as this contains an offset due to the light from the discharge; this level varies over time, contributing 1.4% to the error.

The dominant calibration errors result from understanding the proportionality constant α. In particular, knowledge of the dependence on the pumping cell pressure and on the magnetic field contribute 2.0% each to the total error; this calibration was determined with NMR previously [4].

In addition to the PCP, a new device was used to study the polarization of the target atoms in the storage cell. The Target Optical Monitor (TOM) operates on a similar principle as the PCP. Instead of a RF discharge to excite atomic states, the stored positron beam is used, as it excites these states via the Coulomb interaction. The nuclear polarization is again transferred back to the electronic system via the hyperfine interaction.

The $4^1D_2 - 2^1P_1$ transition is used for the TOM, as its 37 ns lifetime is well suited to the HERA 96 ns beam structure. Furthermore, the relation between the circular polarization and the nuclear polarization at low pressures is well described by results of Pinard and Van Der Linde [1].

The apparatus which directs the photons from the beamline to the detector has been described in detail previously [5] and here we review only the optical components. Circularly polarized photons are directed through a rotating quarter wave plate, then a linear polarizer, then through a beamsplitter where they can pass into one of two phototubes: PMT1, which has an interference filter centered at the $4^1D_2 - 2^1P_1$ wavelength of 492 nm (3 nm bandwidth); and PMT2, with an interference filter at 460 nm (10 nm bandwidth). There is no ^{3}He line at 460 nm and hence PMT2 is used to sample the background, which is assumed to have roughly the same intensity at that wavelength as at 492 nm. Each phototube thus records a sine wave whose amplitude is proportional to the circular polarization of the incident photons. Due to the low rates, single photon counting techniques are used and an asymmetry is constructed:

$$A \equiv \frac{N_+ - N_-}{N_+ + N_-} = \left(\frac{2}{\pi}\right)\left(\frac{1}{1 + B/S}\right) P_\gamma. \qquad (2)$$

$N_{+(-)}$ refers to the number of counts in the positive (negative) lobe of the sine wave and P_γ is the measured circular polarization. S/B is the signal-to-background ratio (typically 1:2), determined by measuring the rates in both PMT1 and PMT2 with the target empty. Most of the background is seen even with a closed shutter in front of the exit window and hence does not come from optical photons, such as synchrotron light, from inside the beampipe.

Preliminary studies [5] indicated that the measured asymmetry was approximately 50% of its expected value, due probably to diffuse reflection of the photons off the cell walls. For this reason, the absolute nuclear polarization cannot be computed directly from this asymmetry. However, the asymmetry can be used to extract a relative polarization measurement when compared to the PCP and hence provides a useful cross-check as well as additional information regarding possible polarization losses between the pumping cell and the storage cell.

The two phototubes' signals are plotted in Figure 1. The target cell is filled with gas at t=0.5 h and the 20 minute time structure of the spin flip for the target is clearly visible in PMT1, where the asymmetry clearly changes sign, whereas PMT2 sees no effect. Both phototubes see a non-zero asymmetry even with the target empty, indicating a background with circular polarization. This is perhaps due to synchrotron radiation which is reflected downstream of the target, where it picks up a phase and becomes partially circularly polarized.

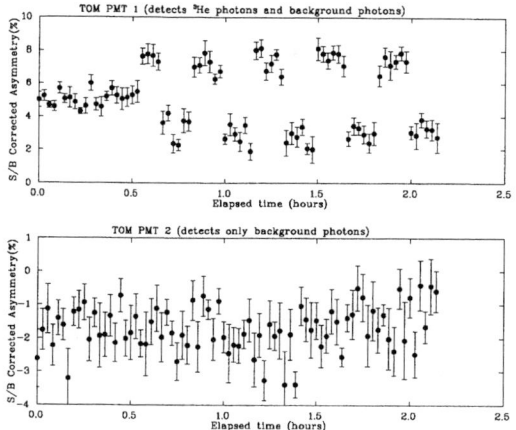

Figure 1: A comparison of the signal seen in each of the two TOM phototubes. Top: PMT1, which sees the 492 nm ^{3}He line and hence clearly sees the structure of the spin flips. Bottom: PMT2, which only sees background light at 460 nm.

In Figure 2, the polarization measured with the TOM and the PCP are compared; the time structure of the spin flips is again clearly visible. The error bars shown on the TOM are purely statistical and can be reduced to approximately 2% over the course of an 8-hour fill, though a systematic error of 6% remains, largely because of the imprecision of the S/B measurement.

Additional Measurements

By comparing the TOM and the PCP, additional information could be gained regarding possible depolarization mechanisms inside the target. Specifically, the dependence of the polarization on the target temperature was examined. Depolarization was expected to be negligible above 12 K, according to measurements

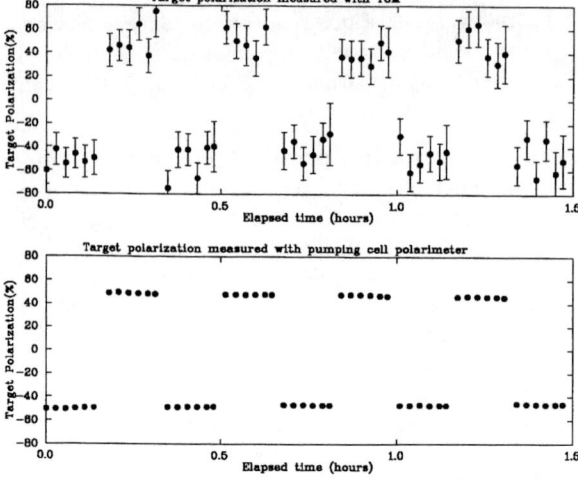

Figure 2: A comparison of the polarization measured in the TOM versus the measurement in the PCP. The time structure of the helicity flip is clearly seen. The TOM measurement has been normalized to the PCP measurement.

made on a similar target [6]. This was confirmed in the range from 18-60 K, as is seen in Figure 3. In addition, the dependence on the stored positron beam current was studied, and the relative change in the polarization in the target was seen to be less than 7% (see Figure 4).

Summary

Two techniques for polarimetry in the internal ^{3}He target at HERMES have been described. In the pumping cell, the nuclear polarization can be determined from analysis of the circular polarization of the decay photons from states excited by an RF discharge. In the target cell, the TOM can be used to analyze photons from states excited by the positron beam and hence provides a relative measurement of the polarization of the atoms which actually interact with the beam. Together these devices can be used to study possible depolarization mechanisms as well as simple polarimetry.

References

[1] M. Pinard and J. Van Der Linde, Can. J. Phys. **52**, 1615 (1974).

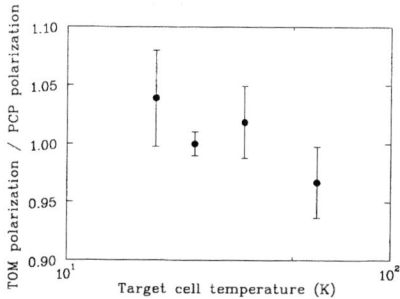

Figure 3: The ratio of the TOM to PCP polarization measurements as a function of cell temperature. No evidence for depolarization at low temperatures is seen.

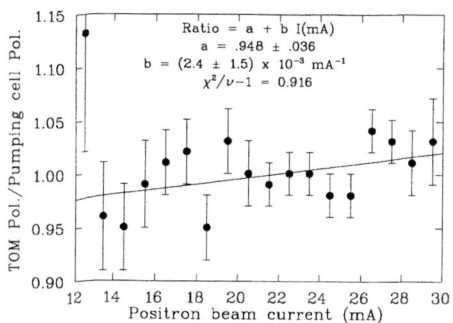

Figure 4: The ratio of the TOM to PCP polarization measurements as a function of positron beam current. No evidence for depolarization by the beam is seen.

[2] C. E. Jones. *A Measurement of the Spin-Dependent Asymmetry in Quasielastic Scattering of Polarized Electrons from Polarized ^{3}He*. PhD Thesis, California Institute of Technology, 1992.

[3] D. De Schepper, *et al.*, *Nuclear Instruments and Methods*, to be submitted.

[4] W. Lorenzon, T.R. Gentile, H. Gao and R.D. McKeown, Phys. Rev. A **47**, 468 (1993).

[5] M. L. Pitt, *et al.*, *International Workshop on Polarized Beams and Polarized Targets*, ed. Hans Paetz gen. Shieck and Lutz Sydow (World Scientific, London 1996) pp. 413-417.

[6] W. Korsch, *et al.*, Nuclear Instruments and Methods A **389**, 389 (1997).

Highlights from HELION97

M. Tanaka

Kobe Tokiwa Jr. College, Ohtani-cho 2-6-2, Nagata, Kobe 653, JAPAN

Abstract

A brief review of the International Workshop on Polarized ^{3}He Beams and Gas Targets and Their Application (HELION97), which was held from 20th through 24th January, 1997 in Kobe, Japan, is presented paying particular attention to the highlights from the workshop. Since topics of the workshop covered almost all of the fields related to polarized ^{3}He and its surroundings, i.e., from particle physics to medical applications, this was really the first meeting through which we could comprehensively discuss a variety of problems raised in the individual field on the common ground. In this respect, the subtitle, *from Quark to Life* seems to hit the mark as another title of HELION97.

AIM OF HELION97 WORKSHOP

The international workshop on polarized ^{3}He beams and gas targets and their application (HELION97) was held on 20th through 24th, January 1997 in Kobe, Japan under the auspices of RCNP, Osaka University. About 100 participants from 9 different countries attended on this workshop. As symbolized by a sub-title of the HELION97, i.e., *from Quark to Life*, topics covered a really broad range of subjects from particle physics to medical applications with polarized ^{3}He from either experimental or theoretical view points. In fact, the discussions covered were;

1. Technology in polarizing ^{3}He and heavier nuclei,

2. Spin structure of nucleon,

3. Symmetry breaking test by polarized ^{3}He,

4. Low and intermediate nuclear physics with polarized ^{3}He,

5. Application of polarized ^{3}He to other fields including;

   (a) Polarized neutron source,

   (b) Atomic physics,

   (c) Fusion problem,

(d) MRI (Magnetic Resonance Imaging) with hyperpolarized ^{3}He, and ^{129}Xe for medical diagnosis,

(e) Astrophysics,

(f) Low temperature Physics,

(g) Fundamental physics, for instance, geometrical phase in quantum mechanics,

In what follows, I will at first touch the history of the polarized ^{3}He to help you understand why the HELION97 workshop should have taken place. Since the first discovery of the ^{3}He isotope in the mid 1930th[1] and the subsequent determination of nuclear moment[2] and spin[3], it had received only a restricted interest for a long time. However, after the invention of optical pumping by Kastler[4], much effort had been paid on the production of a polarized ^{3}He gas only from interest to know the polarization mechanism itself. As a matter of fact, even when the first meeting on the polarized ^{3}He and its application took place at Princeton in 1984, ^{3}He was regarded as a special tool only for material physicists and nuclear physicists.

However, owing to novel discoveries or inventions after that as shown in the following, polarized ^{3}He was gradually getting its status. In 1986 a lot of minery ^{3}He gas was discovered on the moon surface[5]. This fact immediately inspired plasma physicists to realize an aneutronic fusion reactor using the polarized ^{3}He and polarized deuteron whose concept was first suggested by Goldhaber in the beginning of 80th[6]. In 1987 the deep inelastic measurement of lepton on the spin structure function of proton, $g_1^p(x)$ by EMC gave rise to an sensational topic[7] in particle and nuclear physics because the experimental result showed a breakdown of a simple interpretation of nucleon that a static property of nucleon is explained by a naive quark model while high energy scattering is understood with a parton model motivated by QCD. As a counter part of proton the spin structure of neutron was then considered to be of particular importance to disentangle the above difficulty. For this purpose, the construction of polarized ^{3}He targets were initiated at various high energy laboratories in the world since the polarized ^{3}He is approximated by polarized neutron (87%) with a small amount of polarized proton (-2.7%) in a fully polarized ^{3}He target.

Besides these novel discoveries, one of the most epoch-making inventions was recently placed on MRI (magnetic resonance imaging) in vivo with hyperpolarized rare gases (^{3}He[8] and ^{129}Xe). American and European physicists in collaboration with medical people have made a substantial role in this field by using rare gases polarized with either spin exchange or metastability exchange method.

Based on the above aspects, it was timely, we believed, to comprehensively discuss the various problems related to the polarized ^{3}He and its surroundings. In the next section, I will try to describe highlights from topics discussed in the workshop particularly emphasizing on the experimental view points.

HIGHLIGHTS

Technology in Polarizing ^{3}He, and Heavier Nuclei

1. Polarized Target

Since the first demonstration of the transfer of spin angular momentum from optically pumped Rb atoms to ^{3}He 35 years ago, several advances have resulted from detailed study of optical pumping and spin exchange. In HELION97 workshop, four methods have been established or proposed to polarize ^{3}He-gas;

1. spin exchange (SE) with an optically pumped Rb-vapor,

2. metastability exchange scattering (ME) with optically pumped metastable ^{3}He-atom (^{3}He*) in their 1s2s $^3S_1$-state.

3. Solid ^{3}He target,

4. Pomeranchuck cooling.

- **SE** The recent development of high power semiconductor laser has enable to provide high density ^{3}He gas, for example, $\sim 10^{20}$ atoms/cm^3, with high polarization. The disadvantage of this method is the small spin exchange rate, which needs much long time to attain large polarization. Anderson found that K-K spin relaxation rate about a factor of 15 smaller than the measured Rb-Rb spin relaxation cross section, which suggests that optical pumping with K is much better than Rb because of longer relaxation time.

- **ME** Owing to powerful LNA-laser a remarkable development of the latter method could be achieved. Advantage of this method is that the time necessary to polarize is much shorter than the former method, whereas serious disadvantage of this method is that the polarization could be realized only in a low pressure around 1 mbar. Therefore, the gas density should be increased by mechanical compression by means of

Toepler pump. Heil presented ME polarized target successfully operating at Mainz for various remote uses. By choosing a suitable container avoiding paramagnetic impurities, they succeeded in realizing relaxation time longer than 100 hours with a polarization of 0.5 at 1 bar. Schepper presented ME polarized target stored in the cryogenic storage cell used in the HERMES experiment with HERA to increase the target thickness.

- **Solid ^{3}He target** Haase presented a 0.4 mole solid ^{3}He target, cryogenically polarized at 12 mK in a field of 7 T with a polarization of 0.38 for use of absorption measurement on polarized neutron to investigate the excited state of n-^{3}He system. Such a low temperature solid target will hopefully be used for the experiments which do not accompany with much energy dissipation like a neutron.

- **Pomeranchuck cooling** Melting curve for ^{3}He shows a minimum at 320 mK unlike ^{4}He, where the entropy of the liquid phase becomes smaller than that of the solid phase. This means that solidification produces cooling, namely, Pomeranchuck cooling. A method for polarizing liquid ^{3}He was suggested by Castaing and Nozieres; quick melting of polarized solid under the strong magnetic field ($\sim$ 15T) in a time shorter than relaxation time (15000 sec. at 10 mK) of the nuclear polarization gives a liquid with a polarization equivalent to the solid phase. Frossati estimated that production of 100 to 1000 l/day polarized ^{3}He is available. This is enough for a fuel of ^{3}He + D reactor to be used as a 1GW electric power plant.

The polarization degree and density of ^{3}He is another important quantity to be determined. Romalis presented a precision polarimetry of dense polarized ^{3}He target at SLAC to the degree of % by using a standard technique of adiabatic fast passage and a shift of the Rb Zeeman resonance frequency due to the polarization of ^{3}He. ^{3}He density was measured by observing pressure broadening of Rb D_1 and D_2. Another impressing presentation was given on a precision frequency measurement by Chupp. He mentioned a development of spin exchange pumped masers of one and two species which will hopefully enable to measure atomic electric dipole moment (EDM) on ^{129}Xe for T- or CP- violation measurement. Their EDM measurement is accomplished by measuring a change in the Zeeman energy when an electric field is applied. He used polarized ^{3}He as a monitor of the magnetic field variations due to the leakage current induced by the high voltage. He speculated that ^{129}Xe EDM smaller than 10^{-28} e-cm may be possible by measuring a precision frequency shift with this device.

There were presentations on the production and application of the radioactive polarized nuclei. In particular, heavy ion induced reactions have been conveniently used for production of nuclear polarizations.

2. Polarized ^{3}He Beam

As compared with the development of the polarized ^{3}He gas, the development of the polarized ^{3}He beam is still infancy. In fact, though a variety of ideas to polarize ^{3}He ions were offered in the last ^{3}He workshop at Princeton[9], most of them are nowadays interrupted or abandoned. In HELION97 workshop, at least, three types of polarizing methods were presented. Among them, the optical pumping ion source (OPPIS) successfully used for producing polarized H$^-$ beams have been considered to be one of the most reliable methods enabling high current beams of ^{3}He with high polarization and small beam emittance. It is noticed as Zelenski mentioned that the polarized H$^-$ current with 10-30 mA may be possible in a pulsed mode, which is, amazingly, comparable to the beam current available from unpolarized ion sources. Anderson suggested a new prospect to produce nuclear spin polarized ^{3}He ions by the OPPIS. A metastable 2^3S ^{3}He atomic beam selectively formed by ^{3}He$^+$ with a kinetic energy less than 1 keV and a polarized sodium atom optically pumped would be nuclear polarized by using a Sona field. The polarized metastable atom, then, will be ionized by using collisions in a gas target.

The concept of the OPPIS was extended to a novel principle, "Electron pumping"; the optical pumping uses absorption of circularly polarized photons, while the electron pumping uses capture of polarized electron. To realize the electron pumping for ^{3}He, polarized thick Rb vapor ($\sim 10^{15}$ $atoms/cm^2$) is needed, which is an order of magnitude thicker than the value imposed for normal OPPIS ion sources. In RCNP, a new polarized ^{3}He ion source is developed to experimentally prove the electron pumping as presented by Yamagata. He also mentioned that a solid state pulsed laser is developed to produce a super-thick polarized Rb vapor.

The OPPIS can also be used for production of radio-isotopes. Zelenski proposed to produce nuclear polarized ^{8}Li$^-$ ion as one of possible experiments for the ISAC project at TRIUMF; ^{8}Li atom is populated after penetration of ^{8}Li$^+$ ion through a sodium vapor and polarized by means of optical pumping. After polarization be realized, the Li beam is ionized to Li$^-$. It is estimated that about 10% of the primary beam can be optically pumped to a polarization of 80-90%.

Belov presented a new scheme for a polarized ^{3}He^{2+} ion source based on resonant charge-exchange between polarized ^{3}He atoms and unpolarized ^{4}He^{2+} ions, where polarized atoms are produced by optical pumping with metastability exchange.

Spin structure of nucleon

At SLAC and HERA, asymmetry measurements for electron (positron) deep inelastic scattering with polarized ^{3}He targets are undergoing to investigate the neutron spin structure function, $g_1^n(x)$, where x is Bjorken parameter. Of particular interest is $g_1^n(x)$ integrated over all x which gives the parton or quark polarization in the polarized nucleon weighted by the square of the quark charges. Chupp presented a preliminary data at SLAC with good statistics far beyond the data so far. De Schepper presented a preliminary result of deep inelastic scattering of positron from polarized ^{3}He with HERMES.

On the other hand, at MIT-Bates, TJNAF (presented by Gao, Meziani), and AmPS storage ring at NIKHEF (presented by van den Brand), and Mainz (presented by Heil) measurements of electromagnetic form factors corresponding to the distribution of charge and magnetization within the nucleons have been attempted or planned to check a various models based on QCD. In particular, through quasielastic electron scattering with the polarized ^{3}He target the detailed study of the neutron electromagnetic form factors became possible. van den Brand further discussed on the S' (mixed symmetry state) and D-wave parts of the ^{3}He ground-state wave function.

Symmetry breaking test by polarized ^{3}He

Detailed analyses on charged electro-weak currents which provide a possible test of the standard model and beyond were discussed by Govaerts and Souder with a new technique prepared for the asymmetry measurement of muon capture on polarized ^{3}He to the triton channel. The choice of ^{3}He instead of proton is that the proton experiment is too difficult to achieve the desired level of precision.

The most stringent limit of the T-violation will be done by the direct observation of the electric dipole moment (EDM) for polarized rare gas by Chupp as mentioned already. Masuda discussed a rigorous test of P- and T-violation through the neutron transmission experiment with polarized ^{3}He.

Low and intermediate nuclear physics with polarized ^{3}He

A various type of experiments with hadron beams at intermediate energy region have been carried out or planned. Uesaka presented the measurement of polarization correlation coefficient for the $^3He(d,p)^4He$ reaction with polarized deuteron with 270 MeV through which the D-state contribution in ^{3}He can be extracted. Sowinski discussed the spin dependent momentum distributions of protons and neutrons in ^{3}He through ^{3}He(p,pN) reaction with polarized ^{3}He and polarized proton through which the clear evidence for the S' (mixed symmetry) state in ^{3}He was observed.

On the other hand, there was no report on the experiments with polarized ^{3}He beam. This is due to the fact there is no polarized ^{3}He ion source available. Nevertheless, many theoreticians attempted to calculate expected results with it. Sakuragi and Kim, respectively, speculated expected analyzing powers for elastic scattering and for giant resonances excited by the (^{3}He,t) reactions with polarized ^{3}He at intermediate energies.

Application of polarized ^{3}He to other fields

Polarized neutron source

Intense polarized neutron beams are required for a broad range of experiments in condensed matter and nuclear physics. As a one of the most promising polarizer of neutron, a transmission polarization filter based on gaseous nuclear spin-polarized ^{3}He is developed; this uses the fact that the neutron absorption cross section is highly spin dependent due to a broad resonance in the ^{4}He* ($J^\pi=0^+$). Heil presented a project at the High Flux Reactor, Grenoble and Keith discussed a joint IUCF/NIST project in which a polarized neutron in the energy range 10^{-3}-10^1 eV will be available.

Atomic physics

Andersen presented current understanding of electron transfer processes in single collisions of light ions, such as H^+, He^+, and Li^+ with aligned Na atoms in the 3s ground state or 3p excited state at impact energies of a few keV. A particular emphasis was placed on possible polarized ^{3}He ion source. He concluded that every polarization observable at low energy can be predicted correctly by theories based on either AOM (atomic orbital model) or MOM (molecular orbital model).

Fusion problem

Since the discovery of the lunar ^{3}He, fusion reactor based on ^{3}He + D drew much attention. Momota presented use of polarized ^{3}He for the energy production. For this purpose, he offered to construct the D-^{3}He fueled reactor based on a field reversed configuration. Weller presented absolute differential cross section, and the analysing power of the ^{3}He(d,γ)^{5}Li reaction at fusion region.

It was found that the tensor force is of crucial importance, which suggests that the tensor force should be taken into account in evaluating the fusion cross section for ^{3}He + D.

MRI

Since the first success in the magnetic resonance imaging (MRI) of lung tissue filled with hyperpolarized noble gas by Happer et al., hyperpolarized ^{3}He, and ^{129}Xe gases have been used for the MRI of animal and human organs, where a terminology, 'hyperpolarized' comes from the fact that the noble gas is polarized to a degree far beyond the ordinary Boltzmann-equilibrium which is of order 10^{-5} to 10^{-6} smaller than the usual NMR. Heil presented latest results applied on the human lung in collaboration with the NMR group of radiology department at Mainz. They succeeded in adapting their FLASH-3D imaging sequences within 7 seconds which is shorter than T_1 relaxation time of ^{3}He. The image shows a sharp demarcation of the lung against diaphragm, heard, chest wall,and blood bessel, though the conventional ^{1}H-NMR of the thorax only shows images of tissue water and fatty acid protons. Albert and Zhao presented the advantage of ^{129}Xe imaging rather than ^{3}He one by stressing that Xe readily dissolves in blood, and the T_1 of dissolved ^{129}Xe is long enough for sufficient polarization to be carried by the circulation to tissues. He also discussed that the disadvantage of the non-renewable hyperpolarization could be eliminated by a pulse sequence technique allowing high speed imaging techniques (scan time $\sim$ 1-2 seconds). Welsh presented the study of time dependences of the ^{129}Xe magnetization in the whole body and its build-up in the brain.

Astrophysics

Kubono discussed stellar nuclear reactions in explosive hydrogen burning (rp-) process were investigated experimentally using ^{3}He induced reactions. Kajino discussed that ^{3}He and ^{3}H play the critical role in primordial Big-Bang nucleosynthesis of ^{6}Li, ^{7}Li and ^{9}Be

Fundamental physics

Wäckerle reported on the measurement of the geometric (Berry) phase experiments utilizing highly polarized noble-gas atoms, ^{129}Xe (I=1/2), and ^{131}Xe (I=3/2) as sensor nuclei for spatial rotations. The experiment was extended to the non-adiabatic regime for both nondegenerate and degenerate levels and discussed in terms of the non-Abelian gauge kinematics. For this purpose, an experimental device was mounted on a rotating turn table. For realizing degenerate level, an electric quadrupole interaction in the crystal was successfully used.

Asahi shortly touched on a possibility to detect a spin-related Casimir force acting on a nuclear system by using a polarized ^{129}Xe; a tiny but visible energy shift due to an electromagnetic interaction with the vacuum.

CONCLUSION

Our polarized ^{3}He community might be pictorially expressed as the "Polarized ^{3}He Universe" in Fig.1. The abscissa of this universe is a dimension in which each subject is treated, and it spans from quarks to galaxy. The ordinate pointing downward represents the depth of knowledge, and pointing upward represents serendipity, which means an assumed gift for finding valuable or agreeable things not sought for, in other words, an unexpected discovery or invention. In this universe, we can see many small galaxies. The traditional physics, such as nuclear physics, atomic physics, quark-spin physics, etc. is located in the lower area of the universe, while MRI in vivo, lunar ^{3}He, enhanced effect of the P/T violation, and unexpected deficiency of solar neutrinos are good candidates of the subjects to be placed in the upper area. Of course, we know a big difference between the real universe and the "Polarized ^{3}He Universe". There are almost no intragalactic interactions for the real universe, whereas there are strong interactions in the latter universe. e.g., in near future, plasma physics may succeed in igniting the aneutronic ^{3}He + D fueled reactor by using polarized particles in collaboration with nuclear physics at low energy, low temperature physics, atomic physics, and probably astrophysics together with the lunar ^{3}He. This is indicated as a supernova in Fig. 1. Another example of supernova may occur in life science through the NMR of rare gases in vivo.

Lastly, I would close my review by reminding you of an episode of Norman F. Ramsey, a famous experimentalist who made a series of excellent measurements on nuclear moments. He sometimes said to his colleague,

"Intensity is everything".

I believe that this is crucially important for every experimentalist. I understand that poor statistics tells us nothing meaningful and sometimes takes us misleading conclusions. All the experimentalists, therefore, should concentrate their efforts to increase the polarization degree of ^{3}He, the beam intensity, and target thickness. Only through the above efforts, we hope that we will firmly establish our new field and lead us to a discovery of supernovae in *"the polarized ^3He universe"*, in other words, unexpected discoveries and inventions.

References

1. W. Bleakney, G.P. Harnwell, W.W. Lozier, P.T. Smith, H.D. Smyth, Phys. Rev **46** (1934) 81.

2. H.L. Anderson, and A. Novick, Phys. Rev. **73** (1948) 919.

3. A.E. Douglas, and G. Herzberg, Phys. Rev. **76** (1949) 1528.

4. A. Kastler, Journal de Physique et Radium, **11** (1950) 255.

5. L.J. Wittenberg, J.F. Santarius, and G.L. Kulcinski, Fusion Technology **10** (1986) 172.

6. Physics Today, August 1982, p.17.

7. Hai-Yang Cheng, "Status of the Proton Spin Problem", IP-ASTP-3-96, hepph/9607254 and references therein; M. Anselmino, E. Efrremov, and E. Leader, Phys. Rep. **261** 1 (1995).

8. Physics Today, June 1995, p.17.

9. Proceedings of workshop on polarized ^{3}He beams and targets, AIP conf. proc. 131, ed. by R.W. Dunford and F.P. Calaprice, NY 1985.

10. W.E. Burcham, O. Karban, S. Oh, and W.B. Powell, NIM **116** (1974) 1.

11. J. Whiteveen, Nucl. Instr. and Meth. **158** (1979) 55.

12. L.W. Anderson, Nucl. Instr. and Meth. **167** (1979) 363.

13. M. Tanaka, Proceedings of Polarized beams and polarized gas targets, ed. by H.P. gen. Schieck and L. Sydow, World Scientific 1996, p.131.

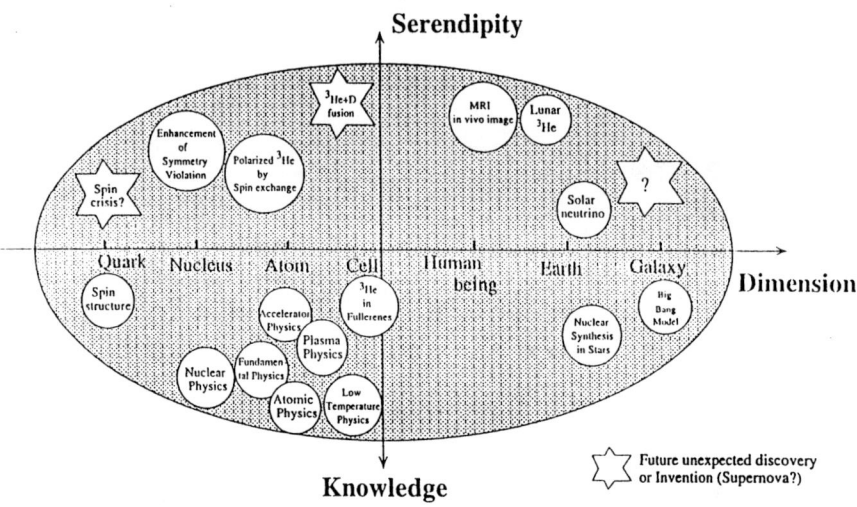

Figure 1: The Universe of polarized ³He

The HERMES Polarized Hydrogen Internal Gas Target

J. Stewart for The HERMES Collaboration

University of Liverpool, Liverpool L69 3BX, UK.

Abstract.
The HERMES polarized hydrogen target is an internal gas target using the storage cell technique. It was installed in the HERA ring at DESY during the 1996 winter shutdown and has been used throughout the 1996 and 1997 running periods. The major components of the target and its operating parameters are described. The performance during data taking is presented along with some studies of the target cell surface. Evidence suggests that the effective HERMES target cell coating is a saturated layer of water on top of the Drifilm coating.

INTRODUCTION

The HERMES experiment at DESY is designed to study the spin structure of the nucleon in inclusive and semi-inclusive deep inelastic scattering. It uses the 27.5 GeV polarized positron beam at the HERA storage ring and polarized internal gas targets. The HERMES hydrogen target is an internal polarized gas target using the storage cell technique. Internal targets have the advantage of a high nuclear polarization, no dilution by spectator nuclei, and a windowless storage cell design. A windowless cell design is required for an internal target since the stored HERA positron beam passes through the target cell. The HERMES detector is a conventional magnetic spectrometer with large acceptance and good particle identification. The combination of a pure target and large acceptance spectrometer make HERMES especially well suited to studying semi-inclusive physics. HERMES is one of several modern experiments studying deep inelastic scattering but it is the only one with such an internal target.

THE HERMES TARGET

The HERMES hydrogen target, shown in Figure 1, consists of a high intensity conventional atomic beam source (ABS) which produces a polarized

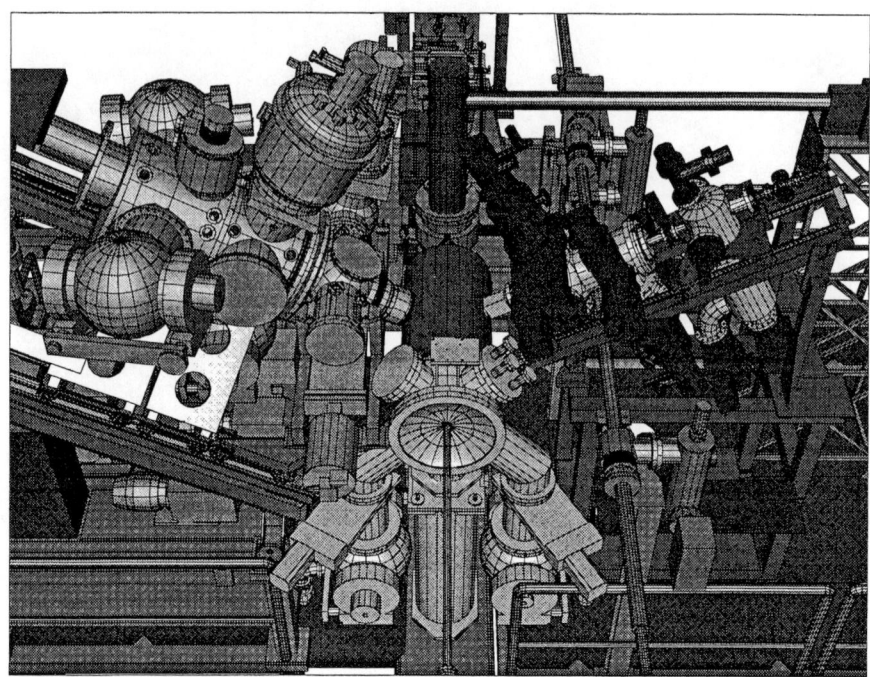

FIGURE 1. The HERMES Target. The HERMES ABS is shown on the left in the figure and the BRP on the right. The cylindrical flux return yoke of the longitudinal holding field magnet is shown in the center. The exit window for the scattered particles is seen in front of the magnet. The HERA proton beam pipe passes through the BRP support frame on the right.

atomic hydrogen beam and focuses it into a storage cell. The storage cell is an elliptical tube mounted coaxially with the HERA beam. It confines the target gas near the high energy positron beam thus increasing the target thickness. A sample of the gas from the middle of the storage cell diffuses into a Breit-Rabi polarimeter (BRP), which measures the atomic polarization, and the target gas analyzer (TGA), which measures the atomic and molecular content of the sample. A superconducting magnet provides a longitudinal holding field around the target cell defining the polarization direction and reducing the depolarization due to hyperfine mixing. The scattered particles exit the target and enter the spectrometer through a a 400 mm diameter 0.3 mm thick stainless steel exit window. A schematic of the target is included elsewhere in these proceedings. [1]

The HERMES storage cell is 29×9.8 mm elliptical tube 400 mm long made from 75 micron thick aluminum. A detailed drawing of the cell is shown in Figure 2. The polarized atomic hydrogen is injected into the cell as a beam through a 100 mm long 10 mm diameter entry tube and then diffuses out of

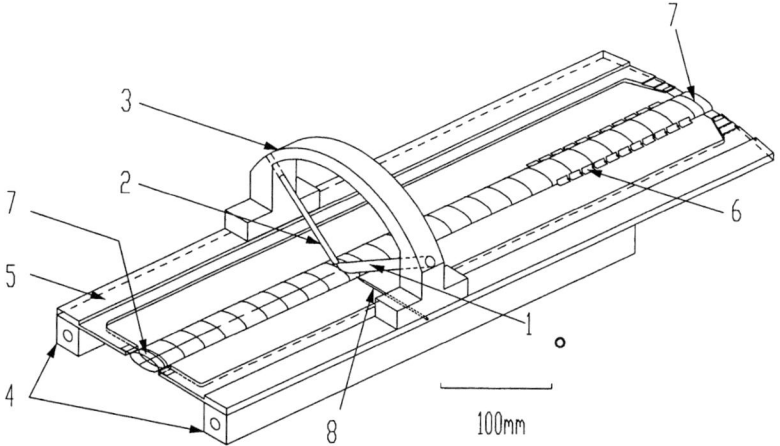

FIGURE 2. The HERMES Target Cell. The various components are as follows: 1) H/D feed tube; 2) Polarization sample tube; 3) Target cell and entry tube support arch; 4) cryogenic-cooling and support rails; 5) Support plates; 6) Cell extension with pumping apertures; 7) Cell end support collar 8) Unpolarized gas feed capillary.

the ends of the cell. By confining the atoms near the beam, a factor of about 100 increase in target density over a free atomic beam is achieved. The target gas is sampled in the middle of the cell through a 100 mm long 5 mm diameter sample tube. A 10 mm diameter 120 mm long extension tube is connected to the sample tube to direct the gas to a region were it can be analyzed. The target thickness increases as $1/\sqrt{T}$ as the target cell temperature is decreased, therefore the storage cell is cooled by flowing cold helium gas through the support structure. Platinum resistance thermometers mounted on the cell support structure monitor the cell temperature. The operating range of the HERMES cell cooling system is 35-260K.

The ends of the cell must be open so the high energy beam can pass freely though the cell. Thus the target atoms can enter the storage ring after leaving the cell. A powerful differential pumping system with a total pumping speed of about 9000 l/s is required to remove the target gas so that it does not significantly degrade the ring vacuum. Two turbomolecular pumps, shown in Figure 1, with a combined pumping speed of 4400 l/s maintain a pressure in the target region of 1.5 $\times 10^{-7}$ mbar. Three additional 1500 l/s pumping stations, two upstream and one downstream reduce the pressure to about 5×10^{-9} mbar at the entrance to the machine.

The hydrogen atoms interact with the walls of the storage cell and have a finite probability of recombining or losing their nuclear polarization. These effects depend on cell temperature, which is therefore optimized to maximize the effective target polarization. The HERMES cell is coated with Drifilm to minimize these effects but they are still present to some degree. The nominal operating temperature of the target cell is 100 K which is the lowest temperature possible without causing substantial degradation in polarization and increased recombination.

The cell coating can be damaged by synchrotron radiation, originating from the positron beam, so a set of two collimator were installed upstream of the target. [2] The first collimator C1 (13.0×5.2 mm^2) is positioned 2 m upstream of the target and shields the cell from direct synchrotron radiation. The second collimator C2 (17.2×6.0 mm^2) is located near the cell and protects the coating from secondary scattered radiation. This collimator is the smallest fixed aperture in the HERA ring. Additionally the beam is capable of ionizing the target gas. The effect of the rf-field of the HERA beam on the ions has been simulated and it was found that the rf-field can accelerate the ions near the beam. [3] At our nominal holding field and target density it was estimated that the surface dose rate due to ion bombardment is approximately 20 Gy/s with beam, and the ion energy is of the order of 100 eV.

The HERA positron beam can also directly alter the proton polarization. A holding field over the target region is required to provide a well defined polarization axis. The HERA beam, because of its bunch structure, produces an rf magnetic field. This combination of a static holding field and an imposed rf magnetic field can under some conditions drive hyperfine transitions, which modify the polarization. Monte-Carlo studies were performed during the design stage of HERMES which indicated that in order to avoid having depolarizing resonances inside the storage cell a relatively high and uniform holding field was required. [4] Based on these studies a holding field magnet with a nominal axial field of 0.35 T and a uniformity of $\pm 1.5\%$ was constructed. It was chosen to make the magnet superconducting because there were very serious space constraints placed on the design by the ABS and BRP. Recent measurements have been made by varying the holding field while a high positron current was circulating in the ring and monitoring changes in the measured atom polarization. These measurements have verified the existence of these resonances at the expected magnetic field settings. Under our normal running conditions the field is set to a value (0.335 mT) where there is no detectable resonance effect, and then no observable effect is seen on the measured proton polarization, due to the beam. A detailed report on these studies is presented elsewhere in this proceedings. [5]

The HERMES ABS shown in Figure 3 is a conventional atomic beam source. [6] Atomic hydrogen is produced in an rf-dissociator and flows through a nozzle into vacuum. The atoms which pass through two collimators form an atomic beam and then enter a 6-pole magnet focusing system. The sextupole magnets focus atoms in the two higher hyperfine energy levels while defocusing the lower two states, thus producing an electron polarized beam. A second set of 6-pole magnets focus this beam into the storage cell. The sextupole magnets are radially segmented permanent magnets with a remnant magnetization of 1.25 T. Adiabatic hyperfine transition units are used to interchange the atom

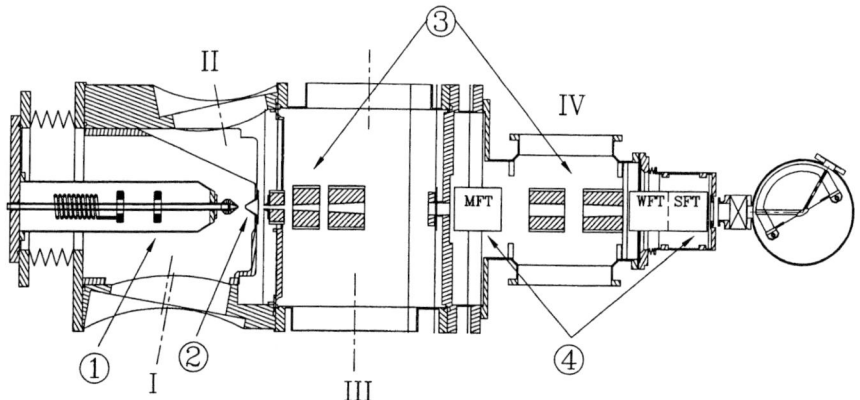

FIGURE 3. The HERMES atomic beam source. The major components are the rf-dissociator 1, the skimmer and collimator 2, the six-pole magnet system 3, and the rf transition units 4. The pumping apertures for the 4 differential pumping stages are denoted with the Roman numerals (I-IV).

state populations producing a nuclear polarized beam. During data taking two spin states are injected into the target cell. For testing and calibration purposes individual spin states can be injected. [1] The atomic flux from the ABS, measured using a compression tube before installation, was 6.4×10^{16} H_1/s which corresponds to a target areal density of 7.0×10^{13} H_1/cm^2 at 100K.

The ABS rf-dissociator [7] operates at a frequency of 13.56 MHz with a nominal power of 300 W. The power was adjusted to maximize the intensity and long term stability of the dissociator. 1.25 mbar l/s of hydrogen is metered into the dissociator along with about 0.14% by volume of oxygen. The oxygen improves the dissociator output and also results in the formation of an ice coating on the nozzle. The nozzle is constructed of 0.995% pure aluminum with a bore of 2mm and is cooled to 100±1 K. Under normal operating conditions the nozzle must be warmed every 3-7 days to prevent clogging with ice. The dissociator tube is typically replaced every 3 months.

The HERMES ABS has a powerful 4 stage differential pumping system with a total pumping speed of over 15,000 l/s. The pressure in the final chamber of the ABS is about 2.7×10^{-7} mbar which is of the same order as the target vacuum chamber pressure of 1.5×10^{-7} mbar. Measurements of the diffusive flow into the target cell from the source indicate that less than 0.5% of the target gas comes from this source.

The target gas analyzer (TGA), shown in Figure 4, measures the relative flux of Mass 1 and Mass 2 exiting the target cell's sample tube. The TGA is mounted off axis, at an angle of 7°, so as not to interfere with the beam entering the BRP. A pair of baffles collimate the beam which enters the TGA, ensuring that no atom exiting the sample tube can strike the metal structure of the quadrupole mass spectrometer. The atoms and molecules are detected

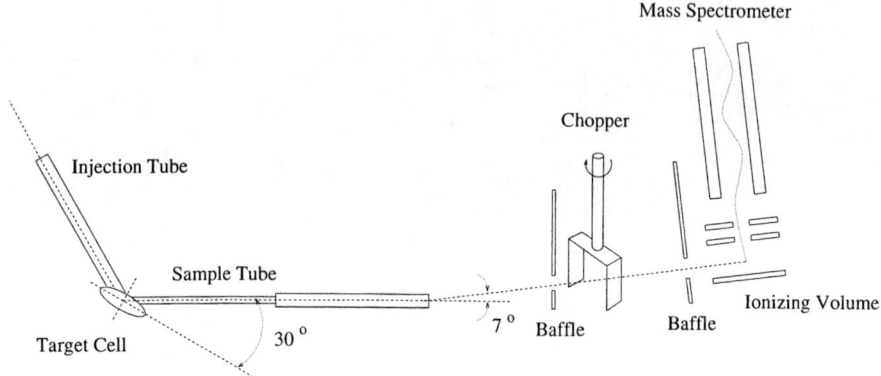

FIGURE 4. The HERMES Target Gas Analyzer TGA.

with a modified Balzers QMG 430 mass spectrometer. The QMG 430 has an open ionizer geometry making it well suited for this application.

The mass spectrometer was modified to make the ion optics remote controlled and a channeltron was used as the ion detector. The channeltron provides a stable response which is linear with changes in emission current. A chopper between the baffles is needed to subtract the count rate caused by the background gas. The TGA's pressure is typically 3×10^{-9} mbar resulting in a signal to noise of 1 to 100. Under these conditions a measurement with 1% statistical precision requires 100 s. The TGA is calibrated by injecting a constant atomic flux into the target cell and varying the amount of recombination. The recombination probability in the cell was changed either by changing the cell temperature or as a result of a HERA beam loss near our experimental hall. When the background corrected Mass 2 rate is plotted versus the Mass 1 rate the points lie on a straight line. The slope of the line is $\sqrt{2}$ times the ratio of the probability of detecting an atom relative to a molecule. The Mass 1 to Mass 2 calibration constant is known to 1-2 % .

The Breit-Rabi Polarimeter measures the polarization of the atoms exiting the sample tube. A polarization measurement with a statistical uncertainty of 0.01 requires 60 s. The present estimate of the systematic error in the atom polarization is less than 1% . The BRP and the calibration of the TGA are discussed in greater detail elsewhere in these proceedings. [1]

TARGET PERFORMANCE

The measured atomic fraction data for 1996 and 1997 up to the present is summarized in Figure 5. The atomic fraction is the ratio of the atomic hydrogen flux to the total atomic and molecular flux seen by the TGA. The quantity (1 − Atomic Fraction) is the molecular content of the sampled beam.

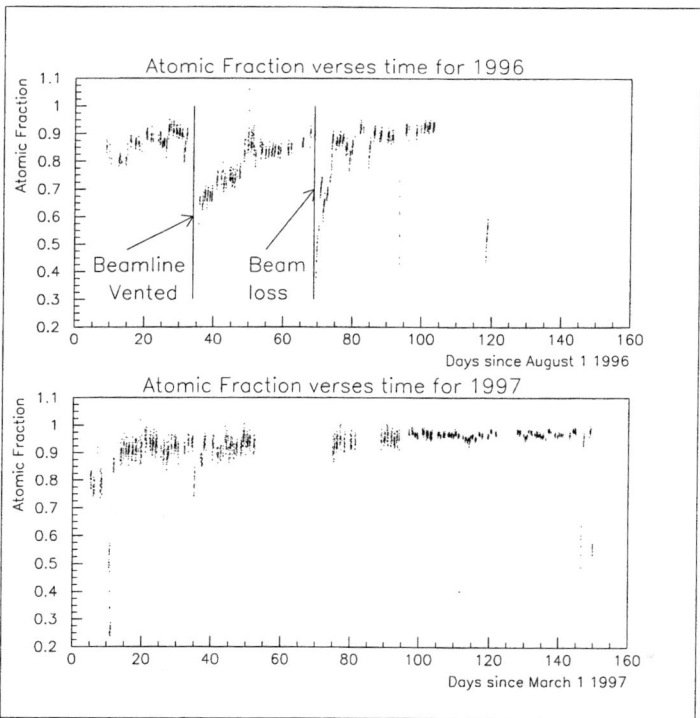

FIGURE 5. The atomic fraction for 1996 and 1997 as measured by the TGA. Only data collected during a polarized physics run is plotted. The gaps correspond to periods of unpolarized running, beam injection, and HERA machine failure. A 2% to 4% correction for the target chamber rest gas and molecular flux from the ABS has been applied; thus the atomic fraction shown reflects the amount of recombination in the target cell.

The average atomic fraction for the 1996 running period was 0.88 ± 0.01. As seen in Figure 5, the atomic fraction dropped sharply twice in 1996, once after the beamline was vented for routine maintenance and once when the positron beam was lost near the experiment. Immediately after these incidents atomic fractions below 0.25 were measured. During the unpolarized physics run the target cell was warmed to determine if the changes in the cell surface quality could be due to water adsorbed on the cold cell surface. The data is shown in Figure 6. The atomic fraction after the water was removed was 0.25, which suggested that the Drifilm coating had been damaged as the value obtained with new Drifilm was much higher. The cell was then cooled to 35 K for an unpolarized physics run. After the unpolarized run and when the cell had been raised to 100K the atomic fraction was 0.3 and improved rapidly as long as the ABS was injecting beam into the cell. In summary, a large change in atomic fraction was observed when the cell temperature was increased to the

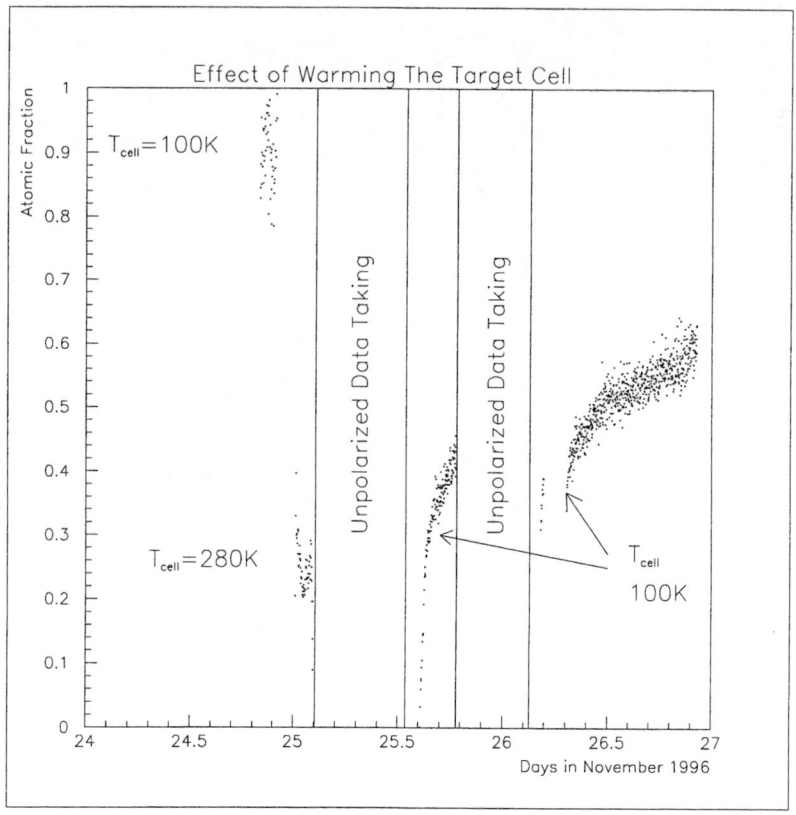

FIGURE 6. Cell Surface studies.

point where the vapor pressure of water becomes appreciable. An increase in vacuum pressure was also seen during the warming process until the water was pumped away. Additionally, the atomic fraction only recovered when the cell was cold and the gate valve between the ABS and the target chamber was open. This is taken as strong evidence that the water produced in the ABS dissociator was coating the surface of the target cell and reducing the recombination.

The target cell was replaced in the 96-97 winter shutdown and measurements were made with undamaged Drifilm. The atomic fraction with the new cell was 0.91 at 140K. Periodically during the positron beam commissioning period the target cell was warmed to remove any ice and the atomic fraction was measured with the bare Drifilm surface. The atomic fraction measured with a warm cell decreased rapidly to 0.80 during this time. Shortly after the start of data taking we stopped removing the water to achieve the highest possible atomic fraction for data taking. The average atomic fraction in 1997 up to

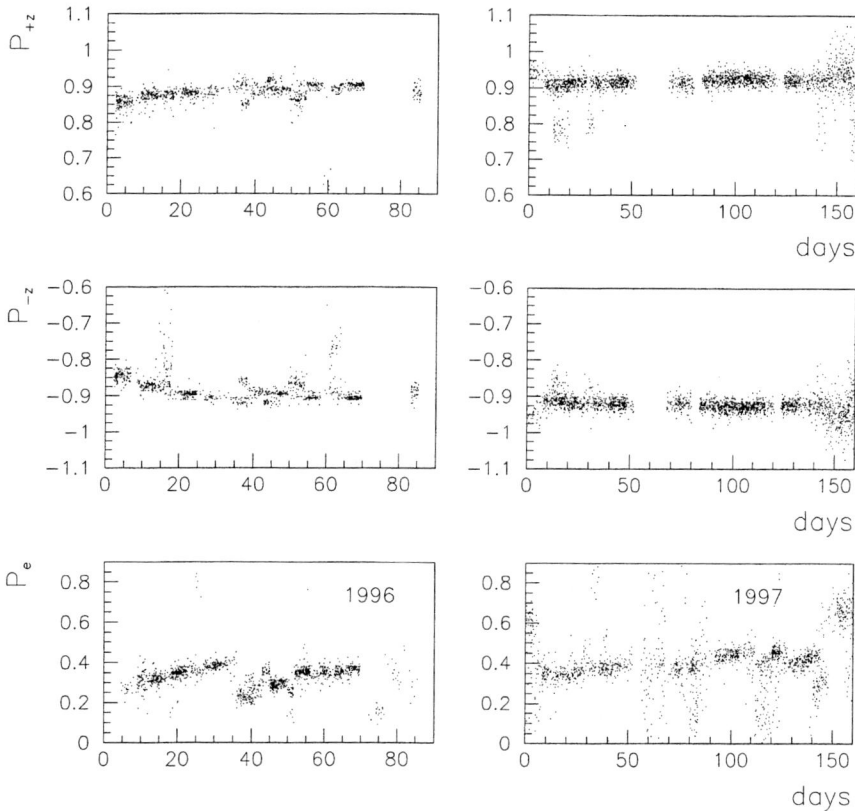

FIGURE 7. The polarization of the sampled atoms for 1996 and 1997 as measured by the BRP.

the present time is 0.95.

The average nuclear polarization of the atoms measured by the BRP was 0.90 ± 0.01 for both polarization directions in 1996 and 0.92 ± 0.01 in 1997 up to the time of this conference. The measured polarization as a function of time is shown in Figure 7. Between positron fills the ABS transitions were set to cycle through both proton polarization directions and to produce electron polarized atoms (with no proton polarization). The measured electron polarization obtained during these times is plotted on the bottom in Figure 7. Much of the scatter in the electron polarization values comes from tests performed between positron fills. The increase in electron polarization with time is not understood.

Both the atomic fraction and the atom polarization measurements are of a sample of the target gas taken from the middle of the cell. There is a systematic uncertainty due to the sampling technique which is still being studied, therefore no average target nuclear polarization values can be quoted at the present. This topic is discussed in more detail elsewhere in these proceedings. [1] Another source of uncertainty in the target polarization is the polarization of the molecules which recombine in the cell. This is more important for the 1996 data, where the atomic fraction is lower, than the 1997 data. A recent measurement of the tensor polarization of deuterium molecules formed by recombination found the molecular polarization to be 0.8 ± 0.30 of the initial atom polarization. [8] However, it is not clear how applicable this data is to the recombination of hydrogen in our target on a different surface. Additional measurements of the molecular polarization are planned at HERMES, NIKHEF, and IUCF.

CONCLUSIONS

In conclusion, the HERMES target was successfully installed in the HERA ring and has operated reliably for the past two years. The beam depolarization effects are understood and no depolarization is observed under optimized running conditions. High polarization and low molecular background are currently obtainable under normal running conditions. The target cell's Drifilm coating damages rapidly in the HERMES experimental environment as evidenced by an increase in recombination when the target cell is operated near room temperature. However, when the cell is cooled to a temperature in the region of 100K, operation with high values of atomic fraction is possible. It is assumed that water molecules formed in the ABS dissociator, due to the deliberate introduction of a small quantity of oxygen, are absorbed onto the cold cell surface and form a coating which inhibits recombination.

REFERENCES

1. Braun, B., contribution to this workshop.
2. Zapfe-Düren, K., et al., *Int. Workshop on Pol. Beams and Pol. Gas Targets*, Cologne, H. Schieck and L. Sydow (ed.), 1996, p.400.
3. Miller, A., Private communication.
4. Kinney, E., *Proceedings of the Workshop on Pol. Ion Sources and Pol. Gas Targets*, Madison 1993, W. Haeberli and L.W. Anderson (ed.), AIP Conf. Proc. 293, (1994) p.278.
5. Kolster, H., contribution to this workshop.
6. Stock, F., et al., *Nucl. Instrum. Methods* A **343**, p.334(1994).
7. Stock, F., et al., *Int. Workshop on Pol. Beams and Pol. Gas Targets*, Cologne, H. Schieck and L. Sydow (ed.), 1996, p.260.
8. van den Brand, J.F.J., et.al., *Phys. Rev. Lett.* **78**, *1235(1997)*.

Polarized Deuterium Internal Target at AmPS (NIKHEF)

M. Ferro-Luzzi[*,†], Z.-L. Zhou[‡,#], J.F.J. van den Brand[*,#],
H.J. Bulten[*,#], R. Alarcon[∥], N. van Bakel[*], T. Botto[*],
M. Bouwhuis[†], L. van Buuren[*], J. Comfort[∥], M. Doets[†],
S. Dolfini[∥], R. Ent[¶,§], D. Geurts[*], P. Heimberg[*],
D.W. Higinbotham[**], C.W. de Jager[§], J. Lang[††], D.J. de Lange[†],
B. Norum[**], I. Passchier[†], H.R. Poolman[*], E. Six[∥],
J. Steijger[†], D. Szczerba[††], O. Unal[#], and H. de Vries[†].

[*] *Dept. of Physics and Astronomy, Vrije Universiteit, 1081 HV Amsterdam, The Netherlands*
[†] *NIKHEF, P.O. Box 41882, 1009 DB Amsterdam, The Netherlands*
[‡] *MIT-LNS, Cambridge, Massachusetts 02139, USA*
[#] *Dept. of Physics, University of Wisconsin, Madison, WI 53706, USA*
[∥] *Dept. of Physics, Arizona State University, Tempe, AZ 85287, USA*
[¶] *Dept. of Physics, Hampton University, Hampton, VA 23668, USA*
[§] *TJNAF, Newport News, VA 23606, USA*
[**] *Dept. of Physics, University of Virginia, Charlottesville, VA 22901, USA*
[††] *Inst. für Teilchenphysik, ETH, CH-8093 Zürich, Switzerland*

Abstract. We describe the polarized deuterium target internal to the NIKHEF medium-energy electron storage ring. Tensor polarized deuterium was produced in an atomic beam source and injected into a storage cell target. A Breit-Rabi polarimeter was used to monitor the injected atomic beam intensity and polarization. An electrostatic ion-extraction system and a Wien filter were utilized to measure on-line the atomic fraction of the target gas in the storage cell. This device was supplemented with a tensor polarization analyzer using the neutron anisotropy of the $^3H(d,n)\alpha$ reaction at 60 keV. This method allows determining the density-averaged *nuclear* polarization of the target gas, independent of *spatial* and *temporal* variations. We address issues important for polarized hydrogen/deuterium internal targets, such as the effects of spin-exchange collisions and resonant transitions induced by the RF fields of the charged particle beam.

INTRODUCTION

Measurements of spin-dependent electron scattering have the potential to enhance our understanding of nucleon and nuclear structure. For example, inelastic spin-dependent electron-proton scattering provides information about the proton spin structure, while spin observables in elastic, quasi-elastic, and deep-inelastic scattering from polarized deuterium are predicted to provide data sensitive to the effects of D-wave components in the deuteron wave function, the largely unknown charge form factor of the neutron, and the neutron spin structure functions. This has prompted development of both polarized ^{1}H and ^{2}H targets for use with internal [1-3] or external beams [4] and polarimeters for measuring the polarization of recoiling hadrons [5].

The internal target facility (ITF) at the Amsterdam Pulse Stretcher (AmPS) at NIKHEF offers the possibility to study in a comprehensive and precise way the spin-dependent electromagnetic response of few-body nuclei at momentum transfers up to 0.8 $(GeV/c)^2$. The experimental program utilizes thin-walled storage cells which are fed by a high intensity source of polarized hydrogen and deuterium atoms. The electrons scatter from a chemically and isotopically pure target of high nuclear polarization. This technique offers rapid polarization reversal and flexible orientation of the polarization axis. For deuterium one has the additional ability to reverse the tensor polarization, P_{zz}, at fixed vector polarization, P_z, and vice versa. Consequently, small systematic errors can be expected, provided that the polarization can be reliably measured. Since there are many mechanisms that can potentially reduce the nuclear polarization of internal hydrogen/deuterium targets (e.g. spin-exchange [6,7], cell-wall collisions [8], electron-beam induced Zeeman transitions [9]), we developed polarimetry that is capable to measure the polarization of the target *in situ* [10]. The method takes advantage of the ionization of atoms and molecules by the stored beam passing through the target cell. The number of ions produced along the storage cell is directly proportional to the product of target density and beam current. The total target polarization can be obtained independent of its spatial and temporal variations by uniformly extracting these ions from the cell, measuring their atomic and molecular fractions, and by directly determining their nuclear polarization. Here, we report on the most recent developments obtained with our tensor polarized deuterium internal target.

INTERNAL TARGET PERFORMANCE

The experiments are performed by using a 100% duty-factor electron beam stored in the AmPS ring. Long beam lifetimes ($\sim$ 15 min) are obtained by compensating synchrotron radiation losses with a 476 MHz cavity in AmPS. Electron beams with energy up to 900 MeV (using ramping) and currents of more than 250 mA (by stacking) have been obtained.

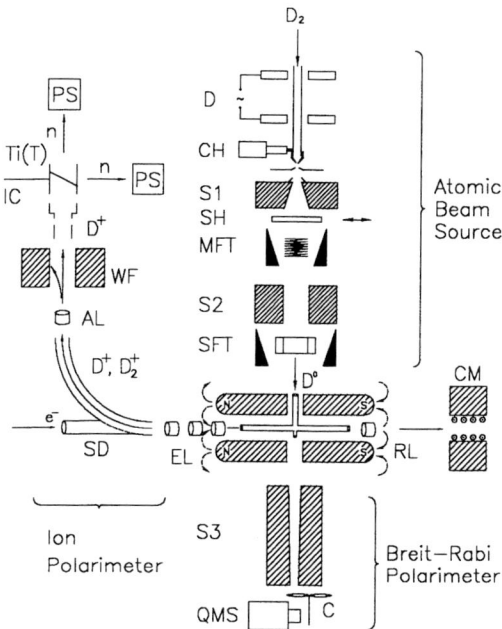

FIGURE 1. Schematic outline of the atomic beam source, Breit-Rabi polarimeter, internal target, and ion-extraction polarimeter. The various components are described in the text.

Fig. 1 shows schematically the experimental setup. Polarized deuterium atoms are produced in an atomic beam source (ABS) and fed into a ø15 mm/400 mm long storage cell that was manufactured from two 25 μm thick ultrapure Al foil coated with Teflon and cooled to about 150 K. Two electromagnets are used to apply a magnetic guide field over the target region to define the target polarization axis. The dissociator (D) was operated in a stable way during the entire run of the experiments at ITF. A degree of dissociation of the atomic beam of $62 \pm 3\%$ (at a gas throughput of ~ 0.6 mbar ℓ/s) was routinely obtained while adding about 0.1% water vapour into the discharge tube. To produce a tensor polarized deuterium beam, a 1-4 transition was induced by a medium-field transition unit (MFT) between the sextupoles (S1, S2) to remove the dilution by hyperfine state 1, while a strong-field transition unit (SFT) produced alternatively a 3-5 or 2-6 transition after the sextupoles. In this way the tensor polarization of the atomic beam was switched every 10 seconds between maximum values. The performance of the RF transition units was studied with a nuclear polarimeter and with the Breit-Rabi polarimeter described below. All transitions were found to be more than 95% efficient [12]. In these circumstances, we achieved a target thickness of $\sim 2 \times 10^{13}$ nuclei/cm^2 (injecting two deuterium hyperfine states) with total target polarizations in the storage cell up to $P_{zz}^{\pm} \simeq +0.5 / -1.0$. The design and performance of the ABS, target cell and polarimeters are described in detail in Ref. [10–12]. Currently, we are implementing an upgraded focusing system that uses rare-earth permanent magnets. Raytrace calculations show that at least a factor of two higher intensity and a substantial improvement of both the unwanted hyperfine state rejection efficiency and of the beam size can be expected.

In order to perform absolute measurements of the tensor analyzing powers in ^{2}H$(e,e'd)$ and ^{2}H$(e,e'p)n$ scattering, two polarimeters were applied. A ø4 mm sample hole in the storage cell allowed tuning and monitoring of the atomic beam intensity and polarization with a Breit-Rabi polarimeter (BRP) which consists of a sextupole magnet (S3) and a quadrupole mass spectrometer (QMS). The BRP allowed us to carry out detailed studies of possible depolarization mechanisms, such as zero-crossings of the magnetic field or electron-beam induced Zeeman transitions (see below). Since there exist many mechanisms that can affect the polarization of the target in the storage cell, we constructed a dedicated polarimeter to determine the total target polarization with the same weighing over gas density as in the e-^{2}H experiment. Ions produced inside the cell by the circulating electron beam were extracted using a set of electrostatic lenses (EL, RL, AL) and a spherical deflector (SD). A Wien filter (WF) was utilized to separate the atomic and molecular contributions. We measured an atomic fraction of 0.71 ± 0.02 and found that about 85% of the molecules in the storage cell were due to residual D$_2$ gas and undissociated molecular beam, while the rest was attributed to recombination of polarized atoms on the cell surface. In case of a precise polarization measurement, the contribution of molecules cannot be neglected. This issue is addressed in detail in Ref. [14] and in another contribution to this conference [15]. Measurements of the deuterium tensor polarization were performed by selecting atomic ions (D$^+$) through the Wien filter and accelerating them to bombard a tritiated titanium foil (Ti(T)). The low energy ^{3}H$(d,n)\alpha$ reaction exhibits a large and well-known tensor analyzing power. By measuring single neutron rates in two plastic scintillators (PS), at 0° and 90° with respect to the polarization axis, we determined the tensor polarization of the impinging deuterons. The ion-extraction polarimeter is described in more detail in Ref. [10] and in another contribution to this conference [13].

Next, we discuss two specific issues of general importance for polarized hydrogen/deuterium internal targets: spin-exchange effects and beam-induced resonant transitions.

SPIN-EXCHANGE EFFECTS

An essential issue for polarized hydrogen/deuterium internal targets is the influence of spin exchange between the atoms. Furthermore, the spin-exchange

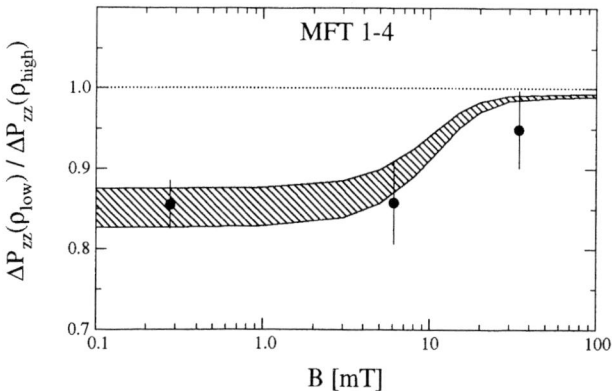

FIGURE 2. Ratio of difference in tensor polarization ΔP_{zz} for measurements at high $(1.1\times 10^{12}\text{ cm}^{-3})$ and low $(0.4\times 10^{12}\text{ cm}^{-3})$ atomic density. ΔP_{zz} was produced by turning on/off the MFT 1-4 transition. The hatched area represents the results of our model. The dotted line indicates the result in the absence of spin-exchange effects.

mechanism enables a novel way to produce polarized atoms [6,16], based on optical pumping of alkali metals. The rate of spin exchange linearly depends on the atomic density and is a function of the external magnetic field. To isolate the effects on the tensor polarization of spin exchange collisions in the storage cell, we measured at two values of the atomic density in the cell center (~ 1.1 and 0.4×10^{12} at/cm^3). This was achieved by adjusting the RF power of the discharge, thus changing the atomic beam intensity. The presence of spin exchange is demonstrated in Fig. 2, which shows the ratio of tensor polarization for measurements at high and low density. In this case ΔP_{zz} was produced by switching on/off the 1-4 MFT. More measurements were made, activating either the MFT, the SFT, or both units, and at three values of the external magnetic field (0.3 mT, 6 mT, and 34 mT). For all combinations of injected states, a higher polarization was found for the measurements at lower target density. We developed a computer code based on a model of the spin exchange rate equations in the storage cell. The shaded band in Fig. 2 shows the results of such a calculation. The width of this band indicates the uncertainty in the calculations, due to the uncertainty in the spin-exchange cross section and the uncertainty in the target density. The results of the calculation for the different transitions in Fig. 2 were averaged in a similar manner as the data. The measurements indicate, that spin exchange lowers the polarization in a storage cell by about 10 % for densities of $\sim 10^{12}$ atoms/cm^3, even for an external magnetic field that is three times larger than the critical field.

BEAM-INDUCED TRANSITIONS

A possible depolarization mechanism in polarized hydrogen/deuterium internal targets originates from the interaction of the atomic spin with the RF fields associated with the bunch structure of the charged particle beam [9,18,19]. To study such a beam-target interaction, we prepared our deuterium beam with selected hyperfine substate populations. The atomic beam was crossed with the AmPS electron beam and subsequently analyzed in the BRP. Measurements were carried out at various target holding fields and electron beam currents. The resonant conditions of individual Zeeman transitions were obtained by scanning the target holding field over the expected resonant fields. The electron-beam induced transitions were identified by changing the hyperfine substate populations of the injected atomic beam.

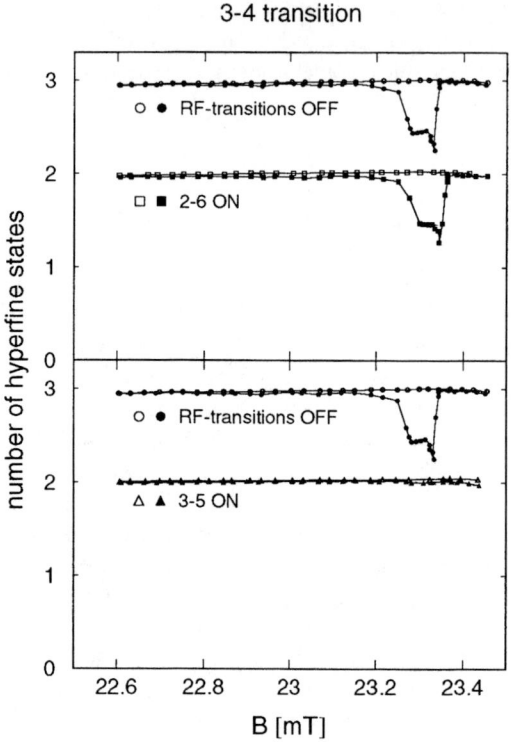

FIGURE 3. Relative atomic beam intensity measured with the BRP as a function of the target holding field for various combinations of injected hyperfine states. The open (solid) markers represent the data obtained with electron beam off (on). The structure around 23.3 mT corresponds to an electron-beam induced 3-4 transition (see text).

Fig.3 shows the occurrence of a 3-4 transition. The BRP signal is shown as a function of the target holding field. In the top graph, we show the data for a first scan with all ABS RF transition units switched off. With no electron beam in the ring, three hyperfine states are detected in the BRP (open circles). When the ring is filled with an electron beam (110 mA in this case), one observes a drop of about 20 % of the QMS signal in the region $B = 23.30 \pm 0.05$ mT (solid circles). If a two-state transition is assumed to occur, then this corresponds to a 60 % transition probability. From the drop in the intensity measured in the BRP it is clear that the electron beam is transfering some of the upper ($m_J = +1/2$) hyperfine states into the lower ones ($m_J = -1/2$), which are then rejected by the BRP sextupole. The data for the second scan shown in the top graph was obtained with the 2-6 SFT turned on, i.e. states 1, 3 and 6 are injected into the electron storage ring. State 6 is rejected by the BRP sextupole, so that the signal with electron beam off (open squares) is two thirds the signal of the previous case (open circles). It is seen that the drop in intensity due to the electron beam remains unchanged (solid squares). Thus, the electron-beam induced transitions 1-6, 2-6, 2-5 and 3-6 are ruled out, while the 3-5 and 3-4 are still compatible with the information obtained so far. In the lower graph, data are shown for the case where the 3-5 SFT in the ABS was turned on. Now the effect of the electron beam on the measured beam intensity is no longer observed. This rules out the electron-beam induced 3-5 transition and leaves the 3-4 transition as the only possible candidate. Indeed, the value of the magnetic field at which the transition is observed precisely matches the resonance field for a 3-4 transition at a frequency of 476 MHz.

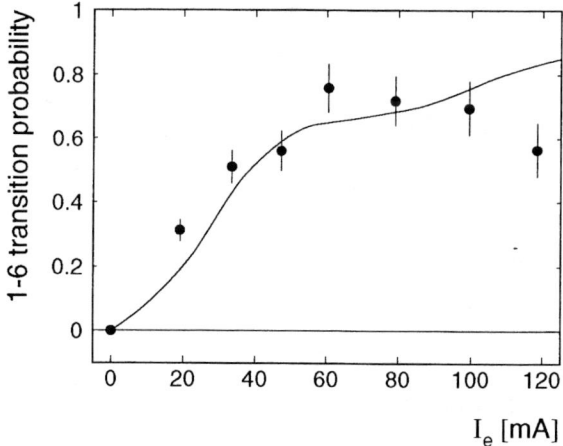

FIGURE 4. Transition probability of the first harmonic (952 MHz) 1-6 transition as a function of the electron beam current. The curve shows the result of our calculations.

We identified in this way all possible electron-beam induced transitions for a frequency of 476 MHz [9] and observed transition probabilities as high as

70 % at large electron currents ($\approx$ 100 mA). In addition, a 1-6 transition resulting from the first harmonic (952 MHz) was identified. Fig. 4 shows the dependence of the transition probability for this transition as a function of beam current. The curve is the result of a calculation which we performed by numerically solving the Schrödinger equation. Beam-induced σ transitions (3-5, 2-6) were not observed when the target holding field was oriented parallel to the electron beam, as expected, since these transitions require the RF field to oscillate parallel to the static field. The effects of beam-induced transitions are of general importance for the polarized internal target technique. These studies enabled us to properly choose the target field during our e-^{2}H nuclear physics experiments, so that such depolarizing effects were negligible.

PHYSICS RESULTS

Elastic electron scattering from the spin-1 deuteron is completely described by three form factors, the charge monopole (G_C), charge quadrupole (G_Q) and magnetic dipole (G_M). Cross section measurements yield the structure functions A(G_C,G_Q,G_M) and B(G_M), which combined with T_{20}(G_C,G_Q,G_M) allow the determination of these form factors. We completed two experiments with our tensor polarized deuterium internal target, in which we measured absolutely and with unprecedented accuracy the tensor analyzing power T_{20} in the range of four-momentum transfer $1.4 < Q < 3$ fm^{-1}. Fig. 5 shows our results [2] in comparison with the world data [1,20] and the prediction of various models [21]. As a consistency check for our experiments, we also performed a precise measurement of T_{22}. This quantity is unambiguously given by the structure functions A and B. In our case, T_{22} was found in agreement with the present data set on A and B. Finally, simultaneous to the elastic scattering data, we obtained first asymmetries for the ^{2}H($e, e'p$)n reaction with a tensor polarized deuterium target. These data provide independent constraints for microscopic models of the deuteron.

In the near future, we will extend our T_{20} measurements up to ~ 4.5 fm^{-1} and obtain data for the quasi-elastic channel at higher missing momenta, in the region where the tensor analyzing powers are predicted to strongly depend on the details of the deuteron spin structure [22].

SUMMARY AND OUTLOOK

A polarized ^{2}H internal gas target has been successfully used in a medium-energy storage ring with an electron beam to carry out spin-dependent scattering measurements. The experiments exploited the unique advantages of internal targets such as purity, high polarization, the ability to manipulate the target spin and clean recoil hadron detection. The total target polar-

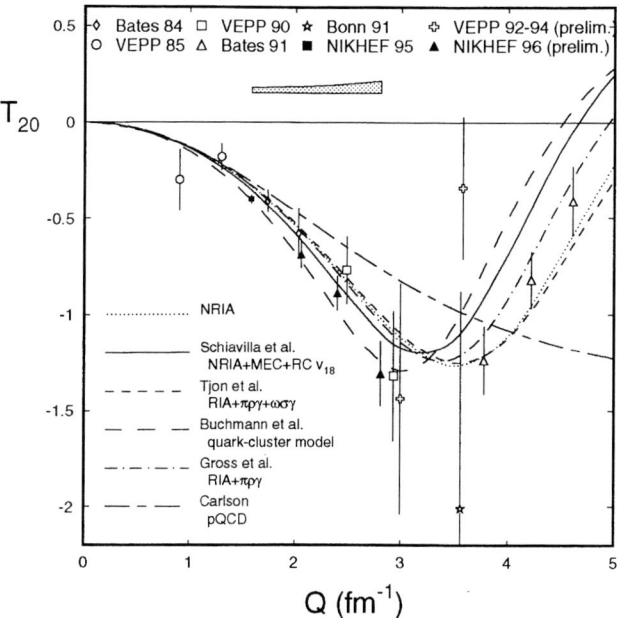

FIGURE 5. The figure shows our T_{20} results in comparison with the world data. The shaded area shows our systematic uncertainty.

ization was measured to an accuracy of 5%, independent of its spatial and temporal variations, by analyzing ions extracted from the storage cell.

Currently, the polarized hydrogen/deuterium internal target is being upgraded. By using new focusing magnets, a longer storage cell, an increased vacuum pumping speed on the target chamber, and a stronger target guide field, we expect to obtain an increase in the figure of merit of more than one order of magnitude for future experiments. This will allow us continuing our study of the spin-dependent electromagnetic response of the nucleon and deuteron. Measurements of tensor analyzing powers will provide information on the S/D-wave structure of the deuteron, while with vector polarized hydrogen/deuterium targets spin-correlation parameters will be measured that are sensitive to small quantities, such as the quadrupole form factors of the N-Δ transition and the neutron charge and magnetic form factors [23].

This work was supported in part by the Stichting voor Fundamenteel Onderzoek der Materie, which is financially supported by the Nederlandse Organisatie voor Wetenschappelijk Onderzoek, the Swiss National Foundation, the National Science Foundation under Grants No. PHY-9316221 (Madison), NATO Grant No. CRG920219, and HCM Grant No. ERBCHBICT-930606 and ERB4001GT931472.

REFERENCES

1. R. Gilman et al., *Phys. Rev. Lett.* **65**, 1733 (1990).
2. M. Ferro-Luzzi et al., *Phys. Rev. Lett.* **77**, 2630 (1996).
3. Ackerstaff et al.(HERMES collab.), *Phys. Lett.* **B 404**, 383 (1997).
4. D.G. Crabb and D. Day, *Nucl. Instrum. Methods* **A356**, 9 (1995)].
5. S. Kox et al., *Nucl. Instrum. Methods* **A346**, 527 (1994).
6. T. Walker and L. W. Anderson, *Nucl. Instrum. Methods* **A334**, 313 (1993).
7. H.J. Bulten et al., accepted for publication in *Phys. Rev.* **A**, (Dec. 1996).
8. J.S. Price and W. Haeberli, *Nucl. Instrum. and Methods* **A326**, 416 (1993) and **A349**, 321 (1994).
9. M. Ferro-Luzzi et al., accepted for publication in *Hyperfine Interactions*, (May 1997).
10. Z.-L. Zhou et al., *Nucl. Instrum. and Methods* **A379**, 211 (1996); Z.-L. Zhou et al., submitted for publication in *Nucl. Instrum. and Methods* (Sep. 1997).
11. Z.-L. Zhou et al., *Nucl. Instrum. and Methods* **A378** 40 (1996).
12. M. Ferro-Luzzi et al., *Nucl. Instrum. and Methods* **A364**, 44 (1995).
13. Contribution to this conference by Z.-L. Zhou et al., "Ion-Extraction Polarimetry for Tensor Polarized Deuterium Internal Targets".
14. J.F.J. van den Brand et al., *Phys. Rev. Lett.* **78**, 1235 (1997).
15. Contribution to this conference by J.F.J. van den Brand et al., "Polarization of Deuterium Molecules".
16. K. Coulter et al., *Phys. Rev. Lett.* **68**, 174 (1992); M. Poelker et al., *Phys. Rev.* **A50**, 250 (1994); J. Stenger et al., *Phys. Rev. Lett.* **78**, 4177 (1997); H. Gao et al., Int. Workshop on Polarized Beams and Polarized Gas Targets, eds. H. Paetz gen. Schiek and L. Sydow, World Scientific (Cologne, June, 1995), p. 67.
17. E. M. Purcell and G. B. Field, *Astrophys. Jour.* **124**, 542 (1956).
18. E. Kinney, HERMES Report Nr. 3-90 (1990); J. Giroux, *Proc. of the Workshop on Pol. Gas Targets for Storage Rings*, in Proc. High Energy Spin Physics, Minneapolis 1988, K.J. Heller ed., AIP Conf. Proc. 187, 1565 (1989).
19. R. Gilman et al., *Nucl. Instrum. and Methods* **A327**, 277 (1993).
20. M.E. Schulze et al., *Phys. Rev. Lett.* **52**, 597 (1984); V.F. Dmitriev et al., *Phys. Lett.* **157B**, 143 (1985); B. Boden et al., *Z. Phys.* **C49**, 175 (1991); M. Garçon et al., *Phys. Rev.* **C49**, 2516 (1994); S.G. Popov et al., in *Proc. of the 8th Intl. Symp. on Polarization Phenomena in Nuclear Physics, Bloomington, Indiana 1994*, AIP Conference Proceedings 339.
21. R. Schiavilla and D.O. Riska, *Phys. Rev.* **C43**, 437 (1990), and R.B. Wiringa et al., *Phys. Rev.* **C51**, 38 (1995); E. Hummel and J.A. Tjon, *Phys. Rev.* **C42**, 423 (1990); A. Buchmann et al., *Nucl. Phys.* **A496**, 621 (1989); J.W. Van Orden, N.Devine and F. Gross, *Phys. Rev. Lett.* **75**, 4369 (1995); C.E. Carlson, *Nucl. Phys.* **A508**, 481c (1990).
22. J.L. Forest et al., *Phys. Rev.* **C54**, 646 (1996).
23. NIKHEF experiment 97-01, spokespersons: M. Ferro-Luzzi and J.F.J. van den Brand, "Spin-Dependent Electron Scattering from Hydrogen and Deuterium".

The Wisconsin-IUCF Polarized Gas Target

Frank Rathmann[1], W. Haeberli, B. Lorentz, P. Quin,
B. Schwartz, and T. Wise

University of Wisconsin-Madison, Madison, WI 53706, USA

H. O. Meyer, R. E. Pollock, J. Doskow, M. Dzemdizic,
J. H. Hardie, B. v. Przewoski, T. Rinckel, F. Sperisen,
and M. Wolanski

Indiana University Cyclotron Facility, Bloomington, IN 47408, USA

P. V. Pancella and P .B. Ugorowski

Western Michigan University, Kalamazoo, MI 49008, USA

W. Daehnick, R. Flammang, and D. Tedeschi

University of Pittsburgh, Pittsburgh, PA 15260, USA

Abstract. This paper will describe the polarized internal gas target installed in the Indiana Cooler and focus to a large extend on operational properties from the experimenters point of view. Measurements of pp elastic spin correlation parameters A_{xx}, A_{yy}, A_{xz} and A_{zz} have been finished and some of the recent results will be presented as well. Studies of systematic effects from background reactions and target beam interactions will be presented also.

INTRODUCTION

The polarized internal gas target at the Indiana Cooler Storage Ring is in operation since about four years. It is installed in the A-region of the Cooler. The focus of the polarized physics experiments performed by the PINTEX-Collaboration (**P**olarized **I**nternal **T**arget **Ex**periments) with the experimental setup in the past has been on measurements of pp elastic spin correlation parameters. Studies in beam physics and non-elastic polarized pp reactions

---

[1] Now at: Physikalisches Institut der Universität Erlangen-Nürnberg, 91058 Erlangen, Germany. Working at: Institut für Kernphysik, 52425 Jülich, Germany.

have been performed as well but will not be emphasized in this paper. These topics are subject of another paper to this conference by P.V. Pancella.

In this paper I will give a brief overview of the experimental equipment and the operating parameters of the polarized source and the internal target. The detection system will be described briefly as well. Systematic effects such as the determination of background and effects from position and angle modulation of the beam on the data will also be discussed. Results from recent measurements of spin correlation parameters will be presented.

INTERNAL TARGET SETUP

The setup of the polarized gas target internal to the Cooler is shown in fig. 1.

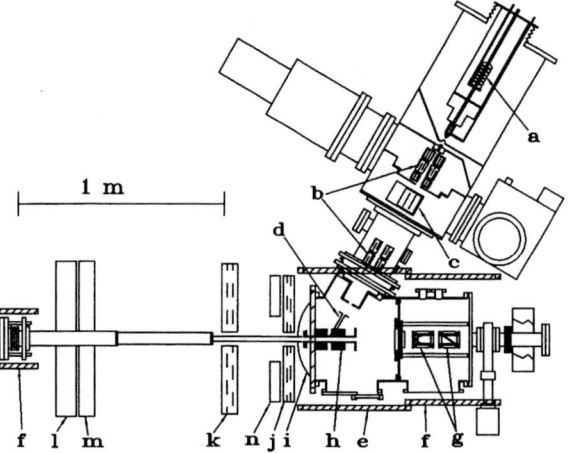

FIGURE 1. Top view of the polarized internal target setup in the Cooler A-region. Hydrogen molecules are dissociated into atoms in the dissociator (a) and electron spin separated in the segmented system of sextupole magnets (b). A medium field rf-transition (c) is used to transfer atoms from hyperfine state 2 into state 3. In the subsequent sextupole magnets, these atoms are then defocussed. The atomic beam then enters the storage cell through the feed tube (d). Three sets of weak magnetic field coils (e), mounted on the outside of the target chamber, are used to orient the target polarization in the desired direction, transverse along the horizontal or vertical (coils not shown) and longitudinally with respect to the beam. For the horizontal guide field direction, additional compensating coils (f) are mounted before the target and behind the detector stack. They are used to reduce closed orbit distortions. (Other segments of the setup will be described in more detail later in the text.)

The Wisconsin Atomic Beam Source [1] delivers an intense polarized hydrogen beam of $3.6 \cdot 10^{16}$ H_1/s, when atoms in hyperfine state 1 are injected into the 13 cm long feed tube of the storage cell (for the labeling of states, see ref. [2]). When two states $(1+2)$ are injected, by switching off the $2-3$ medium field RF transition unit (c in fig. 1), the atomic beam intensity reaches $6.7 \cdot 10^{16}$ H_1/s. Incomplete defocussing of unwanted state 3 atoms behind the transition unit and a non-ideal transition efficiency lead to a calculated maximum achievable polarization of 0.87 for the polarization of the atoms that are injected into the cell. Three sets of guide field coils (e) provide a field of about 0.3 mT. They are used to orient the target polarization along the vertical, horizontal or longitudinal direction with respect to the beam. For the horizontal guide field, additional compensating coils (f) are located before the storage cell and behind the detector stack in order to reduce closed orbit distortions of the beam.

The assembly of the storage cell through which the stored proton beam passes is shown in figs. 2 and 3.

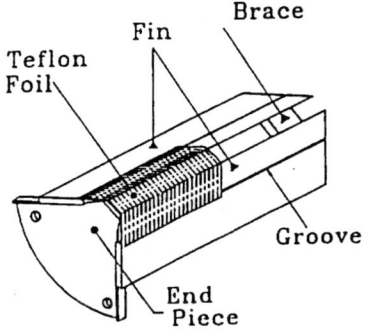

FIGURE 2. One quadrant of the storage cell target. Thin teflon foil is stretched over fins and held in place by a wire pressing into a groove.

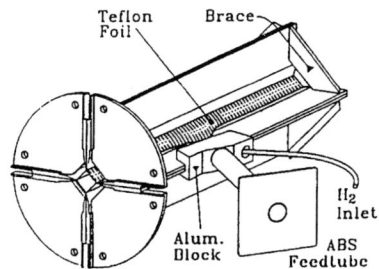

FIGURE 3. Fully assembled storage cell with feed tube for the atomic beam and unpolarized inlet for H_2 or other gases.

In fig. 2, one quadrant is shown in detail. Thin teflon foil with a thickness of 0.43 mg/cm^2 is stretched over two fins and held in place by a thin aluminum wire that presses into a groove. Four of these quadrants are joined together and form a storage cell of a quadratic cross sectional area of about 1 cm^2 as shown in fig. 3. Teflon as a material for storage cells for polarized targets has been shown to prevent depolarization in collisions of the polarized atoms with the walls of the cell [3]. The number of wall bounces in the cell is estimated from a Monte Carlo simulation to be about 200. The energy loss in the teflon foil for recoiling protons at the smallest scattering angles of 4° in the lab at 200 MeV incident proton energy is less than 100 keV. During operation the

25 cm long cell remains at room temperature. With the above given atomic beam intensities for the injection of one hyperfine state, a target thickness of $3.1 \cdot 10^{13}$ H_1/cm^2 has been measured by comparison to a calibrated flux of unpolarized hydrogen gas injected into the center of the cell. A detailed description of the storage cell target assembly can be found in refs. [5,6].

DETECTOR SYSTEM AND PARTICLE IDENTIFICATION

The system of recoil detectors is shown in fig. 4. The eight detectors have an active area of 4×6 cm^2 each and a thickness of about 1 mm. They are located 50 mm from the stored Cooler beam. Each detector contains 28 strips of width 2 mm. The position resolution is approximately 1 mm.

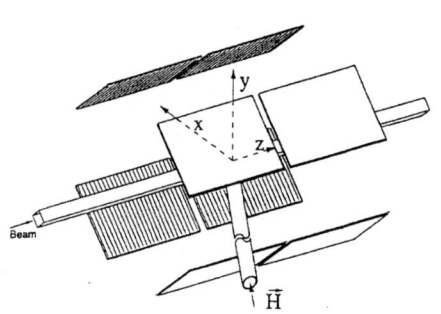

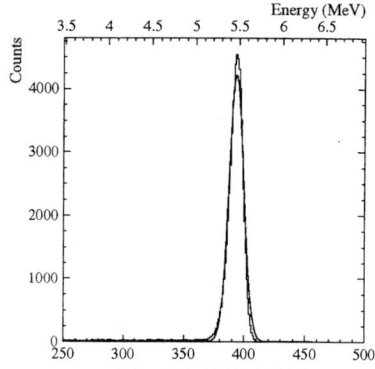

FIGURE 4. The eight recoil detectors that surround the storage cell are located at azimuthal angles of $\pm 45°$ and $\pm 135°$. Four detectors are located upstream, the other four downstream of the cell center. The atomic beam enters the storage cell at an azimuthal angle of $0°$ in order to not interfere with the detection system.

FIGURE 5. Spectrum of the recoil pulse height from the 10 Hz ^{241}Am α-sources. The spectrum is recorded under normal running conditions with beam through the cell and illustrates that even a slow source can be resolved nicely.

In order to calibrate the recoil energy pulse height ^{241}Am α-sources (5.5 MeV) are mounted on the fins of the cell support structure. A pulse height spectrum of the detectors with trigger on single recoils in the presence of the stored beam is shown in fig. 5. The peak corresponds to a gaussian of width 75 keV and illustrates the low rate of background, which allows to observe even a 10 Hz source. A more detailed description of the recoil detector system can be found in ref. [4].

The detector system, shown in fig. 1, is used to trigger elastic pp coincidences. There is either a coincidence between any of the eight recoil detectors (h in fig. 1) and the forward scintillators (l,m) or, at larger angles, a coincidence between two of the four scintillators (n) on opposite sides that are located at azimuthal angles of ±45° and ±135°. The particle tracks are determined from the position information of the wire chambers (j,k) and if available the positions in the recoil detectors are used as well. In most cases there is more information available than necessary to determine the tracks of the ejectiles in pp elastic scattering. Therefore kinematical fitting is used to find the most probable values for scattering and azimuthal angles and the vertex. The resulting distribution of χ^2's is in good agreement with the expected probability distribution.

Additional constraints for the identification of pp elastic scattering events can be obtained from the pulse height distribution of the recoil energy deposit vs forward scattering angles. Such a spectrum is shown in fig. 6.

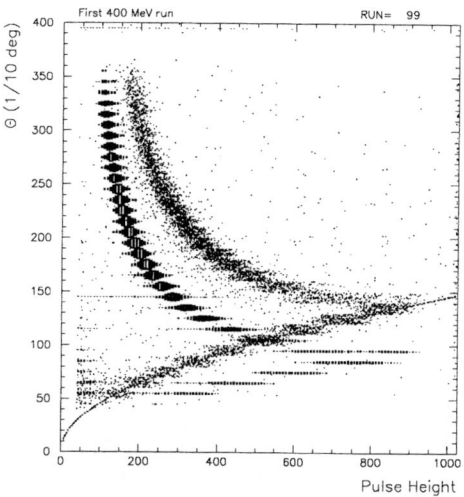

FIGURE 6. Recoil energy pulse height vs forward scattering angle. The boxed spectrum corresponds to 450 MeV and the dotted one to 200 MeV incident proton energy. The dotted line underneath the 200 MeV pp elastic locus is the result of a calculation of the forward scattering angles from the calibrated recoil energy information (see fig. 5).

The figure actually shows two different incident proton energies, 200 MeV (dotted) and 450 MeV (boxed). The characteristic shape comes about from the energy deposit in conjunction with the finite thickness of the recoil detectors of 1 mm. At 200 MeV recoil protons at forward scattering angles up to 14° are stopped in the detectors, while for larger scattering angles they pass through the detectors. Below 14° it is possible to determine the forward scattering angles using the calibration of the recoil energy from the α-sources shown in fig. 5. This leads to the dotted curve underneath the 200 MeV locus in fig. 6. The spectrum taken at 450 MeV beam energy (boxed) reveals the same characteristics, except that the punch-through of the recoil protons occurs at smaller scattering angles.

OPERATION AND PERFORMANCE OF THE TARGET

During operation the target polarization is reoriented every 2 s in a sequence $\pm x$, $\pm y$ and $\pm z$. This is illustrated in fig. 7. ¿From the known analyzing power A_y the target polarization for x- and y-guide field is deduced. For z-guide field it is assumed that the target polarization is equal to the average of the target polarizations with x- and y-field.

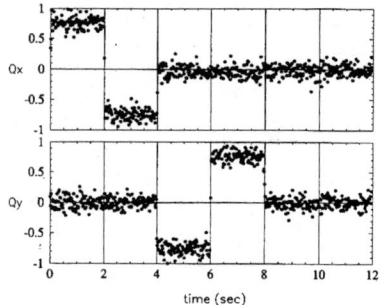

FIGURE 7. Polarized subcycle of the gas target. The sequence is $\pm x$, $\pm y$ and $\pm z$. The direction of the guide field is reversed or reoriented every 2 s.

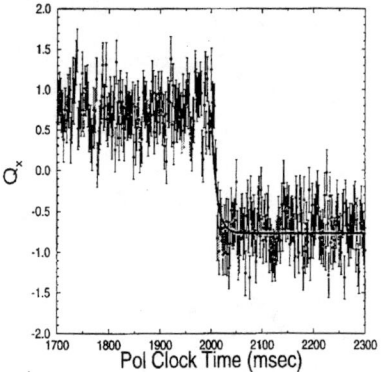

FIGURE 8. Reversal of the target polarization from $+x$ to $-x$. After about 50 ms the target polarization has reversed sign.

In fig. 8 the transition region when the target field is reversed from $+x$ to $-x$ guide field is shown in more detail. The target polarization has completely reversed sign after about 50 ms. It should be noted that the atoms spend only about 5 ms in the cell.

The target polarizations Q_x and Q_y are in perfect statistical agreement. In fig. 9 the average target polarization $\frac{1}{2} \cdot (Q_x + Q_y)$ is shown as a function of running time. There is no deterioration of the target polarization over time visible. From this one can conclude that there is no radiation damage to the teflon walls during operation over the course of about a week. However, the values of the target polarization are lower by about 9% compared to the expected polarization of the atoms in the beam from the source. It is at present not clear which effects account for this difference.

The very precise vertex determination due to the good resolution of the recoil detectors makes it possible to determine also the target polarization as a function of position along the cell. This is actually only possible for the central portion of the storage cell, since the recoil detectors extend only over about half the length of the storage cell. In fig. 10 the target polarization is plotted as a function of the z-coordinate along the storage cell. There is no z-dependence of the target polarization visible. The same plot also shows negligibly small values for the unwanted components of the target polarization.

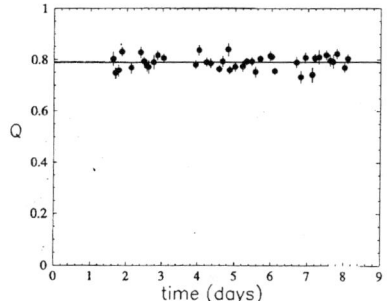
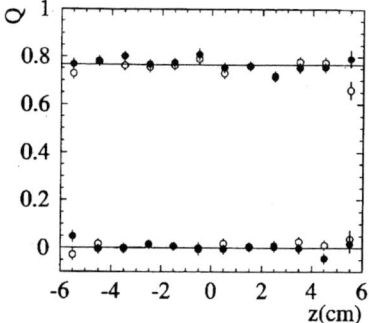

FIGURE 9. Average target polarization $\frac{1}{2} \cdot (Q_x + Q_y)$ as a function of time during a one week run.

FIGURE 10. Polarization of the target as a function of the longitudinal coordinate z along the storage cell. Open circles stand for Q_x, closed circles for Q_y. The points near zero show the unwanted components of the target polarization.

SYSTEMATIC EFFECTS

Although the magnetic guide field for the alignment of the target polarization is weak, reversal of the target guide field nevertheless affects the position of the stored proton beam. The question arises how large these guide field associated beam and angle modulations are and what impact they have on the resulting data. For that purpose the raw position information from the wire chambers is used to find the interception of a particle track with a plane perpendicular to the beam at the reconstructed vertex position. Repetition of this procedure for many events leads to a hitpattern from which horizontal and vertical position of the beam with respect to the wire chamber coordinates can be determined. This information can be prepared separately for each of the six guide field states of the target and therefrom the position modulation of the beam induced by reversal of the guide field can be obtained. In a very similar way it is also possible to determine the angle modulation of the beam from fluctuations of the pulseheight of the recoil detectors at certain angles. Details regarding this procedure can be found in [7,4]. The values for the beam position modulation are small and reach at most about 20 μm and about 100 μrad for the angle modulation. The accuracy is astonishingly high, 4 μm for the position and 50 μrad for the angle modulation. The largest position modulation are found for the vertical guide field state of the target, where no compensation coils are used. The largest tilt is found for the x target state. The resulting modulations were used as input to a Monte Carlo simulation. It was found that the effect on spin correlation data determined by the experiment is less than 10^{-3}.

The walls of the storage cell are about 10^9 times heavier than the polarized gas stored inside. Therefore it is relevant to study the contribution from undesired reactions. The background mechanism assumed is quasifree scattering of beam particles from light nuclei in the walls (F, C) of the storage cell. In order to study these reactions unpolarized nitrogen gas was injected into the cell. The fractional background was determined by comparing the number of accepted events with nitrogen to those accepted with a hydrogen target. This procedure is illustrated in fig. 11.

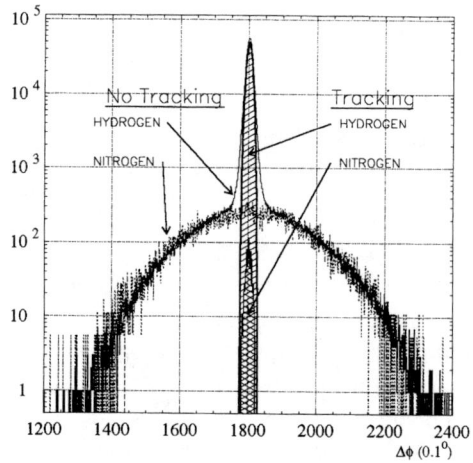

FIGURE 11. The background was studied by injection of N_2 into the storage cell. In pp elastic scattering the outgoing protons span a plane that includes the beam direction, $\Delta(\phi) = 180°$. For quasifree pp scattering there is no such constraint, the Nitrogen spectrum therefore shows a broad distribution. The ratio between the number of events accepted after processing through the tracking code with N_2 as target gas compared to H is determined by scaling the number of counts with N_2 to those with H. The extracted ratio is $P_{\text{back}} < 0.2\%$.

The upper limit for the fractional background is determined to be 0.2%.

Also here the impact of background on the spin correlation data is negligible compared to the statistical errors.

RECENT RESULTS AND CONCLUSIONS

Unprecedented small statistical and systematical uncertainties in pp elastic measurements of spin correlation parameters and analyzing powers have been achieved. *Preliminary* results of A_{xx}, A_{yy}, A_{xz} and A_y in the measured angular range $5° \leq \theta_{\mathrm{Lab}} \leq 45°$ for seven incident proton energies 250, 280, 295, 310, 350, 400 and 450 MeV are shown in fig. 12. The graphs also show results from partial-wave analysis by the VPI group [8]. All the data shown were acquired in about two weeks of beam time. The high statistical precision and small systematic uncertainties are rather astonishing and underline the great advantages of thin polarized gas targets for precision measurements.

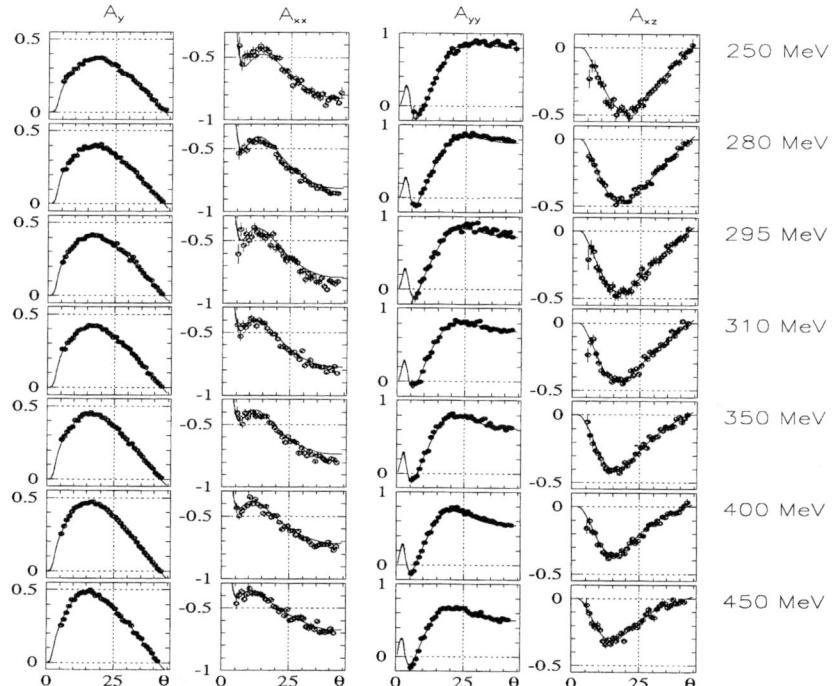

FIGURE 12. Angular distributions of spin correlation parameters A_{xx}, A_{yy}, A_{xz} and of analyzing power A_y determined in the angular range $5° \leq \theta_{\mathrm{Lab}} \leq 45°$ at energies of 250, 280, 295, 310, 350, 400 and 450 MeV. The data shown are still preliminary.

The IUCF polarized internal hydrogen target has shown a remarkably stable operation over the course of the last four years. Target polarizations close to 0.8 and target densities of about $3.1 \cdot 10^{16}$ H_1/cm^2 have been achieved regularly. Average pp elastic event rates of up to 87k/h have been recorded with the apparatus at beam energies between 200 and 450 MeV.

There has been no indication of damage of the storage cell walls due to radiation in the proton machine environment of the Cooler. Experimentally there is no indication of a loss in polarization along the axis of the storage cell. The depolarization probability per wall bounce is estimated to be on the order of 1/5000 or less. The impact of small background contributions of less than 0.2% on the final data is negligibly small.

Recently, a consistency test of the working assumption that the target polarization with longitudinal guide field is equal to the transverse target polarization could be carried out. For the first time longitudinally polarized protons were stored in the Cooler, thereby making accessible measurements of spin correlation parameter A_{zz}. Details regarding this interesting new development can be found in ref. [9]. At a scattering angle of 90° CM theory predicts that $A_{xx} - A_{yy} - A_{zz} = 1$ [10]. At 200 MeV the measured result of 1.004 ± 0.03 is in perfect agreement with this prediction and underlines once again how well understood polarized internal gas targets are.

REFERENCES

1. T. Wise, A.D. Roberts and W. Haeberli, Nucl. Instrum. Meth. A **336** 410 (1993).
2. W. Haeberli, Ann. Rev. Nucl. Sci. **17** 373 (1967).
3. J.S. Price and W. Haeberli, Nucl. Instrum. Meth. A **326** 416 (1993).
4. W. Haeberli, B. Lorentz, F. Rathmann, M.A. Ross and T. Wise, W.A. Dezarn, J. Doskow, J.G. Hardie, H.O. Meyer, R.E. Pollock, B. von Przewoski, T. Rinckel, and F. Sperisen, P.V. Pancella, Phys. Rev. C **55** 597 (1997).
5. W.R. Lozowski, J. Hudson and W.A. Dezarn, Nucl. Instrum. Meth. A **362**, 189 (1995).
6. M.A. Ross, A.D. Roberts, T. Wise, W. Haeberli, W.A. Dezarn, J. Doskow, H.O. Meyer, R.E. Pollock, B. von Przewoski, T. Rinckel, F. Sperisen, P.V. Pancella, Nucl. Instrum. Meth. A **344** 307 (1994).
7. F. Rathmann et al., *Event reconstruction for an extended internal polarized gas target, including beam position analysis and vertex determination*, Proc. of the Intl. Workshop on Pol. Beams and Pol. Targets, Cologne 1995, Eds. H. Paetz gen Schieck and L. Sydow, World Scientific, 376 (1996).
8. R.A. Arndt et al., Solution C200 of the VPI SAID program.
9. B. Lorentz et. al, Proceedings of the Symposium on High Energy Spin Physics SPIN96 (Amsterdam, September 1996), to be published.
10. J. Bystricky, F. Lehar and P. Winternitz, J. de Physique **39** (1978); G.G. Ohlsen, Rep. Progr. Phys. **35 717** (1972).

The EDDA Experiment at COSY

H. Rohdjess for the EDDA Collaboration[1]

Institut für Strahlen- und Kernphysik, Universität Bonn, D-53115 Bonn, Germany

Abstract. Polarized and unpolarized proton-proton elastic scattering is investigated with the EDDA-experiment at the Cooler Synchrotron COSY at Jülich to significantly improve the world data base in the beam energy range 500-2500 MeV. Measurements during beam acceleration with thin internal targets and a large acceptance detector produce excitation functions over a broad angular and energy range with unprecedented internal consistency. Data taking with an unpolarized CH_2 fiber target and an unpolarized beam have been completed and the derived differential cross sections demonstrate the benefit of this technique. With a polarized atomic beam target recently installed in COSY and a polarized COSY beam – currently under development – the measurements will be extended to analyzing powers and spin correlation parameters.

INTRODUCTION

Nucleon-nucleon (NN) elastic scattering is fundamental to the understanding of the NN force, being the basis for many models in nuclear and heavy ion physics. Having been studied for decades, an enormous experimental database has been acquired. When compared to theoretical models of the NN force the experimental information generally enters through empirical phase shifts, which are fit to the data assuming a smooth energy dependence (e.g. [2]). Below about 1 GeV the phase shifts are well known owing to the many experiments performed in this region. Above 1 GeV the data points are less numerous – especially for polarization observables – and scatter considerably. Until recently global phase shifts did not extend beyond 1.6 GeV. For a substantial improvement of the NN database experiments are needed which provide internally consistent data over a wide kinematic range and/or with small uncertainties. Storage rings for nuclear physics offer novel experimental possibilities [3]: low background, easy changes of beam energy and the possibility to use pure polarized internal hydrogen targets in the recirculating beam.

These novel options are exploited by the EDDA experiment [4], designed to measure unpolarized cross sections, analyzing power and the spin correlation parameters A_{yy}, A_{xx}, and A_{xz} between 500 and 2500 MeV in narrow energy

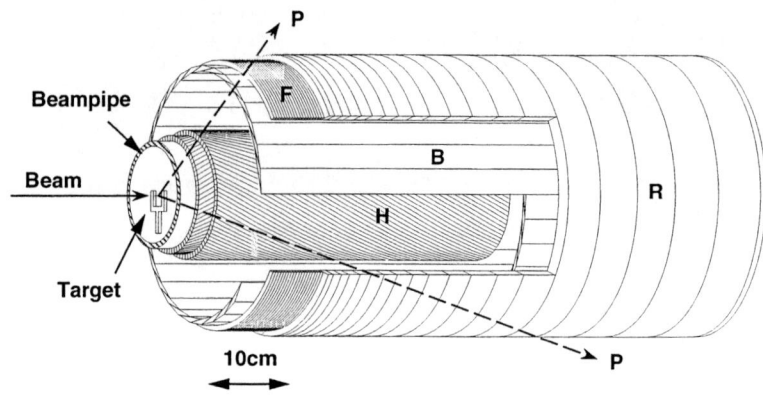

FIGURE 1. The EDDA detector (not to scale): **Target** fiber (CH_2) or a polarized atomic hydrogen beam; **B** scintillator bars ; **R** scintillator semi-rings; **F** semi-rings and **H** helix (4 layers) made from scintillating fibers.

steps and over a wide angular range ($35° < \theta_{c.m.} < 90°$). It uses the internal proton beam of the Cooler Synchrotron COSY in Jülich on a thin 5x4 μm^2 CH_2 fiber target for unpolarized and an atomic beam target [5] for polarized measurements.

Data are taken **during synchrotron acceleration**, such that a low-statistics excitation function is acquired in a single machine cycle (injection, acceleration, deceleration, reset). Statistics is gained by repetition of this cycle. This technique was first used for fixed angle measurements of differential cross sections at SATURNE [6] and analyzing powers at KEK [7]. Because all beam energies are sampled within a few seconds, systematic errors due to drifts in luminosity monitors, target and beam performance are greatly reduced.

Excitation function data of this type, being quasi-continuous in beam energy, is ideally suited to look for energy-dependent structures in elastic scattering observables. Such structures have been proposed (e.g. [8]) as a signature of a coupling of dibaryons – genuine six quark states – to the NN channel but, despite large efforts, no experimental evidence has been found as yet. Searching a large energy range with many observables and good resolution will hopefully clarify if such states exist with an appreciable coupling to the NN channel.

DETECTOR

The EDDA detector is shown schematically in Fig. 1. Its design is based on a fast triggering of coplanar two–prong events of charged particles that fulfill the kinematic relation between the laboratory angles for elastic proton–proton scattering. The detector consists of a cylindrical double layer that surrounds the thin walled beampipe downstream from the internal target (In Fig. 1 a horizontally oriented fiber target) . The angular range covered extends from $\Theta_{lab} = 10°$ to $72°$ subtending 85% of the solid angle. The outer layer consists of 32 scintillator bars (B in Fig. 1) parallel to the beam axis, surrounded by scintillator semi–rings (R,F). The scintillators are in both layers partially overlapping; their cross sections have been designed such that each particle from the vertex traversing the outer layer deposits energy in two adjacent bars and two semi–rings. Analysis of the fractional light output is then used [9] to determine scattering angles with a resolution of $1.0°$ $(1.9°)$ (FWHM) in $\Theta_{c.m.}(\Phi_{c.m.})$.

For unpolarized measurements the reaction vertex is well determined by the beam width and the size of the fiber target. Here, the outer layer suffices to measure scattering angles. Use of the atomic beam target requires vertex reconstruction. For this purpose an inner detector made from 4 layers of scintillating fibers (160 each) wound helically around the beam pipe in alternating directions has been installed. (H in Fig. 1).

UNPOLARIZED MEASUREMENTS

Measurements were performed with about 10^7 protons circulating in COSY, data were collected during beam acceleration (1.15 GeV/c/s) between 500 and 2500 MeV proton energy. A total of $4 \cdot 10^7$ elastic scattering events have been accumulated.

Background originating from the carbon content of the CH_2 target was measured with a pure carbon target and subtracted statistically. Contributions from inelastic pp reactions are small and were investigated with Monte-Carlo simulations.

The relative change of luminosity as a function of beam energy was monitored by concurrent measurements of the total yield of secondary electrons emanating from the target and the δ electron yield from e-p elastic scattering. They are described in [10]. Both methods agree within 2.5% for all energies. Absolute normalization of the whole data set was achieved with reference to data at 793 MeV from LAMPF [11] where an absolute normalization uncertainty of 1% was achieved.

Unpolarized differential cross sections based on $2 \cdot 10^7$ elastic pp events have been published [10] and are shown in Fig. 2. They add 2121 entries to the NN elastic scattering data base with typically 5% error (statistical and systematic). Two excitation functions are compared to previously published data in Fig. 3. The impact of the new EDDA data on phase shifts [2] is evinced in the difference of the dashed (solution SM94, without EDDA data) and solid (solution SM97, with EDDA data) lines. Note, that the phase shift analysis

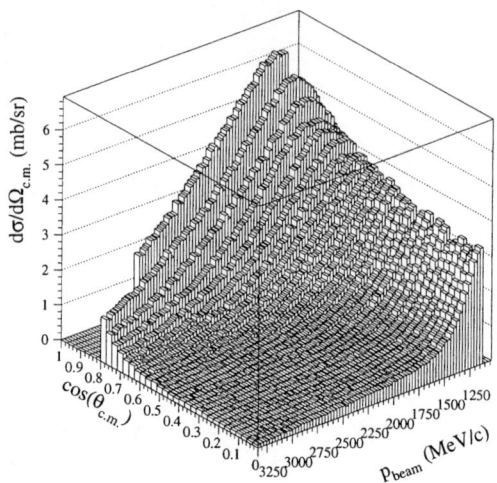

FIGURE 2. Differential proton-proton elastic scattering cross section as a function of c.m. scattering angle and proton beam momentum [10].

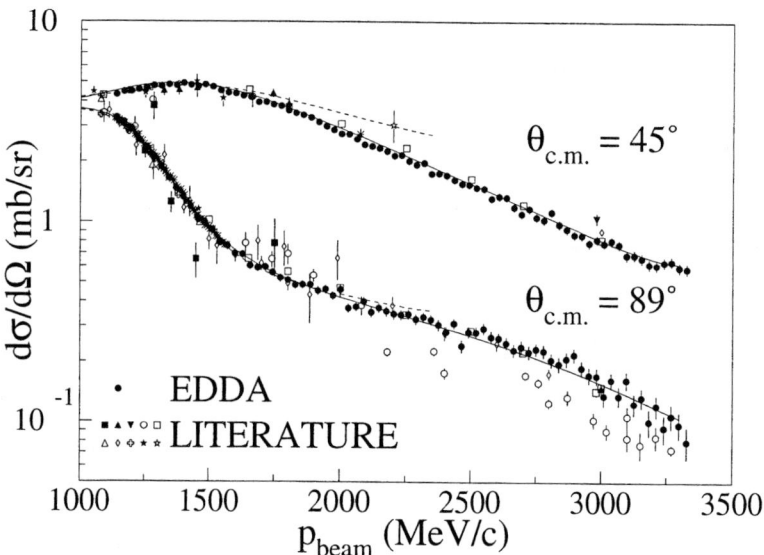

FIGURE 3. Excitation functions of unpolarized differential cross sections at two selected c.m. scattering angles. The EDDA data from [10] are compared to previously published data (extracted from the SAID [2] database) and two phase shift solutions [2] excluding (dashed) and including (solid line) this data.

could be extended to 2500 MeV. No indication for structures in the present data have been observed [10].

Currently, a refined analysis of all acquired data is underway in order to further reduce the error bars. Improvements are expected due to increased statistics, better background suppression by use of self-organizing maps for event classification and an additional luminosity monitor.

The quality of the unpolarized cross sections acquired by the EDDA experiment show the potential of the experimental technique. The EDDA collaboration now strives to achieve the same level of precision on polarization observables.

PREPARATIONS FOR POLARIZED MEASUREMENTS

For the experimental program of the EDDA experiment on polarization observables the detector has been upgraded by the inner helix detector (cf. Fig. 1) by the end of 1996. With the polarized atomic beam target – installed in May this year – and an unpolarized COSY beam, analyzing powers will be measured first. For spin correlation parameters both a polarized target and beam are needed and will commence when a polarized COSY beam over the full energy range and with sufficient intensity is available.

Polarized Target

The design of the atomic beam target [5] had to meet constraints imposed by the EDDA experiment. First, the space close to the interaction region is limited: target components must be outside the angular acceptance of the EDDA experiment. This allows only weak holding fields such that pure spin states must be used. Secondly, measurements during beam acceleration makes the use of a storage cell unfavorable. The reduction of the acceptance, and thus the injected current at 40 MeV beam energy, and possible beam movements during acceleration to 2500 MeV would at least partly offset the benefit of higher target densities. Since elastic scattering has a large cross section (2.5 -25 mb in the angular range covered by the EDDA detector) target densities of a few 10^{11} H/cm^2, as they are obtained by atomic beams, suffice. With a recirculating (1.5 MHz) COSY beam of $> 5 \cdot 10^{10}$ protons luminosities of $> 10^{28}$ cm^{-2}s^{-1} will produce rates of $> 50 - 200$ elastic events per second.

The atomic beam target as it was used in first tests during June 1997 is shown in Fig. 4. Hydrogen is dissociated in a 350 W RF discharge and passes through an aluminum nozzle cooled to 30 K. The low temperature of the nozzle leads to a decreased lower velocity (most probable velocity 1.1 km/s) and thus an increased target thickness. Hydrogen atoms in the $m_j = +1/2$

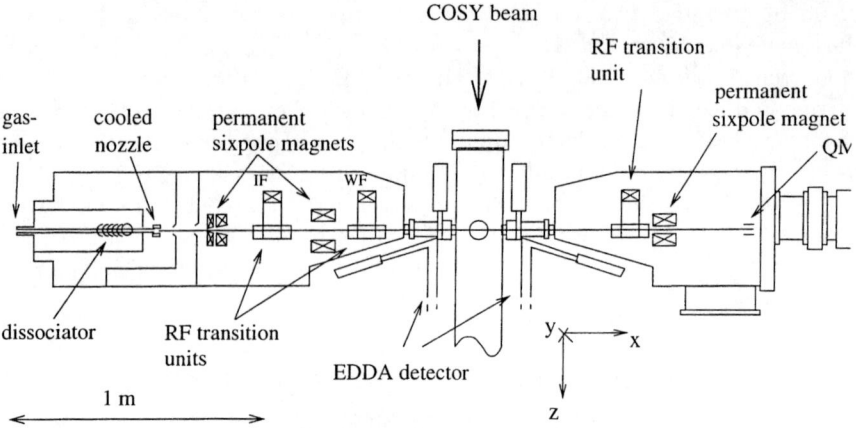

FIGURE 4. Simplified drawing of the atomic beam target [5]. Only the upstream light-guides and phototubes of the EDDA detector (cf. Fig. 1) are shown to show the space constraints imposed on the target chambers.

spin states are focused by a set of three permanent sixpole magnets into the interaction region while the $m_j = -1/2$ states are defocused.

The proton spin can be prepared by two RF-transition operating at intermediate (IF, 2→ 4) and weak fields (WF, 1→ 3). Nuclear polarization in a pure state is obtained when the IF transition is in operation. The proton spin is aligned by a 0.9 mT holding field in the interaction region and can be flipped by either the WF transition or by changing the sign of the holding field. In the beam dump (right side of Fig. 4) a sixpole magnet, RF transition and a QMS is installed for beam optimization and monitoring.

Right after installation the atomic beam target was tested with a 1455 MeV/c internal COSY beam, i.e. data were acquired after synchrotron acceleration. The preliminary result of this very first test is presented here, with the analysis still going on. It should be noted that due to limited time for optimization of the atomic beam, performance with the design values [5] is not yet to be expected.

The vertical beam profile of the atomic beam has been measured by sweeping the COSY beam through steerer magnets across the target and measuring the vertex distribution of elastic proton-proton scattering events. In Fig. 5 the movement of the COSY beam (top) and the distribution of the scattering vertex along the COSY beam (bottom) is shown as a function of cycle time. From the modulation of scattering rate from the atomic beam at z=0 with the vertical position of the COSY beam, an atomic beam width of 13 mm (FWHM) is derived. This is in agreement with the width extracted from the z-distribution of the scattering vertex when the resolution of vertex reconstruction of 2.5 mm is considered. Scattering rates from the atomic beam

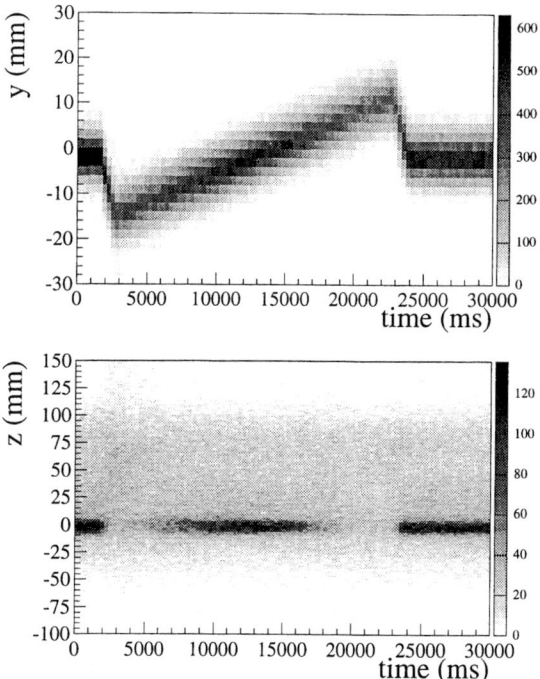

FIGURE 5. Top: Distribution of the reconstructed scattering vertex perpendicular to both the COSY and atomic hydrogen beam as a function of cycle time. The COSY beam is swept across the target beam by steerer magnets. **Bottom:** The resulting modulation of the scattering vertex distribution along the COSY beam (z). From the change of the scattering rate with time at the atomic beam location (z=0, bottom) the vertical atomic beam profile is deduced. For the choice of coordinates refer to Fig. 4.

part of the target thickness distribution are in line with densities of about 10^{11} H/cm^2.

In the tests only the IF transition was used to prepare a pure $m_j = +1/2$, $m_I = +1/2$ state. The spin was flipped by reversing the holding field. The target polarization was measured by observing the asymmetry of elastic proton-proton scattering events with known analyzing power [2]. The scattering rate to the left ($-70° < \phi < 70°$) and right ($-110° < \phi < 250°$) are plotted in Fig. 6 (a) and (c) as a function of the reconstructed scattering vertex position along the COSY beam axis. The derived polarizations are shown in (b) and (d). Apart from the polarized beam at z=0, a diffuse cloud of unpolarized hydrogen extends up and downstream. The decrease in rate below z<-20mm and z>80mm is due to the acceptance of the EDDA hardware trigger. Polarizations of about 60% (when the diffuse background is subtracted of around

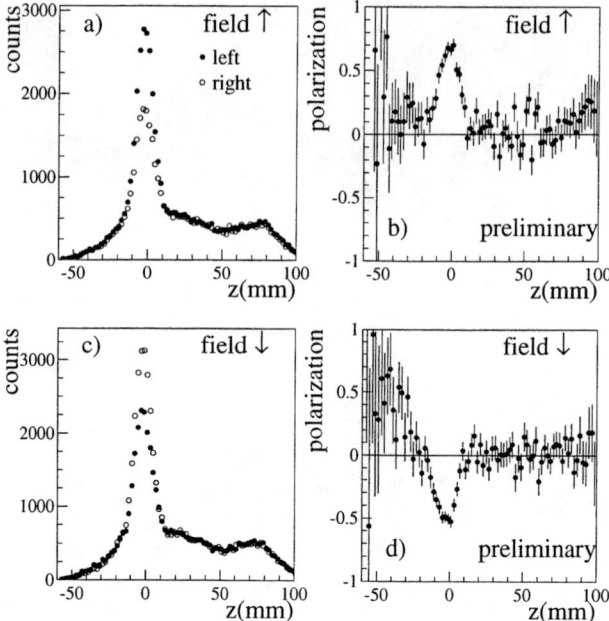

FIGURE 6. Elastic proton-proton scattering rate as a function of the position z along the unpolarized COSY beam to the left and right of the detector (left column) and the derived polarization (right column). The polarized atomic hydrogen beam stands out at z=0 on a unpolarized background. A holding field in the interaction region of 0.9 mT pointing up (top row) or down (bottom row) was used to align the target spins.

80%) have been reached in this first test. The difference in polarization between the two field directions – measured separated in time – may reflect true changes in polarization or be due to false asymmetries in this preliminary analysis and needs to be investigated. Technical improvements now focus on a reduction of the unpolarized hydrogen background by vacuum improvements and optimization of the atomic hydrogen beam transport and are discussed in [5].

Polarized Beam

Polarized beam development using the polarized source at COSY [12] has started [13]. Here, the EDDA detector serves as a fast polarimeter for beam development by acquiring continuously the asymmetry of scattering events from a CH_2 fiber target as a function of beam energy. Within a few minutes, relative changes of the beam polarization can be measured. Up to now the COSY team has successfully accelerated polarized protons up to 2200 MeV/c crossing 7 depolarizing resonances. This was achieved by dynamic changes of ion optics without recourse to fast quadrupoles. Polarized beam development will continue later this year.

CONCLUSION

The quality of the unpolarized cross section data produced by the EDDA experiment have demonstrated that the experimental method of data acquisition during synchrotron acceleration works well. Here, the main challenge has been and is the consistent normalization.

Preparations for continuing the experiment with a polarized atomic beam target and a polarized beam are well under way. A detector upgrade has been successfully completed and the target has produced both density and polarization not too far from its design values. However, unpolarized background observed in these first tests needs to be reduced, and first steps in this direction have been taken. Development of the polarized beam at COSY, i.e. preserving polarization when crossing depolarizing resonances, is under way and has led to promising results.

ACKNOWLEDGMENT

The excellent beam support and collaboration of the COSY operating team, headed by R. Maier, is greatfully acknowledged. This work is supported by the BMBF and by the Forschungszentrum Jülich.

REFERENCES

1. M. Altmeier[1], F. Bauer[2], J. Bisplinghoff[1], T. Bissel[1], R. Bollmann[2], M. Busch[1], K. Büßer[2], P. Cloth[3], R. Daniel[1], O. Diehl[1], F. Dohrmann[2], H.P. Engelhardt[1], J. Ernst[1], P.D. Eversheim[1], O. Felden[1], J. Flammer[2], M. Gasthuber[2], R. Gebel[3], J. Greiff[2], A. Groß[2], R. Groß-Hardt[1], K. Hebbel[2], F. Hinterberger[1], T. Hüskes[1], R. Jahn[1], I. Koch[2], R. Langkau[2], T. Lindemann[2], J. Lindlein[2], R. Maier[3], R. Maschuw[1], T. Mayer-Kuckuk[1], M. Pfuff[2], D. Prasuhn[3], H. Rohdjeß[1], D. Rosendaal[1], P. von Rossen[3], N. Schirm[2], M. Schulz-Rojahn[1], V. Schwarz[1], W. Scobel[2], S. Steinbeck[2], G. Sterzenbach[3], S. Thomas[1], H.J. Trelle[1], M. Walker[1], E. Weise[1], A. Wellinghausen[2], K. Woller[2], R. Ziegler[1]
   (1) Inst. f. Strahlen- und Kernphysik, Univ. Bonn
   (2) I. Inst. f. Experimentalphysik, Univ. Hamburg
   (3) Inst. f. Kernphysik, KFA Jülich
2. R. A.. Arndt et al., *Phys. Rev.* **C 50**, 2731 (1994) and preprint nucl-th/9706003 (1997); Program SAID solutions SM94 and SM97.
3. H. O. Meyer "Nuclear Physics with Light-Ion Beams" to be published in *Ann. Rev. Nucl. Part. Sci.*

4. J. Bisplinghoff and F. Hinterberger, AIP Conf. Proc. **221**,312 (1991); W. Scobel, Phys. Scripta **48**,92 (1993); H. Rohdjeß, Proc. Int. Conf. on Physics with GeV-Particle Beams (Jülich 1994), p.334.
5. P.-D. Eversheim, Contribution to this Conference
6. M. Garcon et al., *Nucl. Phys.* **A445**, 669 (1985)
7. H. Y. Yoshida et al., *Nucl. Phys.* **A541**, 443 (1992)
8. P. Gonzalez et al. *Phys. Rev.* **D35**, 2142 (1986) ; R. Vinh Mau et al. *Phys. Rev. Lett.* **67**, 1392 (1987)
9. J. Bisplinghoff et al. (EDDA Collaboration), Nucl. Instr. and Meth. **A329**,151 (1993);
10. D. Albers et al. (EDDA Collaboration), *Phys. Rev. Lett* **78**, 1652 (1997)
11. A. J. Simon et al. *Phys. Rev.* **C48**, 662 (1993) K. Ackerstaff et al., Nucl. Instr. and Meth. **A335**,113 (1993).
12. R. Gebel, Contribution to this Conference
13. A. Lehrach et al., *Proceedings of the 12th International Symposium on High-Energy Spin Physics*, Amsterdam 1996, World Scientific 1997, pp. 416-418.

CRYOGENIC ATOMIC BEAM SOURCE AT VEPP-3

L.G.Isaeva*, B.A.Lazarenko*, S.I.Mishnev*, D.M.Nikolenko*,
A.N.Osipov[†], S.G.Popov*[1], I.A.Rachek*, Yu.V.Shestakov*,
A.A.Sidorov[†], V.N.Stibunov[†], D.K.Toporkov*,
D.K.Vesnovsky* and S.A.Zevakov*

*Budker Institute of Nuclear Physics, Novosibirsk 630090, Russia
[†]Institute for Nuclear Research, Tomsk 634050, Russia

Abstract. The experiment on elastic and inelastic scattering of 2 GeV electrons by internal polarized target is in progress at the VEPP-3 storage ring in Novosibirsk. It's carried out by Novosibirsk/St.-Petersburg/Tomsk/Argonne/Illinois/NIKHEF collaboration. A cryogenic Atomic Beam Source having five superconducting sextupoles is under manufacturing to feed by polarized deuterium atoms an internal storage cell target. All the magnets have been manufactured and tested. The magnetic poletip field up to 4.8 T was measured at the cylindrical magnets having 44 mm inner diameter while 3.1 T and 4.0 T were measured for the tapered magnets. The dissociation degree of about 90% has been achieved for a gas throughput 1 mb×l/sec. The expected flux of polarized deuterium atoms into the storage cell is 1.0×10^{17} at/sec (in two substates). The geometry of the magnetic system, results on the dissociation measurements, testing of the superconducting magnet and expected parameters of the target are presented.

INTRODUCTION

At present a number of intermediate and high-energy facilities are carried out experiments with polarized internal target technique [1-3]. In spite of a fast progress of laser-driven spin-exchange technology an atomic beam method is the most widely used technique for producing polarized hydrogen or deuterium. This is especially true if high degree of polarization and purity of the beam are required. The classical Stern-Gerlach separation scheme is the base of this method. A number of groups have reported about the construction of the intense atomic beam sources [4,5]. The intensity of the beam from the

---

[1] Deceased

source depends on many factors: the degree of dissociation in the discharge tube, the beam formation system, the temperature of the nozzle, the velocity distribution in the beam, the attenuation of the beam in the source due to the scattering off background atoms and molecules and especially the properties of the focusing magnet system.

The magnets having axial symmetry are widely used to obtain large acceptance of the beam focusing system. For typical geometries of the atomic beam sources the accepted beam solid angle and thus the flux of the selected particles is proportional to the magnetic field strength at the largest usable radius of the magnet (B_0). When considering the focusing properties of the magnet the one with a higher multipolarity is preferable, because it allows to get higher acceptance and/or flux of polarized atoms at the same value of the magnetic poletip field. However, only quadrupole or sextupole magnets are used because of the difficulty to satisfy mentioned above requirement for higher multipolarity - to keep the same magnetic field strength B_0 at the magnet. Also, due to their harmonic focusing field sextupole magnets are preferable to focus atoms at a given position.

In the commonly used sextupole electromagnets, the saturation in the mild iron pole pieces limits B_0 to about 1 T. The use of permanent magnets is limited by the demagnetization of the material. With neodymium-iron-boron alloys as a permanent magnetic material (high resistance against demagnetization) a usable sextupole field strength of 1.44 T has been obtained at the inner aperture of 24.5 mm in diameter [6].

Substantial increase of the magnetic field might be obtained in a superconducting magnet. At the same time the cold surfaces of the wall of the magnet might be used as a cryopump to get good vacuum conditions along the beam path. The attraction to use high B field superconducting magnets for the atomic beam sources has been discussed many times [7], however the magnets providing only 1.8 T magnetic field were constructed thus far [8].

DESIGN OF THE MAGNETS

One of the principal innovation of the source is the magnetic-optical system including five strong superconducting sextupoles. Figure 1 shows a vertical cut through the ABS along the atomic beam. The dimensions of the magnets, their positions along the source axis and expected values of the magnetic pole tip field, which were used for ray tracing calculations, are shown in Table 1.

These placement and parameters of the magnets allow to provide an achromatic focusing of the atoms into the storage cell feed tube and to get a good substate selection with a medium field RF transition. Figure 2 shows the velocity distribution of the atoms transmited into the storage cell. The calculation has shown that about 67% of the atoms which reach the entrance of the first magnet and should be focused are transported into the storage cell.

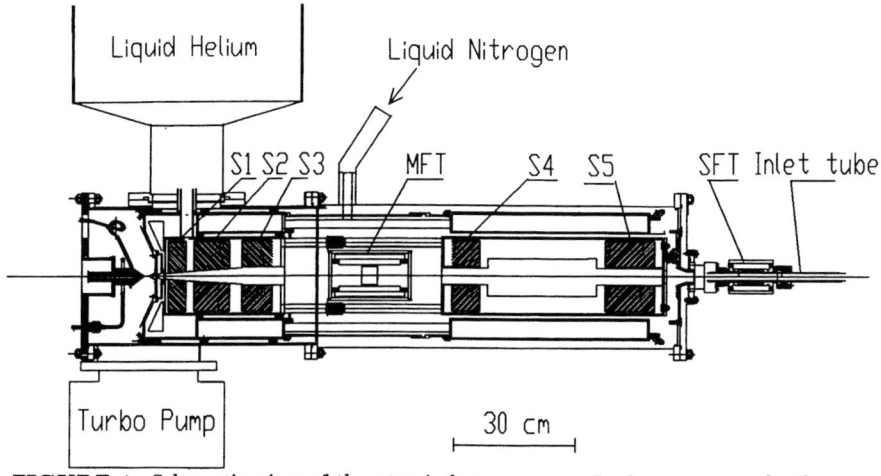

FIGURE 1. Schematic view of the atomic beam source. S1–S5 – superconducting sextupoles, MFT – medium field transitions, SFT – strong field transitions.

The flux of the atoms undergoing a medium field transition is about 5% of the total flux into the storage cell.

The cross section of the cylindrical magnet with an aperture 44 mm and the result of the two-dimensional field calculation using code MERMAID [9] are presented in Figure 3. For these calculations the total current of 120 kA was distributed uniformly in the rectangular area 3×0.5 cm^2. Figure 4 shows the results of these calculations. For the cylindrical magnets the magnetic field distribution doesn't change along the magnet, but that not longer valid for the tapered magnets, because the ratio between the dimensions of coil and iron differs for the entrance and the exit of the magnet. The dimensions of the coil and iron were chosen to get high quality magnetic field for the mid position of the magnet.

TABLE 1. The positions of the magnets along the source and their parameters. Two values of the radius and the magnetic field correspond to the entrance and the exit of the magnet.

	inner diameter [cm]	B field [T]	length [cm]	position [cm]
nozzle	0.2	—	—	0
1 magnet	1.4–2.2	3.2–3.7	4.5	6
2 magnet	2.5–3.6	4.2–4.3	9.0	12.5
3 magnet	4.4–4.4	4.6–4.6	7.5	24.5
4 magnet	4.4–4.4	4.6–4.6	7.0	76.0
5 magnet	4.4–4.4	4.6–4.6	12.5	114.0
inlet tube	2.0	—	35.0	137.5

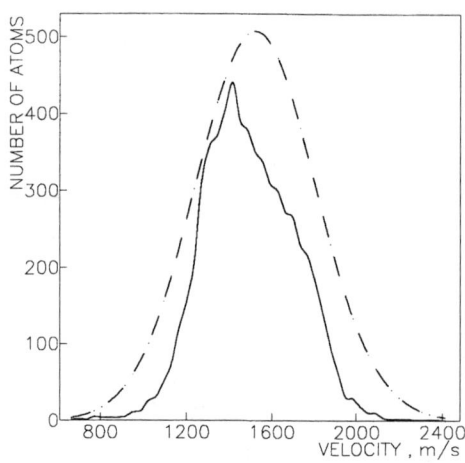

FIGURE 2. Velocity distribution of the atoms: dashed line–at the entrance of the magnet system, solid line–at the exit of the inlet tube.

CONSTRUCTION OF THE MAGNET AND RESULTS OF THE TEST

All five magnets have been manufactured using mild iron Steel KP08 for the pole tips and for the yoke. The outer diameter of the magnets 170 mm was chosen to get small magnetization in the body of the magnet. A multifilament NbTi wire with a diameter 0.85 mm was used for the coil of the cylindrical magnets. The characteristic parameters of the wire, specified by the manufacturer is a critical current $I_c \leq$ 594 A at 4.2 K and $B_c \leq$ 5 T. The number of turns around each pole is 243 and the smallest radius of winding around the poletip is about 0.5 cm.

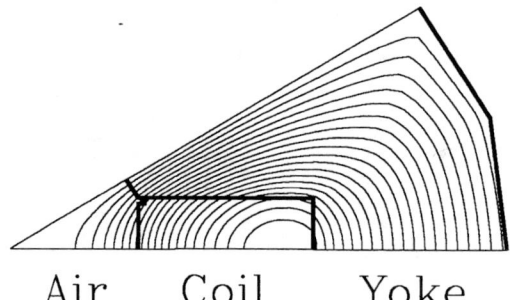

FIGURE 3. MERMAID calculation for a segment of 30 °. The figure shows flux lines of the magnetic field.

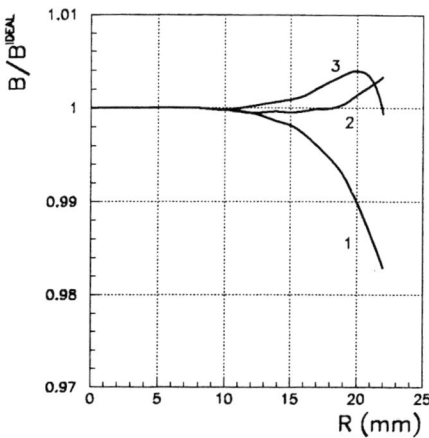

FIGURE 4. The ratio of the calculated magnetic field to the field of ideal sextupole magnet $| B^{ideal} | = B_o \times (R/R_o)^2$ as a function of the radius for the different azimuthal angle. Curve 1 corresponds to the direction on the pole, 2 - direction on the coil, 3 - in between.

The epoxy glue with titanium oxide filler was used to fix the wires. The gluing was done by "wet winding", applying the glue directly to the coil and winding the wire into the glue. After the winding the coils were pressurized to fix the configuration. Each pole was wound separately and later they were assembled into the magnet. The coils were connected with the use of indium solder along the length about 10 cm. A thinner NbTi wire with a diameter 0.50 mm was used for winding of the tapered bore magnets because of the smaller radius of the coils and higher current density in the coils. To avoid small radius of the winding around the pole, which could limit the current, the wires were glued to the pole with a special form provided a large curvature. The cross section of the coil in the second magnet is a rectangle 0.55×1.5 cm^2 which is a constant along the magnet and consists of 210 turns. For the first

TABLE 2. Results of the field measurements for three magnets. The gradient was calculated from the magnetic field measured at a single point assuming a quadratic dependence of the magnetic field on radius.

	max. current [Ampere]	B field [T]	radius [cm]	gradient [T/cm]	cur.density [kA/cm^2]
1 magnet	270	3.1	0.9	6.9	64
2 magnet	250	4.0	1.5	5.3	64
3 magnet	513	4.8	2.2	4.4	49

magnet the coil is a rectangle 0.35×1.5 cm^2 and consists of 125 turns.

All magnets were tested in a cryostat filled with liquid helium. The magnetic field was measured by the miniature Hall sensor, placed under the pole. The accuracy of the measurements were limited by the uncertaincy of the position of the sensor. The results of the measurements are presented in the Table 2. These results were obtained after many (20-50) coil quenches are occurred. Figure 5 shows the dependence of the magnetic field measured by sensors on the current through the coil.

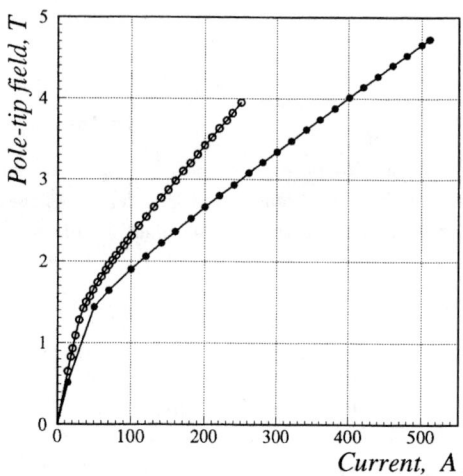

FIGURE 5. The magnetic field versus the current in the coil: full circles–magnet 3 (cylindrical), open circles–magnet 2 (tapered).

DISSOCIATOR

A thermal atomic deuterium beam is produced by RF dissociator followed by a liquid nitrogen cooled aluminum nozzle, see Figure 6. The construction of the dissociator is very similar to that of the Wisconsin ABS [4,10]. The discharge is induced inside a water cooled quartz tube 10 mm inner diameter and 1.5 mm wall thickness. A 5kW RF generator operated at 16 MHz was used to get a discharge, supplying 250 - 400 W power depending on the gas throughput. This power was measured by the temperature rise of the water cooling the discharge. The nozzle had a diameter of 2.5 mm and an internal half angle of 10 °, an external half angle of 25 ° and the length of about 25 mm. The nozzle terminated in a knife edge. A certain amount of oxygen was added to the discharge to get higher degree of dissociation.

The measurements of the degree of dissociation were performed at the test

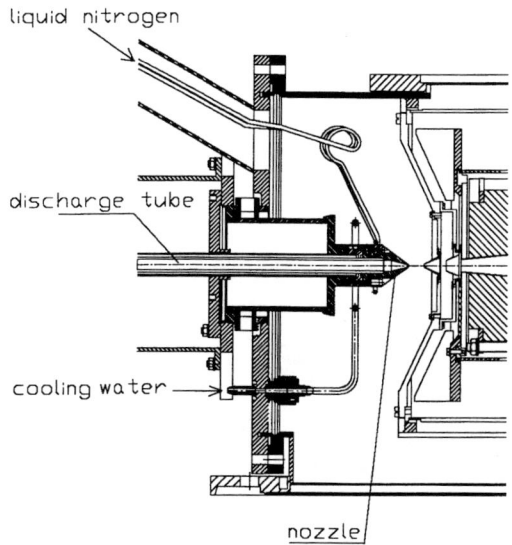

FIGURE 6. Schematic of the dissociator arrangement.

bench, contained a straight section of the VEPP-3 storage ring and Breit-Rabi atomic beam polarimeter [11]. The polarimeter consisted of three permanent quadrupole magnets (8 mm diameter, 90 mm length, pole tip field strength 1.1 T) and quadrupole mass analyser. A chopper and paddle were used for background signal suppression. During these measurements the nozzle chamber was pumped by a cryopump, having a pumping speed about 5000 l/s. A small, 1.17 mm diameter, knife edge skimmer was used to separate nozzle chamber from skimmer chamber because of a small (40 l/s) pumping speed in the skimmer chamber.

Figure 7 shows the lockin amplifier signals measured by QMA for atomic and molecular fractions as a function of the gas throughput. The dissociation degree derived from the measured signals is also shown. At the small gas throughput, when the attenuation of the molecular beam is small, the dissociation degree was determined with the use of the molecular signals only, corresponding the situation when the discharge was switched on and off. This value of the degree of dissociation was used for normalization. As it is seen from the Figure 7 the degree of dissociation is almost constant for the range of a gas flux up to 1.5 mbar×l/sec.

Some phenomenon was found during the increase of the RF discharge power in the range of 1 mbar×l/sec gas flux: it was observed that a luminous gas comes out from the nozzle into the vacuum and simultaneously a substantional gain (almost double) of the atomic beam density measured by QMA was found. Time of flight measurements were performed to obtain the velocity distributions parameters for different conditions of the discharge. The Figure 8

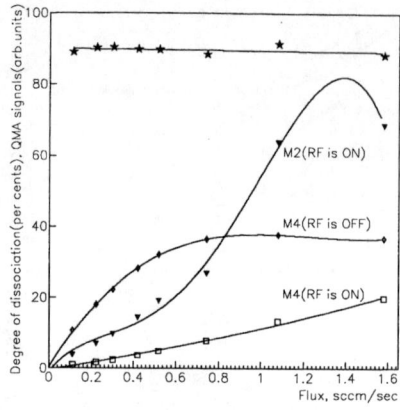

FIGURE 7. Degree of dissociation and QMA signals versus the gas throughput.

shows the time of flight signals obtained for the different level of RF discharge

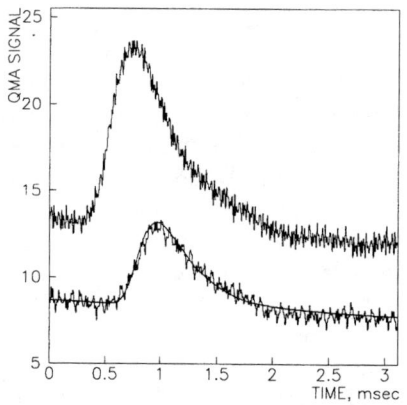

FIGURE 8. Time of flight signals measured for different RF power.

power. The lower curve was found for the regular condition of the discharge (power 290 W) and the upper one was obtained for increased power (380 W), when the phenomenon is on. The visible increase of the velocity of the atoms (about 25%) was detected for the latter case (see Figure 8).

With the purpose to explain the phenomenon the distance between the nozzle and skimmer was enlarged to 50 mm and a movable wire 0.5 mm diameter was placed at the distance 17 mm from the nozzle. The Figure 9 shows the pressure in the skimmer chamber versus the position of the wire for different level of the RF power. From these measurements it was concluded

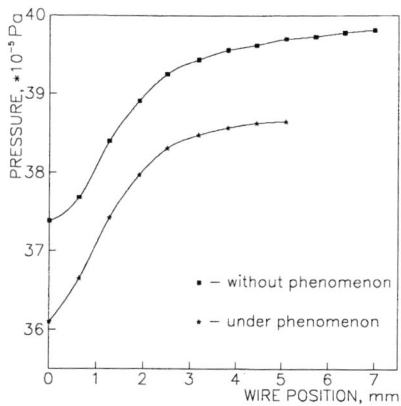

FIGURE 9. The skimmer pressure versus the position of the shadowing wire.

that when the discharge is coming out from the nozzle the geometry of the flow is changed and higher forward intensity is occured as a result of this change.

CONCLUSION

Up to now all superconducting magnets were fabricated and tested. The tests showed that the measured magnetic field are close to the expected ones. The dissociator is also ready for operation. The dissociation degree of the beam is sufficiently high at the gas throughput about 1 mbar×l/sec. It is expected that the operation of the dissociator at higher RF power will gain in final intensity, in spite of the velocity increase. The RF transition units were manufactured and tested to operate at given frequencies. The expected flux of polarized atoms into the storage cell should exceed 1.0×10^{17} at/sec (in two substates) providing the target thickness more than 1.0×10^{14} at/cm^2. It is planed that the source will be tested in a full scale in this year.

ACKNOWLEDGEMENTS

We are grateful to D. Geesaman, the Argonne National Laboratory and R. Holt, the University of Illinois for the many equipments used during the measurements. We greatly appreciate discussions with and support by Yu.M. Shatunov, V.V. Nelyubin, I.N. Nesterenko and N.A. Mezentsev. We are also grateful to A.M. Efimov who has done the excellent work on winding of the magnets.

REFERENCES

1. R.Gilman et al., Phys. Rev. Lett. v.**65** (1990) 1733.
2. M.Ferro-Luzzi et al., Phys. Rev. Lett. v.**77** (1996) 2630.
3. K.Coulter et al., Proposal DESY PRC-90/01 (1990).
4. T.Wise, A.D.Roberts and W.Haeberli, Nucl. Instr. and Meth.**A 336** (1993) 410.
5. F.Stock et al., Nucl. Instr. and Meth.**A 343** (1994) 344.
6. P.Schiemenz, A.Ross and G.Graw, Nucl. Instr. and Meth.**A 305** (1991) 15.
7. D.K.Toporkov, Proc. of the Inter. Workshop on Polarized Ion Sources and Polarized Gas Jets, Feb. 12-17,1990. KEK Report 90-15, p.208.
   L.Dick and W.Kubischta, Proc. of the Inter. Workshop on Polarized Ion Sources and Polarized Gas Jets, Feb. 12-17,1990. KEK Report 90-15, p.232.
8. A.V.Evstigneev, S.G.Popov and D.K.Toporkov, Nucl. Instr. and Meth. **A 238** (1985) 12.
9. A.N.Dubrovin. MERMAID, user guide.
10. T.Wise, personal communication.
11. D.Toporkov et al., Proc. of Inter. Workshop on Polarized Gas Targets. June 6-9, 1995, Cologne, Germany, p.80.

Ultra-Cold Methods for Polarized Atomic Hydrogen

V.G. Luppov[1], J.D. Arnold[1], B.B. Blinov[1], M.A. Bychkov[1,3],
S.E. Gladycheva[1], A.D. Krisch[1], A.M.T. Lin[1], R.S. Raymond[1],
V.V. Fimushkin[2], V.V. Mochalov[3], P.A. Semenov[3]

(1) Randall Lab. of Physics, University of Michigan, Ann Arbor, MI 48109-1120, USA
(2) Joint Institute for Nuclear Research, Dubna, RU-141980, Russia
(3) Institute for High Energy Physics, Protvino, RU-142284, Russia

Abstract. Using the ultra-cold electron-spin-polarized atomic hydrogen technique, one can produce a slow monochromatic beam for use as a polarized jet target. We will first review the development of the ultra-cold technique and then discuss the recent progress on Michigan's Mark-II ultra-cold proton-spin-polarized hydrogen jet target.

INTRODUCTION

Electron-spin-polarized hydrogen atoms in the lower hyperfine states $|3\rangle$ and $|4\rangle$ can be stabilized for a long time in a magnetic field above 5 T at a temperature below 500 mK (see Fig.1). Under these conditions, the electron-spin magnetic energy difference between states with the electron spin "up" and "down" ($2\mu_e B$) is much larger than the thermal energy (kT); therefore, only the two lowest states are significantly populated. For a magnetic field of 8 T and an atomic hydrogen temperature of 300 mK, the ratio of electron-spin-down to electron-spin-up states is $\exp(2\mu_e B/kT) = 3.6 \cdot 10^{15}$.

The first successful long-term stabilization of atomic hydrogen gas (1) used a high magnetic field and an ultra-cold cell coated with superfluid ^{4}He to suppress surface depolarization and recombination into molecular hydrogen. Atomic hydrogen densities of up to $3 \cdot 10^{17}$ atoms cm^{-3} have been achieved using this technique (2). It was then proposed to use high density stabilized ultra-cold electron-spin-polarized atomic hydrogen in polarized sources and targets (3,4,5). We will discuss the general methods and the main results of the ultra-cold technique.

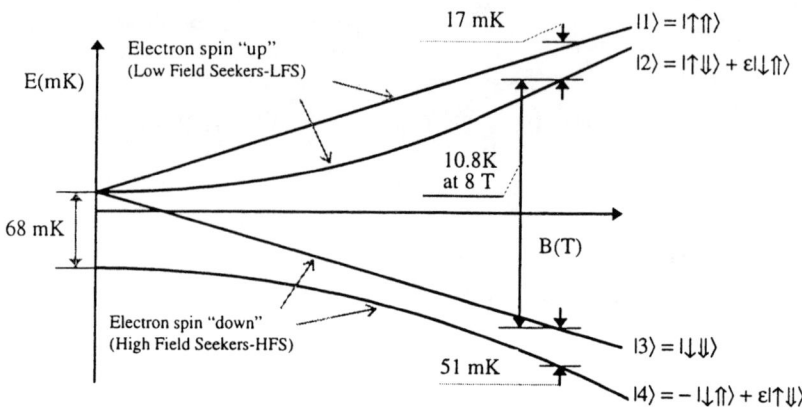

FIGURE 1. Hyperfine energy level diagram of a hydrogen atom in a magnetic field B. (Not to scale)

MICROWAVE EXTRACTION OF STABILIZED ELECTRON-SPIN-POLARIZED ATOMIC HYDROGEN

One early proposal involved electron-spin resonance (ESR) pumping and the subsequent extraction of stored hydrogen atoms from a magnetic bottle (3,4). The atoms pumped from a lower hyperfine state $|4\rangle$ or $|3\rangle$ to an upper state $|1\rangle$ or $|2\rangle$, were focused and driven by a magnetic field gradient into the low magnetic field region, where they formed a slow beam with a small divergence and a small velocity spread. Several authors studied spin-polarized atomic hydrogen ESR pumping (6,7) and one observed atoms ejected from a microwave cavity into a closed storage cell (8). The first electron-spin-polarized atomic hydrogen beam was formed by microwave extraction around 1991 (9). In this experiment the ESR transition at a 5 T field was driven by up to 60 mW of 140 GHz microwave radiation in a storage cell, which was produced by a Varian 20 W Extended Interaction Oscillator.

Atomic hydrogen was later microwave extracted from a 7.6 T magnetic field (10). The external magnetic field and the 300 mK stabilization cell, of 36 cm length and 2.5 cm diameter, are shown in Fig. 2. Atomic hydrogen at 20 K was fed into this cell, which was coated with superfluid ^{4}He; the atoms passed through baffles which cooled them to 300 mK; they were then electron-spin-separated by the magnetic field gradient. The electron-spin-up atoms were repelled by the gradient and did not enter the cell, while the electron-spin-down atoms were attracted into the central cell volume and stabilized. The spin-up atoms could either recombine or depolarize into spin-down atoms and then enter the cell.

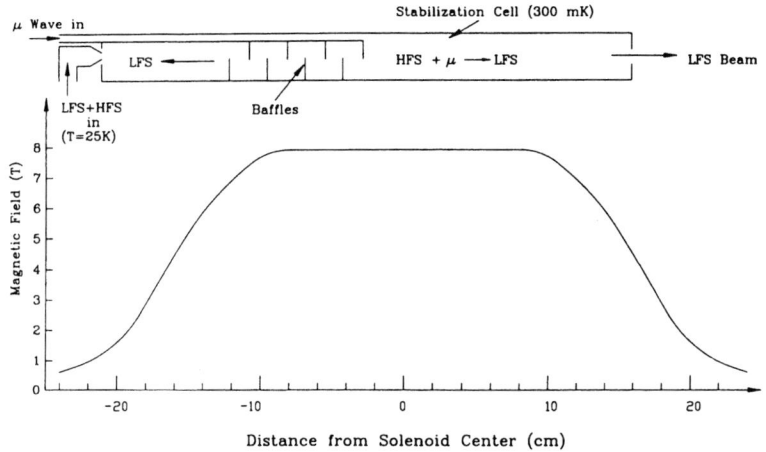

FIGURE 2. The stabilization cell used in (10,11) is displayed relative to the solenoid field.

The ESR transition was driven at the resonance frequency, by 213 GHz microwaves, produced by a 71 GHz Klystron and a microwave tripler; however, the maximum microwave power available at the storage cell was only 5 mW.

We later studied the same microwave-driven extraction of stabilized electron-spin-polarized atomic hydrogen with a 212 GHz microwave source of 2 W (11). The extracted beam was focused by a sextupole. We observed the extracted atomic hydrogen beam using both a compression tube detector and a thermal ring detector at 5 K. Fig. 3 shows a typical thermal detector signal obtained by ramping the magnetic field, with 9 mW of microwave power going into the stabilization cell. The two peaks, corresponding to the $|4\rangle \rightarrow |1\rangle$ and $|3\rangle \rightarrow |2\rangle$ transitions, are separated by 510 ± 20 G, which is in good agreement with the calculated hyperfine splitting of 507 G or 68 mK. A typical measured compression tube signal is plotted against the microwave power at the stabilization cell in Fig. 4. The maximum observed continuous atomic hydrogen flow into the compression tube of $7\,10^{14}$ atoms sec^{-1} corresponded to a jet density of about $5\,10^{10}$ atoms cm^{-3}.

Note that the proton polarization of the extracted beam is near zero. This occurs because the stabilization cell atoms are almost equally in the $|1\rangle$, $|2\rangle$, $|3\rangle$ and $|4\rangle$ states, which is due to spin-exchange processes, such as $|4\rangle + |2\rangle \leftrightarrow |3\rangle + |1\rangle$, having very large rates (6).

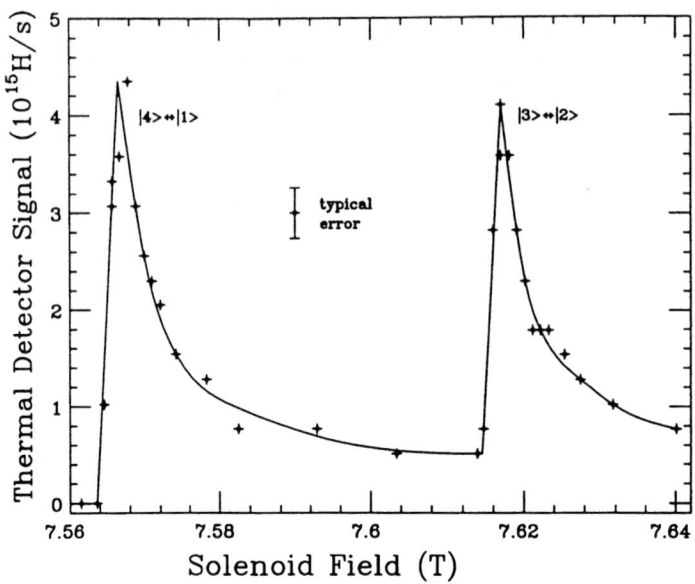

FIGURE 3. The thermal detector signal is plotted versus the ramped magnetic field (11).

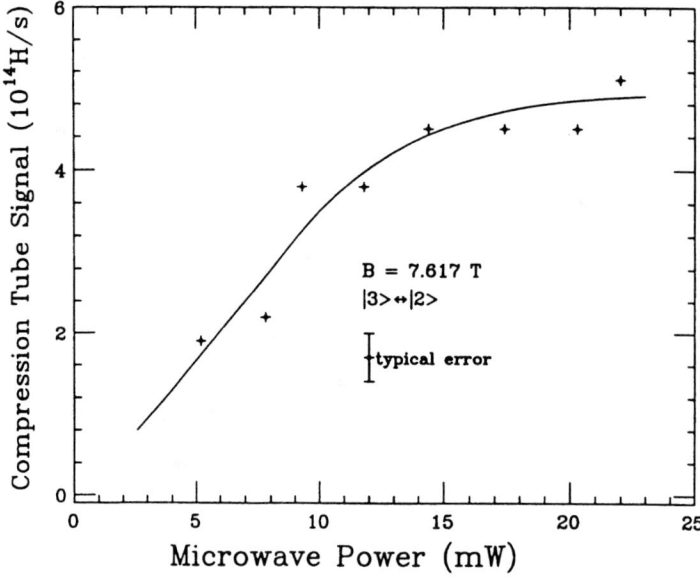

FIGURE 4. The compression tube signal is plotted versus the microwave power into the stabilization cell (11).

MAGNETIC-FIELD-GRADIENT ELECTRON-SPIN-SEPARATION OF ULTRA-COLD ATOMIC HYDROGEN

It was proposed (5) to use the steep magnetic field gradient itself to separate the ultra-cold hydrogen atoms of different electron-spin states. The Michigan prototype jet was then modified to study the continuous production of ultra-cold electron-spin-polarized hydrogen atoms (10), as shown in Fig. 5.

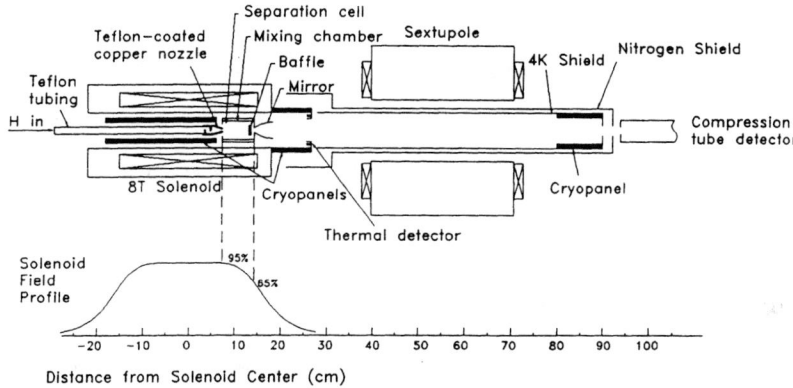

FIGURE 5. The Michigan Prototype ultra-cold electron-spin-polarized atomic hydrogen jet.

Atomic hydrogen was guided into the 300 mK ultra-cold separation cell through a Teflon-coated copper nozzle held at about 20 K. The cell was covered with a superfluid ^{4}He film to suppress the atomic hydrogen surface recombination. The cell's entrance and exit apertures were respectively located at 95% and 65% of the central 8 T solenoid field.

After the hydrogen atoms were thermalized by collisions with the cell surface, the magnetic field gradient separated the atoms according to their electron-spin states. The atoms in the two lowest hyperfine states in Fig. 1 (high field seekers) were attracted toward the high field region. Most of these atoms eventually escaped from the cell through an annular gap around the entrance nozzle; then they recombined on bare surfaces and were pumped away by cryopanels. The atoms in the two higher hyperfine states (low field seekers) were repelled toward the low field region, where they exited through a 5-mm-diam aperture. After emerging from the aperture, these electron-spin-polarized atoms were magnetically accelerated by the remaining field gradient. About 3 10^{15} electron-spin-polarized atoms per second reached the detectors; however, note that the sextupole had a small acceptance; thus, only 3% of the effused atoms reached the compression tube. This observed flux corresponded to a target density of about $1.3\ 10^{11}$ atoms cm^{-3}.

A pulsed ultra-cold polarized atomic hydrogen source using the original magnetic field gradient separation idea (5) is under development in Dubna. They have extracted a pulsed atomic beam with a 300 ms duration and a 0.25 Hz repetition frequency (12).

FOCUSING AN ATOMIC HYDROGEN BEAM WITH A HELIUM-FILM-COATED QUASIPARABOLIC MIRROR

The quantum reflection of cold hydrogen atoms from a superfluid-helium-film-covered surface was first demonstrated by Berkhout *et al.* (13). They measured about 80% specular reflectivity for normal incidence on a hemispherical optical-quality concave quartz mirror coated with a 100 mK ^{4}He film. The quantum reflection occurs because hydrogen atoms are light and interact very weakly with the helium surface.

The first formation of an external beam of ultracold electron-spin-polarized hydrogen atoms, using a highly polished quasiparabolic copper mirror coated with ^{4}He film, was reported in 1993 (14). The Michigan prototype jet (10), using the magnetic-field-gradient electron-spin-separation method, was used to perform these measurements. The mirror was located in the gradient of an 8 T solenoid magnet and mounted on an ultra-cold separation cell at 350 mK (see Fig. 5). After the formation by the mirror, the beam was focused with a sextupole magnet. The mirror, which was especially designed for operation in the magnetic field gradient of our solenoid, increased the focused beam intensity by a factor of about 7.5 for a fixed atomic hydrogen feed rate. The $3.7\ 10^{15}$ ultra-cold low-velocity atoms per second, focused into the compression tube, corresponded to a jet target density of about $3\ 10^{11}$ atoms cm^{-3}. Note that this kind of mirror could be used for any ultra-cold atomic hydrogen beam, including a beam obtained with microwave extraction.

MARK-II ULTRA-COLD POLARIZED HYDROGEN JET.

At the University of Michigan High Energy Spin Physics Lab, we are developing an ultra-cold high-density jet target of proton-spin-polarized hydrogen atoms. The Mark-II jet, which is shown in Fig. 6, uses the magnetic-field-gradient electron-spin-separation method described above.

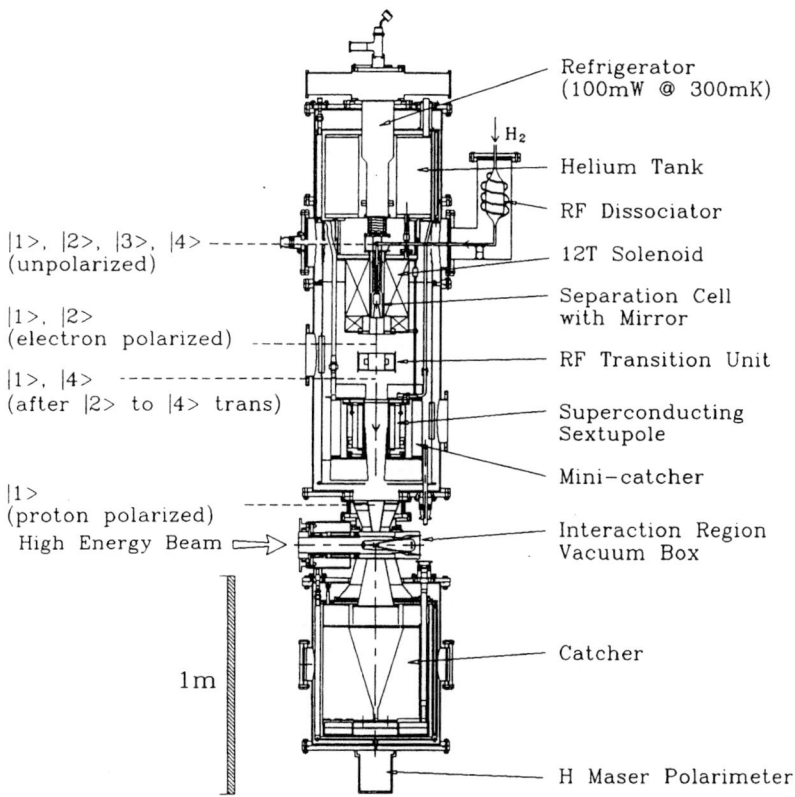

FIGURE 6. Layout of the Mark-II ultra-cold jet

Atomic hydrogen is first produced in a room-temperature rf dissociator and then guided through a Teflon transport tube and a Teflon-coated copper nozzle into the 0.3 K separation cell, which is coated with superfluid ^{4}He. The double walls of the cell form the mixing chamber of the dilution refrigerator. The cell's entrance and exit apertures are respectively located at about 95% and 50% of the superconducting solenoid's 12 T central magnetic field. After the hydrogen atoms are thermalized by collisions with the cell surface, the magnetic field gradient separates the atoms according to their electron-spin states. The two higher hyperfine state atoms |1⟩ and |2⟩ are repelled toward the low field region and effuse from the exit aperture, forming an electron-spin-polarized beam. We plan to use a focusing mirror, with a polished surface covered with superfluid ^{4}He, similar to the prototype mirror (14). The mirror should significantly increase the jet density. After the rf transition unit changes the state |2⟩ atoms into state |4⟩

atoms, a superconducting sextupole focuses the state |1⟩ atoms into the interaction region while defocusing the state |4⟩ atoms, which are then cryopumped. The proton-spin polarized beam then passes through the interaction region and is caught below by the cryopumping catcher. A hydrogen maser polarimeter is placed below the catcher to monitor the proton polarization.

The ultra-cold method has several advantages compared to other methods:

• The ultra-cold beam divergence is much less. As shown in Fig 7a, for molecular effusion from a thin-walled aperture, the atoms' angular distribution is proportional to $\cos \theta$, where θ is measured normal to the aperture. The ultra-cold atoms are then accelerated by a longitudinal magnetic field, which turns their trajectories toward the axis and narrows their angular distribution. Monte-Carlo simulations of the angular distributions before and after the acceleration, respectively, are shown in Fig. 7a and 7b for the 12 T Mark-II with the separation cell exit aperture at 6 T and a field gradient of about 1 T cm^{-1}. As shown in Fig. 7c, a focusing quasiparabolic mirror should further decrease the beam divergence and thus significantly increase the jet density.

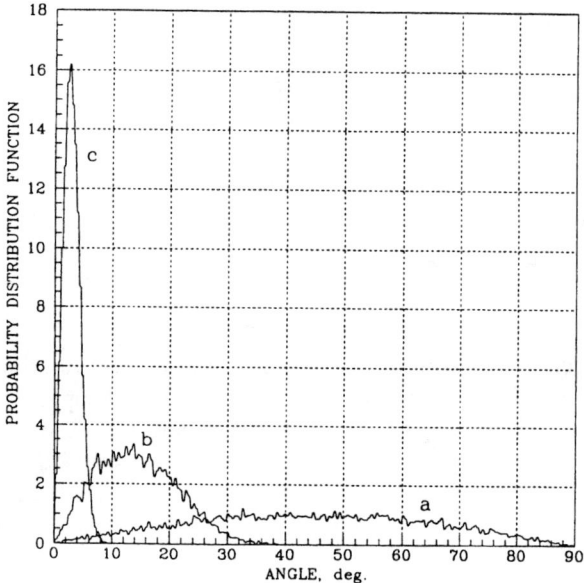

FIGURE 7. Calculated angular distributions of the |1⟩ and |2⟩ states: a) emerging from the cell, b) after acceleration by the magnetic field gradient with no mirror, and c) after acceleration with a quasi-parabolic mirror.

- The ultra-cold beam is much more monochromatic due to the large magnetic-field-gradient acceleration. This monochromaticity allows the beam to be focused into the interaction region with much higher efficiency. The simulations indicate that about 45% of the beam emerging from the separation cell should be focused into a 1 cm x 2 cm target region.
- A free jet's density varies inversely with the atoms' velocity. Since the ultra-cold atoms' average kinetic energy corresponds to about 4 K, the target density should be 4.5 times higher than at liquid nitrogen temperature of 80 K.

Most of the Mark-II parts have been fabricated and successfully tested. This hardware includes: a 12 T superconducting solenoid; a dilution refrigerator with a cooling power of about 50 mW at 300 mK; its pumping system; a 20-cm-long superconducting sextupole magnet with 10.5 cm diameter bore iron poles; a cryocondensation pump with a measured pumping speed of about $1.2 \; 10^7$ liters sec^{-1} which is $4.2 \; 10^{26}$ atoms Torr^{-1} sec^{-1} (15); a hydrogen maser polarimeter capable of monitoring the polarization to about ± 2% precision in a few minutes; and the Mark-II computer-controlled instrumentation.

The electron polarization will be converted into proton polarization by adiabatic passage through an rf transition unit with a novel ring dielectric resonator that accepts a 6-cm-diameter beam. A room temperature prototype rf unit was recently tested; its measured transition efficiency was about 95% (16). We recently started the first beam tests of Mark-II. Fig. 8 shows the measured atomic hydrogen transport line efficiency plotted against the nozzle temperature. This efficiency includes both the dissociation efficiency and the recombination losses on the Teflon transport tube and the nozzle.

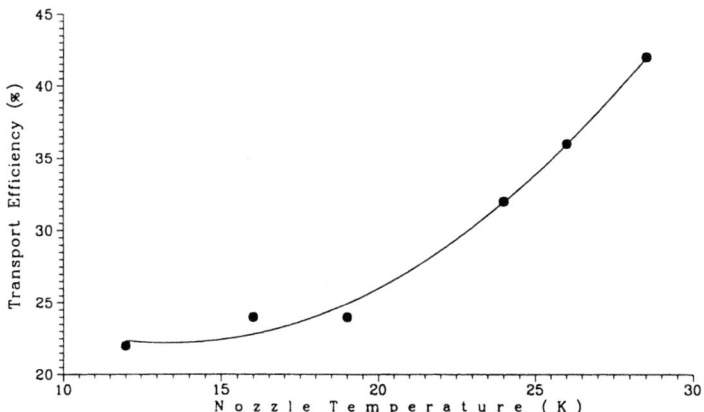

FIGURE 8. Atomic hydrogen transport efficiency versus the Teflon-coated copper nozzle's temperature.

Extrapolating our prototype jet experimental results (14), we expect to achieve a Mark-II target thickness of about 10^{13} atoms cm^{-2} with a 2-cm-thick jet crossing the accelerator beam.

ACKNOWLEDGEMENTS

We would like to thank Prof. D. Kleppner (MIT) for helpful discussions.
This work was supported by the U.S. Department of Energy.

REFERENCES

1. Silvera, I.F., and Walraven, J.T.M., *Phys.Rev. Lett.* **44**, 164 (1980).
2. For a review of the physics of spin stabilized atomic hydrogen, see Silvera, I.F., and Walraven, J.T.M., in *Progress in Low Temperature Physics* (Elsevier Science Publisher B.V., Amsterdam), Vol. **X**, 139-370 (1986).
3. Niinikoski, T.O., in *Proceedings of the International Symposium on High-Energy Physics with Polarized Beams and Polarized Targets,* Lausanne, 1980 (Birkhauser EXS 38, Basle, 191 (1981)).
4. Kleppner, D., in *Proceedings of the Workshop on Polarized Proton Ion Sources,* Ann Arbor, 1981 (AIP Conference Proceedings, No. 80, 111 (1982)).
5. Mertig, M., Levkovich, A.V., Luppov, V.G., and Pilipenko, Yu.K., in *Proceedings of the 9-th International Symposium on High-Energy Spin Physics,* Bonn, 1990 (Springer-Verlag, 164 (1991)).
6. Niinikoski, T.O., Penttila, S., Rieubland, J.-M., and Rijllart, A., in *Proceedings of the Workshop on High Intensity Polarized Proton Ion Sources,* TRIUMF, Vancouver, 1983 (AIP Conference Proceedings, No.117, 139 (1984)).
7. Matthey, A.P.M., van Zwol, J., Walraven, J.T.M., and Silvera, I.F., *Phys. Rev.* **B 37**, 4831 (1988).
8. Hurlimann, M.D., Hardy, W.N., Berlinsky, A.J., and Cline, R.W., *Phys. Rev.* **A 34**, 1605 (1986).
9. Roser, T., et al., *Nucl. Instr. and Methods*, **A 301**, 42 (1991).
10. Kaufman, W.A., Roser, T., and Vuaridel B., *Nucl. Instr. and Methods,* **A 335**, 17 (1993).
11. Kaufman, W.A., Krisch, A.D., Luppov, V.G., and Raymond, R.S., *High power microwave extraction of stabilized electron-spin polarized atomic hydrogen*, to be submitted to *Nucl. Instr. and Methods*.
12. Fimushkin ,V.V., and Pilipenko, Yu.K., in *Proceedings of the 6th International Workshop on High Energy Spin Physics*, Protvino, 1995 (Institute for High Energy Physics, 167 (1996)).
13. Berkhout, J.J., et al., *Phys. Rev. Lett.* **63**, 1689 (1989).
14. Luppov, V.G., Kaufman, W.A., Hill, K.M., Raymond, R.S., and Krisch, A.D., *Phys. Rev. Lett.* **71**, 2405 (1993).
15. Arnold, J.D., et al., *Nucl. Instr. and Methods*, **A 391,** 398 (1997)
16. Raymond, R.S., *Tests of a Prototype Large-Bore, Low Power |2⟩ to |4⟩ RF Transition Unit*, in Proceedings of this Workshop.

The Argonne Laser-Driven D Target: Recent Developments and Progress.

J. A. Fedchak, K. Bailey, W. J. Cummings, H. Gao,
C. E. Jones[1], R. S. Kowalczyk, J. Magnes[2], and M. Pipes[3]

Argonne National Laboratory
Argonne, IL 60439-4843

Abstract.
  The first direct measurements of nuclear tensor polarization p_{zz} in a laser-driven polarized D target have been performed at Argonne. We present p_{zz} and electron polarization P_e data taken at a magnetic field of 600 G in the optical pumping cell. These results are highly indicative that spin-temperature equilibrium is achieved in the system. To prevent spin relaxation of D and K atoms as well as the molecular recombination of D atoms, the walls of the laser-driven D target are coated with organosilane compounds. We discuss a new coating technique, the "afterwash", developed at Argonne which has yielded stable atomic fraction results when the coating is exposed to K. We also present new coating techniques for glass and Cu substrates.

I INTRODUCTION

Preliminary results of the first direct measurement of nuclear polarization in a laser-driven D target were reported nearly two years ago by Fedchak, Jones, and Kowalczyk [1]. Since then, the Argonne group has performed a systematic study of the nuclear tensor polarization p_{zz} and has developed a new coating technique, the "afterwash". A laser-driven H and D internal target has been installed in the Cooler ring at IUCF by the CE66/68 collaboration, representing the first use of a laser-driven H and D target in a nuclear physics experiment [2]. Here we will summarize the afterwash coating technique and also present new coating techniques for Cu and glass substrates which are currently being developed at Argonne. Additionally, we will present results of nuclear tensor polarization measurements in a laser-driven target with $B = 600$ G

---

[1] Present address: Kellogg Lab 106-38, Caltech, Pasadena, CA 91125
[2] Dept. of Physics, U. of Delaware
[3] Dept. of Industrial Technology, Western Illinois U.

in the optical pumping cell. A detailed report of the p_{zz} measurements is forthcoming [3].

II PHYSICS OF THE LASER-DRIVEN TARGET

In the laser-driven D target, K atoms are polarized by optical pumping with polarized laser light in a high magnetic field. The electron polarization is transferred to D atoms in D-K spin-exchange collisions. Deuterium electron polarization is subsequently transferred to the deuterium nucleus in D-D spin-exchange collisions. Walker and Anderson [4] have examined the consequences of D-D spin-exchange collisions in a laser-driven polarized D target. In the limit of many spin-exchange collisions, the population distribution of the six magnetic ground substates of deuterium will come to an equilibrium distribution characteristic of the total angular momentum pumped into the system by the laser. At equilibrium, the relative population of a given deuterium magnetic substate is given by

$$n(m_I) = e^{\beta m_I}, \tag{1}$$

where β^{-1} is a parameter known as the spin-temperature and m_I is the projection of the nuclear spin on the spin quantization axis. Eq. (1) can be used to calculate both the p_{zz} and the electron polarization P_e at equilibrium. The timescale for attaining spin-temperature equilibrium is inversely proportional to the D density n_D, and sets the timescale for polarizing the nucleus. Spin-temperature equilibrium in a laser-driven target has been discussed in several previous publications [1,5–7].

III EXPERIMENTAL APPARATUS

A schematic of the laser-driven target is shown in Fig. 1. Details of the experiment can be found in Refs. [3,5,6] and will only be summarized here. Deuterium flows into a rf discharge which dissociates the molecular deuterium into atoms. Deuterium atoms and molecules flow from the dissociator into a cylindrical optical pumping cell which lies between two pole faces of an electromagnet. The optical pumping cell has a diameter of 2.2 cm and a length of 4.6 cm. Potassium enters the cell through a 0.9 mm aperture connected to a K reservoir heated to 180°C, corresponding to a K density of $3.8 \pm 0.8 \times 10^{11}$ cm^{-3}. The $4^2S_{1/2} - 4^2P_{1/2}$ resonance of K is optically pumped by about 2 W of circularly polarized light from a Ti:sapphire ring laser broadened by an electro-optical modulator to match the Doppler width of the σ_+ or σ_- transition. Atoms experience about 700 wall bounces in the optical pumping cell and enter the transport tube through a 3.1 mm aperture. The pumping cell and Pyrex transport tube are heated to between 200°C and 250°C to

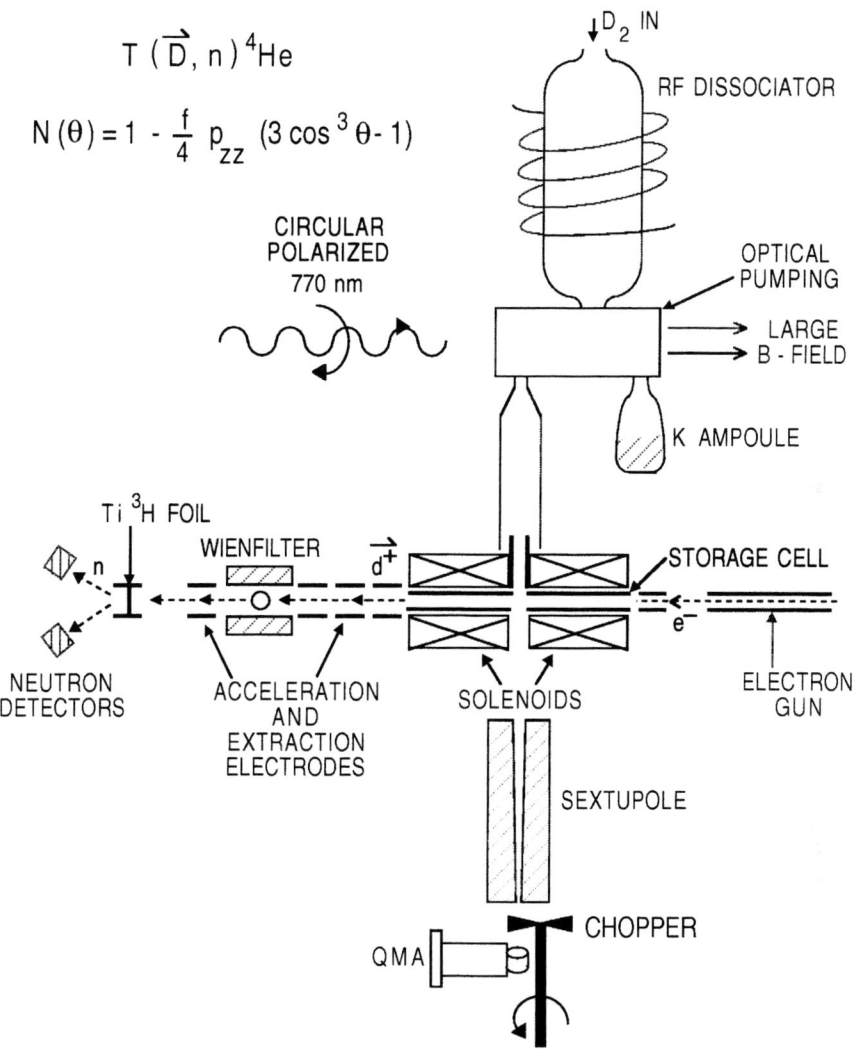

FIGURE 1. Schematic of the laser-driven D target.

prevent K from condensing on the walls. The Pyrex transport tube extends to a length of 30 cm, has a diameter of 19 mm, and is joined to a 6 cm length of aluminum tube by a viton O-ring. Atoms exiting the extended transport tube flow into an open ended aluminum storage cell 48 cm in length and 23 mm in diameter. Both the storage cell and aluminum section of the transport tube are heated to over 200°C. A 5 mm hole in the center of the storage cell allows the P_e polarimeter to sample atoms leaving the transport tube.

The P_e polarimeter is located downstream of the transport tube and has an acceptance such that the polarimeter samples atoms which have made many wall collisions in the transport tube. A sextupole magnet focuses atoms with spin up and defocuses those with spin down. Focused atoms pass through a chopper wheel and are detected by a quadrupole mass analyzer (QMA). A removable shutter placed before the sextupole allows one to measure the background signal in the QMA. Atomic polarization can be measured by blocking and unblocking the laser beam. The atomic polarization is determined by:

$$P_e = \frac{n_\uparrow - n_\downarrow}{n_\uparrow + n_\downarrow} = \frac{n_{unblocked}}{n_{blocked}} - 1, \qquad (2)$$

where $n_{\uparrow(\downarrow)}$ denotes deuterium atomic states with electron spin up (down) and $n_{(un)blocked}$ is the signal measured with the QMA with the laser beam (un)blocked.

The atomic fraction, f_a, defined as the total number of deuterium nuclei in the form of atoms over the total number of deuterium nuclei (in the form of D or D_2) is also determined using the P_e polarimeter. This is determined with the QMA by measuring the mass 4 (amu) signal with the dissociator rf power on and off:

$$f_a = 1 - \frac{n_{on}}{n_{off}}. \qquad (3)$$

With the rf power off the mass 4 signal n_{off} is entirely due to D_2, but when the dissociator rf power is turned on the mass 4 signal, n_{on}, measures the molecular flow from D_2 molecules that are not dissociated before entering the optical pumping cell and from D atoms which recombine on surfaces between the time they leave the dissociator and when they reach the QMA. For both f_a and P_e, the error in the measurements are dominated by systematic uncertainties and is less than 2%.

In addition to P_e, we can also measure the nuclear tensor polarization p_{zz}, defined by

$$p_{zz} = 1 - 3n_0, \qquad (4)$$

where n_0 represents the fractional population density of the nuclear spin substates with $m_I = 0$. The p_{zz} polarimeter is based upon the technique developed by Price and Haeberli [8]. To determine p_{zz} the polarimeter employs the low energy $^2\text{H} + ^3\text{H} \to \text{n} + ^4\text{He}$ reaction in which the angular distribution of the outgoing neutrons is anisotropic if the incident deuterium ions are tensor polarized. The expression for the differential cross section for a tensor polarized incident deuteron beam is

$$\sigma(\theta) = \sigma_0(\theta)\left(1 - \frac{f}{4}p_{zz}(3\cos^2\theta - 1)\right), \tag{5}$$

where σ_0 is the unpolarized cross section, which is isotropic in the center-of-mass frame, θ is the angle between the direction of the outgoing neutron and the spin of the deuteron in the center-of-mass system, and the dilution factor f accounts for the small admixture of reaction channels which have no tensor analyzing power. For ion energies in the range used for the polarimeter, the dilution factor has been measured and found to be near unity ($f \approx 0.96$) [9]. The neutron anisotropy R, defined as

$$R = \frac{\sigma(0°)}{\sigma(90°)} = \frac{1 - \frac{f}{2}p_{zz}}{1 + \frac{f}{4}p_{zz}}, \tag{6}$$

is measured with and without the optical pumping light incident on the pumping cell. The ratio of the measured R values for polarized and unpolarized ions is used to determine the tensor polarization of the deuterons in the storage cell. Differences in the detector efficiencies and angular acceptance largely cancel in the ratio.

A 2 keV electron beam directed along the length of the storage cell ionizes the deuterium atoms to produce D^+. A solenoidal coil surrounding the storage cell creates the magnetic holding field of about 300 G that serves both to separate the deuterium magnetic substates and act as a guide field for the ions to prevent them from hitting the cell walls. Ions are extracted from the storage cell and accelerated towards a tritiated foil target mounted inside an electrostatic lens maintained at a potential of 50 kV. A Wein filter ($\vec{E} \times \vec{B}$) located between the storage cell and tritiated foil acts as a velocity selector to separate deuterium atoms and molecules. The deuteron spin direction, which is parallel to the solenoidal holding field in the storage cell, precesses by 45° in the B-field of the Wein filter. Neutrons resulting from the fusion reaction $^3\text{H}(d,n)\alpha$ are detected in two NE110 scintillators placed parallel and perpendicular to the spin direction, thus maximizing the neutron anisotropy in the two detectors.

Tests of the polarimeter with unpolarized deuterium show that the measurements are reproducible to better than $\Delta R/R = 0.01$. Corrections for the variation of p_{zz} due to the angular acceptance contribute an uncertainty of $\Delta p_{zz}/p_{zz} < 0.005$. The overall systematic error is dominated by the uncertainty in the spin rotation angle in the Wien filter, which gives a contribution of $\Delta p_{zz}/p_{zz} < 0.065$.

IV COATING TECHNIQUES AND TESTS

In order to reduce the recombination on surfaces as well as preserve D and K polarization, the Pyrex glassware is coated using SC-77 plus a trimethylmethoxysilane "afterwash", while the aluminum storage cell and short connecting tube are coated with a mixture of dimethyldimethoxysilane and methyltrimethoxysilane. Both the afterwash technique and aluminum coating techniques are discussed in ref. [10]. Below we will summarize the afterwash technique developed at Argonne, and present measurements relevant to the Pyrex glassware used during the p_{zz} measurements. We will also present new developments in the technique of coating Cu surfaces with drifilm and of coating Pyrex with SC-77.

It is believed that standard coating techniques with organosilane compounds such as SC-77 will leave OH sites on the coated surfaces [10]. Consider, for example, the process of coating Pyrex with SC-77. SC-77 is a mixture of $(CH_3)_2SiCl_2$ and $(CH_3)SiCl_3$. During the coating process, the SC-77 is combined with H_2O resulting in the Cl atoms being replaced by OH groups. The OH groups attached to the silane compounds will combine in such a way as to produce chains of Si-O bonds with methyl groups (CH_3) attached to the Si. These chains bond to the Pyrex surface to produce a surface of methyl groups. The methyl groups have a low dielectric constant and are fairly inert, therefore H or D atoms will tend not to stick to a surface of methyl groups. Consequently, these surfaces do not promote recombination of atoms to form molecules nor the spin relaxation of the polarized atoms. Occasionally, an OH group bonded to Si will remain in the final coating. As OH is a fairly polar and reactive molecule, the OH will be sites for relaxation and recombination of H or D atoms. Additionally, these sites will be subject to attack from K atoms, which are known to bond with OH.

To reduce the number of OH sites left in the final surface, a second coating, or afterwash, of a mixture containing trimethylmethoxysilane $((CH_3)_3Si(COH_3))$ is washed over the SC-77 coating. Ideally, the methoxy group (COH_3) will attach to the OH site to form a Si-O bond and the remaining surface will then contain only methyl groups. We obtain excellent results using this technique.

We use the atomic fraction as a measure of surface quality rather than the electron polarization because the atomic fraction samples all of the flow from the source, while measurements of the polarization select only the atoms in the flow, which in the event of significant molecular recombination on the surface is heavily weighted by the particles which experience the least collisions. Our experience is that f_a is a more sensitive monitor of the surface quality than P_e.

Fig. 2 shows f_a measurements made with the Pyrex glassware coated using the afterwash technique. The measurements were made without the storage cell in place, but the glassware and coating are the same as were used for the p_{zz} measurements presented in section V. These do not represent "best ever" data, but represent the performance of the glassware used for the p_{zz} measurements. As seen in the figure, the atomic fraction is relatively stable for the three days data was taken. This is a marked improvement over f_a results for glassware coated with SC-77 alone which, in our experience, tends to yield a low f_a after being exposed to K for a similar amount of time.

Recently, the group at Argonne has been developing better coating techniques for Cu substrates. In the past, Cu substrates have been coated using a recipe similar to that used to coat aluminum substrates [10,1,11]. Both the work of Swenson and Anderson [11] and observations made at Argonne indicate that drifilm coatings on Cu rapidly deteriorate when exposed to alkali. We now have a new recipe for coating Cu substrates: The Cu substrate to be coated is cleaned with Alconox followed by washes of HCl and deionized water to both clean the surface and remove the oxide layer. This process is performed several times, finishing with the rinse of deionized water. Immediately following the cleaning procedure, the Cu piece is immersed in deionized water. To this is added equal amounts of dimethyldimethoxysilane and methyltrimethoxysilane, so that the final solution consists of 10% dimethyldimethoxysilane, 10% methyltrimethoxysilane, 80% deionized water,

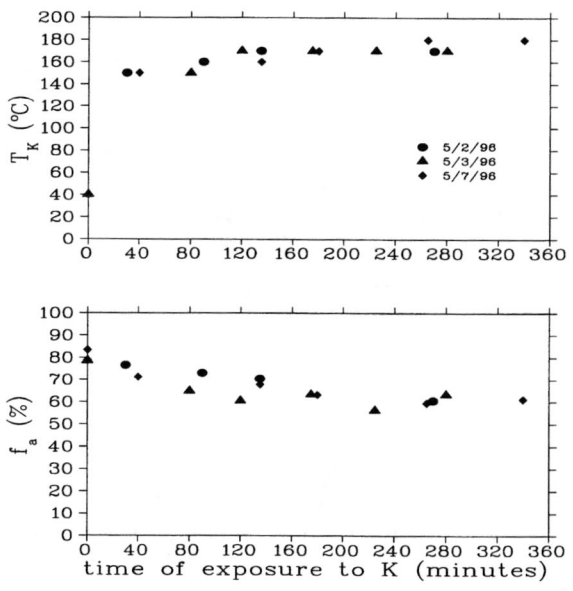

FIGURE 2. The atomic fraction f_a measured as a function of time exposed to K. The three symbols represent three different days of data. The upper graph shows the K ampule temperature as a function of the time corresponding to the f_a measurements, and thus is an indication of the K density in the optical pumping cell.

and enough HCl to raise the pH of the solution to between 2.7 and 3.1. The solution is not agitated. After 48 hours in the solution, the substrate is baked in vacuum for 24 hours at 225°C.

Researchers have found it difficult to consistently prepare SC-77 glass coatings which are clear and of some desired thickness. We have developed a method of coating glass with SC-77 which consistently yields a clear coating and allows one to control the coating thickness. The SC-77 coating procedure is performed in a hood filled with N_2. This prevents atmospheric water from reacting with the SC-77 before the coating is applied. The glass piece to be coated is suspended over a container holding $10-20$ ml of SC-77. Deionized water is dripped into the container at a rate of about 0.02 ml/s. By steadily dripping the deionized water into the SC-77, the SC-77 is vaporized at a constant rate thus allowing one to control the amount of coating deposited on the surface. The glass piece is exposed to the SC-77 vapor for 2 to 20 minutes, depending on how thick a coating is desired, after which the glass piece is left in air for 30 to 60 minutes. An afterwash can then be performed or the piece can be directly baked in vacuum for 24 hours.

V POLARIZATION RESULTS

With the storage cell in place, both the electron polarization and nuclear tensor polarization can be measured as a function of flow. Fig. 3 shows P_e and p_{zz} measurements with $B = 600$ G in the optical pumping cell. The relative independence of P_e on flow is indicative that the system is in equilibrium over the entire range of flows investigated. Once n_D is great enough to achieve spin-temperature equilibrium, the dominant flow dependent mechanism for losing polarization is K-D spin-exchange collisions, which occur at a rate proportional to n_D. In this case, the 16% decrease in P_e should be reflected in a decrease in the alkali polarization P_K. This does not necessarily mean that P_K is a sensitive measurement of P_e, as a small change in P_K could result in a large change in P_e. Additionally, we have neglected other depolarization mechanisms that are dependent on D_2 density and hence flow, such as spin relaxation due to $D - D_2$ collisions and the loss of P_K due to K-D_2 collisions. To the authors knowledge, these spin relaxation rates have never been determined.

The helicity of the circularly polarized laser light was reversed to check for systematic effects. Other than the expected sign change, no dependence of $|P_e|$ or p_{zz} on laser light helicity can be seen within the error of the measurements. This is also a strong signature of spin-temperature equilibrium.

The measured p_{zz} is less than 0.035 for all the flows we tested. Assuming spin-temperature equilibrium, Eq. (1) can be used to calculate the electron polarization that one expects from the measured p_{zz}. The ratio of the electron polarization inferred from p_{zz} to that of the measured P_e is shown in Fig.

3(c). This ratio should be fairly free of experimental fluctuations such as laser frequency and power. The relative independence of this ratio with flow is a signature of spin-temperature equilibrium. On the average, the electron polarization inferred from spin-temperature, P_e^{ST}, is about 40% less than the measured P_e. This is not surprising since the electron polarimeter samples atoms from the transport tube, whereas the nuclear polarimeter samples along the length of the storage cell. Using Monte Carlo techniques, we estimate that the average D atom leaving the storage cell experiences about 650 more wall bounces than those detected in the P_e polarimeter. Therefore, we can use the formula $P_e^{ST} = P_e e^{-\frac{N}{N_{relax}}}$ to estimate the probability of relaxing per wall bounce, N_{relax}^{-1}. We calculate that $\langle N_{relax} \rangle = 1370$ wall bounces. It must be noted that the electron beam of the nuclear polarimeter ionizes atoms along the length of the storage cell, therefore 650 wall bounces represents an upper limit to the difference in wall bounces between the two polarimeters. $\langle N_{relax} \rangle$ could also be anomalously low due to poor coupling of the Pyrex transport tube to the aluminum tube, so that some of the atoms may interact with the viton O-ring or enter the O-ring groove. Therefore $\langle N_{relax} \rangle$ is not representative of wall conditions alone.

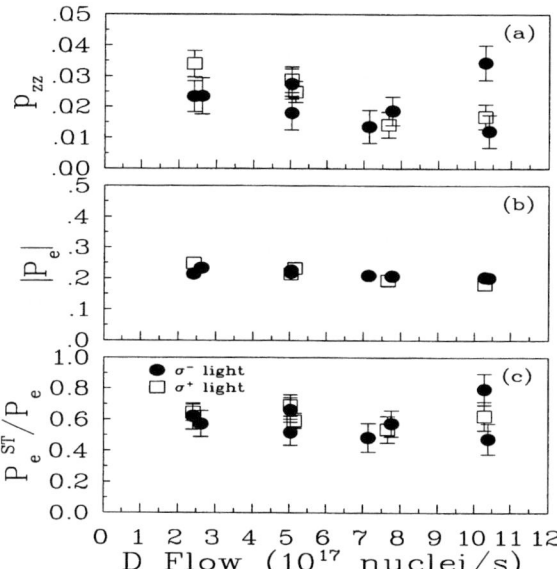

FIGURE 3. (a) p_{zz} vs. flow of D nuclei; (b) $|P_e|$ vs. flow of D nuclei; (c) the ratio of the electron polarization inferred from the p_{zz} data based upon spin-temperature equilibrium to the measured P_e as a function of D nuclei flow. All data is taken with of field of 600 G in the optical pumping cell.

VI CONCLUSION

In a laser-driven D target, wall relaxation is the dominant mechanism for loss of polarization and atomic fraction. We have developed new coating techniques which show improvements over previous methods used, both in terms of atomic fraction and the ability to obtain a consistent coating. We have also performed the first direct measurement of nuclear polarization in a laser-driven D target. This verifies that it is possible to obtain polarized deuterium nuclei without the use of rf transitions. The p_{zz} measurements provide further understanding of the operation of a laser driven target. A full report of the p_{zz} measurements is in preparation.

ACKNOWLEDGEMENTS

We are grateful to W. Haeberli, J. van den Brand, and the U. of Wisconsin for the loan of many parts of the nuclear polarimeter, and also to R. Holt and M. Poelker for their preliminary work on the polarimeter. We appreciate the skillful glassblowing of Joe Gregor, who made the laser-driven target glassware. This work is supported by the U.S. Department of Energy, Nuclear Physics Division, under contract No. W-31-109-ENG-38.

REFERENCES

1. J. A. Fedchak, C. E. Jones, and R. S. Kowalczyk, in *International Workshop on Polarized Beams and Polarized Gas Targets, Cologne, 1995* edited by H. P. gen. Schieck and L. Sydow (World Scientific, Singapore, 1996), pp. 72-79.
2. See M. Miller *et al.* in these proceedings.
3. Manuscript in preparation.
4. T. Walker and L. W. Anderson, *Nucl. Instrum. Methods A* **334**, 313 (1993).
5. M. Poelker *et al.*, *Phys. Rev. A* **50**, 2450 (1994).
6. M. Poelker *et al.*, *Nucl. Instrum. and Methods A* **364**, 58 (1995).
7. J. Stenger *et al.*, *Phys. Rev. Lett* **78**, 4177 (1997).
8. J. S. Price and W. Haeberli, *Nucl. Instr. Methods A* **326**, 416 (1993).
9. G. G. Ohlsen, J. L. McKibben, and G. P. Lawrence in *Polarization Phenomena in Nuclear Reactions, Proceedings of the Third International Symposium, Madison, 1970,* edited by H. H. Barschall and W. Haeberli (The University of Wisconsin Press, Madison, 1971), pp. 503-505.
10. J. A. Fedchak *et al.*, *Nucl. Instrum. Methods A* **XX**, xxx (1997).
11. D. R. Swenson and L. W. Anderson, *Nucl. Instrum. and Methods B* **12**, 157 (1985).

Nuclear spin polarized H and D by means of spin–exchange optical pumping

Jörn Stenger[1], Carsten Grosshauser, Wolfgang Kilian, Wolfgang Nagengast, Bernd Ranzenberger, Klaus Rith, Frank Schmidt

Universität Erlangen–Nürnberg
91058 Erlangen, Germany

Abstract. Optically pumped spin–exchange sources for polarized hydrogen and deuterium atoms have been demonstrated to yield high atomic flow and high electron spin polarization. For maximum nuclear polarization the source has to be operated in spin temperature equilibrium, which has already been demonstrated for hydrogen. In spin temperature equilibrium the nuclear spin polarization P_I equals the electron spin polarization P_S for hydrogen and is even larger than P_S for deuterium. We discuss the general properties of spin temperature equilibrium for a sample of deuterium atoms. One result are the equations $P_I = 4P_S/(3+P_S^2)$ and $P_{zz} = P_S \cdot P_I$, where P_{zz} is the nuclear tensor polarization. Furthermore we demonstrate that the deuterium atoms from our source are in spin temperature equilibrium within the experimental accuracy.

INTRODUCTION

Optically pumped spin–exchange sources for polarized hydrogen and deuterium atoms have been demonstrated to yield high electron polarization of around 70 % at a flow of several 10^{17} $atoms/sec$ [1], [2]. In such a source atomic hydrogen or deuterium is electron spin polarized by means of spin–exchange collisions with optically pumped potassium atoms. The nuclei become polarized mainly due to spin–exchange collisions among the hydrogen or deuterium atoms themselves. With a sufficiently large number of spin–exchange collisions the atoms approach the spin temperature equilibrium. Spin temperature equilibrium is the most probable relative substate population for a given total angular momentum [3]. In spin temperature equilibrium the nuclear polarization is highest, if no RF–transitions are applied in the source, which would complicate the apparatus substantially.

---

[1] now at MIT, Cambridge, MA 02137, USA

Spin temperature equilibrium in a spin–exchange source operated with hydrogen at a flow of $5 \cdot 10^{17}$ $atoms/sec$ has already been demonstrated [4]. In this article we mainly discuss the properties of deuterium and report experimental results.

SPIN TEMPERATURE EQUILIBRIUM

The deuterium atoms represent a system of coupled spin–1/2 electrons and spin–1 nucleons. The deuterium groundstate consists of six hyperfine substates, labeled i, which can be characterized by the total angular momentum F and the z-component m_F: $|F = 3/2, m_F = 3/2, 1/2, -1/2, -3/2 \rangle$ ($i = 1, 2, 3, 4$) and $|F = 1/2, m_F = -1/2, 1/2 \rangle$ ($i = 5, 6$), respectively. Spin exchange collisions among the deuterium atoms allow the atoms to change the substate under conservation of the sum of the z–components $m_F^{(a)} + m_F^{(b)}$. E.g. $|a\rangle|b\rangle \longrightarrow |a'\rangle|b'\rangle = |3/2, -1/2\rangle|1/2, 1/2\rangle \longrightarrow |1/2, 1/2\rangle|1/2, -1/2\rangle$.

The vector polarization P of a sample of spin–1/2 particles, such as electrons, is defined as

$$P_S = \frac{N_{+1/2} - N_{-1/2}}{N_{+1/2} + N_{-1/2}}, \tag{1}$$

where $N_{+1/2}$ and $N_{-1/2}$ denote the numbers of particles with spin orientation parallel and antiparallel to the quantisation axis. A sample of spin–1 particles, such as deuterons, may have vector or tensor polarization, being defined as

$$P_I = \frac{N_{+1} - N_{-1}}{N_{+1} + N_0 + N_{-1}},$$
$$P_{zz} = \frac{1 - 3N_0}{N_{+1} + N_0 + N_{-1}}. \tag{2}$$

Possible values for P range between -1 and $+1$, those of P_{zz} between -2 and $+1$.

The polarizations for deuterium are [5]

$$P_S = n_1 + (n_2 - n_6)\cos 2\theta_+ - n_4 + (n_3 - n_5)\cos 2\theta_-, \tag{3}$$
$$P_I = n_1 + n_2 \sin^2\theta_+ - n_3 \cos^2\theta_- - n_4 - n_5 \sin^2\theta_- + n_6 \cos^2\theta_+, \tag{4}$$
$$P_{zz} = n_1 + n_2(1 - 3\cos^2\theta_+) + n_3(1 - 3\sin^2\theta_-)$$
$$+ n_4 + n_5(1 - 3\cos^2\theta_-) + n_6(1 - 3\sin^2\theta_+). \tag{5}$$

P_S denotes the electron polarization, P_I the nuclear vector polarization and P_{zz} the nuclear tensor polarization, respectively. n_i are the normalised occupation numbers; $\sum_{i=1}^{6} n_i = 1$. The mixing angles θ_+ and θ_- obey:

$$tan 2\theta_{\pm} = \frac{\sqrt{8}}{3B/B_c \pm 1}, \tag{6}$$

where B is the external magnetic field and $B_c = E_{HFS}/(g\mu_B) = 11.7\,mT$ is the critical magnetic field of deuterium. $E_{HFS} = 1.35 \cdot 10^{-6}\,eV$ is the hyperfine energy splitting in zero magnetic field, g the electron g–factor and μ_B the Bohr magneton.

In the case of spin temperature equilibrium the relative substate population can be characterized by a spin temperature parameter β:

$$n_i = A e^{m_F \beta}. \qquad (7)$$

The factor A is given by the normalisation condition.

With this distribution of hyperfine substate populations ($n_2 = n_6, n_3 = n_5$), the polarizations become independent of the external magnetic field

$$P_S = n_1 - n_4,$$
$$P_I = n_1 + n_2 - n_3 - n_4,$$
$$P_{zz} = n_1 - n_2 - n_3 + n_4. \qquad (8)$$

By combining eqs. (7) and (8), the nuclear polarization P_I and the nuclear tensor polarization P_{zz} can be derived as functions of the electron polarization P_S:

$$P_I = \frac{4 P_S}{3 + P_S^2}, \qquad (9)$$

$$P_{zz} = \frac{4 P_S^2}{3 + P_S^2}, \qquad (10)$$

and therefore

$$P_{zz} = P_S \cdot P_I. \qquad (11)$$

The electron spin polarization can easily be measured by means of a sextupole magnet. Once the deuterium atoms have been measured to be in spin temperature equilibrium, the nuclear polarizations can immediately be calculated.

Fig. 1 shows a plot of eqs. (9) and (10). It can be seen, that in spin temperature equilibrium the tensor polarization is always positive and the nuclear spin polarization is larger than the electron spin polarization, although the polarization transfer from the optically pumped potassium atoms to the deuterium atoms is a purely electronic process.

The spin–exchange source has to be designed in such a way that the number of spin–exchange collisions is high enough to reach spin temperature equilibrium but the number of depolarizing wall collisions is as small as possible. Whereas the substate distribution in spin temperature equilibrium is independent of the external magnetic field, the number of spin–exchange collisions needed to approach the equilibrium increases proportional to $1 + (B/B_c)^2$ [5]

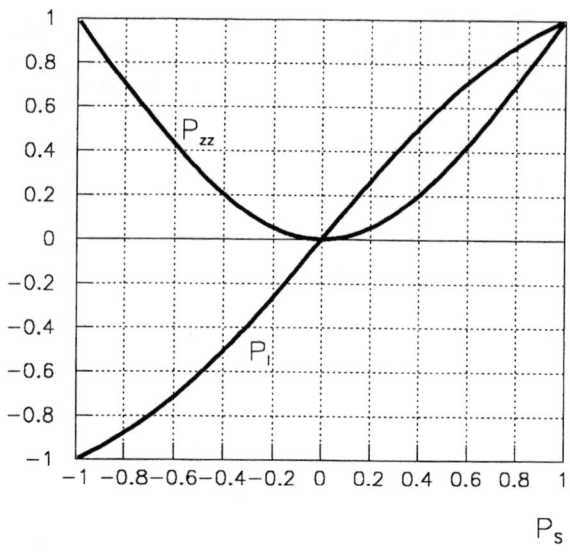

FIGURE 1. P_I and P_{zz} versus P_S in spin temperature equilibrium.

in large fields. The number of spin–exchange collisions is $T_{dwell} \cdot \rho_D \cdot \langle \sigma v \rangle$, where $\langle \sigma v \rangle$ is the averaged product of the spin–exchange cross section and the thermal relative velocity [5], [6]. The mean dwell time T_{dwell} and the density ρ_D in the cell in which the collisions take place, must be chosen according to the external field strength. At a magnetic field of $100\,mT = 8.5 \cdot B_c$ around 150 spin–exchange collisions per spin–exchange collision with a potassium atom are necessary to approach spin temperature equilibrium.

A magnetic field of $100\,mT$ or more is necessary to reduce radiation trapping [7]. In these fields the various transitions of the potassium D1–line are separated by more than the Doppler–broadened linewidth of $1\,GHz$.

DETERMINATION OF SPIN TEMPERATURE EQUILIBRIUM

The spin–exchange source apparatus and the atomic polarimeter are described in [2], [4] and [8]. Only brief descriptions are given here.

Deuterium atoms from a RF–dissociator enter the optical pumping cell, together with a small potassium admixture (0.3 %) from a side arm. Light from a Ti:Sapphire laser with an optical power of up to $4\,W$ is used for optical pumping. The dissociator tube, the optical pumping cell and the exit tube to the atomic beam polarimeter are made of glass and are connected by capillaries with defined conductances. The ratios of these conductances and the volume V of the pumping cell define the numbers of spin–exchange collisions and wall collisions at a given flow. The potassium density ρ_K and the deuterium density ρ_D in the pumping cell have been chosen to $\rho_K = 6 \cdot 10^{11}\, cm^{-3}$ and $\rho_D = 1.8 \cdot 10^{14}\, cm^{-3}$ at an atomic deuterium flow of $F_D = 3 \cdot 10^{17}\, 1/s$. A magnetic field of $100\, mT$ was applied.

With the mean dwell time $T_{dwell} = \rho_D V/F_D = 30\, ms$ each deuterium atom undergoes 32 spin–exchange collisions with potassium atoms and around 120 times more with other deuterium atoms in the pumping cell. The atoms are therefore expected to be close to spin temperature equilibrium.

The atomic beam polarimeter consists of high frequency transitions, sextupole magnets and a quadrupole mass spectrometer. P_S can be derived from the intensity ratio of the atomic deuterium in the quadrupole mass spectrometer with pumping laser on resonance, I^{L-on}, and blocked, I^{L-off} [4]

$$P_S = \frac{I^{L-on}}{I^{L-off}} - 1. \qquad (12)$$

Due to the six hyperfine substates of deuterium, in general five high frequency transitions together with the normalization condition are necessary to completely determine the nuclear spin polarizations. However, as we will show only one high frequency transition is necessary to demonstrate spin temperature equilibrium.

We applied a high frequency transition which allowed us to interchange the occupation numbers of the deuterium states 3–4, 2–4 and 1–4. Intensity ratios can be measured with high frequency transition on and off. The pump laser must be on resonance in both cases.

The sextupole magnet in the atomic beam polarimeter, placed in between the high frequency transition unit and the quadrupole mass spectrometer, deflects the deuterium states 4, 5 and 6. Therefore, e.g. the ratio

$$R^{(3-4)on/off} = \frac{I^{(3-4)on}}{I^{(3-4)off}} = \frac{n_1 + n_2 + n_4}{n_1 + n_2 + n_3} \qquad (13)$$

can be measured, where $I^{(3-4)on}$ is the signal in the quadrupole mass spectrometer when the transition unit 3–4 is switched on and $I^{(3-4)off}$ with the transition unit switched off.

The expected ratios $R^{(i-j)on/off}$ depend on the nuclear polarization. Limits can be calculated for the assumption $P_I = 0$ and for the assumption of spin temperature equilibrium, where P_I is maximum. With an efficiency of one of the high frequency transitions the measured ratios must be within these limits.

The limits for spin temperature equilibrium for the various transitions can be calculated by combining the corresponding eqn. (13) and eqn. (7)

$$R^{(1-4)on/off} = \frac{1 - P_S}{1 + P_S}, \tag{14}$$

$$R^{(2-4)on/off} = \frac{3 - P_S + 5P_S^2 + P_S^3}{3 + 3P_S + P_S^2 + P_S^3}, \tag{15}$$

$$R^{(3-4)on/off} = \frac{3 + P_S + 5P_S^2 - P_S^3}{3 + 3P_S + P_S^2 + P_S^3}, \tag{16}$$

$$R^{(1-6)on/off} = \frac{3 - 2P_S - P_S^2}{3 + P_S^2}, \tag{17}$$

$$R^{(2-6)on/off} = 1, \tag{18}$$

$$R^{(3-5)on/off} = 1. \tag{19}$$

The limits for the assumption that the nuclear spin is completely unaffected by spin–exchange collisions and therefore $n_1 = n_2 = n_3$ and $n_4 = n_5 = n_6$, are

$$R^{(i-j)on/off} = \frac{1 + \frac{1}{3}P_S}{1 + P_S}; \quad P_I = 0 \tag{20}$$

for all of the above transitions. This ratio is also measured when a calibration sextupole magnet in between the spin–exchange source and the atomic beam polarimeter is used, while the pump laser is blocked.

Fig. 2 shows a plot of the intensity ratios $R^{(i=j)on/off}$ versus the electron spin polarization P_S.

A high frequency transition unit consists of a homogeneous magnetic field, a superimposed gradient field, and a high frequency field [8], [9]. For an interchange of the occupation numbers of two specific states, the high frequency must match the corresponding transition frequency for the magnetic field.

The transitions 3–4, 2–4, and 1–4 can be operated in the same high frequency transition unit. Only the static magnetic field is changed in order to Zeeman–shift the transition frequencies to the external high frequency. The transitions 2–4 and 1–4 are manifold transitions. The transition 2–4 corresponds to the sequential transitions 2–3 and 3–4, whereas the transition 1–4 corresponds to the sequential transitions 1–2, 2–3 and 3–4. It depends on the strength of the gradient field and the length of the unit, whether all three transitions can be operated with high efficiency.

Scans of the D_1 signal I in the mass spectrometer as a function of the magnet current, which is proportional to the magnet field strength, are shown in fig. 3. For the scan shown in the upper part of the figure a calibration sextupole magnet was placed in between the spin–exchange source and the polarimeter, while the pumping beam was blocked. Assuming a complete substate separation in the sextupole magnet, $P_S = 1$ and $P_I = 0$. Therefore

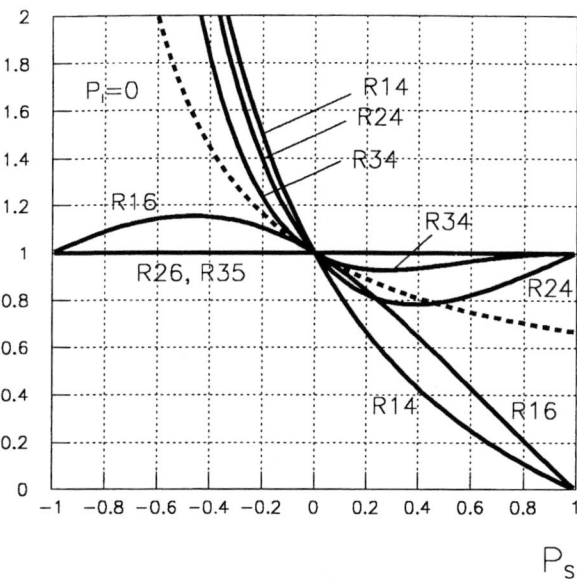

FIGURE 2. Expected intensity ratios for the various transitions plotted versus P_S The solid lines correspond to spin temperature equilibrium and the dashed line to $P_I = 0$, respectively.

the expected ratio is $R^{(i-j)on/off} = (1 + \frac{1}{3}P_S)/(1 + P_S) = 2/3$. The measured ratio is $R^{(i-j)on/off} = 0.68 \pm 0.01$. With the calibration magnet the different transitions can not be distinguished, since $n_1 = n_2 = n_3$. By comparing the measured and the calculated ratios an efficiency E of the transition 3-4 can be determined [8] to $E = 0.96 \pm 0.03$.

The lower part of fig. 3 shows the same scan while the spin–exchange source is operated (without calibration magnet). The electron polarization was measured to 0.35 ± 0.03. The expected ratios are $R^{(3-4)on/off} = 0.930 \pm 0.003$, $R^{(2-4)on/off} = 0.784 + 0.003 - 0.002$, $R^{(1-4)on/off} = 0.48 \pm 0.03$. The measured ratios are $R^{(3-4)on/off} = 0.97 \pm 0.03$ and $R^{(2-4)on/off} = 0.81 \pm 0.02$ The smallest ratio for the transition 1-4 within the current range for this transition is $R^{(1-4)on/off} = 0.61 \pm 0.03$. Including the efficiency measured with the calibration magnet, the first two ratios are consistent with the assumption of spin temperature equilibrium within the errors. The large value for the transition 1-4 is probably due to the low efficiency for the 'subtransition' 1-2 (which has no influence when using the calibration magnet). Therefore P_I and P_{zz} can be calculated by using eqs. (9) and (10), $P_I = 0.45 \pm 0.04$ and

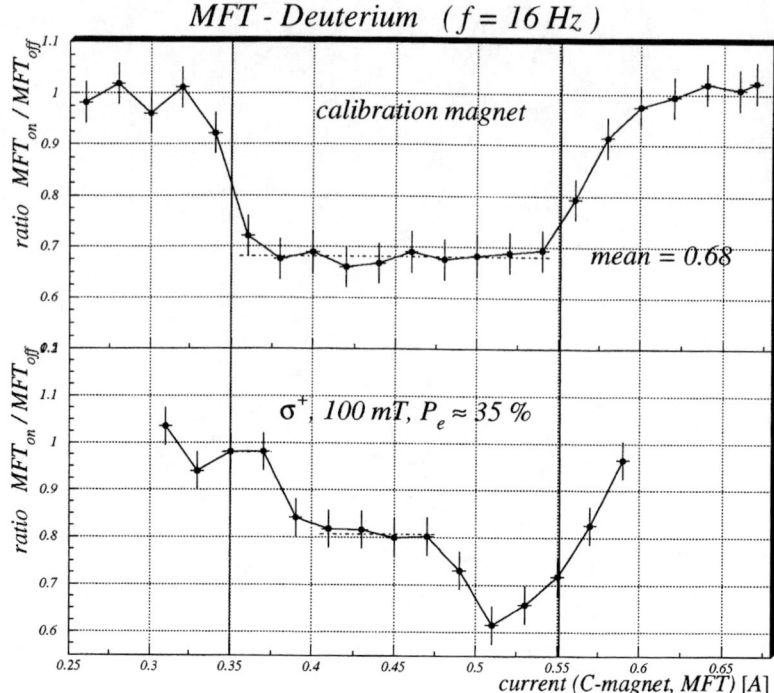

FIGURE 3. The D_1 signal versus the current in the magnet for the homogeneous field. The upper part of the figure corresponds to a sample beam from the calibration magnet, the lower part corresponds to a beam from the spin–exchange source.

$P_{zz} = 0.16 \pm 0.03$.

CONCLUSIONS

We discussed the general properties of a sample of deuterium atoms in spin temperature equilibrium and calculated the intensity ratios in the quadrupole mass spectrometer for various high frequency transitions for the assumption of spin temperature equilibrium (spin–exchange source) and completely unpolarized nuclei (polarization by the calibration sextupole magnet). Whereas in general five high frequency transitions are necessary to completely determine the nuclear spin polarizations of deuterium, only one high frequency transition unit is necessary to demonstrate spin temperature equilibrium using the calculated intensity ratios.

The atomic density ρ_D and the dwell time τ_{dwell} in the pumping cell were chosen to $\rho_D = 1.8 \cdot 10^{14}\, cm^{-3}$ at a flow of $3 \cdot 10^{17}\, 1/s$ and $\tau_{dwell} = 30\, ms$, respectively. With these parameters the atoms undergo enough spin–exchange collisions to approach spin temperature equilibrium in the high magnetic field of $100\, mT = 8.5 \cdot B_c$, in which nucleon and electron spins are strongly decoupled. The deuterium atoms from the spin–exchange source could experimentally be demonstrated to be close to spin temperature equilibrium within the errors of the measurements. Presently a high frequency transition unit for the transitions 1–6, 2–6 and 3–5 is under construction, which allows more precise measurements.

Future work aims to increase flow and polarization of the source.

This work was supported by the Bundesministerium für Bildung, Wissenschaft, Forschung und Technologie, BMBF, through Grant No 06ER358.

REFERENCES

1. M. Poelker et al.; Nucl. Instr. Meth. **A364** (1995) 58.
2. C. Grosshauser et al.; Proceedings of the 12th Int. Symp. on High Energy Spin Physics, Amsterdam, Sep. 10–14 1996, in press, J. Stenger et al.; ibid.
3. W. Happer; Rev. Mod. Phys. **44** (1972) 169.
4. J. Stenger, C. Grosshauser, W. Kilian, B. Ranzenberger, K. Rith; Phys. Rev. Letters **78** (1997) 4177.
5. T. Walker and L.W. Anderson; Nucl. Instr. and Meth. **A334** (1993) 313.
6. H.R. Cole, R.E. Olsen; Phys. Rev. **A31** (1985) 2137.
7. C. Coulter et al.; Phys. Rev. Letters **68** (1992) 174.
8. C. Grosshauser et al.; to be published.
9. M. Ferro–Luzzi et al.; Nucl. Instr. Meth. **A364** (1995) 44.

First Use of a Laser-Driven Polarized H/D Target at the IUCF Cooler

M. A. Miller[1a], K. Bailey[b], J. Brack[c], R. V. Cadman[a],
W. J. Cummings[b], J. Fedchak[2b], B. Fox[c], H. Gao[3b],
C. Grosshauser[d], R. J. Holt[a], C. Jones[e], E. Kinney[c],
R. Kowalczyk[b], Z.-T. Lu[b], W. Nagengast[d], B. Owen[a],
K. Rith[d], F. Schmidt[d], E. Schulte[a], J. Sowinski[f],
F. Sperisen[f], J. Stenger[4d], E. Thorsland[a],
and S. Williamson[a]

[a] *University of Illinois at Urbana-Champaign, Urbana, Illinois 61801*
[b] *Argonne National Laboratory, Argonne, Illinois 60439*
[c] *University of Colorado at Boulder, Boulder, Colorado 80309-0446*
[d] *Universität Erlangen-Nürnberg, 91058 Erlangen, Germany*
[e] *California Institute of Technology, Pasadena, California 91125*
[f] *Indiana University Cyclotron Facility, Bloomington, Indiana 47408*

Abstract. The HERMES Laser-Driven Target Task Force (Argonne, Erlangen, Illinois and Colorado) is charged with developing a polarized H/D target for use in the HERA ring at DESY. Rapid progress was made in the beginning of 1996, leading us to the decision to test the target in a realistic experimental environment. In particular, polarizations of 0.6 and flows above 10^{18} atoms·s^{-1} have been achieved on the bench. The laser-driven target and a simple detector system are currently installed in Cooler storage ring at the Indiana University Cyclotron Facility in order to test its applicability to nuclear physics experiments. Target polarizations are being measured using the $\vec{H}(p,p)$ and $\vec{D}(p,p)$ reactions. Initial tests were reasonably successful and the target is well along toward becoming viable for nuclear physics

---

[1] Corresponding author, `miller5@uiuc.edu`
[2] Current address: University of Wisconsin-Madison, Madison, Wisconsin
[3] Current address: Laboratory for Nuclear Science, Massachusetts Institute of Technology, Cambridge, MA 02139
[4] Current address: Massachusetts Institute of Technology, Cambridge, MA 02139

INTRODUCTION

Polarized targets are an important tool for nuclear physics, used both for investigating spin-dependence in nuclear reactions and structure and for studying the spin structure of the nucleon. The laser-driven polarized hydrogen and deuterium target, originally described in Ref. 1, has been developed to provide a high density polarized gas target for storage ring experiments. Since that time, development has continued at several laboratories[5] and has reached a point where the target is ready to be tested in a realistic nuclear physics environment.

In order for a target to be useful for nuclear physics, several criteria have to be met. The target thickness has to be sufficient to provide high enough luminosity to carry out an experiment, the nuclear polarization also must be high enough to provide a measurable spin-dependent signal and the target must be robust enough to maintain thickness and polarization continuously over the days, weeks, or even months, of an experiment. The laser-driven source has an advantage in this first area. Typical operating flows are 10^{18} nuclei/second. At present, the limitation on flow comes from the affect of the target on the lifetime of a stored beam in the IUCF ring, rather than from the source itself.

The nuclear polarization in the source is produced in several steps. Atoms are polarized by spin exchange collisions between D (or H) atoms, produced by an RF dissociator, and polarized potassium atoms. A small fraction of potassium atoms is introduced into the source from a potassium ampoule which is heated to provide the desired density. The potassium atoms are optically pumped by polarized 770.1 nm laser light from standing-wave Ti-sapphire lasers pumped by two Ar^+ lasers. Polarization is transferred from the K atoms to H/D atoms in spin-exchange collisions and then to H/D nuclei by HH or DD spin-exchange collisions which bring the system to spin-temperature equilibrium, as discussed elsewhere in this proceedings [7] and in Ref. 8. In order to maintain the polarization of the nuclei in the source and as they move to the target, the source is coated with drifilm [9], which minimizes recombination into molecules and depolarization. Atomic polarizations of $p_e \sim 0.6$ have been achieved with atomic fraction ~ 0.5. Typical polarization behavior on the bench is shown in Fig. 1, with $p_z = 0.6$, and an atomic fraction of 0.35. The source is described in more detail in Refs. 1–5 and elsewhere in these proceedings [6].

The effective nuclear polarization of the target, p_{eff}, depends several quantities: the electronic polarization of the atoms, the atomic fraction, whether the

---

[5] The HERMES laser-driven target task force consists of research groups at the University of Illinois at Urbana-Champaign, Argonne National Laboratory, the Universität Erlangen-Nürnberg and the University of Colorado at Boulder.

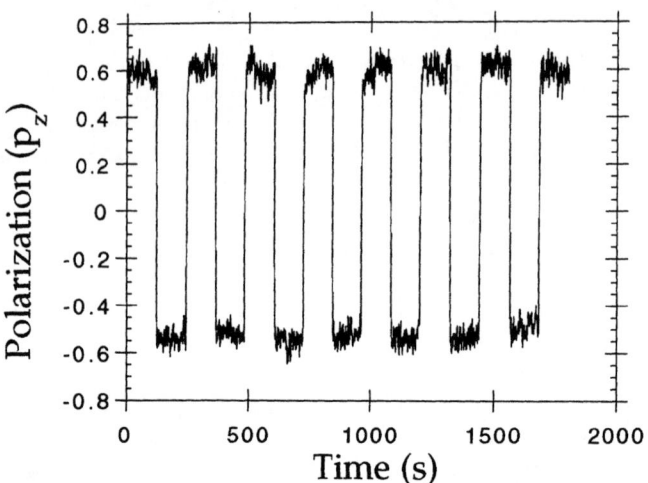

FIGURE 1. Polarization performance of the laser-driven source. This is the polarization p_z of deuterons, assuming that the system is in spin temperature equilibrium, plotted versus time. The sign of the polarization is changed by switching the helicity of the laser light incident on the target.

system has achieved spin-temperature equilibrium and depolarization caused in the source (from collisions with surfaces and gas, from recombination effects). In addition, if the target is used in an experiment which probes the nucleon spin (for example, deep-inelastic electron scattering), the largely unpolarized nucleons in small fraction of the potassium atoms in the target dilutes the overall nuclear polarization (typical potassium densities are $\sim 0.1\%$ of the H/D density). It should be noted that this assumes that nuclei in atoms that have recombined into molecules are depolarized in the process. If these nuclei retain some or all of their original polarization, the effective polarization would be significantly larger than that predicted from the atomic fraction.

INSTALLATION IN THE COOLER

The Cooler light ion storage ring [10,11] at the Indiana University Cyclotron Facility (IUCF) is well suited to demonstrate that the target is viable under realistic experimental conditions. The main goals of this project are to install the target in the Cooler, to measure the nuclear polarization of the target and to measure the polarization of nuclei in molecules. After this program is

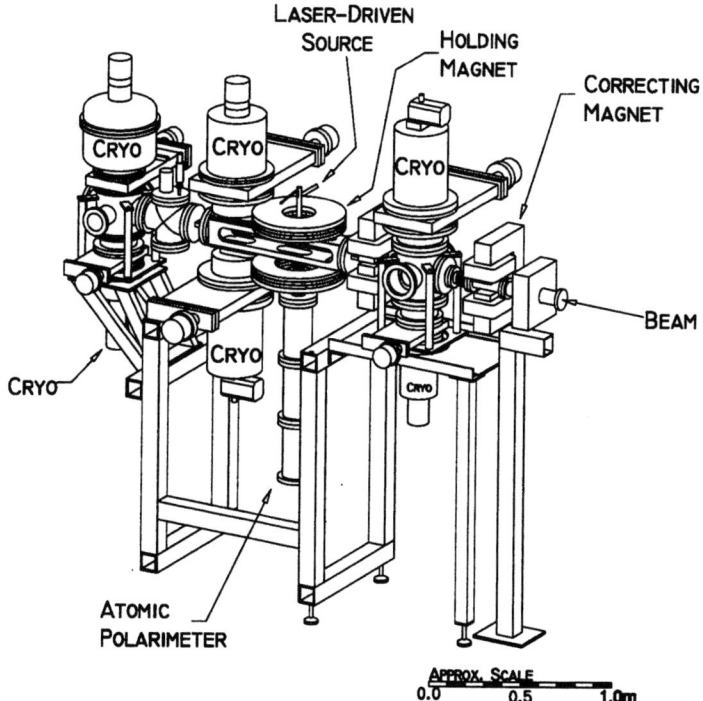

FIGURE 2. A view of the polarized internal H/D target and beam line as installed in the Cooler G-region. The Cooler beam passes from right to left in the figure. The two correcting magnets compensate for the deflection of the beam due to the target holding field.

complete, the target will be used study the spin-dependence of the deuterium wave function via the $\vec{D}(\vec{p}, 2p)$ and $\vec{D}(\vec{p}, pn)$ reactions. The target installation in the Cooler G-region is shown in Fig. 2.

The target cell dimensions and the flux of polarized nuclei from the source define the target thickness. In the present experiment, the horizontal and vertical dimensions of the target cell are chosen such that the phase space acceptance of the target nearly matches the design acceptance of the accelerator. The target cell is made from a single piece of aluminum 400 mm long, 32 mm wide by 13 mm high. Its sides have been machined to provide exit windows approximately 0.2 mm thick. The ends of the cell are open to allow the Cooler beam to pass through. The polarized source is mounted above the target cell with a vertical polarization axis defined by a vertical holding field ~ 1 kG. The inner surfaces of both the pumping cell and the target cell are coated with drifilm to minimize the wall depolarization and recombination

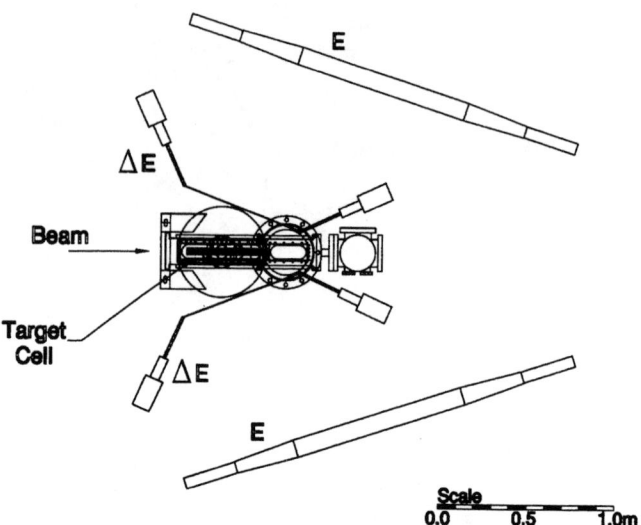

FIGURE 3. An overhead view of the target and detector arrangement used to measure the target polarization. The 200 MeV polarized proton beam enters from the left. ΔE and E scintillator detectors are placed symmetrically on either side of the target.

effects. The source and target are heated to ~ 250 C° to prevent potassium atoms from condensing onto the wall surfaces.

The detector arrangement, shown in Fig. 3, consists of thin plastic scintillators (one on either side of the target, 3 mm thick) and scintillator bars (two per side, 1 m long by 100 mm thick and 150 mm high) acting as ΔE-E pairs The extended target and large detectors provide coverage for approximately 40-100° in the p+p center of mass. Particle identification provided by time-of-flight between ΔE and E and dE/dx allows the separation of protons and deuterons. Time differences between photo-tubes at each end of the ΔE and E detectors are used to measure the angles of detected particles. A coincidence between both ΔE detectors and at least one E detector is required for a trigger.

The laser system for optical pumping of the source consists of two Ar^+/Ti-sapphire pairs with the Ar^+ lasers acting as pump lasers for the tunable Ti-sapphire lasers. They are installed outside the Cooler approximately 60 m from the target area with the laser light transported to the target on multi-mode fiber optics, as described elsewhere in this proceedings [12]. This fiber optic transport allows us to place the lasers in an easily accessible location where they can be tuned while the Cooler is operating. The net efficiency of the light transport is $\sim 70\%$, with the losses dominated by input and output coupling to the fibers. The light is split into two polarization states at the target and recombined after the polarizations have been rotated so that all of the light is circularly polarized. The helicity can be switched by changing between two $\lambda/4$ plates in a sliding stage.

An atomic beam polarimeter, consisting of a permanent quadrupole separator magnet and mass spectrometer is installed beneath the target chamber. This is used to measure the polarization and atomic fraction of the atoms coming from the source. A second mass spectrometer (the dissometer) looks at the downstream end of the target cell through a series of apertures. This is used to measure the atomic fraction of atoms coming out of the end of the target cell. The combination of these two atomic fraction measurements, at the center and at the end of the target cell, gives us an important diagnostic of recombination inside the target cell. In addition to the pump lasers, two beams from diode probe lasers are brought from the laser room to the target on separate fibers. These are used to measure relative potassium density in the source (through photo-absorption) and the atomic polarization of the potassium (via Faraday rotation [13]).

PRELIMINARY POLARIZATION MEASUREMENTS

The elastic $\vec{H}(p,2p)$ and $\vec{D}(p,pd)$ reactions are used to measure the polarizations of the target when running hydrogen and deuterium, respectively. The reactions have relatively large analyzing powers ($A_{n0} \sim 0.3$ at $\theta_{cm} \sim 40°$ for $\vec{H}(p,2p)$ and $it_{11} \sim 0.35$ at $\theta_{cm} \sim 40°$ for $\vec{D}(p,pd)$) and differential cross sections of a few mb/sr. This allows us to measure the asymmetries with a relative uncertainty of a few percent in 15 minutes or so.

We have run the target with beam twice to date. The measured spin asymmetry for the $\vec{H}(p,2p)$ reaction with the target both polarized and unpolarized is plotted versus center of mass scattering angle in Fig. 4. The running conditions here were not the best — the atomic fraction as measured by the atomic polarimeter was 0.50 while that measured with the dissometer was consistent with zero. This was a consequence of potassium migrating from the source to the target cell and damaging the drifilm coating. The curve shown in Fig. 4 is a fit of the known $\vec{H}(p,2p)$ analyzing power [14] to the data with a polarization parameter p_{eff}. The value of p_{eff} is 0.09 ± 0.01, consistent with the assumption of a linearly decreasing atomic fraction, falling from 0.5 at the target center to zero at the ends of the target cell.

As mentioned above, the migration of potassium from the source is a concern with the LDS. In our first measurements at IUCF, potassium contaminated the target cell (probably due to an argon bubble formed in the ampoule in assembly), resulting in high recombination in the cell. Such problems will be avoided in the future by distilling the potassium before use. The potassium can also affect the dissociator, resulting in a gradual degradation of atomic fraction. Adding a small fraction of oxygen to H/D dissociators has long been known to improve dissociator lifetime and efficiency. Studies begun to investigate the affect of oxygen on the LDS dissociator have yielded promising initial results.

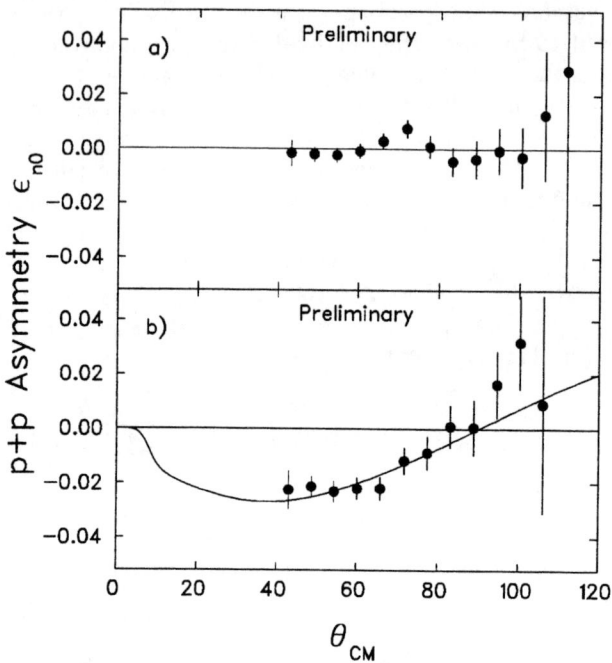

FIGURE 4. Preliminary measurements of the $\vec{H}(p,p)$ nuclear asymmetry ($\epsilon_{n0} = p_{\text{eff}} A_{n0}$) plotted versus center of mass angle: a) the measured asymmetry with the target unpolarized, b) the asymmetry with the target polarized. The curve in b is a fit of the $\vec{H}(p,p)$ analyzing power [14] to the data. The atomic fraction in the target cell is undetermined for this measurement, but varied between 0.5 and zero. The running conditions and polarization are discussed in the text.

SUMMARY

A laser-driven polarized H/D target has been installed in the IUCF Cooler. First tests with proton beam have been started in order to measure the nuclear polarization of the target. The first measurements have shown that the principle of the target is sound and that it produces polarized nuclei. A number of areas have been identified where further improvements can be made. Over the coming months, further studies will be carried out in order to improve reliability and performance of the source and target. The target will be also used to study the spin-dependence of the deuteron wave function via the $\vec{D}(\vec{p}, 2p)$ reaction (IUCF experiment CE68). The coupling of the high flux LDS and the high cross sections of the (p,2p) reaction will allow us to measure the effective polarization of the nucleons in the deuteron for momenta to 350 MeV/c.

ACKNOWLEDGEMENTS

The authors would like to thank D. Tupa for the loan of two Ti-sapphire lasers, W. Lawrence, our glass blower at UIUC, A. Kenyon for his help with assembly and maintenance of the target, J. Doskow for maintaining the vacuum system and the rest of the IUCF operators and staff for their assistance. This work was supported in part by the U.S. National Science Foundation under grants PHY94-20787 and PHY94-20470 and by the U.S. Department of Energy under contract No. W-31-109-ENG-38 and grant DE-FG02-86ER40269.

REFERENCES

1. K. P. Coulter et al., Phys. Rev. Lett. **68**, 174 (1992).
2. M. Poelker et al., Phys. Rev. A **50**, 2450 (1994).
3. H. Gao, R. V. Cadman, R. J. Holt, and E. Thorsland, in *Proceedings of the International Workshop on Polarized Beams and Polarized Gas Targets*, edited by H. P. gen. Scheick and L. Sydow (World Scientific, Singapore, 1996), p. 67.
4. M. Poelker et al., Nucl. Instrum. Meth. **A 364**, 58 (1995).
5. B. Owen et al., in *Spin 96 Proceedings*, edited by C. de Jager et al. (World Scientific, Singapore, 1996), pp. 490–494.
6. R. V. Cadman et al., in *Proceedings of the Seventh International Workshop on Polarized Gas Targets and Polarized Beams*, edited by R. J. Holt (American Institute of Physics, New York, 1997), in preparation.
7. J. Stenger et al., in *Proceedings of the Seventh International Workshop on Polarized Gas Targets and Polarized Beams*, edited by R. J. Holt (American Institute of Physics, New York, 1997), in preparation.
8. T. G. Walker and W. Happer, Rev. Mod. Phys. **69**, 629 (1997).
9. D. R. Swenson and L. W. Anderson, Nucl. Instrum. Meth. **B 29**, 627 (1988).
10. H. O. Meyer, *Indiana Cooler User Guide*, 2nd ed., 1988.
11. R. E. Pollock, Annu. Rev. Nucl. Part. Sci. **41**, 357 (1991).
12. W. Cummings et al., in *Proceedings of the Seventh International Workshop on Polarized Gas Targets and Polarized Beams*, edited by R. J. Holt (American Institute of Physics, New York, 1997), in preparation.
13. J. Stenger, M. Beckman, W. Nagengast, and K. Rith, Nucl. Instrum. Meth. **A 384**, 333 (1997).
14. R. Arndt and R. Workman, *The SAID program: A Guide for Users*, 1994, available through anonymous ftp at ftp://clsaid.phys.vt.edu/pub/said/said_manual.

The HERMES Polarimeter

B. Braun,
for the HERMES Collaboration

Phys. Inst. Univ. Erlangen–Nuernberg [1],
Erwin-Rommelstr. 1, D-91058 Erlangen, Germany

Abstract. The HERMES hydrogen target was operated during 1996 and 1997 in the HERA electron ring at DESY to study the spin structure of the proton by deep inelastic lepton proton scattering. An **A**tomic **B**eam **S**ource (ABS) injects an atomic hydrogen beam of high intensity and high proton polarization into a storage cell, made of thin aluminum and coated with Drifilm. A sampling tube attached on the side of the cell forms a sample beam whose polarization is analyzed by means of a **B**reit-**R**abi **P**olarimeter (BRP). A **T**arget **G**as **A**nalyser (TGA) measures its degree of dissociation. During running with the 27.6 GeV positron beam the target ran smoothly at nuclear polarization of the atoms above 0.9 and molecular fraction below 10 %.

INTRODUCTION

The formation of the polarized atomic hydrogen beam by the ABS is based on Stern-Gerlach separation in a system of five sextupole magnets (Fig. 1). The electron polarization of the atomic beam after the magnet system is transfered to the protons by means of two adiabatic RF transitions. A weak field transition (WFT 1–3) after the magnet system provides negative and a strong field transition (SFT 2–4) positive proton polarization. A beam of high electron polarization is obtained when all the transition units are switched off. A medium field transition (MFT 1–3/2–3) after the third sextupole can be used to transfer either hyperfine state $|1\rangle$ or $|2\rangle$ into state $|3\rangle$ which is then deflected in the second part of the magnet system. Using in addition the two transition units at the end of the ABS, atomic beams consisting predominantly of one single hyperfine state can be prepared. Seven beams of different hyperfine compositions and polarization states can be injected by the ABS (Table 1).

---

[1] Supported in part by the BMBF, Germany, contract number 057ER12P(2) and 056MU22I(1).

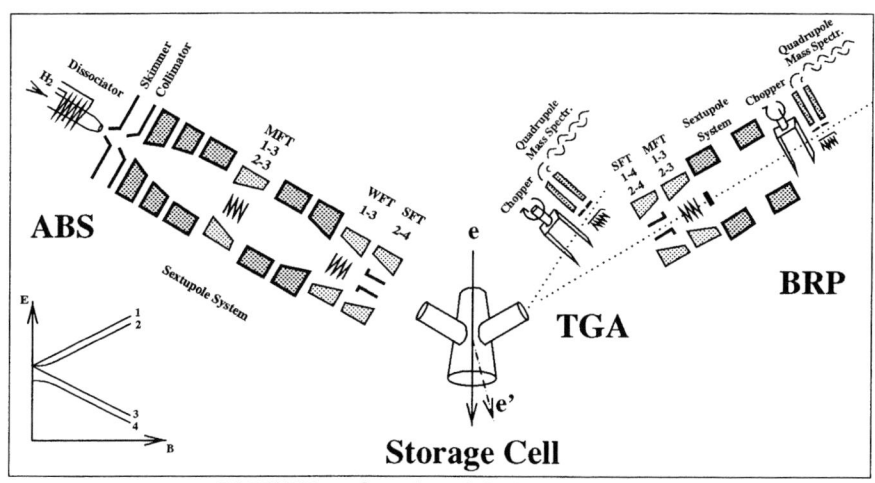

FIGURE 1. Overview of the Hermes Target.

The storage cell is located in a vacuum chamber surrounded by a super conducting magnet which provides a quantization axis for the proton and electron spins. The field is produced by 4 coils resulting in a non-uniformity of $\frac{\Delta B}{B} = \pm 1.5\%$ along the cell axis at the working point of $B = 335\,\mathrm{mT}$ [1].

A small fraction of the target gas leaves the cell via an additional sample tube, attached on the side of the cell. A sample beam is formed and directed into TGA and BRP which measure its degree of dissociation and polarization, respectively.

THE TARGET POLARIZATION

The evaluation of the target polarization must consider the polarization of atoms P^T_{atom}, the degree of dissociation α_T, a possible polarization of molecules β and various sources of molecules:

$$P_T = \alpha_0 \left[\alpha_r^T + \left(1 - \alpha_r^T\right) \beta \right] P^T_{\mathrm{atom}}, \qquad (1)$$

where $\mu_0 = 1 - \alpha_0$ is the fraction of protons entering the cell in molecules. These protons are unpolarized. The dominant sources of the molecules quantified by μ_0 are the residual gas pressure inside the target chamber and ballistic flow from undissociated gas in the ABS. These contributions are small and result in a small error on α_0 of 0.01, with α_0 values ranging from 0.96 to 0.98. Another fraction of molecules $\mu_r^T = 1 - \alpha_r^T$ is formed by recombination inside the cell. These molecules may carry a significant proton polarization. This is specified by the parameter β which is the ratio of the proton polarization in the molecules to the proton polarization of the atoms from which the molecules originated. P^T_{atom} is the polarization of the atomic fraction.

TABLE 1. Injection modes of the ABS.

ABS status	inj. states	P_e	P_z
off	$|1\rangle, |2\rangle$	+1	0
WFT 1–3	$|2\rangle, |3\rangle$	0	-1
SFT 2–4	$|1\rangle, |4\rangle$	0	+1
MFT 2–3, SFT 2–4	$|1\rangle$	+1	+1
MFT 1–3, WFT 1–3	$|2\rangle$	+1	-1
MFT 2–3, WFT 1–3	$|3\rangle$	-1	-1
MFT 1–3, SFT 2–4	$|4\rangle$	-1	+1

The first row shows RF transitions switched on to select a certain composition of hyperfine states quoted in the second row. The third and the forth row indicate ideal values for electron and proton polarization. Second and third mode are used for DIS data taking, the injected beam consists of two hyperfine states. The other modes are essential for the the calibration of the BRP and for polarized Bhabha scattering.

The degree of dissociation α_r^T and the atomic polarization P_{atom}^T of the target as seen by the beam will in general differ from the corresponding quantities α_r^{TGA} measured by the TGA and P_{atom}^{BRP} measured by the BRP. Two sampling correction factors c_α and c_p relate the measured quantities to the one seen by the beam:

$$\alpha_r^T = c_\alpha \left(\alpha_r^{TGA}\right) \cdot \alpha_r^{TGA}, \qquad P_{atom}^T = c_p \left(P_{atom}^{BRP}\right) \cdot P_{atom}^{BRP}. \qquad (2)$$

Both sampling corrections are not constant, but depend on the degree of dissociation and polarization, respectively. They can be determined either by means of Monte Carlo simulations or by indirect measurements like Bhabha scattering or a luminosity measurement (see below).

THE TARGET GAS ANALYZER

The TGA consists of a quadrupole mass spectrometer alternately measuring mass 1 and 2. A chopper in front of it allows for the subtraction of background originating from residual gas. The degree of dissociation in the sample beam α_r^{TGA} [2] can be calculated from the two resulting beam count rates S_1^{TGA} and S_2^{TGA}:

$$\alpha^{TGA} = \frac{S_1^{TGA}}{S_1^{TGA} + \kappa \cdot \sqrt{2} \cdot S_2^{TGA}}. \qquad (3)$$

The calibration constant κ corrects for the relative sensitivity of the TGA for mass 1 and mass 2. This calibration constant can be measured [2] knowing the total flux injected by the ABS into the cell is constant as represented by

$$S_1^{TGA} + \kappa \cdot \sqrt{2} \cdot S_2^{TGA} = \text{const.} \qquad (4)$$

---

[2] S_2^{TGA} has to be corrected for the beam count rate caused by μ_0 to derivate α_r^{TGA}.

In a graph (Fig. 2) where the two beam count rates S_1^{TGA} and S_2^{TGA} are plotted vs. each other, the points for different values of α^{TGA} are on a line whose slope is related to κ as

$$\frac{dS_2^{TGA}}{dS_1^{TGA}} = -\frac{1}{\kappa \cdot \sqrt{2}}. \tag{5}$$

Because the recombination is strongly temperature dependent, a variation of α can be forced by changing the cell temperature between 35 and 100 K. However, this method requires a temperature correction of the measured beam count rates which introduces additional systematic errors.

The data in figure 2 was taken after the beam was dumped accidentally close to the target region. Due to this event the cell surface was modified suddenly and α^{TGA} reduced to about 0.2. It recovered to its original value of 0.95 within 24 h. During this time all other conditions including the cell temperature remained constant, resulting in a very small systematic error $\Delta\alpha_r^{TGA}$ of 0.01.

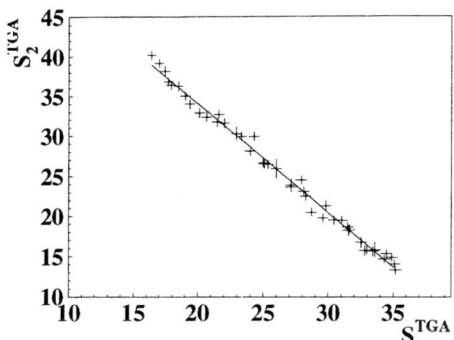

FIGURE 2. Calibration of TGA. Count rate mass 2 is plotted vs. count rate mass 1 while α was varying inside the cell after an accidental beam dump close to the target region. The slope of the line defines the calibration constant κ. Its distance to the origin is a measure for the total flux injected by the ABS. Points closer to the vertical axis belong to lower α, those closer to the horizontal to higher α.

THE BREIT-RABI POLARIMETER

A combination of RF transitions followed by a sextupole system and a beam detection system which is the same as the one used in the TGA allows the polarization of the atoms in the sample beam to be measured (Fig. 1). Switching on one of the transition units the beam count rate will change, if the two involved hyperfine states are not equally occupied. The measurement of 4 beam signals, i.e. one measurement without any transition and three with three different transitions active, provide a system of 4 linear equations for the 4 unknown occupation probabilities for the hyperfine states. The electron and proton polarization can be calculated from the hyperfine occupation knowing the guide field. Under data taking conditions all four possible transitions are switched providing four different evaluations and online quality monitoring of the BRP.

To calibrate the BRP one has to measure 10 transition efficiencies [3] and one ratio of transmission probabilities of the sextupole system. Four transition combinations involving the SFT and MFT being on can be set up in addition to the normal operating modes of the BRP. Altogether 11 different BRP states are possible for each ABS state, leading to a restriction not only for the occupation probabilities, but also partially for the transition efficiencies.

The system becomes completely constrained, if the ABS is also switched through its 7 possible states (Table 1). In addition to the 11 calibration constants one has 28 occupation numbers, i.e. 39 unknowns. From 11 BRP states for each of the 7 ABS states one gets 77 equations, which is resolvable for the 39 unknowns. The system is not linear and a χ^2–minimization procedure is used to solve this equation system leading to a systematic error of 0.01 for the proton polarization in the sample atom beam.

POLARIZATION OF MOLECULES

Information concerning the polarization of molecules can be obtained by performing an asymmetry measurement sensitive to the nucleon polarization at two different values for α^T. The first measurement of this kind has been made at NIKHEF using electron scattering from tensor polarized deuterons indicating high polarization of molecules [3]. Recently at HERMES some data has been taken at low α by raising the cell temperature to 260 K to increase the recombination. The evaluation is in progress and will give the first experimental information concerning the proton polarization of molecules under HERMES conditions. Further measurements are planned at NIKHEF and IUCF.

SAMPLING CORRECTIONS

Presuming a homogeneous cell surface, the sampling corrections (2) are uniquely defined and can be determined by a Monte Carlo simulation of molecular flow inside the storage cell. However, information concerning the sampling corrections can also be obtained experimentally.

The conductance of the storage cell depends on the mass of the gas particles leading to a higher target density, if the molecular fraction is increased. As a consequence a luminosity measurement can be regarded as a direct measurement of α_T as long as one can make a measurement either $\alpha_T = 1$ or 0 and normalize to the luminosity at this point. However, the effect is small and this method works well only for low α_T.

---

[3] A MFT is operating with two separated resonances to be described by two efficiencies. Additionally it is operated in two different modes depending on the setup of the SFT close to it, whose magnetic fields influence also the fields in the MFT.

The luminosity monitor of the HERMES detector can be used to measure an asymmetry in Bhabha scattering of polarized positrons from polarized target electrons [4]. Reasonable electron polarization of the target with both signs can be achieved by injecting only one hyperfine state with the ABS (Table 1). The target polarization formula (1) simplifies, because molecules do not carry electron polarization, i.e. $\beta_e = 0$. The measured asymmetry can be related to the electron polarization measured by the BRP and α_r^{TGA} measured by the TGA. Hence the product of the two sampling corrections $c_p \cdot c_\alpha$ can be extracted. The observed c_p is related to a relatively low electron polarization. The extrapolation of c_p to the high values of the proton polarization is not trivial, but straight forward, because the relaxation of electrons and protons are related to each other only by hyperfine coupling. The technique to separate the two factors, which is required to quote the correct target polarization, is still being developed. The errors on these sampling corrections will probably dominate the systematic error for the target polarization.

CONCLUSION

The target diagnostic system consisting of the TGA and the BRP works very well. The degree of dissociation and the atomic polarization of the sample beam are measured with systematic errors of 0.01, which are better than the design values. Both devices were operating smoothly and reliably throughout the whole running period, showing as well the stable and good performance of the ABS and the target itself [5]. The error contributions from the sampling corrections and the polarization of molecules to the target polarization are constrained by the high polarization and the high degree of dissociation. The techniques for their determination are presently being worked out.

REFERENCES

1. Kolster H., contribution to this workshop.
2. Baumgarten C., *Diploma thesis*, Univ. of Hamburg (1996).
3. Van den Brand, J.F.J. et al., *Phys. Rev. Lett.* **78**, 1235 (1997).
4. Benisch T., *Dissertation*, Univ. of Erlangen, in preparation.
5. Stewart J., contribution to this workshop.

Beam-Induced Resonant Depolarization Effects in a Polarized Atomic Hydrogen Gas Target at HERA

H. Kolster,
for the HERMES Collaboration.

LMU München, Sektion Physik[1]*,
Am Coulombwall 1, D-85748 Garching, Germany*

Abstract. In the HERMES experiment at HERA deep inelastic scattering of polarized positrons from internal polarized gas targets is used to study the spin structure of the nucleons.

For the hydrogen target, a strong magnetic field parallel to the positron beam direction is used to maintain the polarization of the atoms inside the target cell. Depolarization of the atoms due to time dependent magnetic fields resulting from the time structure of the high energy positron beam have been studied using a Breit-Rabi type polarimeter. Resonant effects are observed as function of the magnetic holding field. They are in agreement with model expectations. At the working point at 335 mT in between two resonances depolarization is negligible.

INTRODUCTION

In the HERMES experiment the spin structure of the nucleon is investigated by inclusive and semi inclusive observation of positron - proton scattering using the polarized positron beam of the HERA accelerator and an internal polarized hydrogen gas target (Diagram 1). The positron beam is bunched to provide high luminosity for the collider experiments and there are nominally 220 bunches circulating in the HERA positron ring. At the beam energy of 27.5 GeV the average beam current after injection is 35 mA. The high peak currents and the time structure of this beam lead to intense transient magnetic fields transverse to the beam direction which have a typical peak value of 80 mT at a distance of 0.15 mm from the beam axis. The periodicity of these transient fields is seen by the hydrogen atoms which are moving with thermal

---

[1] Work is supported by BMBF Germany, 056MU22I(1) and 057ER12P(2).

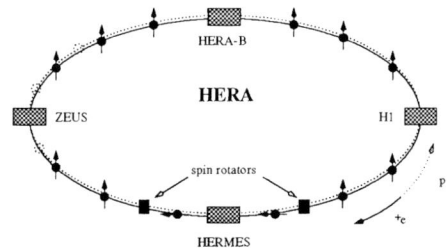

FIGURE 1. Schematic of the HERA accelerator at DESY.

The transversal polarization of the positrons in the ring results from the Sokolov Ternov effect. The HERMES experiment is located in between two spin rotators to provide a longitudinally polarized positron beam.

velocity within the interaction region where there is a static holding field which provides the quantization axis for the proton spins. This leads to depolarizing resonances at certain values of this field. There are a large number of closely spaced resonances in the usable range of holding field values. The holding field must therefore be rather uniform and the working point chosen carefully. The basic features of depolarizing resonances had been calculated by Kinney [1–3], they gave the design criteria for the target holding field magnet.

THE HERMES TARGET

The components of the internal target which are important in the present context are shown in Figure 2. In the atomic beam source (ABS) [4–6] the

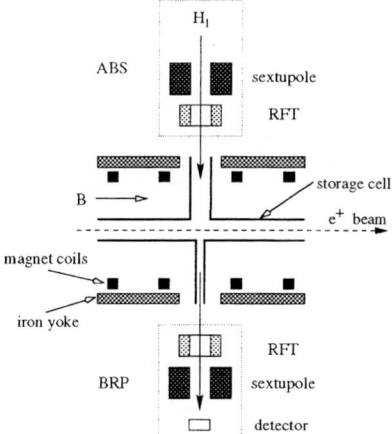

FIGURE 2. Schematic of the Hermes Target.

From top to bottom: ABS (sextupole magnets and RF transition units); target cell and positron beam; target magnet coils to provide the holding field; BRP (RF transition units, sextupole magnets and quadrupole mass spectrometer (QMS) for single ion detection).

atomic hydrogen electrons are polarized using Stern-Gerlach-separation in a sextupole system. Adiabatic RF transitions [6] interchange sub-state populations of hyperfine states to allow the selection of different proton polarization states in the injected target gas [7]. A super conducting magnet [5,8] provides the quantization axis parallel to the positron beam axis. The field is produced by 4 coils giving a non uniformity of $\frac{\Delta B}{B} = \pm 1.5\%$ along the interaction region axis. The target cell [5,8] increases the target density by a factor of hundred compared to a free jet. The achieved target thickness is $7 \cdot 10^{13}$ nucl cm^{-2}. The atoms in the cell are sampled and their polarization measured using a Breit-Rabi type polarimeter (BRP) [4,5,7]. A combination of adiabatic RF transitions [4] followed by a sextupole system allows the magnetic sub-state population and thus the polarization of the atoms to be measured.

HYPERFINE SPLITTING

The Hamiltonian of hydrogen is given by

$$\mathbf{H} = \mathcal{A} \cdot \mathbf{IS} + \omega_S \mathbf{S}_Z - \omega_I \mathbf{I}_Z. \tag{1}$$

where S and I indicate the electron and proton spin operators, respectively, and ω is the Larmor frequencies according to the external magnetic field B. The dependence of the energies of the hyperfine sub-states on the magnetic field strength is shown in Figure 3. At the working point of several hundred mT the field is well above the critical field $B_c = 50.7$ mT.

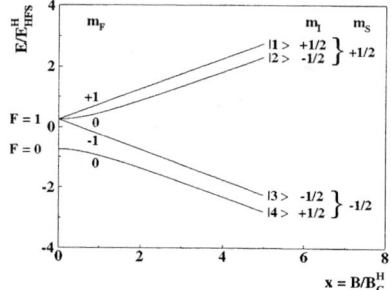

FIGURE 3. Hyperfine splitting of hydrogen. The energy of the sub-states of hydrogen is plotted versus the strength of the magnetic field in units of the critical field $B_c = 50.7$ mT.

A periodic magnetic field may induce transitions between two sub-states $|a\rangle$ and $|b\rangle$, if the perturbation provides the transition (resonance) frequency ν_{ab}:

$$h\nu_{ab} = \Delta E_{ab}. \tag{2}$$

Because of the transverse direction of the perturbing fields relative to the holding field only π-transitions with $\Delta m_F = 1$ are allowed. These are

proton transitions: $|1\rangle \leftrightarrow |2\rangle$, $|3\rangle \leftrightarrow |4\rangle$,
electron transitions: $|1\rangle \leftrightarrow |4\rangle$, $|2\rangle \leftrightarrow |3\rangle$.

In the strong field limit the transitions (12) and (34) change the proton polarization and the (23) and (34) transitions the electron polarization only. In a transverse holding field one has to consider in addition the σ transition

$|2\rangle \leftrightarrow |4\rangle$ with $\Delta m_F = 0$, which changes both the electron and proton polarization.

TIME STRUCTURE OF THE POSITRON BEAM

The frequency spectrum of the perturbing field is the Fourier transform of the time structure of the positron beam field. Each bunch in the HERA ring has a length of approximately $\sigma_z = 8$ mm and an elliptical shape with $\sigma_x = 0.26$ mm and $\sigma_y = 0.07$ mm. The revolution time of a single bunch at 27.5 GeV is T= 21.14 μs and the time interval between subsequent bunches is $\Delta t = 96$ ns. The magnetic field of a single bunch is modified due to reflections at the beam tube. Thus the resulting bunch field is longer than the bunch length. A schematic of the time structure is given in Figure 4.

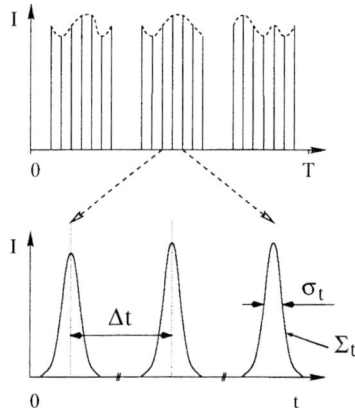

FIGURE 4.
Upper: Schematic of the time structure of the positron beam. The beam current I is plotted versus time. Some of the buckets are left empty.
Lower: The time Δt between single bunches is 96 ns. The shape of single bunches is is approximated by a Gaussian function Σ_t of width σ_t.

A schematic of the Fourier transform of the beam is plotted in Figure 5. The frequency interval $\Delta \nu$ between the harmonics is $\Delta \nu = 10.4$ MHz $= \Delta t^{-1}$. The shape function of the envelope Σ_ν is given by the average field of the single

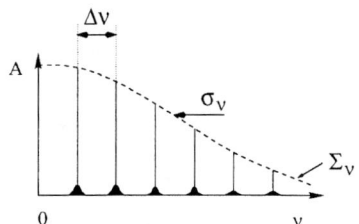

FIGURE 5.
Schematic of the Fourier transform of the positron beam. The frequency of the harmonics is $\nu = k \cdot \Delta \nu$. The envelope Σ_ν is the Fourier transform of the field of the single bunches Σ_t, $\sigma_\nu = \sigma_t^{-1}$.

bunches Σ_t, which is approximated by a Gaussian function. The short length of the bunches leads to a large number N of harmonics. For an 8 mm bunch length, which is a lower limit, the $1\sigma_\nu$ range of the envelope Σ_ν is expected at a frequency of $\nu = 5\,\mathrm{GHz}$.

RESONANCE CONDITION

The energy difference ΔE_{ab} between the hyperfine states is a non linear function of the magnetic field. It is plotted in Figure 6 for transitions between the states $|1\rangle$ and $|4\rangle$, $|1\rangle$ and $|2\rangle$, and $|3\rangle$ and $|4\rangle$. The field values, where a beam harmonic leads to a resonance are shown schematically in the graph for both resonance types (12) and (34).

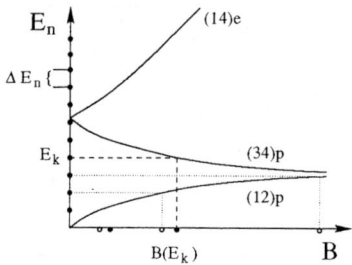

FIGURE 6. Schematic of the position of the harmonics versus holding field. The energy differences for the transitions (14), (12), and (34) are plotted. E_k is the energy value for the beam harmonic no. k given by the condition $\Delta E_{ab} = kh \cdot \Delta\nu$.

It should be noted that precise calculation of the resonances at high fields requires the use of the full Hamiltonian. Significant errors will occur for the calculated positions of the resonances if the proton term is neglected (Equation 1).

MEASUREMENT TECHNIQUES

Two different techniques had been used to observe the resonances.

- "Flip -in" measurements. In this technique the ABS and BRP are operated continuously. The RF transitions are chosen so that in the absence of resonant transitions in the cell no intensity in the BRP detector is expected. Then, the ABS is injecting sub-states $|1\rangle$ and $|4\rangle$ into the target cell. The (12) or (34) proton resonance inside the cell depopulates the states $|1\rangle$ or $|4\rangle$, respectively. With the BRP MFT transition (13) the injected sub-states $|1\rangle$ and $|4\rangle$ are removed from the detector acceptance. Thus only atoms which experienced the bunch field resonances (12) or (34) in the cell contribute to the signal in the BRP, which is proportional to the strength of the resonance.

injected   BRP detector

$$\begin{pmatrix} n_1 \\ 0 \\ 0 \\ n_4 \end{pmatrix} \xrightarrow{\text{Reson.:}} \begin{pmatrix} n_1 - \epsilon_1 \\ \epsilon_1 \\ \epsilon_4 \\ n_4 - \epsilon_4 \end{pmatrix} \xrightarrow{\text{MFT13}} \begin{pmatrix} \epsilon_1 \\ \epsilon_4 \\ n_1 - \epsilon_1 \\ n_4 - \epsilon_4 \end{pmatrix} \xrightarrow{\text{Sextup.}} \begin{pmatrix} \epsilon_1 \\ \epsilon_4 \\ 0 \\ 0 \end{pmatrix}$$

The advantage of this technique is the short time needed to obtain statistically relevant information. The data is obtained by scanning the magnetic field and gives information about the position and shape of the resonances.

- Polarization measurement. The BRP is operated to measure the occupation numbers $n_1...n_4$ of the sample. The polarization results as:

$$P_e = n_1 - n_3 + (n_2 - n_4)\sqrt{\tfrac{x}{1+x^2}},$$
$$P_z = n_1 - n_3 - (n_2 - n_4)\sqrt{\tfrac{x}{1+x^2}}. \qquad (3)$$

using $x = \tfrac{B}{B_c}$. This measurement requires stable magnetic field for a single measurement and gives detailed information about the resonances inside the cell for different injected states from the ABS.

MEASUREMENTS

First measurements were taken at beam currents between 28 mA and 38 mA after injection and ramping up the positron beam energy to 27.5 GeV. During a scan of the magnetic field the H_1 signal in the BRP was measured with the RF transitions in the ABS and the BRP running continuously ("flip in"). The lower part of Figure 7 shows the measured H_1 signal as a function of the magnetic field. The observed peaks are compared to the expected positions of the resonances calculated from the Fourier spectrum of the beam (upper part of Figure 7). The position of the resonances are at the expected magnetic field values.

In the second measurement the proton and electron polarization of the hydrogen atoms have been measured in the magnetic field range near the (12) resonance of the 62nd harmonic of the beam (see Figure 8). The measurements clearly indicate a change in the proton polarization caused by beam induced resonances. As expected, no polarization transfer from proton to electron is seen.

The transition strength of the electron resonances, e.g. (14), depends on much higher Fourier components of the bunch field (see Figure 5). Because of their narrow spacing and the non-uniformity of the holding field inside the cell they cannot be resolved. A first measurement of the electron resonances,

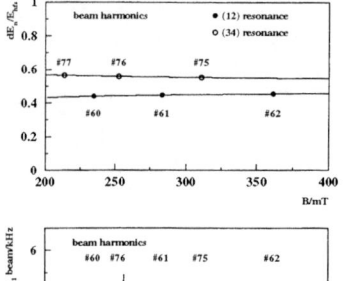

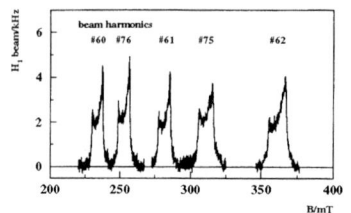

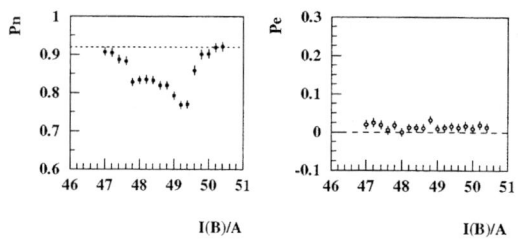

FIGURE 7.
Upper graph:
The expected position of the resonances derived from the Fourier spectrum of the beam. The number of the beam harmonics is plotted next to the data points.
Lower graph:
Measured positions of the resonances versus magnetic field. The HERMES operating point is $B = 335\,\mathrm{mT}$ in between two resonances.

FIGURE 8. Measured depolarization in the range of the 62nd beam harmonic for injected states $|1\rangle$ and $|4\rangle$. The left picture shows the proton polarization, the right picture shows the electron polarization. ($B = 7.42 \cdot I_{\mathrm{mag}} \cdot \frac{\mathrm{mT}}{\mathrm{A}}$).

comparing data with and without positron beam has been performed scanning the magnetic field from $B = 0\,\mathrm{mT}$ to $335\,\mathrm{mT}$. This measurement gives information about the length of the bunch fields.

RESONANCE SHAPE

All measured resonances show a characteristic shape which is the same for the (12) and (34) resonances. This indicates that the measured resonance shape results from broadening of an intrinsic resonance and is determined by the geometry and the density distribution in the cell and the non-uniformity of the magnetic holding field. The influence of these effects is discussed below.

Beam-Induced Resonances inside the Target Cell

The dependence of the amplitude B_{rf} of the bunch field on the distance from the beam is given by a function $\propto r^{-1}$. The transition probability is proportional B_{rf}^2. The average transition probability of an atom moving in the cell depends on the distance of the particle from the beam (see Figure 9). The bunch field may be approximated by a constant value within a certain radius, then the strong interaction between the gas particles and the perturbing field takes place inside a cylinder with radius r_{int} along the beam axis. The thermal

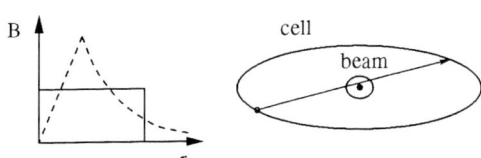

FIGURE 9. Left: Radial dependence of the perturbing field (dashed line). Left and Right: Approximation of the bunch field by a constant field value within a cylinder which is passed by the particle with the probability ε.

motion of the particles (see Figure 9, right) give a finite interaction time $\tau = 2 r_{\text{int}} v_{\text{therm}}^{-1}$ which leads to an averaged intrinsic width $\Gamma_E = \hbar \tau^{-1}$ of the resonance with a spectral density distribution given by:

$$\rho(\Delta \nu) = \frac{\tau}{\pi} \frac{1}{1 + (\nu_{\text{rf}} - \nu_{ab}(B))^2 \tau^2}, \quad (4)$$

where ν_{rf} is the frequency of the perturbing field and $\nu_{ab}(B)$ is the frequency according to the energy gap between states $|a\rangle$ and $|b\rangle$ at the actual magnetic holding field B. The transition probability W_{ab} between states $|a\rangle$ and $|b\rangle$ is then given by

$$W_{ab}(\nu_{\text{rf}}) = |\langle b | \mathbf{A}(S, I) | a \rangle|^2 \, \rho(\Delta \nu) \, \mathcal{F}_{e^+}(\nu_{\text{rf}})^2 \quad (5)$$

where $|\langle b | \mathbf{A}(S, I) | a \rangle|$ is the matrix element and $\mathcal{F}_{e^+}(\nu_{\text{rf}})$ is the calculated Fourier transform of the bunch field of the actually measured positron beam time structure.

The motion of particles inside the cell is diffusive and the probability to cross the cylinder ε_0 is independent on the z position along the cell axis. In addition, particles are injected into the cell center as ballistic particles. This leads to an additional probability ε_1 to cross the cylinder. The total probability $\varepsilon(z)$ to cross the cylinder around the beam depends on z.

Holding Field Uniformity

For a given setting of the magnetic holding field current the magnetic field inside the cell shows a non-uniformity of $\frac{\Delta B}{B} = \pm 1.5\%$ (see Figure 10). The resonance condition for a certain frequency of the perturbing field is met in a few points only. Thus the resonance condition appears locally inside the cell at different positions depending on the holding field current. The spectral density distribution $\rho(\Delta \nu(B))$ of the resonance depends on the position z inside the cell (see Equation 4).

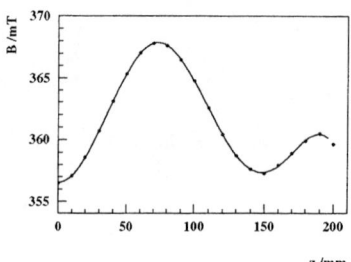

FIGURE 10. Magnetic field map of the holding field. The dots show the measurements of the field along the cell axis. The center of the cell is at $z = 0$ mm. The maxima at $z = 75$ mm and $z = 190$ mm correspond to the position of the magnet coils.

Particle Sampling

The BRP detects a sample of the particles near the center of the cell. The sampling probability for a particle is given by the probability for a particle at a certain z position to move to the center of the cell. It is linearly dependent on z. The density distribution $n(z)$ of the target gas is given by the condition $n = \Phi C$, where Φ is particle flux through the cell and C is the conductance of the cell. This again shows a linear dependence on z. Their product is the total weight function $S(z)$. It gives the sensitivity of the BRP to detect a change in the proton polarization depending on the position inside the cell.

RESULTS

For polarized gas the measured polarization in the BRP is given by

$$P_{\text{meas}} = P_0 \cdot (1 - R), \qquad (6)$$

where R is a function describing the polarization loss measured by the BRP. The function R is calculated by multiplying Equation 5 with the weight functions $S(z)$ and $\varepsilon(z)$ and integrating along the cell and over all frequencies of the perturbing field

$$R = \int_0^\infty \int_{\text{cell}} W_{ab}(\nu_{\text{rf}}) \, \varepsilon(z) \, S(z) \, dz \, d\nu_{\text{rf}}. \qquad (7)$$

Fit parameters are the intrinsic resonance width and the probabilities ε_0 and ε_1 to pass the cylinder near the beam. The resulting function is plotted in Figure 11. It shows good agreement with the measured data.

The interaction time of the particles with the perturbing field can be calculated from the fitted intrinsic resonance width as

$$\Gamma_E = 5.2 \cdot 10^{-7} \text{eV} \Rightarrow \tau = \frac{h}{\Gamma_E} = 8 \cdot 10^{-7} s.$$

Assuming thermal velocity of the atoms at 100 K this gives an interaction path length of 0.8 mm much larger than the average beam diameter of 0.15 mm, but much smaller than the minimum cell diameter of 100 mm.

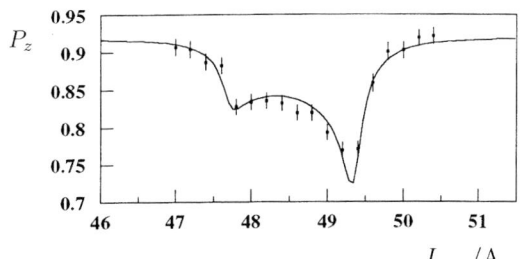

FIGURE 11. Measured polarization versus magnetic field for the (12) resonance at the 62nd beam harmonic. The solid line represents the calculated depolarization due to the resonance according to Equations 6 and 7. ($B = 7.42 \cdot I_{mag} \cdot \frac{mT}{A}$).

CONCLUSION

Beam induced resonances at HERA have been observed for the first time. The physics of the resonances is discussed, the measured data are well fitted by a model based on the cell geometry and three parameters. The quality of these data also shows that the BRP works as a versatile and reliable tool.

The longitudinal magnetic field of the HERMES target is sufficiently uniform to operate in between two resonances. At the working point at a high magnetic field of $B = 335\,\mathrm{mT}$ no depolarizing effects are observed, in agreement with the model discussed above.

REFERENCES

1. E.R. Kinney, "Simulations of the Atomic Polarization and Density in the HERMES Polarized Internal Target", HERMES Internal Report 3/90, (1991)
2. E.R. Kinney, Proc. Pol. Gas Targets and Storage Rings, Heidelberg 1991, pp. 155
3. E.R. Kinney, Proc. Pol. Ion Sources and Pol. Gas Targets, Madison 1993, pp. 78
4. B. Braun, PHD-thesis, München 1995
5. J.A. Stewart, contribution to this workshop
6. F. Stock, Proc. Pol. Beams and Pol. Gas Targets, Cologne 1995, pp. 260
7. B. Braun, contribution to this workshop
8. J.A. Stewart, Proc. Pol. Beams and Pol. Gas Targets, Cologne 1995, pp. 408

Novel Polarization Measurements at IUCF: Recent Work and Future Plans of the PINTEX Collaboration

Paul V. Pancella* for the PINTEX[1] collaboration

Western Michigan University, Kalamazoo, Michigan 49008

Abstract. This talk is a general overview of several recent measurements made by our group using the Wisconsin ABS target and the IUCF Cooler ring. The target and its operation have been described by Frank Rathmann in another talk at this workshop. We have used this facility to study beam depolarization by measuring polarization lifetimes under various conditions. We have used the ramping capability to transport calibration standards to different beam energies, and to increase average luminosity, thereby decreasing systematic and statistical uncertainties in our results. We are currently using spin observables to sort out the contributions of various partial waves to single pion production in proton-proton collisions. Finally, our future plans include a study of the three-body force using pd breakup, and an investigation of whether molecules made from polarized atoms retain any of their polarization.

INTRODUCTION

The PINTEX collaboration (**P**olarized **IN**ternal **T**arget **EX**periments) formed in the last few years to exploit new capabilities available at the IUCF Cooler ring in Bloomington, Indiana. With the installation of the Wisconsin ABS target, it is now possible to do nuclear physics measurements with stored, cooled beams of polarized protons or deuterons on pure polarized targets of hydrogen or deuterium gas. The first measurements of spin correlation parameters in pp elastic scattering [1] have shown that very precise measurements of spin-spin observables are possible with this system.

---

[1] H. O. Meyer, R. E. Pollock, W. A. Dezarn, J. Doskow, M. Dzemidzic, J. H. Hardie, B. v. Przewoski, T. Rinckel, F. Sperisen, and M. Wolanski; Indiana University Cyclotron Facility, Bloomington, IN 47408; W. Haeberli, B. Lorentz, P. Quin, F. Rathmann, B. Schwartz, and T. Wise; University of Wisconsin–Madison, WI 53706; W. W. Daehnick, R. Flammang, D. Tedeschi; University of Pittsburgh, Pittsburgh, PA 15260; P. B. Ugorowski, Western Michigan University.

After a brief summary of the capabilities of the beam, target, and detector systems, the bulk of this talk will concentrate on measurements we have made recently, and some which are currently underway. With the help of the polarized target, we can measure beam polarization with reasonable accuracy in less than a minute. This has allowed us to trace the decay rate of beam polarization as a function of betatron tune in the vicinity of an intrinsic depolarizing resonance. Speculation on the actual *mechanism* of depolarization has been tested by investigating the effect of increasing internal target thickness on the polarization lifetime.

The ability to accelerate *and* decelerate the stored beam, as well as to reverse the polarization direction without significant loss of polarization in either case, is very useful for reducing uncertainties in nuclear physics experiments involving spin observables. In particular, calibration standards can now be transported to any energy in the Cooler range with high precision. We are in the process of extending our earlier studies of single pion production near threshold a little higher in energy. The use of polarization observables will allow us to sort out the contributions from additional partial waves.

The last part of this talk concerns some of our plans for the immediate future. These include a study of deuteron breakup caused by interaction with a proton. Calculations indicate that certain polarization observables in certain configurations of phase space are very sensitive to the three-body nuclear force. In another approved experiment, we plan to study the polarization of hydrogen and deuterium *molecules*, formed by the recombination of polarized atoms from the ABS.

CAPABILITIES

Polarized beams

Polarized protons from the High Intensity Polarized Ion Source at IUCF are accelerated by the cyclotrons and stored in the Cooler at $T_{beam} = 200$ MeV. From there they can be ramped to any energy up to 450 MeV, and back down again, if desired, with almost no loss of polarization. Vertical polarization is easiest to store, and is typically 70%. The direction of the polarization of the stored beam can be reversed in a matter of seconds with a demonstrated efficiency of $0.984 \pm .004$ [2]. By adding solenoidal fields and compensating the tune of the machine, proton beams can be stored with similar total polarization and a sizable longitudinal component. The magnets which are presently available limit the longitudinal component to 0.95 of the total polarization at 200 MeV, and 0.80 at 375 MeV. The polarization direction of this beam can also be flipped rapidly. Fill rates of over $100\mu A$ per minute have become routine.

Deuterons with vector and tensor polarizations can also be stored, with a vertical quantization axis, although we have less experience with these beams. Early indications are that the vector and tensor components may have different depolarization rates.

Polarized targets

The polarized target now in use was designed and built at the University of Wisconsin–Madison. It has been well described in Frank Rathmann's report elsewhere in these proceedings ("The IUCF Polarized Gas Target"). It can provide protons with a polarization of 0.75 in any direction. The guide fields used are relatively weak ($\leq$1mT) and are compensated to make their effect on the beam position almost undetectable. When the beam is well positioned, the target in its windowless cell is extremely pure, ie., nearly all of the scattering is from target atoms. The best stable thickness achieved is about $4 \times 10^{13}/\text{cm}^2$. Combined with the average stored beam current, this yields an average luminosity of about $10^{29}/(\text{s·cm}^2)$. The target has very little effect on the beam storage lifetime. Relatively minor modifications are required to produce polarized deuterons with complete flexibility in state selection.

RECENT MEASUREMENTS

Beam Depolarization

The polarized proton target allows us to make useful measurements of the beam polarization in as little as 30 seconds. This is because we have very accurate measurements of the spin correlation coefficients A_{xx} and A_{yy} for 200 MeV pp elastic scattering. In an angular range accessible to our detector system (5° to 43° in the lab) the quantity $A_{xx} - A_{yy}$ takes a rather large average value, -1.475. By reversing the horizontally transverse target polarization every two seconds, and storing a few hundred μA of beam, yields are sufficient to do a cross-ratio type measurement of the vertical beam polarization. [3]

This has allowed us to study the polarization lifetime of a stored beam in detail. For previous experiments, the synchrotron was tuned in such a way that no loss of polarization was detectable during the storage times which were used (15 minutes or so). However, various depolarization resonances in circular machines are known to exist. We chose to vary the betatron tune near one of the so-called intrinsic resonances, a place in tune space where the betatron oscillations are in phase with the spin precession of the protons. For each value of the tune, the beam was stored for a time comparable to the polarization lifetime, and the polarization was measured every 30 seconds. These cycles were repeated to get sufficient statistics. The polarization as a function of time was fitted to a decaying exponential to get the lifetime parameter, with an error estimate determined by the quality of the fit. Figure 1 shows the results of these fits for several values of the betatron tune, along with some simple model calculations.

Simple theoretical models of depolarization require some mechanism to incoherently mix the phase space of the particles forming the stored beam. It is natural to assume that scattering from an internal target or residual gas could provide such a mechanism. If this were the case, one would expect that a thicker target would decrease both the beam current lifetime and its polarization lifetime. Quantitatively, the simple models we have used predict that these two lifetimes should remain in the same ratio. Figure 2 is a dramatic illustration that this hypothesis is not correct. We chose a tune where the polarization lifetime was approximately 1000 seconds. Then in the middle of each cycle, we turned on an internal nitrogen target at another location in the ring. This additional target thickness was sufficient to reduce the beam current lifetime by a factor of four. Several runs with different tune values failed to show *any* effect of target thickness on beam polarization lifetime. It appears that some other as yet unidentified mechanism besides scattering from target gas dominates the depolarization process.

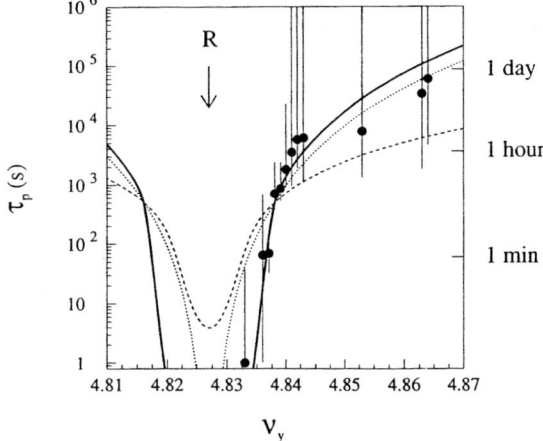

FIGURE 1. Beam polarization lifetime τ_p as a function of betatron tune ν_y. The dotted line is a model based on tune mixing only, caused by scattering. The solid line is the same folded with a Gaussian distribution of tunes among the beam particles. The dashed line is a model based on mixing of betatron amplitudes only. The value of ν_y at resonance (R) is fixed at 4.872, a value predicted independently of this measurement. The strength or width of the resonance is also fixed, leaving only one free scale parameter to fit for each curve.

Transporting Calibration Standards

Since a perfectly symmetric detector is impossible to realize in practice, high-precision measurements of spin observables benefit from calibration against reactions of known asymmetry the same way that cross section measurements benefit by direct comparison with reactions of known cross section. Such calibration standards are more difficult to come by in the spin realm. In the course of our early spin measurements, we have developed a technique for transporting such standards from energies where they exist (from previous, careful measurements) to new, generally higher energies. [4]

At the root of this process is the technical feat of accelerating *and* decelerating the polarized beam without appreciable loss of polarization. Once this is accomplished, a reaction of known asymmetry can be measured at one energy, then in the same cycle the beam energy increased, and the asymmetry measured at a new energy where it is not so well known. This measurement is made with the assumption that the beam polarization did not change during the energy ramp. This assumption can be checked by reducing the energy back to the original calibration point in the same cycle, and measuring again. If this latter measurement indicates a small reduction in beam polarization, then the average of the polarization before and after the energy changes can be used for the new (higher energy) asymmetry measurement, with an error given by half the change in beam polarization.

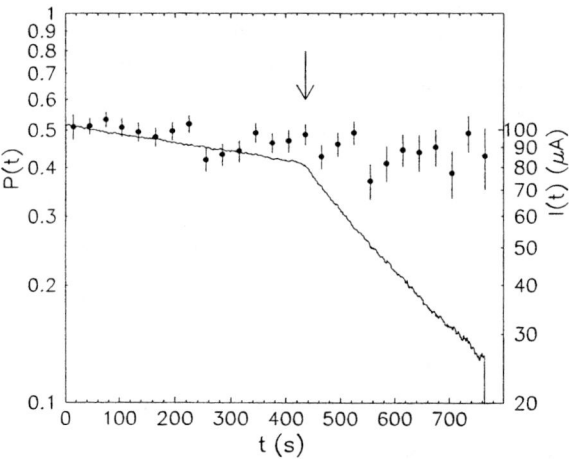

FIGURE 2. Beam current (solid line, right-hand scale) and polarization (points, left-hand scale) as a function of time during a typical run for which the tune $\nu_y = 4.842$. The arrow indicates the time during each cycle at which an additional nitrogen target was turned on. The reduction in beam lifetime by a factor of four is clearly seen, with no corresponding effect on the beam polarization.

Figure 3 shows this difference in beam polarization resulting from ramps up and down to various energies. The small size of this difference (0.022 out of about 0.7) indicates the relative uncertainties in the new standards. By using a polarized target as well as a polarized beam, and a reaction like pp scattering where the particles are identical, we can also determine the beam polarization directly at the higher energy, based on the well-established assumption that the target polarization is stable. Figure 4 shows the results of this measurement for several energies, indicating that at this stage of development, most of the polarization loss occurs on the ramps which reduce the beam energy.

This technique has ramifications for statistical as well as systematic uncertainties. It is desirable to alternate the direction of the beam spin fairly often to reduce systematic errors. Without the ability to flip the spin of the stored beam, one would normally store alternate spin states on each fill cycle of the storage ring. Since our polarized targets are so thin, this inevitably means discarding a lot of beam at the end of every cycle, so as not to dilute the polarization of the new beam injected with opposite spin. However, by ramping back down to the injection energy at the end of each cycle (preserving most of the beam polarization) and flipping the spin of the stored beam an odd number of times during each cycle, we have more than doubled our statistical yields, simply by not throwing away the old beam. [2]

Single Pion Production

Our current project is a followup to an earlier successful study of pp→ppπ° very close to threshold. In this reaction, there is an intersting range of excitation energies so close to threshold that only one (non-resonant) partial

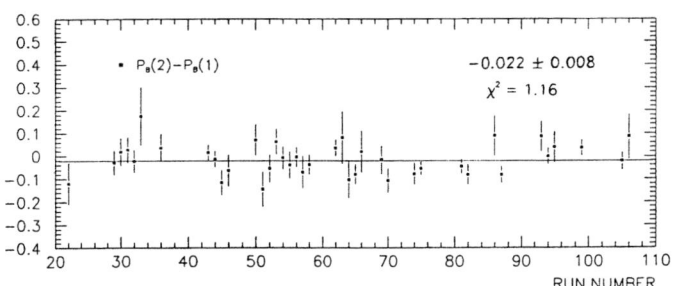

FIGURE 3. Beam polarization after ramping up and down minus the beam polarization before energy ramps for several runs. Beam polarization is measured at 200 MeV in every case, where the asymmetry in pp scattering is well-known. The intermediate energies vary from 250 to 450 MeV. Only statistical errors are shown. Also shown is the best fit constant difference.

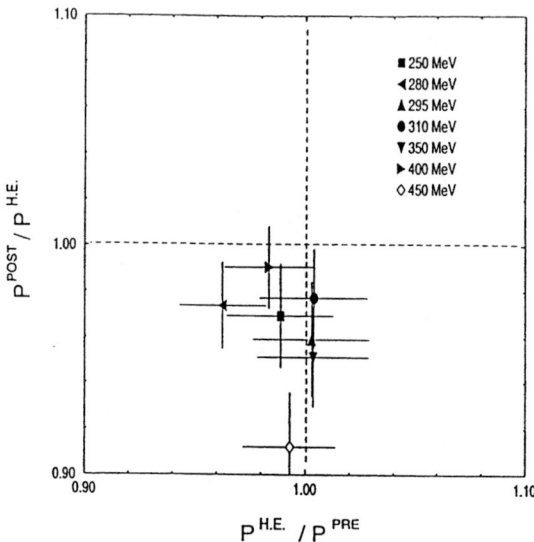

FIGURE 4. Ratio of the beam polarization before and after each of the two energy ramps, for the ramp up (horizontal axis) and down (vertical axis). Only statistical errors are shown. The data are grouped with respect to the energy after ramping up (see inset). A ratio of less than one indicates polarization loss. For most energies, the greatest loss occurred during the down ramping.

wave can contribute to the reaction amplitude. Thus it was possible to confront theoretical calculations directly with only total cross section data, with interesting results. [5]

By employing a polarized beam and a polarized target in this same reaction, we can continue this study at somewhat higher energies with similar theoretical impact. This is because a "spin sorting" technique allows us to determine the relative contributions of individual partial waves in the energy regime where only a few of them contribute. The appropriate combinations of spin observables take on different values for the partial waves with different quantum numbers, so that they can be separated without the difficulty of fitting angular distributions of limited resolution.

Specifically, certain combinations of the total cross section, $\Delta\sigma_L$ and $\Delta\sigma_T$ will be used to unambiguously measure the fraction of singlet contribution relative to the triplet, and to separate the three possible triplet P (initial pp orbital angular momentum) contributions. $\Delta\sigma_L$ is the cross section difference between parallel and anti-parallel spins when both beam and target are polarized longitudinally, while $\Delta\sigma_T$ is the same difference for transverse spins. This requires only total cross section measurements, accomplished in this case by measuring the vector momenta of both protons in the final state, for which this detector has good phase space coverage up to nearly 400 MeV beam kinetic energy. The pnπ^+ final state will be measured simultaneously, by employing a large-area neutron hodoscope, with similar physics goals. The required degree of longitudinal beam polarization has been demonstrated.

FUTURE PLANS

Three-body Force in pd Breakup

Future plans include an approved experiment to study the breakup of polarized deuterons induced by polarized protons. Faddeev calculations show that some observables are very sensitive to contributions of a three-body nuclear force. We will use polarized protons and polarized deuterons in both normal and reversed kinematics, ie., H(d,2p)n at 280 MeV and D(p,2p)n at 140 MeV, to access various areas of final state phase space with the same energy in the center of mass system. The calculations show that certain spin correlation coefficients as a function of scattering angle, in certain final state configurations, show a large effect depending on whether or not a three-body nuclear force is included. Other configurations are almost completely insensitive to the inclusion of a three-body force. We will make measurements in configurations of *both* types in order to test these theories.

Polarized Molecules

It has long been assumed that the recombination of polarized atoms of hydrogen or deuterium must result in the loss of nuclear polarization in a target cell. If recombined molecules remained polarized, it would raise many interesting practical possibilities beyond the theoretical interest. Recent work by van den Brand et al. [6] has suggested that this is indeed the case for deuterium molecules.

By modifying our existing target cell, we will provide a target with a variable amount of recombination. Since atoms and molecules exhibit conductances which are significantly different in the density regime where we operate, we will be able to measure the degree of recombination in all cases simply by measuring the target thickness and knowing the flow rate of atoms into this open-ended cell. Our existing detector system will then be able to measure the nuclear target polarization very accurately. We will be able to do this for both hydrogen and deuterium targets. The only major addition required is a pair of superconducting coils to provide a longitudinal holding field stronger than the critical field. This is considered necessary to decouple the nuclear spins from the electron spins, the latter being of course anti-parallel in the molecule.

REFERENCES

1. W. Haeberli, et al., *Phys. Rev.* **C55**, 597 (1997).
2. B. von Przewoski, et al., *Rev. Sci. Inst.* **66**, 12 (1996).
3. H. O. Meyer, et al., *Phys. Rev.* **E56**, 3578 (1997).
4. R. E. Pollock, et al., *Phys. Rev.* **E55**, 706 (1997).
5. H. O. Meyer, et al., *Nucl. Phys.* **A539**, 633 (1992).
6. J. F. J. van den Brand, et al., *Phys. Rev. Lett.* **78**, 1235 (1997).

Beam Polarimetery at HERA

W. Lorenzon

(on behalf of the HERMES collaboration and the HERA polarimeter group)

Randall Laboratory of Physics, University of Michigan, Ann Arbor, MI 48109-1120, USA
and
Deutsches Elektronen-Synchrotron, Notkestrasse 85, 22603 Hamburg, Germany

Abstract. The polarization of the 27.5 GeV electron/positron beam of the HERA *ep* collider ring is routinely measured with two independent Compton polarimeters. The transverse Compton polarimeter was developed as a tool for the HERA operators to tune the machine for high beam polarization. Currently it is also used to provide reliable beam polarization measurements for the HERMES experiment which investigates the spin structure of the nucleon. An additional Compton polarimeter, which was commissioned over the last twelve months, provides an independent measurement of the longitudinal beam polarization between the spin rotators at the HERMES experiment. The operation of the two polarimeters is presented and compared to each other.

INTRODUCTION

In high energy storage rings the electron (positron) spin can become transversely polarized through the Sokolov-Ternov effect [1]: the synchrotron radiation process contains a small asymmetric spin-flip amplitude that enhances the polarization state antiparallel (parallel) to the magnetic bending field. The polarization increases in time according to

$$P(t) = P_{\max}(1 - e^{-t/\tau}), \qquad (1)$$

where τ is the polarization build-up time and $P_{\max}$ is the asymptotic polarization.

In an ideal flat machine and in the absence of depolarizing effects the maximum polarization theoretically achievable (P_{ST}) is 0.924. In real machines, there are also depolarizing effects which can counteract the Sokolov-Ternov effect. The strength of these depolarizing mechanisms can be quantified with the time constant τ_D. These mechanisms compete with the Sokolov-Ternov build-up, and the value of the asymptotic polarization $P_{\max}$ is determined by the relative strengths of the two processes according to

$$P_{\max} = P_{\text{ST}} \frac{\frac{1}{\tau_{\text{ST}}}}{\frac{1}{\tau_{\text{ST}}} + \frac{1}{\tau_D}}. \tag{2}$$

The effective build-up time τ is also reduced to

$$\frac{1}{\tau} = \frac{1}{\tau_{\text{ST}}} + \frac{1}{\tau_D}. \tag{3}$$

The Sokolov-Ternov time constant, τ_{ST}, for HERA at 27.5 GeV is 37 min. The actual build-up time scales with $P_{\max}$ and will thus be shorter than τ_{ST}. This can be seen by rearranging Eq. 2 and 3, which gives

$$P_{\max} = \tau \left(\frac{P_{\text{ST}}}{\tau_{\text{ST}}}\right). \tag{4}$$

This important feature can be exploited to obtain an independent determination of $P_{\max}$ and thus a scale calibration of the polarization measurement from the actually observed build-up time. For example, $\tau = 20$ min would correspond to $P_{\max} = 0.50$. Therefore, assuming that the Sokolov-Ternov calculation of the build-up time is exact, the polarization scale can be calibrated using the characteristic rise-time behavior.

For a non-flat machine (i.e. a machine with spin rotators), however, Eq. 2 is not exact. There is an extra factor $(1+\delta)$ in the numerator. This can be seen by inspecting the Derbenev-Kondratenko formula [2] which generalizes the Sokolov-Ternov formula to take depolarizing effects into account. When the spin rotators are off, δ is so small that it can be neglected. With the spin rotators on, δ cannot be estimated reliably but must be determined experimentally.

HERA Spin Rotators

The HERMES experiment, located in the East section of the HERA ep collider ring, requires longitudinal beam polarization. This is accomplished with two spin rotators [3] located at the entrance and exit of the East straight section, that precess the spin direction from vertical to longitudinal at the HERMES target position. Currently, only HERMES uses spin rotators. The spin rotators for the H1 and ZEUS collider experiments will be installed in the 1999/2000 shutdown period.

The transverse beam polarization is measured in the HERA West section whereas the longitudinal beam polarization is measured near the HERMES target in the HERA East section. The degree of polarization is invariant over the entire ring; thus the two Compton polarimeters must measure the same polarization. A prominent feature of the HERA spin rotators is that they can be arranged to reverse the longitudinal spin direction [3]. This is an important tool for any experiment at HERA that takes data with polarized beams to reduce systematic uncertainties in the measured physics observables.

POLARIZATION MEASUREMENT

The method for measuring the beam polarization with either polarimeter installed in the HERA storage ring is based on the Compton scattering process. Here, we summarize briefly Compton scattering of circularly polarized laser light on a polarized electron beam, and describe how the transverse and longitudinal polarization of the beam is measured. The transverse polarimeter has been described in great detail in the past [4,5]; therefore, more emphasis will be put on the new longitudinal polarimeter.

Compton Scattering

Compton polarimeters utilize the spin-dependent cross section for Compton scattering of polarized photons on electrons. The differential Compton cross section can be written as a function of the initial electron and photon polarizations **P** and **S**. Since only vertical and longitudinal polarization components of the electron beam are measured, averaging over the horizontal component x leads to [4]

$$\frac{d\sigma_c}{d\Omega}(\mathbf{S},\mathbf{P}) = \frac{1}{2}r_0^2 \left(\frac{k_f}{k_i}\right)^2 [\Sigma_0(\theta) + S_1\Sigma_1(\theta,\phi) + S_3\{P_Y\Sigma_{2Y}(\theta,\phi) + P_Z\Sigma_{2Z}(\theta)\}], \tag{5}$$

with S_1 and S_3 the linear and circular components of the initial photon polarization, P_Y and P_Z the vertical and longitudinal components of the initial electron polarizations, r_0 the classical electron radius, and k_i, k_f the initial and final photon momenta in the electron rest frame. Σ_0 contributes to the unpolarized cross section, Σ_1 and Σ_{2Y} depend on the azimuthal scattering angle ϕ and are used for the vertical polarization measurement, and Σ_{2Z} is used for the longitudinal polarization measurement.

Measurement of Transverse Polarization

The transverse polarization (P_Y) is measured in a position sensitive calorimeter using the up-down asymmetries in the backscattered Compton photons. The "single-photon method" is used where the laser intensity is chosen such that the probability for multiple Compton photons per bunch crossing is about 1% and the energy and position of each single photon is used for the analysis. This is achieved with a 10 W continuous beam of 2.4 eV photons from an argon-ion laser. The laser beam is transported through remotely controlled mirrors and lenses over about 200 m to the $e\gamma$ interaction point in the HERA tunnel. The photon polarization is switched with a frequency of 84 Hz by a Pockels cell. The backscattered Compton photons are detected in a tungsten/scintillator sandwich calorimeter about 65 m from the interaction point. The electron beam optics at the interaction point is such that the Compton photons converge to a vertical focus near the location of the calorimeter. This choice, together with the very small electron beam emittance, make possible the measurement of the Compton photon angle to a precision of a few μrad. The calorimeter is split into an optically decoupled upper and lower part. The energy of the incoming photon is the sum of the energies of the two halves, $E_\gamma = E_u + E_d$, and the vertical distance of the Compton photons to the center of the calorimeter is measured using the asymmetry of the energies $\eta(y) = (E_u - E_d)/(E_u + E_d)$ [4].

There are two ways to measure the vertical polarization [4]:

1. The vertical polarization can be obtained from measurements of the asymmetry

$$\mathcal{A}(y, E_\gamma) = \Delta S_1 \Sigma_1 + \Delta S_3 P_Y \Sigma_{2Y}, \quad (6)$$

with e.g. $\Delta S_1 = \frac{1}{2}(S_{1,L} - S_{1,R})$, where $S_{1,L}$ and $S_{1,R}$ are the degrees of linear polarization of the laser light. Note that the asymmetry measurement not only depends on the circular, but also the linear component of the laser light.

2. The polarization P_Y can also be obtained from the shift of the mean vertical positions $\langle y \rangle$ measured with left and right circularly polarized light

$$\Delta \langle y \rangle (E_\gamma) = \frac{1}{2}(\langle y \rangle_L - \langle y \rangle_R) = P_Y \Delta S_3 \Pi(E_\gamma). \quad (7)$$

The analyzing power $\Pi(E_\gamma) \leq 170\mu m$ is equal to the shift measured when $\Delta S_3 P_Y = 1$. Tis is the standard method used to measure P_Y.

The statistical error in the transverse polarization δP_Y obtained in one minute is about 0.01 to 0.02 depending whether the data is taken in the beginning or towards the end of a fill. The beam currents can vary between 40 to 10 mA during a fill.

Measurement of Longitudinal Polarization

In contrast to the small spatial asymmetry measured with the transverse calorimeter, the measurement of the longitudinal polarization (P_Z) is based on large asymmetries in the energy distributions of the backscattered photons, with $\mathcal{A}(E_\gamma) = \Delta S_3 P_Z \Sigma_{2Z}$.

The energy-dependent analyzing power Σ_{2Z} is shown in Fig. 1. The longitudinal polarization P_Z can be obtained from the measurement of the energy-dependent asymmetry under reversal of the laser photon helicity using the "single-photon method". However, because the HERMES gas target produces intense bremsstrahlung background at the calorimeter of the longitudinal polarimeter, the "multi-photon method" is preferred. In contrast to the "single-photon method" utilized in the transverse polarimeter, where the energy of every individual Compton photon is analyzed, the "multi-photon method" is based on a measurement of the total energy deposited in the detector by about 1000 Compton photons per bunch crossing. This is achieved with a frequency doubled, pulsed YAG laser. It produces 2.3 eV photons with a repetition rate between 0.1-100 Hz and pulse energies from 1 to 250 mJ. The pulsed laser beam is guided with remotely controlled mirrors and lenses in a 72 m long stainless steel vacuum pipe to the $e\gamma$ interaction point. The laser is pulsed at 100 Hz and a Pockels cell switches the photon polarization before each laser pulse.

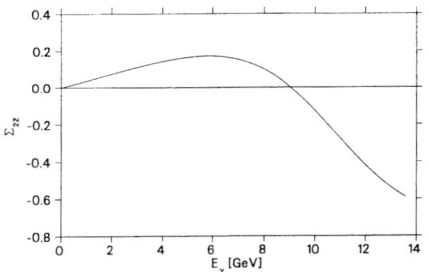

FIGURE 1. The energy-dependent analyzing power Σ_{2Z} plotted versus the photon energy E_γ for a 2.3 eV photon scattered off a 27.5 GeV electron. At the Compton edge, at 13.6 GeV, the asymmetry is close to -0.60.

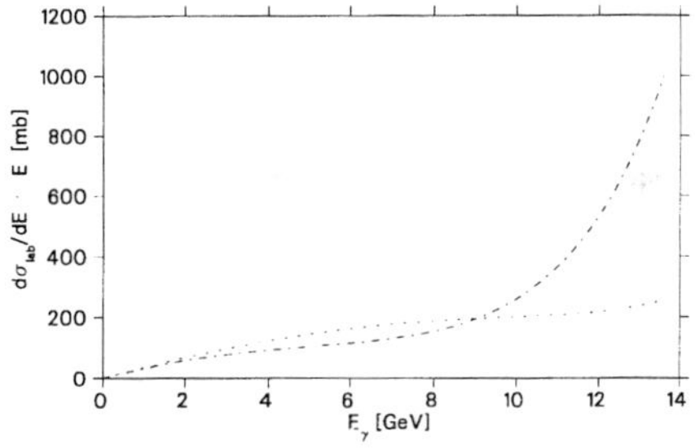

FIGURE 2. Energy weighted spectra of the scattered photons in the lab frame for scattering with longitudinally polarized electrons for the cases of $S_3 P_Z = +1$ (dotted curve) and $S_3 P_Z = -1$ (dash-dotted curve).

Due to the enormous kinematic boost from the beam (Lorentz factor is $E/m_e \approx 54,000$), most backscattered photons are contained in a narrow cone centered along the initial electron beam direction. They are detected in a calorimeter 54 m from the interaction point. It consists of four optically decoupled, radiation hard NaBi(WO$_4$) Cerenkov crystals, arranged in a 2 × 2 array. The energy of the incoming photons is the sum of the energies of all four crystals, and the 2 × 2 configuration allows a precise determination of their horizontal and vertical positions. The spatial distribution of the incoming photons are mainly determined by the beam optics; therefore, at the calorimeter position, the cone has a radius of 7 mm for photon energies above 500 MeV. Compton photons below that energy threshold contribute less than 0.5% to the asymmetry and can be neglected. Therefore, in the "multi-photon method" the longitudinal polarization is obtained from measurements of the energy weighted asymmetry

$$\mathcal{A}(\Sigma E_\gamma) = \Delta S_3 P_Z \Sigma_{Z_{lr}}, \qquad (8)$$

with

$$\Sigma_{Z_{lr}} = \frac{\Sigma_l - \Sigma_r}{\Sigma_l + \Sigma_r}, \quad \text{and} \quad \Sigma_i = \int_{E_{min}}^{E_{max}} \left(\frac{d\sigma}{dE}\right)_i \cdot E \cdot dE, \quad \text{where} \quad i = l, r.$$

The energy weighted analyzing power $\Sigma_{Z_{lr}}$ is 0.184. The energy weighted spectra with longitudinal electron polarization $P_Z = 1$ and photon polarization $S_3 = \pm 1$ are shown in Fig. 2. Assuming 1000 backscattered Compton photons are produced per laser pulse, 6820 GeV is deposited in the calorimeter for an unpolarized electron beam, since the average energy deposited per Compton photon is 6.8 GeV. For the two polarization states, one measures two peaks separated by 1770 GeV for a 0.70 polarized electron beam, as shown in Fig. 3. (In comparison, for the transverse polarimeter, a spatial shift of 100μ has to be detected). The measured energy peak positions depend on the luminosity, and the electron and laser polarizations, P_Z and S_3.

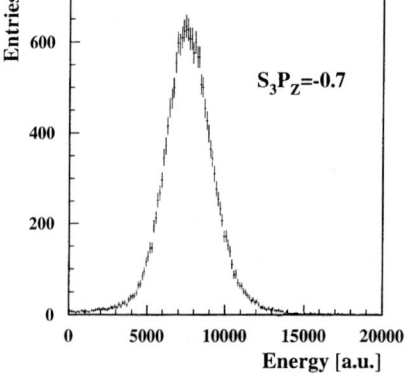
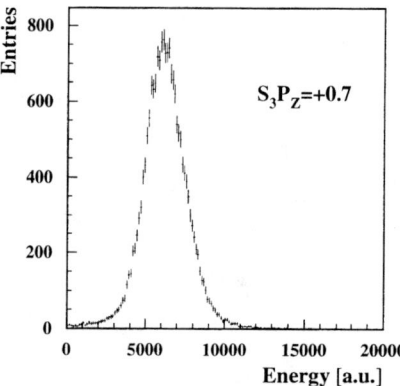

FIGURE 3. Histogram for 11,000 events collected in "multi-photon mode". The polarization is determined from the centroid difference of these two peaks, with an extracted value of 70%. The two peaks are the results of changing the laser helicity.

The polarization is determined from the centroid difference of these two peaks divided by the energy weighted analyzing power (0.184) under the assumption that $\Delta S_3 = 1$. Note that it is not necessary to know the absolute energy scale, it is sufficient to know that the energy scale is linear. This can be checked by varying the laser power.

To achieve a statistical error in δP_Z of 0.01, a measurement takes about one minute. The laser power can be adjusted easily to keep the luminosity constant. The background is measured between each laser pulse and is negligible in the "multi-photon method".

ABSOLUTE POLARIZATION SCALE

Within a factor of two, both Compton polarimeters deliver statistical errors in the polarization of about 0.01 per minute. The main goal for building the newly commissioned longitudinal polarimeter was to have a second, independent measurement of the beam polarization, and to reduce the overall systematic uncertainty in the absolute polarization value. Experience has shown, each polarimeter is available for reliable polarization measurement about 90% of the HERMES data production time. The overall efficiency to deliver valid polarization values to HERMES is close to 100%.

As discussed in Section II.B and II.C, the polarization of the HERA beam can be determined from basic principles by starting from the spin-dependent Compton cross section. This requires a very detailed understanding of all the components involved as well as an accurate determination of the analyzing power, $\Pi(E_\gamma)$, for the transverse polarimeter, or the energy weighted analyzing power, $\Sigma_{Z_{lr}}$, for the longitudinal polarimeter, which are determined from Monte Carlo simulations.

A much more straightforward way of obtaining the absolute polarization scale is by using the rise-time calibration described in Section I.

Rise-Time Calibration

In 1994, rise-time curves were taken to get an absolute calibration of the transverse beam polarization. This was done with the spin rotators deactivated, i.e. with a flat machine. Analysis of those eleven rise-time curves which passed quality cuts indicated that the polarization scale predicted by Monte Carlo simulations resulted in values that were 6% too high. The systematic uncertainty associated with that scale calibration was 3.2%, a considerable improvement from the 9% systematic error assigned to the Monte Carlo simulations [4].

In 1997, after completion of the longitudinal polarimeter, another set of rise-time calibration runs were taken to compare the performance of the longitudinal and transverse polarimeters, and to study the effect of the spin rotators on the absolute polarization scale. One of these rise time curves, taken simultaneously with the transverse and longitudinal polarimeters, is shown in Fig. 4. There, the polarized beam was depolarized about six hours into the fill by activating a kicker magnet at a depolarizing resonance frequency. After about ten minutes this depolarizer was turned off and the polarization started to rise again. At the end of the rise-time curve the electron beam was depolarized again to check for stability in the beam conditions. In case of unstable machine conditions, the depolarizing frequency would have drifted. This was not the case for the curves shown. The agreement in the fitted polarization build-up time τ for both polarimeters is better than 1% which is not surprising since the polarization is expected to be invariant over the entire ring. Using τ, both polarimeters can be calibrated absolutely up to the factor $(1+\delta)$ discussed in Section I. The scale correction needed for the longitudinal polarimeter, which was already applied in Fig. 4, lowered the measured values by 20%. It is believed that this large number can be reduced in the near future after more detailed Monte Carlo studies and more extensive systematic studies are completed.

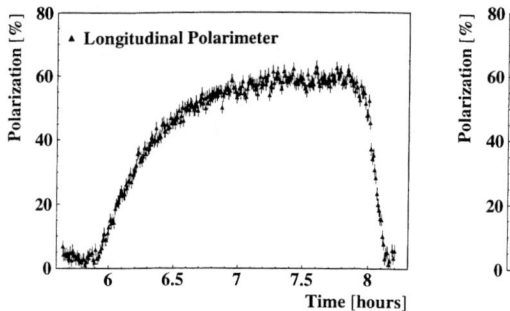 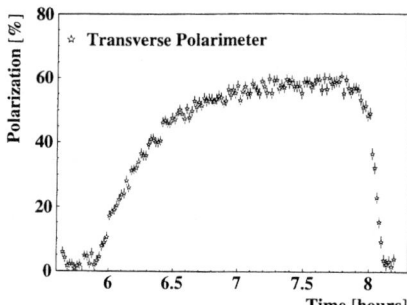

FIGURE 4. Rise-time curve taken simultaneously with the longitudinal and transverse polarimeters to compare build-up time and general performance of both polarimeters. At the beginning and the end of the spectrum, the beam was depolarized on purpose.

In 1997, the machine conditions were much more stable than in 1994. Therefore, new rise-time calibration data were taken successfully which yielded a factor 1.01 ± 0.01 as compared to the calibration taken in 1994. In order to improve the overall polarization scale uncertainty, another set of measurements has to be taken, again with the spin rotators turned off, therefore re-establishing a flat machine. If a similar accuracy can be obtained, then the absolute polarization scale is known to better than 2% and the unknown correction δ is known to the same precision. Current plans forsee that those measurements will take place in late fall 1997.

Polarization Uncertainties

The overall error in $\Delta P/P$ for the transverse polarimeter is shown in Table 1. The independent statistical and systematic errors are added in quadrature, whereas the scale uncertainty obtained from the rise-time calibration is added linearly. In order to reduce the overall error in the polarization measurement to the 3% level, rise-time studies with flat spin rotators have to be performed.

TABLE 1. Fractional errors in $\Delta P/P$ for transverse polarimeter.

	1995	1996	1997 expected
Statistical (10 min.)	0.8%	0.5%	0.5%
Light polarization	0.5%	0.5%	0.5%
Calorimeter Calibration	2.0%	0.5%	0.5%
Quadratic Sum	2.2%	0.9%	0.9%
Rise-time correction	3.2%	3.2%	$\sim 2\%$
Typical Uncertainty (linear Sum)	5.4%	4.1%	$\sim 3\%$

OUTLOOK

There is a very interesting feature of the longitudinal polarimeter. Due to the high backscattered photon rates (10^3) and the high repetition rate of the YAG laser (100 Hz), the polarization of individual bunches can be measured to 2% statistical precision in approximately twenty seconds. Thus in about ten minutes, the polarization of each bunch is known to 5%, as shown in Fig. 5. This enables us to study beam-beam interaction and beam orbit effects on the polarization. Not all the electron/positron bunches collide with proton bunches. Those so-called pilot bunches which do not collide can have significantly different polarization values than the bunches that do collide with protons. This has been very valuable information for the machine physicists.

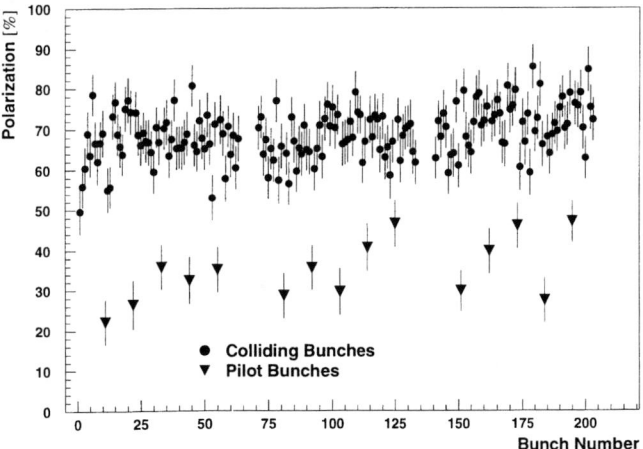

FIGURE 5. Single bunch polarization values for colliding and non-colliding (pilot) bunches are shown.

One possible explanation of this effect is that in fact colliding bunches have different tunes, and therefore different orbit corrections than pilot bunches due to beam-beam interactions. During polarization tuning by the machine operators the average beam polarization is optimized. This means that the polarization for colliding bunches is preferentially optimized, since the ratio of pilot to colliding bunches is about 12:1.

The ability to measure the polarization of individual bunches can also be used to better monitor the stability of the positron beam in conjunction with the concept of partial depolarization of the beam. In this concept only a fraction of the 189 filled bunches are depolarized, after all the bunches have reached asymptotic polarization values. Rise-time measurements can then be performed on these few depolarized bunches, while the rest of the bunches stay at constant levels. This scheme allows us to monitor the stability of the positron beam during rise-time curves, because any instability could be observed as a non-statistical change of the asymptotic polarization values. Stable beam conditions are essential to extract meaningful rise-time curves. More importantly, it also allows us to check the calibration of the polarization scale continuously by sacrificing e.g. one single bunch which can be depolarized every few hours.

ACKNOWLEDGEMENTS

I wish to thank my colleagues in the HERMES collaboration and the HERA polarimeter group. I acknowledge P. Schüler, H. Fischer and C.A. Miller for critical reading of the manuscript. The author's research is supported in part by the U.S. National Science Foundation, Nuclear Physics Division under grant No. PHY-9724838. I thank the University of Michigan and DESY for support while on research leave.

REFERENCES

1. A.A. Sokolov and I.M. Ternov, Sov. Phys. Doklady **8** (1964) 1203.
2. Y.S. Derbenev and A.M. Kondratenko, Sov. Phys. JETP 37, N°6 (1973) 968.
3. J. Buon and K. Steffen, Nucl. Instr. Meth. **A256** (1986) 248.
4. D.P. Barber et al., Nucl. Instr. Meth. **A329** (1993) 79.
5. D.P. Barber et al., Nucl. Instr. Meth. **A338** (1994) 166.

Radio-Frequency Polarimetry

Ya. S. Derbenev

Randall Laboratory of Physics
University of Michigan
Ann Arbor, MI 48109-1120, USA

Abstract. A method of fast non-destructive absolute spin monitoring for a bunched beam in an accelerator ring based on use of the RF techniques is considered. The coherent spin of the beam is driven by RF magnets in the spin echo regime. A passive superconducting resonator is proposed to respond to the flipping spin. A possibility is established to enhance the spin-related excitation of the resonator using the charge-resonator dipole interaction and spin-orbit coupling induced by the quadrupoles. It is shown that the spin impedance can be gauged via measurements of the beam dipole impedance. The noise demands are evaluated. Numerical examples are given.

INTRODUCTION

The idea of the radio-frequency method of spin monitoring for a polarized beam circulating in an accelerator or storage ring was proposed in recent years [1–3]. It is based on a consideration that, if the coherent spin of the beam is declined from the equilibrium direction, $\vec{n}$ (vertical or, in general case, periodical along the beam orbit), then the electromagnetic field of the beam "observed" by a cavity located at a point of the orbit has to have a modulation with periodicity of the spin precession in the accelerator, which is different from the periodicity of the particle orbital motion. Apparently, this signal is very small. However, if the spin tune spread, $\delta\nu$, is small enough, the beam current, J, is sufficiently high, and the quality Q_R of the cavity tuned in resonance with the spin free precession is also high enough, then the beam could excite the cavity resonance mode to a measurable level while an unpolarized beam would not be able to do this.

The attractiveness of the RF method is increased by the fact that its efficiency does not drop with the beam energy. It should be also noted that it is equally efficient for proton and electron beams although in the case of electron beams there are such well-developed and tested non-destructive methods as Compton backscattering [4] and "spin light" [5]. The most critical issue of the RF polarimetry is spin decoherency due to the spin tune spread.

Below we will discuss the principles and recent progress in the RF polarimetry concepts.

GENERAL SCHEME OF THE RF SPIN MEASUREMENT

RF arrangement for non-touching beam spin monitoring involves the following basic elements:

- the spin-resonance RF magnets to cause spin precession around a horizontal (rotating) axis, similarly to spin motion in NMR;

- a passive superconducting resonator (single or a few in a row) to be excited by the precessing coherent spin by mean of the electromagnetic interaction of the polarized beam with a resonance standing wave;

- monitoring RF circuit (a loop, filter, amplifier, scope) to determine the spin-related high frequency voltage accumulated in the resonator;

- optionally, the spin-orbital amplifier (SOA), i.e. RF quadrupoles can be introduced in the ring to induce the spin-correlated beam dipole moment for an enhancement of spin-related excitation of the resonator.

THE DYNAMIC BACKGROUNDS

There are shown in Table 1 the quasiclassical Hamiltonian and ground equations those have been used to calculate the spin-related effects in polarized beam dynamics and electrodynamics.

The Hamiltonian is simply combined as a sum of the conventional spinless Hamiltonian and spin term defined to reproduce the Thomas-BMT equations for spin [6–8]. This composition coincides with the Hamiltonian that can be derived from Dirac-Maxwell equations for a particle with an anomalous magnetic moment, when splitting the Dirac equation in two branches with positive and negative energies according to Foldi-Wouthuysen [9,10]. Thus, the treatment presented in Table 1 can be considered as adequate to the fundamental theory, complete, and sufficient for quasiclassical conditions.

Earlier, this approach was used [11] to reproduce the Sokolov-Ternov polarization effect for ultrarelativistic electrons; in this case, the contribution of the Lorentz force term in $\vec{W}$ is negligible. In contrary, this term dominates in case of ultrarelatistic particles ($\gamma G \gg 1$) in external fields, static or alternating.

TABLE 1. Hamiltonian and basic equations (c=1).

(1) Total Hamiltonian: $H = \Sigma(H_o + \vec{W} \cdot \vec{S}) + H_R$
(2) Spinless energy: $H_o = \sqrt{(\vec{p} - e\vec{A})^2 + m^2} + eA_o$
(3) Resonator energy: $H_R = \frac{1}{2}(P^2 + \omega_R^2 Q^2)$
(4) Spin precession velocity ($\gamma^{-2} = 1 - v^2$):
 $\vec{W} = -\frac{e}{m}[(G + \frac{1}{\gamma+1})(\vec{E} + \vec{v} \times \vec{B}) \times \vec{v} + \frac{1+G}{\gamma}(\vec{B}_v + \frac{1}{\gamma}\vec{B}_{tr})]$
(5) Electromagnetic field: $\vec{A} = \vec{A}_{ext} + Q(t)\vec{e}(\vec{r})$
 $\vec{E} = -\frac{\partial \vec{A}}{\partial t} - \vec{\nabla}A_0; \vec{B} = \vec{\nabla} \times \vec{A}; \int \vec{e}^2 d^3 r = 4\pi$
(6) Non-zeroth Poisson's brackets: $\{\hat{p}_\alpha, r_\alpha\} = 1; \{P, Q\} = 1; \{S_\alpha, S_\beta\} = -\epsilon_{\alpha\beta\gamma}S_\gamma$
(7) Orbit equations ($\vec{p} \equiv \hat{\vec{p}} - e\vec{A}, \vec{v} \equiv \vec{p}/\sqrt{p^2 + m^2}$):
 $\dot{r}_\alpha = v_\alpha + \frac{\partial}{\partial P_\alpha}(\vec{W}\vec{S}); \dot{p}_\alpha = -e\frac{dA_\alpha}{dt} - e\vec{v}\frac{\partial \vec{A}}{\partial r_\alpha} - \frac{\partial}{\partial r_\alpha}\vec{W}\vec{S}$
(8) Resonator field equations: $\dot{Q} = P + \frac{\partial}{\partial P}\Sigma(\vec{W}\vec{S}); \dot{P} = -\omega_R^2 Q + e\vec{v}\vec{e} - \frac{\partial}{\partial Q}\Sigma(\vec{W}\vec{S})$
(9) Thomas-BMT equation: $\dot{\vec{S}} = \vec{W} \times \vec{S}$
Here the symbol Σ means summation on particles of a polarized beam.

SPIN DRIVER

In the earlier RF polarimetry concept [1], the free precession of the coherent spin in magnetic field of an accelerator ring was supposed to be excited (by a pulse RF magnet) to effect a superconducting cavity. Then, the maximum time of RF voltage accumulation in the cavity was limited by the spin tune spread, $\delta\nu_o$, where ν_o is number of spin precessions per particle turn in a ring. The $\delta\nu_o$ value is defined by the beam emittances and magnetic lattice imperfections, it grows rapidly with particle energy. It is relatively small in electron storage rings (after averaging on energy oscillation), thank to a small value of the vertical beam emittance. The $\delta\nu_o$ value becomes essentially reduced in a ring with Siberian snakes [12]. However, it seems difficult to reduce $\delta\nu_o$ in hadron colliders to a value less than $(1-0.3) \cdot 10^{-3}$. The RF polarimetry with this coherent spin quality value seems possible but not highly efficient.

A simple step can improve the spin quality: apply the spin-resonance RF field of a relatively large strength $\varepsilon \equiv GeBl/2\pi m \gg \delta\nu_o$ and let the spin precess around this field during a possibly long time (Fig. 1). Then the effective spin tune spread becomes reduced by a factor $\sim \delta\nu_o/2\varepsilon$, with the correspondent increase of the spin coherency time, τ. Optionally, a SC resonator tuned to frequency $(k + \varepsilon)\omega_o$, or $(k + \nu_o + \varepsilon)\omega_o$, can be choosen. Note, that the reduction of the spin tune shift and spread also helps to stabilize the spin against the weak depolarizing resonances.

Next, consider the spin echo regime when the RF field changes the phase on $\pi/2$ at the moments when the spin is close to the vertical direction, twice a single precession (Fig. 2). This cycle can be perfectly synchronized with the beam revolution. The spin dynamics here is similar to spin motion in a

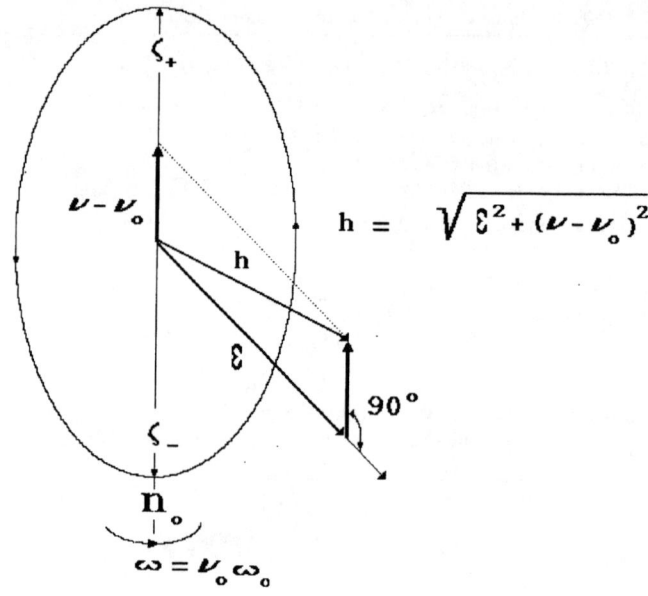

FIGURE 1. Reduction of the spin tune dispersion in resonance RF field.

ring with two Siberian snakes [13]. Thus, the spin precession in the RF field becomes stable (periodical of a period $T = 2\pi(2q + 1)/\omega_o\varepsilon$, where q is an integer) motion with no phase divergence: the free precession now moved in the plane transverse to the periodical spin, with effective tune $\hat{\nu} = 1/2\omega_o T$ and tune divergence $\triangle\hat{\nu} \sim (\triangle\nu_o)^2/2\varepsilon$, at $q \ll (\varepsilon/\triangle\nu_o)^2$.

Now, we nominate the vertical spin oscillating (with infinite quality) with frequency $(\omega_o\bar{\varepsilon}/2\pi) \equiv (2q + 1)/T$ to be in charge to excite the SC resonator. This frequency effectively becomes a characteristic spin tune, inspite it is a harmonic (high order, $2q + 1$) of the RF field phase switching frequency. The demands related to the parasitic beam excitation (of period T) because of the RF field misalignments are discussed in section 7.

SPIN-INDUCED CHARGE-RESONATOR INTERACTION

The precessing or flipping coherent spin will effect the beam orbit via the spin-orbit interaction in quadrupoles according to the general orbital equations shown in Table 1; the reduced equations are given in Table 2.

Here $\vec{\zeta}(t)$ is the coherent beam polarization, n is the focusing field index, and g_x, g_y are the factors due to the coherent beam dipole interaction. The spin-

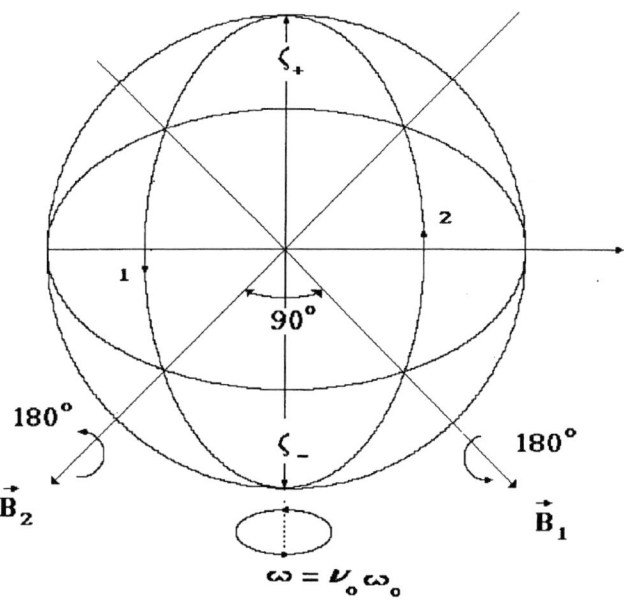

FIGURE 2. Spin motion schematic in RF spin echo regime.

correlated beam dipole moment, $\hat{\vec{d}}$, contributes to the resonance excitation of the SC cavity, effectively renormalizing the spin-related interaction. The effect grows near the spin-orbit resonances $k \pm \nu_{sp} = \pm\nu_b$, where ν_b is one of two tunes of beam coherent oscillations; this mechanism can be used to raise the RF polarimeter efficiency. Use of this effect is complicated by the fact that it alternates with energy. But, observation of the resonance behavior of the spin-orbital effect leads to idea to introduce the RF quadrupoles of the bridge frequency $\omega_q = \omega_R - k\omega_o \pm \omega_{sp}$, at $\omega_R \approx (k_1 + \nu_b)\omega_o$. The RF quadrupoles can be skew, if needed. With the RF spin-orbital amplification (SOA), the immediate spin-resonator interaction becomes non-resonant, hence, negligible. The SOA is limited by the depolarization near spin-orbital resonances due to the beam emittances. Other limiting factors are betatron tune spread and (or) beam damping, coherent and incoherent. Estimations show that the maximum gain varies from a factor of about 50 in high energy hadron rings to $\sim 10^3$ in $e^\pm$ storage rings and low energy proton synchrotrons with cooled beams, although one can meet a lower practical limit because of necessity to control the beam tunes with a high precision.

TABLE 2. RF polarimetry equations at $\gamma G \gg 1$.

(1) $\hat{\vec{d}} = GNe\hbar\hat{\vec{\rho}}/2m; \quad \hat{\vec{\rho}} = (\hat{x}, \hat{y}) \equiv \hat{g}\vec{\zeta}(t)$

(2) $\ddot{\hat{y}} + 2\lambda_y \dot{\hat{y}} + (n + g_y)\hat{y} = \frac{e}{\gamma m} \frac{\partial \vec{B}_{1r}}{\partial y} \vec{\zeta}$

(3) $\ddot{\hat{x}} + 2\lambda_x \dot{\hat{x}} + (K^2 - n + g_x)\hat{x} = \frac{e}{\gamma m} \frac{\partial \vec{B}_{1r}}{\partial x} \vec{\zeta}$

(4) Spin-induced charge-resonator interaction:

$(-e\vec{v}\vec{A})_{sp} = -Q\vec{d}\frac{\partial \vec{e}_v}{\partial \vec{\rho}}$

(5) Resonator equation:

$\ddot{Q} + 2\lambda_R \dot{Q} + \omega_R^2 Q = -\frac{NeG\hbar}{2m}(\vec{v} \times \vec{\nabla}\vec{e}_v + \vec{\nabla}\vec{e}_v \cdot \hat{g})\vec{\zeta}(t)$

SPIN IMPEDANCE

<u>Spin responding RF modes.</u> There are three basic types of standing waves to interact with spin:

a) solenoidal TE_{011} mode of an axial resonator. Use of this mode may be reasonable at $\gamma G \leq 1$.

b) magnetic dipole TM_{110} mode.

c) standing (TEM $\frac{\lambda}{4}$) wave of feeder type (Fig. 3).

In the area $\gamma G \gg 1$, spin interaction with modes b) and c) does not drop with energy. An additional advantage of this modes (at all energies) is that they also interact with the beam dipole motion (excited by the coherent spin). The specific feature of the feeder mode is that the transverse size of the resonator, hence, the interaction strength, is not limited by the wave length (i.e., by the bunch length).

<u>Spin impedance gauging.</u> Spin impedance can be precisely calculated by mapping the RF modes. At high energies ($\gamma G \gg 1$) and short bunches ($\ell \ll \beta$, where β is the beam focussing parameter) it also can be gauged via measurement of the beam dipole impedance [14]. To prove this possibility, note that the Lorentz force in $\vec{W}$ can be represented as $\vec{F} = -e(d\vec{A}/dt) - e\vec{\nabla}(\vec{v}\vec{A})$; since the spin does not precess in the resonator, the first term can be omitted. Then, one can see from the comparison with the charge interaction Hamiltonian, $-e\vec{v}\vec{A}$, that the ratio between spin and beam dipole impedances is simply $G + \frac{1}{\gamma}$. The final equation for the resonator field is shown in Table 2.

<u>The accumulated HF voltage.</u> Table 3 illustrates the estimations of the accumulation rate and maximum value of spin-related voltage in the feeder type $\frac{\lambda}{4}$ resonator based on two simple formulas: $\dot{V} = 2\pi\sqrt{2}(\hbar/m)(J/\ell d)(G + \frac{1}{\gamma})(1 + g)$, and $V_{max} = \dot{V}/\lambda_R$, at polarization degree $\zeta = 1$, $d = 2cm$, $\lambda_R^{-1} = 2 \cdot 10^{-2} s (Q_R = (1-3) \cdot 10^7)$, and spin-orbital gain $g = 30 - 250$ (optional). At number N_R resonators in a row, the integrated voltage is increased by a factor N_R.

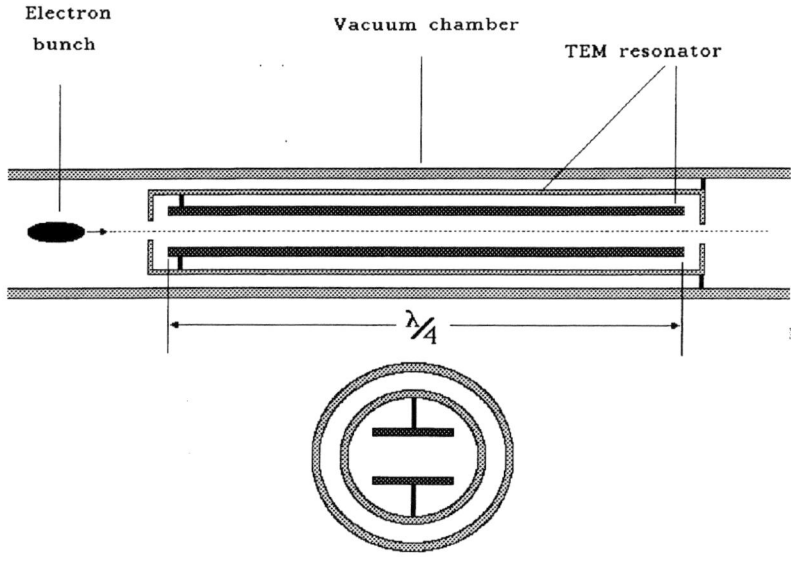

FIGURE 3. Superconducting $\lambda/4$ spin-responding resonator.

MEASUREMENT AND NOISE DEMANDS

There is a number of different kind of noise signals which can overlap with the spin-related signal. They have to be filtered or reduced to a sufficiently low level.

Thermal noise of the resonator [1] does not present a serious problem for a precision spin monitoring in most cases shown in Table 3 ($V_T \sim \sqrt{T_R/\ell} \sim 10^{-3} mV$ at $T_R = 1°K$), especially at use of the spin-orbital amplification.

The input thermal voltage of a voltmeter can be as low as $10^{-5} mV$ or less value, i.e. negligible [1].

The incoherent beam noise can limit closeness to the spin-orbital resonance that is used for an enhancement of the spin-resonator interaction. However, the dipole noise of an intense bunched beam is absent if the coherent tune shift exceeds the tune spread.

Estimations of the beam overtonal thermal noise indicates that all the non-linear harmonics of order less then 5-6 have to be tuned off the voltmeter frequency band, $\triangle\omega$.

Coherent noise due to the orbit-resonator misalignments can be filtered at condition $\triangle\omega \ll \varepsilon\omega_o$.

A special issue is filtering of the transient signal of the beam caused by the

TABLE 3. Estimations of HF voltage accumulated in $\lambda/4$ resonator.

Machine particles	Energy γ	Beam current mA	Resonator length cm	Voltage rate mV/s SOA	no SOA	Maximum voltage mV SOA	no SOA
IUCF Cooler p	1.3	1	100	0.6	$2.4 \cdot 10^{-3}$	0.01	$0.4 \cdot 10^{-4}$
RHIC p	270	50	30	36	0.36	0.7	$0.7 \cdot 10^{-2}$
HERA e±	$6 \cdot 10^4$	30	10	70	0.7	1.4	$1.4 \cdot 10^{-2}$
HERA p	900	100	30	20	0.7	0.4	$1.4 \cdot 10^{-2}$
Tevatron p	900	100	30	20	0.7	0.4	$1.4 \cdot 10^{-2}$

misalignments of the spin driver (see section 4) and (in product) the spin-orbital amplifier (RF quadrupoles). To eliminate the transient echo from the observation, the beam, resonator, and filter have to relax with exponent λ^{-1} short with respect to the spin driver switching period ($\lambda \gg 2\pi q\varepsilon/\omega_o$). The shunting measures can be undertaken, if needed.

CONCLUSIONS

Recent investigations of the RF polarimetry method resulted in a few important improvements.

The most significant step is that the RF spin echo technique is proposed to drive the coherent spin. It removes the spin decoherence and allows, at the same time, the spin to manifest in a signal of a characteristic frequency, in accordance with the basic principle of the RF polarimetry.

The proposal of the feeder $\lambda/4$ superconducting resonator increases the polarized beam-resonator interaction and makes the RF polarimeter efficient and practical at relatively long bunches (0.5 - 1 m in length), as well.

Involvement of the resonance spin-orbit interaction, intrinsic or RF-induced, makes it possible to frequently increase the spin-related excitation of the superconducting resonator.

Today estimations of the RF polarimeter efficiency encourage us to call for practical design of the involved RF elements and test experiments.

ACKNOWLEDGEMENTS

I acknowledge my colleagues A. Krisch, P. Cameron, and T. Roser for stimulating discussions of the RF polarimetry issues.

This work was supported by grants from the U.S. Department of Energy and DESY.

REFERENCES

1. Derbenev, Ya. S., "RF-resonance Beam Polarimeter, Part 1. Fundamental Concepts", NIM A **336** (1993), 12-15.
2. Derbenev, Ya. S., "RF-resonance Beam Polarimeter", 11th International Symposium on High Energy Spin Physics, Bloomington, IN 1994, ISBN 1-56396-374-4, AIP Conference Proceedings **343** (1995) 264-272.
3. Cameron, P.R., Goldberg, D.A., Luccio, A.U., Shea, T.J., and Tsoupas, N., "Squids, Snakes, and Polarimeters: A New Technique for Measuring the Magnetic Moments of Polarized Beams", Proceedings of the 1996 Beam Instrumentation Workshop, Argonne, 1996.
4. "DESY Polarized Beams at PETRA", CERN Courier, 20(1980)5.
5. S.A. Belomesthnykh, A.E. Bondar, M.N. Yegorychev, V.N. Zhilitch, G.A. Kornyukhin, S.A. Nikitin, E.L. Saldin, A.N. Skrinsky and G.M. Tumaikin, "An Observation of the Spin Dependence of Synchrotron Radiation Intensity", NIM A**227** (1984)173.
6. L. H. Thomas, Nature, 1926, v.117, p. 514.
7. V. Bargmann, L. Michel, V. L. Telegdi, Phys. Rev. Lett., 1959, v.2, p.435.
8. L. D. Landau and E. M. Lifshits, Course of Theoretical Physics, v.4 Quantum Electrodynamics, 2nd ed. (Pergamon, 1982).
9. J. Costella and B. McKellan, UM-P-95/12 University of Melburn, 1995.
10. K. Heinemann, Thesis. DESY, Hamburg, 1997.
11. Ya. S. Derbenev and A. M. Kondratenko, Sov. Phys. JETP **367**, 968 (1973).
12. Ya. S. Derbenev and A. M. Kondratenko, AIP Conf. Proc. 51, 292 (1978).
13. Ya. S. Derbenev and A. M. Kondratenko, Sov. Phys. Tech. Phys. 34(10), 1152 (1989).
14. M. M. Blaskiewicz, P. R. Cameron, Ya. S. Derbenev, D. A. Goldberg, A. V. Luccio, F. G. Mariam, T. J. Shea, M. J. Syphers, and N. Tsoupas, "Absolute Calibration and Beam Background of the Squid Polarimeter", Proc. of the 12th International Symp. on High Energy Spin Physics, p.777. World Scientific, Singapore 1997.

Magnetic Resonance Imaging with Laser Polarized ^{129}Xe

Scott D. Swanson, Matthew S. Rosen, Bernard W. Agranoff,
Kevin P. Coulter, Robert C. Welsh, Timothy E. Chupp

University of Michigan
Ann Arbor, Michigan 48109

Abstract. Magentic Resonance Imaging with laser-polarized ^{129}Xe can be utilized to trace blood flow and perfusion in tissue for a variety of biomedical applications. Polarized xenon gas intoduced in to the lungs dissolves in the blood and is transported to organs such as the brain where it accumulates in the tissue. Spectroscopic studies combined with imaging have been used to produce brain images of ^{129}Xe in the rat head. This work establishes that nuclear polarization produced in the gas phases survives transport to the brain where it may be imaged. Increases in polarization and delivered volume of ^{129}Xe will allow clinical measurements of regional blood flow.

I INTRODUCTION

Advances in laser polarization of ^{3}He and ^{129}Xe, motivated in large part by polarized target experiments, has led to significant activity in NMR and MRI of laser-polarized noble gases for medical applications. This is because the NMR signal per atom for laser polarized gases is sufficient to make up for the small concentrations compared to protons in tissue used in conventional NMR and MRI. Gas phase images of ^{3}He or ^{129}Xe have been obtained from excised mouse lungs [1], in vivo guinea pig lungs [2], and humans [3–5]. Gas phase imaging may provide detailed, tomographic information about lung ventilation as dicussed in other contributions in this session. With xenon, it is possible to create tissue phase images in addition to gas phase images. Xenon dissolves in blood [6], accumulates in the brain [7], and may be treated as a freely diffusible tracer in vivo [8]. These properties have allowed xenon to be used by SPECT [9] and CT [10] to measure regional cerebral blood flow (rCBF) by analyzing uptake and washout curves of xenon in the brain [11]. A number of spectroscopic studies have also been performed of ^{129}Xe dissolved in blood in vitro [12,13] and in blood and/or tissue in vivo [14]. Given high levels of nuclear polarization and desirable biological properties for measuring rCBF, we set out to determine the feasibility of using laser-polarized ^{129}Xe as a magnetic tracer to measure rCBF and perfusion in other tissue such as the heart. For this method to be successful, the nuclear polarization created in the gas phase must survive transport from the lungs, into the blood, through the heart, and into the brain.

II TECHNIQUES

We have constructed an optical pumping and polarized ^{129}Xe delivery system shown in figure 1.

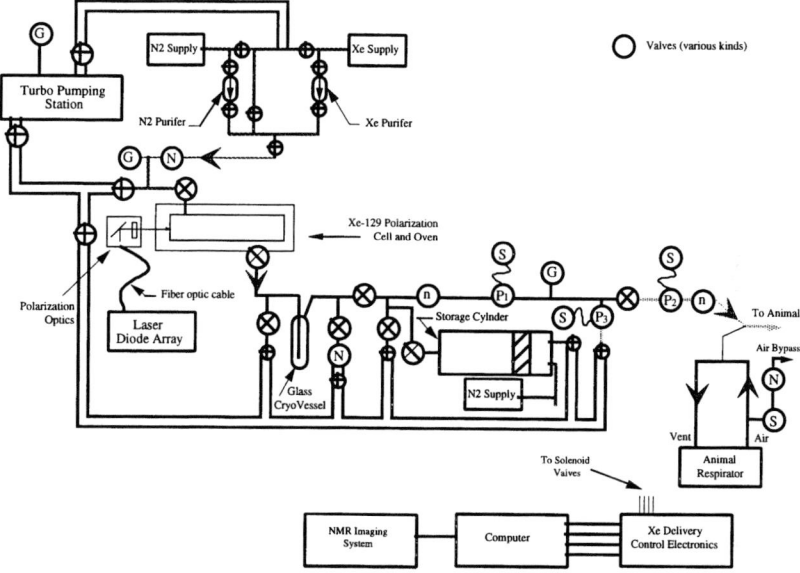

Figure 1. Schematic diagram of laser polarization and delivery system for ^{129}Xe MRI. The optical pumping takes place in the fringing field of the 2 T MRI magnet. The ^{129}Xe is polarized in the pumping chamber, transfered to a storage volume and delivered to the animal in single breath doses.

High nuclear polarization of noble gases is achieved by spin-exchange with optically-pumped Rb [15–19]. The results shown below were obtained with ^{129}Xe polarized to approximately 3.5% in an optical pumping cell containing 1700 Torr of xenon (26.4% ^{129}Xe) and 200 Torr of N2 to quench radiation trapping. The optical pumping cell, a 75 cc Pyrex cylinder with hemispherical end windows, is coated with octadecyltrichlorosilane to reduce depolarizing ^{129}Xe-wall interactions [20]. High vacuum techniques are essential to minimize contaminants such as paramagnetic O_2 which decrease nuclear spin T1 and limit absolute polarization [21]. Two Teflon valves mounted on the cell allow evacuation, filling, and delivery of the gases. Optimum density of rubidium vapor was achieved by heating the cell to 95°C. Two laser diode arrays (Optopower, Tucson, AZ), each providing at 15 Watts of continuous laser light, were used for optical pumping. Commercial availability of these high powered, efficient, GaAlAs laser diode arrays at the Rb D1 wavelength has made a tremendous impact on the field [22]. The entire apparatus including lasers, optics, valves, heating and cooling, and transport tubing lies in the peripheral magnetic field of the 2 T solenoid magnet with the optical pumping cell collinear with the solenoid axis. Typical polarization times in these studies were 10 to 15 minutes, producing sufficient amounts of polarized ^{129}Xe to ventilate a 250 gm rat for 40 to 50 seconds. NMR and MRI data were obtained with a three-turn, doubly-tuned proton-^{129}Xe surface coil [23] with 3.5 cm diameter. Excitation and response of the surface coil were determined by phantom studies to be sufficiently homogenous over the region of interest.

Spectra and images obtained with rats are shown in figures 2 and 3. The ^{129}Xe spectrum shown in figure 2 was obtained with 64 averages and a 50 microsecond RF pulse with an estimated tip angle of 10°.

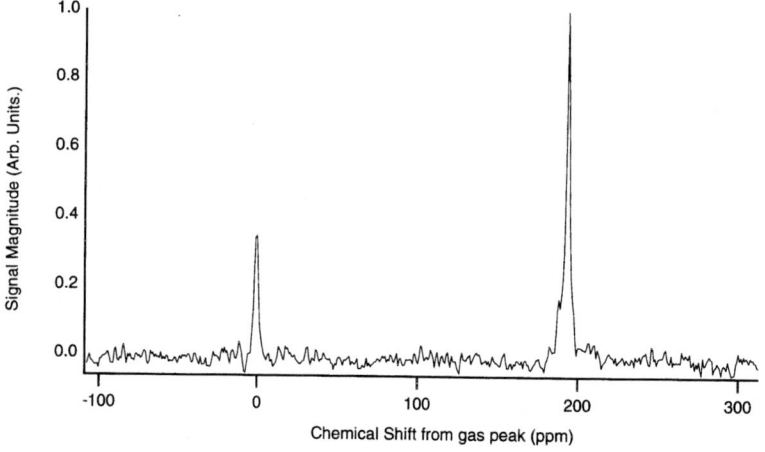

Figure 2. Magnetic resonance spectra of ^{129}Xe obtained with the head coil. The blood tissue resonance is near 200 ppm. The small gas peak near 0 ppm is due to gas in the intubation tube, sinuses, and mouth.

The ^{129}Xe tissue image shown in Fig. 3 was acquired with a 2D chemical shift imaging pulse sequence that uses phase encoding to separate the blood/tissue spectral component from the gas phase peak [24] The sequence produced a slice thickness of 10 mm, a field-of-view of 50 mm x 50 mm, a block size of 256 complex points and an estimated flip angle of 20°. The total imaging time was 73 s. The 256 x 16 x 16 CSI dataset was zero filled along both phase-encode dimensions to produce a final matrix size of 256 x 32 x 32. These data were Fourier transformed along each dimension with the magnitude of the spectrum calculated in each voxel. A Lorentzian function was fitted to the ^{129}Xe tissue phase resonance in each of the 1024 spectra and the amplitudes of the fit used to generate the ^{129}Xe image shown in Fig. 2A. The proton image shown in Fig. 2C was acquired with a conventional spin-echo pulse sequence with a slice thickness of 10 mm, a field of view of 50 x 50 mm. Precise registration of the ^{129}Xe image with respect to the proton image is important to interpretation of information contained in the ^{129}Xe image.

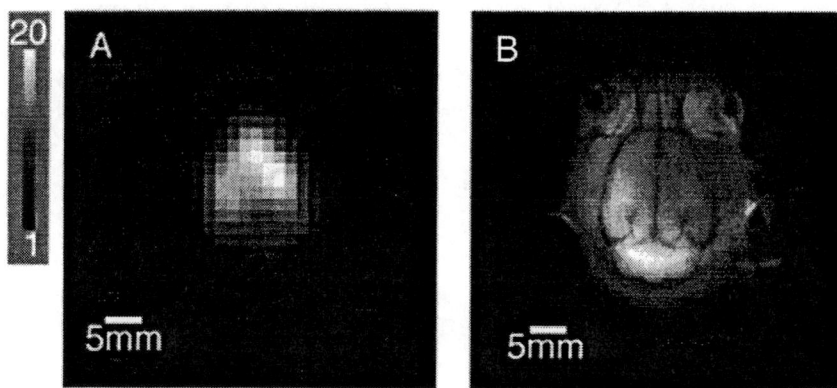

Figure 3. Magnetic resonance images of the rat head with
A) ^{129}Xe and B) proton imaging.

III RESULTS AND DISCUSSION

The NMR spectrum acquired from the rat head (figure 2) has two resolvable resonances: a small gas resonance due to ^{129}Xe in the intubation tube and a single, dominant, blood/tissue resonance at 194.5 ppm relative to the gas peak. The NMR image of ^{129}Xe in the head is presented in figure 3A. The gray scale for indicates signal amplitude relative to the RMS noise fluctuations of the background (SNR). The maximum SNR for a voxel is approximately 20. The blood/tissue resonance seen in Fig. 2 is assigned to ^{129}Xe in brain tissue based on the following argument: the rat brain cerebral blood volume is relatively low ($\approx$ 5%); cerebral blood flow very high ($\approx$ 1.0 ml/g per min.); and the partition coefficient between brain and blood is approximately 1.0 [7]. Since brain tissue has a much larger capacity for xenon than the brain blood and since xenon will freely diffuse into brain tissue, the NMR signal observed will be dominated by ^{129}Xe in the brain. This reasoning is supported by the image of ^{129}Xe in the rat head (Figs. 2A and B) which shows that the ^{129}Xe signal is localized to the brain and not seen the surrounding fat or muscle. ^{129}Xe magnetization observed in the rat brain is not uniform. ^{129}Xe signal in the cerebellum is less than the signal in the cerebrum. The signal difference could be due to partial volume effects or to lower blood flow in the cerebellum than in the cerebrum. Sakurada *et al.* found blood flow for cerebellar gray matter similar to or somewhat lower than that of cerebral gray matter [?]. In addition, inhalation of xenon has been shown to increase cerebral but not cerebellar regional blood flow [28]. Figures 3A and B also show that the two hemispheres of the cerebrum are marginally separated in the ^{129}Xe image.

The time dependence of ^{129}Xe magnetizatio in the brain has been studied with a model of ^{129}Xe magnetization uptake in the brain [25,26]. The data and results are shown in figure 4, where time dependence from the model are shown for several values of ^{129}Xe T1. We use this to estimate the T1 relaxation time of polarized ^{129} in the brain, indicating that T1 is approximately 30 seconds or greater.

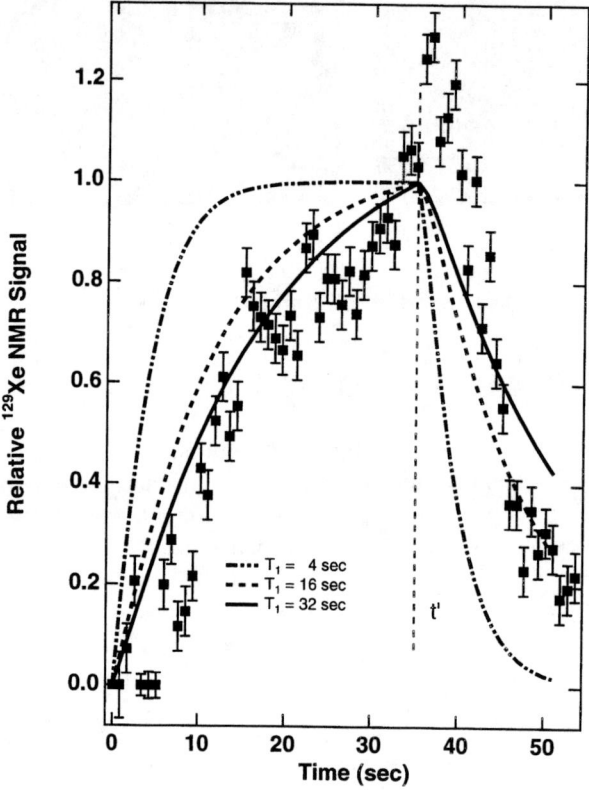

Figure 4. Uptake of blood/tissue spectral component in the head. The solid lines are model results for several T1 ^{129}Xe relaxation times.

IV CONCLUSION

The work described here is part of the large world-wide effort to use laser polarized noble gases for biomedical NMR and MRI. This effort has engaged many physicists with expertise in polarized targetry. Our work represents the first use of laser-polarized ^{129}Xe for NMR spectroscopy and imaging of the brain, in vivo. The importance of laser-polarized ^{129}Xe in studies of regional brain activation by measurement of rCBF follows from the limitations of currently employed techniques and the complementary use of radioactive and magnetic tracers.

In order to extend these techniques to humans, greater volumes and polarizations of laser-polarized ^{129}Xe will be required. In our current apparatus, we accumulate and store laser-polarized ^{129}Xe by freezing [29] for long-duration animal and human rCBF experiments. We also mix oxygen and reduce the Xe concentration to approximately 30% to avoid anoxia and anesthetic effects of xenon. Signal loss must be compensated for by increases in polarization, or the use of expensive isotopically enriched ^{129}Xe.

Though useful experiments can be performed with ^{129}Xe polarization of 1 - 10%, higher polarization will be necessary to increase the spatial and/or temporal resoulution of ^{129}Xe brain imaging. Polarizations of 50% or more have been achieved in small volumes with more suitable lasers. Advances in laser technology and the polarization delivery system should yield significantly higher ^{129}Xe polarization in the future.

Laser-polarized ^{129}Xe imaging need not be limited to the brain. Since xenon dissolves in blood and tissue, combined gas phase/tissue phase pulmonary imaging of ^{129}Xe may provide tomographic ventilation/perfusion information not available with ^{3}He. In addition, measurement of cardiac perfusion may be feasible with laser-polarized ^{129}Xe if the blood and tissue components resonate at different frequencies. Finally, since the polarization produced by laser optical pumping is magnetic field independent, it is interesting to consider of MRI at low fields (0.1 T or less). Low field techniques may provide less expensive, lighter, and more compact scanners for medical purposes.

V ACKNOWLEDGMENT

We gratefully acknowledge essential discussions with Drs. Paul Carson, Kirk Frey, Robert Koeppe, and Martin Prince and Philip Sherman for assistance with animal care and use. This work was supported in part by NSF and the University of Michigan's Office of Vice President for Research, Frontiers in Neuroscience Fund, the Department of Physics, and the Department of Radiology.

REFERENCES

1. M.S. Albert, G.D. Cates, B. Driehuys, W.Happer, B Saam, C.S. Springer, Jr., and A. Wishnia, *Nature*, **370**, 199-201 (1994).
2. R.D. Black, H.L. Middleton, G.D. Cates, G.P. Cofer, B. Driehuys, W. Happer, L.W. Hedlund, G.A. Johnson, M.D. Shattuck, and J.C. Swartz, *Radiology*, **199**, 867-70 (1996).
3. J.R. MacFall, H.C. Charles, R.D. Black, H. Middleton, J.C. Swartz, B. Saam, B. Driehuys, C. Erickson, W. Happer, G.D. Cates, G.A. Johnson, and C.E. Ravin. *Radiology*, **200**, 553-8 (1996).
4. H.U. Kauczor, D. Hofmann, K.F. Kreitner, H. Nilgens, R. Surkau, W. Heil, A. Potthast, M.V. Knopp, E.W. Otten, M. Thelen, *Radiology*, **201**, 564-8 (1996).
5. J.P. Mugler, III, B. Driehuys, J.R.Brookeman, G.D. Cates, S.S. Berr, R.G. Bryant, T.M. Daniel, E.E. de Lange, C.J. Erickson, W. Happer, D.P. Hinton, T. Maier, B.T. Saam, K.L. Sauer, M.E. Wagshul, *Proc., ISMRM, 5th Annual Meeting, Vancouver*,2113 (1997).
6. S.Y. Yeh, R.E. Peterson,. *Appl. Physiol.*, **20**, 1041-7 (1965).
7. R.Y.Z. Chen, F.C. Fan, S. Kim, K.M. Jan, S. Usami, S. Chien, *J. Appl. Physiol.*, **49**, 178-181 (1980).
8. S.S. Kety, *Pharmacol. Rev.*, **3**, 1-41 (1951).
9. N.A. Lassen, in "*Cerebral Metabolism and Neural Function*" (J.V. Passonneau, R.A.Hawkins, W.D. Lust, F.A. Welsh, Eds.) pp. 144-150, Williams and Wilkins, Baltimore, (1980).
10. K. Nambu, R. Suzuki, K. Hirakawa. *Radiology*, **195**, 53-57 (1995).
11. S.S. Kety, C.F. Schmidt, *Am. J. Physiol.*, **143**, 53 (1945).
12. A. Bifone, Y.Q. Song, R. Seydoux, R.E. Taylor, B.M. Goodson, T. Pietrass, T.F. Budinger, G. Navon, A. Pines,*Proc. Natl. Acad. Sci.*, **93**, 12932-6, (1996).
13. M. Albert, V.D. Schepkin, T.F. Budinger,*J. Comput. Assist. Tomogr.*, **19**, 975-978 (1995).
14. M.E. Wagshul, T.M. Button, H.F. Li Z. Liang, C.S. Springer, K. Zhong, and A. Wishnia, *Magn. Reson. Med.*, **36**, 183 - 91 (1996).
15. T. G. Walker, W. Happer, *Rev. Mod. Phys.*, **69**, 629 - 42 (1997).
16. T.E. Chupp, K.P. Coulter, *Phys. Rev. Lett.*, **55**, 1074-1077 (1985).
17. X. Zeng, Z. Wu, T. Call, E. Miron, D. Schreiber, W. Happer, *Phys. Rev. A*, **31**, 260 - 78 (1985).
18. T.E. Chupp, M. Wagshul, K.P. Coulter, A.B. McDonald, W. Happer,*Phys. Rev. C*, **36**, 2244-2241 (1987).
19. G.D. Cates, R.J. Fitzgerald, A.S. Barton, P. Bogorad, M. Gatzke, N.R. Newbury, B. Saam, *Phys Rev. A*, **45**, 4631-9 (1992).
20. E.R. Oteiza, *Ph.D. Dissertation*, Harvard University (1992).
21. B. Saam, W. Happer, H. Middleton, *Phys. Rev. A*, **52**, 862-865 (1995).
22. M.E. Wagshul, T.E. Chupp, *Phys. Rev. A*, **40**, 4447-4454 (1989).
23. V. Cross, R. Hester, J. Waugh, *Rev. Sci. Instrum.*, **47**, 1486-1488 (1976).
24. T.R. Brown, B.M. Kincaid, K. Ugurbil, *Proc. Natl. Adac. Sci. U.S.A.*, **79**, 3523-3526 (1982).

25. S. Peled *et al. Mag. Res. Med.* **37**, 809-815 (1997).
26. K.P. Coulter *et al.* private communication – to be published (1997).
27. O. Sakurada, C. Kennedy, J. Jehle, J.D. Brown, G.L. Carbin, L. Sokoloff, *Am. J. Physiol.*, **234**, H59-H66 (1978).
28. L. Junck, V. Dhawan, H.T. Thaler, D.A. Rottenberg, *J. Cereb. Blood Flow Metab.*, **5**, 126-132 (1985).
29. G.D. Cates, D.R. Benton, M. Gatzke, W. Happer, K.C. Hasson, N.R. Newbury, *Phys. Rev. Lett.*, **65**, 2591-2594 (1990).

Provision of hyperpolarized $^3\vec{\text{He}}$ and its application in MRI

P. Bachert[b], J. Becker[a], J. Bermuth[a], M. Bock[b], A. Deninger[a], M. Ebert[a], T. Grossmann[a], W. Heil[a], D. Hofmann[a], H.U. Kauczor[c], M.W. Knopp[b], K.F. Kreitner[c], L. Lauer[a], M. Leduc[d], H. Nilgens[c], E.W. Otten[a], L.R. Schad[b], R. Surkau[a], T. Roberts[c], M. Thelen[c]

for the

[a] Institut für Physik, Johannes Gutenberg Universität Mainz, D-55099 Mainz, Germany;
[b] Deutsches Krebsforschungszentrum (DKFZ), D-69120 Heidelberg, Germany;
[c] Klinik mit Poliklinik für Radiologie, Johannes Gutenberg Universität Mainz, D-55131 Mainz, Germany;
[d] Ecole Normale Supérieure, F-75231 Paris, France.

Abstract. Magnetic Resonance Imaging (MRI) usually relies on magnetization of hydrogen nuclei (protons) in water or molecules in tissue as source of the signal. Biological environments with low proton content, notably the lungs, are difficult to image. Inhaling of hyperpolarized ^{3}He gas opens the possibility to investigate ventilated spaces by MRI. To overcome the loss in signal due to the low density of the gas the nuclear polarization of the ^{3}He spins is greatly enhanced by laser Optical Pumping.

For more than three decades Optical Pumping of noble gases has been investigated, using spin exchange scattering (SE) [1] or metastability exchange scattering (ME) [2]. Since powerful resonant laser light is available for Optical Pumping, large quantities of ^{3}He gas can be operated [3,4]. The original interest was the development of dense spin polarized targets for fundamental research in physics [5–8]. As a spin off, the possibility of MRI of lung tissue filled with hyperpolarized ^{129}Xenon was demonstrated in 1994 [9]. Later ^{3}He was used for MRI in a guineapig [10]. While these authors have used the SE method to polarize noble gases, more recently ^{3}He MRI in human lungs was reported by our group [11] where the ME method is in use.

PROVISION OF $^3\vec{\text{He}}$

Optical Pumping of noble gases is a two step process. First an intermediary gas is polarized, for instance an alkaline vapor for SE (mostly Rubidium) or metastable ^{3}He (^{3}He*) for ME. By absorbing light, which is circularly polarized along a weak guiding field of about 0.8 mT, angular momentum is gained in the intermediary system. By spin exchange this angular momentum is transferred to the ^{3}He nucleus. The favourable aspect of the SE method is its feasibility at a pressure of several bars. ME, by contrast, takes place in a discharge plasma at 1 mb pressure. A polarization preserving compression is needed after polarizing. The advantage of ME is the ability to polarize larger quantities of ^{3}He.

For this purpose we have built a ^{3}He polarizer and compressor. The gas is polarized by resonant light at the $\lambda = 1083$ nm ^{3}He* transitions provided by two home made LNA-lasers [12,13] with an outpit power of 16 W. Thereafter a two stage piston compressor, built from nonmagnetic material (titanium), compresses the gas by nearly a factor of 10000 into a glass cell, where a polarization of 50% is reached at an average flux of 0.5 bar liter / h [14,8]. Increasing the flux to 2 bar liter per hour, still 30 % polarization is optained – by far sufficient for ventilation measurements.

The glass cells can be closed by glass cocks and disflanged from the polarizer after filling. A weak guiding field around the cell, provided by Helmholtz coils, serves to maintain the magnetization during transportation. Polarization losses inside the cell are mainly due to wall relaxation. Cells for medical application made from special Supremax glass with low iron content (provided by Schott Glaswerke, Mainz) show relaxation times up to 70 h. At the scanner itself a fast transfer of the cell through its stray field causes no significant polarization loss.

MAGNETIC RESONANCE IMAGING (MRI)

Magnetic Resonance Imaging (MRI) allows one to yield information about the three dimensional interior structure of a biological environment, for example animals or humans, by Nuclear Magnetic Resonance (NMR).
A macroscopic magnetization can be disalined against an external guiding field B_0 by an angle $\alpha = \gamma * B_1 * \Delta t$ by applying an oszillating magnetic field B_1 perpendicular to B_0 with an oszillating frequency ω_1 for a short time interval Δt. The frequency has to obey the resonance condition

$$\omega_1 \cong \omega_L = \gamma * B_0, \tag{1}$$

i.e. ω_1 of the RF pulse has to matche the Larmor frequency ω_L determined by the gyromagnetic ratio γ ($2 * \pi * 32.4$ MHz / T for ^{3}He, and $2 * \pi * 48.4$ MHz / T for ^{1}H). After excitation of the disaligned magnetization precession is observed by monitoring the Free Induction Decay (FID) using a system of a purpose built RF coil and a broadband amplifier. By help of the resonance condition (eq. (1)), spatial information is obtained. Variation of the field strength B_0 over a certain direction z as $B = B_0 + G_z * z$ using a gradient $G_z = \partial B_0 / \partial z$ enables one to excite the magnetization selectively in a thin slice by emitting only RF with appropriate ω_1. Vice versa looking at the FID with a gradient $G_x = \partial B_0 / \partial x$ switched on, the detected signal S(t) of an integral of different FIDs is charaterized by its $\omega_L(x) = \gamma * (B_0 + G_x * x)$ and its amplitudes. The density of the excited spins $\varrho(\omega_L(x))$ is responsable for the contrast. Hence a spectrum of the Fourier trasformed time dependend signal mirrors direct the spatial information in x direction.

The relative phase of the different FIDs can be used to encode the y direction. A gradient G_y is applied during a time interval Δt_y accumulating a y dependent phase

$$\Delta\varphi(y) = \Delta t_y * \gamma * (B_0 + G_y * y) = \Delta t_y * \omega(y) \tag{2}$$

prior to analysis of the FID. Again, spatial information $S(\varphi(\omega_p))$ is obtained and finally a 2D Fourier transform creates an image from the raw data. Since a single excitation can only measure one line of the $\varrho(\omega_p)$ spectrum, the number of neccessary NMR measurements with different G_y is determined by the desired image resolution in y direction.

SIGNAL INTENSITIES

To explain the major differences between MRI of Boltzmann polarized hydrogen and $^3\vec{He}$ with polarizations far above the Boltzmann equilibrium (hyperpolarization) it is worth to compare the signal intesities under typical MRI conditions.

Neglecting the time course of the FID, the primary signal amplitude is proportional to the density ϱ, the polarization P, the Larmor frequency ω and the sine of the flipping angle α (eq. (3)).

$$S \propto \varrho * P * \omega^2 * sin(\alpha) \tag{3}$$

	$^1\vec{H}$	$^3\vec{He}$
$\varrho [1/cm^3]$	$6.70 * 10^{22}$	$1.35 * 10^{19}$
P	$8 * 10^{-7}$	30 %
$\omega[MHz]$ (@ 1.5T)	64.0	48.4
α	18^0	1^0
Dilution		1/8 (@ 0.5 bar liter $^3\vec{He}$)

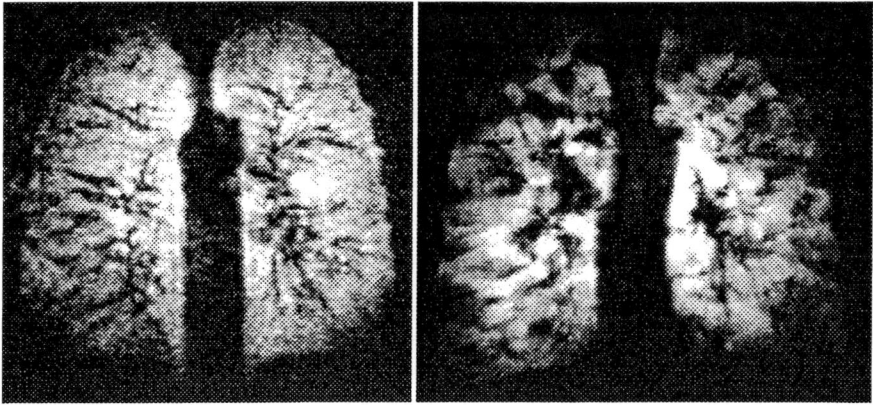

FIGURE 1. Comparison of two $^3\vec{He}$ MRI pictures of humans lungs. Left: The homogeneously ventilated lungs of a volunteer. Right: The lungs of a patient suffering from chronical obstructive pulmonary disease.

While the density of ^{3}He is 3 magnitudes lower, this lack is more than compensated by its huge hyperpolarization, whereas the Larmor frequencies differ only by a factor of 1.32. In the case of hydrogen one chooses higher flipping angles because the polarization is recovered by T_1 relaxation towards the Boltzmann equilibrium. In the case of $^3\vec{He}$ its hyperpolarization depletes with every NMR excitation. Also the hyperpolarizaton is destroyed by T_1 relaxation, leading towards thermal (Boltzmann) equilibrium. Thus one is restricted to very small flipping angles to preserve hyperpolarization sufficiently long to acquire all data for a 2D or 3D image. Moreover for inhaled $^3\vec{He}$ dilution effects have to be taken into account as shown in the tabulary. Nevertheless the signal amplitudes for hydrogen MRI and hyperpolarized helium MRI are of the same order of magnitude. Since no time is needed for polarization recovery image taking of hyperpolarized gases is quite fast and restricted by the scanning proces itsself. Thus a 3D image containing 14 slices with a spatial resolution of 128 x 256 pixels needs 22 seconds to be taken. Figure 1 shows two examples from a study on volunteers and patients carried out in Mainz in 1996 [15]: on the left a homogeneously ventilated lung of a volunteer; on the right a lung of a patient suffering from chronical obstructive pulmonary disease. Both pictures were taken with 0.5 bar liter $^3\vec{He}$ with a polarization of about 30 %. The differences in the signal distribution in both pictures are due to the patient's disease.

This predicts $^3\vec{He}$ MRI to be a promising tool for investigating the ventilation of the air spaces as the oral cavity or the lung.

CONCLUSION

Large amounts of ^{3}He in the order of one bar liter per hour could be hyperpolarized and transported to the MR scanner without significant polarization losses. ^{3}He imaging opens a new modality to determine pulmonary ventilation and its abnormalities.

This work was supported in parts by the Bundesministerium für Bildung, Wissenschaft, Forschung und Technologie under contract number 03 OT3MAI and 03OT4MAI and the Deutsche Forschungsgemeinschaft (DFG) under grant number (Th 315/8-1).

REFERENCES

1. M.A. Bouchiat et al.; Phys. Rev. Lett. $\underline{5}$ (1960) 373 - 375
2. F.D. Colegrove et al.; Phys. Rev. Lett. $\underline{132}$ (1963) 2561 - 2572
3. T.E. Chupp et al.; Phys. Rev. C $\underline{36:6}$ (1987) 2244 - 2251
4. G. Eckert et al.; Nucl. Inst. & Meth. A $\underline{320}$ (1992) 53 - 65
5. P. Anthony et al.; Phys. Rev. Lett. $\underline{71}$ (1993) 959 - 962
6. M. Meyerhoff et al.; Phys. Lett. B $\underline{327}$ (1994) 201 - 207
7. H. Sato et al.; Hyperfine Interactions $\underline{84}$ (1994) 205 - 209;
8. R. Surkau et al.; Nucl. Instrum. & Meth. A $\underline{384}$ (1997) 444-450
9. M.S. Albert et al.; Nature $\underline{370}$ (1994) 199 - 201
10. H. Middleton et al.; Magnetic Resonance Med $\underline{33}$ (1995) 271-75
11. M.Ebert et al.; THE LANCET $\underline{347}$ (1996) 1297 - 1299
12. L.D. Schearer and P. Tin; J. App. Phys. $\underline{68}$ (1990) 943-949
13. C.G. Aminoff et al.; Opt. Commun. $\underline{86}$ (1991) 99
14. J. Becker et al.; Nuc. Instr. & Meth. A $\underline{346}$ (1994) 45 - 51
15. H.U. Kauczor et al.: Radiology $\underline{201}$ (1996) 564-568 $\underline{201}$ (1995) 337 - 343

Polarized Noble Gas MRI

James R. Brookeman*, John P. Mugler III*,
Paul Bogorad§, Thomas M. Daniel†, Eduard E. de Lange*,
Bastiaan Driehuys§, Jack Knight-Scott*, Therese Maier*, Jonathon D.
Truwit††, Gordon Cates§, and William Happer§

Departments of Radiology, Surgery†*
and Internal Medicine††, University of Virginia Health Sciences Center,
Charlottesville, Virginia 22908, and Department of Physics§, Princeton
University, Princeton, New Jersey 08544

Abstract. The development of convenient methods to polarize liter quantities of the noble gases helium-3 and xenon-129 has provided the opportunity for a new MRI method to visualize the internal air spaces of the human lung. These spaces are usually poorly seen with hydrogen-based MRI, because of the limited water content of the lung and the low thermal polarization of the water protons achieved in conventional magnets. In addition, xenon, which has a relatively high solubility and a sufficiently persistent polarization level in blood and biological tissue, offers the prospect of providing perfusion images of the lung, brain and other organs.

Techniques employed for creating laser-polarized spin targets for high-energy physics have been adapted (1) to polarizing liter quantities of the noble gases ^{3}He and ^{129}Xe, that when inhaled make possible magnetic resonance (MR) images of the human lung spaces. Although the noble gas density achieved in the lung is considerably less than the water density in most human tissues, the gain in magnetization achieved by the laser polarization process is so great ($>10^5$) that excellent quality lung space images can be obtained with ^{129}Xe and ^{3}He MRI.

MR STUDIES WITH ^{129}XE

Early MRI studies of rats and mice with ^{129}Xe (2) showed the potential for polarized noble gas MRI and provided the incentive to develop laser gas polarizers suitable for human studies. In April 1996 a prototype laser diode-array polarizer for ^{129}Xe was assembled by the Princeton physics group at the University of Virginia. Enriched xenon gas (71% ^{129}Xe), polarized via spin exchange with an optically pumped rubidium vapor, was accumulated as a solid in a cold finger from a gas stream of 1% Xe, 1% N$_2$, and 98% ^{4}He. The frozen xenon was then sublimed and collected in a 500 cm^3 plastic bag for delivery to the subject positioned in a 1.5 Tesla whole-body MR

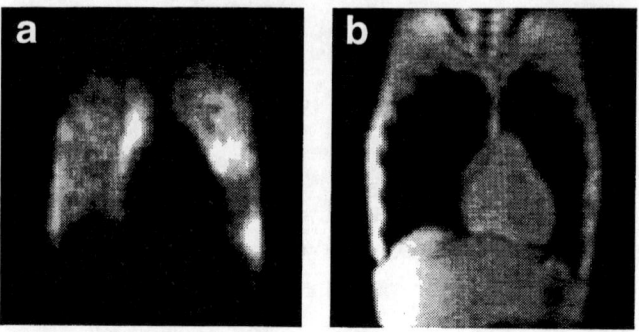

Figure 1: (a) Xe-129 lung MR image, and (b) corresponding H-1 MR image of a healthy human volunteer. The Xe-129 image is one of eleven 2-cm thick sections acquired during a 12-sec breathhold.

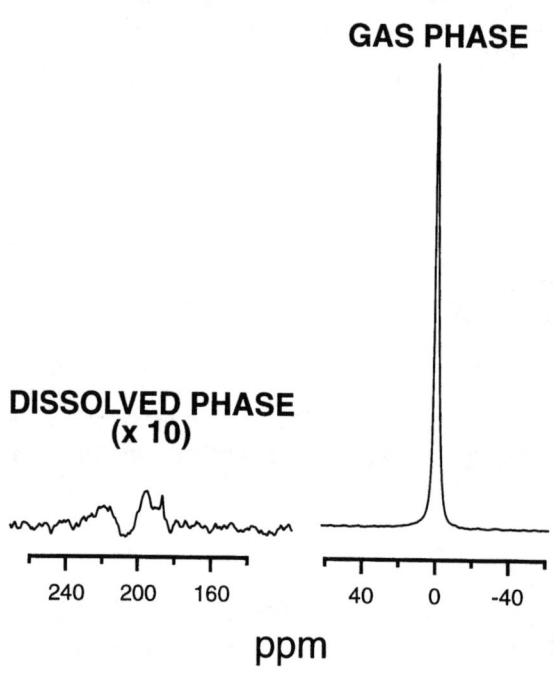

Figure 2: Xe-129 NMR spectra from the chest of a healthy human volunteer showing dissolved-phase peaks 185, 196, and 216 ppm with respect to the gas peak at 0 ppm.

scanner (Magnetom Vision, Siemens Medical Systems, Iselin, NJ). With polarization levels of approximately 2% cross-sectional images of the lung gas-spaces were acquired during a 12 second breathhold (see Figure 1). In separate MR spectroscopy experiments, three dissolved-phase peaks were detected from the chest (see Figure 2), and a single prominent peak, shifted 196 ppm from the gas peak, was observed in the human brain (3).

MR STUDIES WITH ^{3}HE

In April 1997 lung images with ^{3}He were obtained at the University of Virginia (see Figure 3), employing an investigational noble gas polarizer (Magnetic Imaging Technologies Inc., Durham, NC). This system, which is similar in operation to the ^{129}Xe polarizer, is capable of producing 2-3 liters of 15-25% polarized ^{3}He gas in a

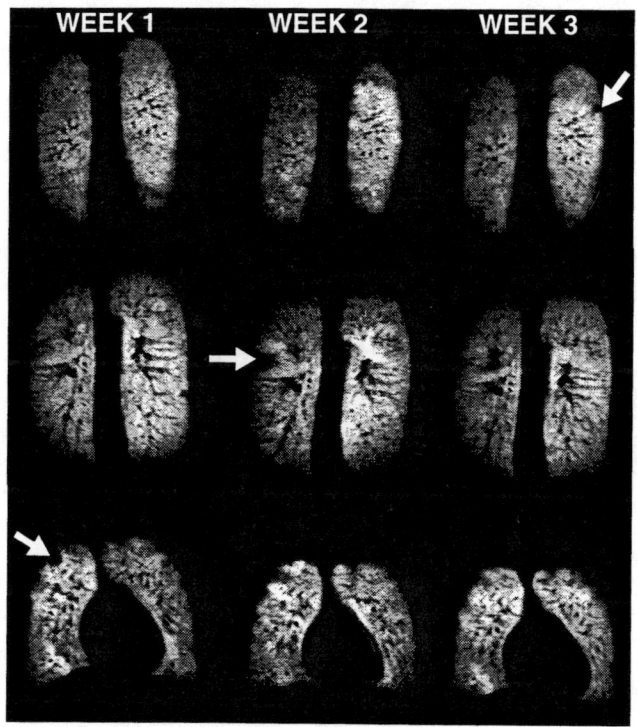

Figure 3: He-3 lung MR images of a volunteer taken at three different times, one week apart. Images in each row are at the same anatomic level. Each column of 3 images was acquired as part of a set of 13 contiguous 1-cm thick sections in a 22-sec breathhold with a FLASH acquisition (TR/TE 13/5 ms). The arrows indicate some variable subtle ventilation defects in this volunteer probably associated with the presence of small mucus plugs.

3-4 hour accumulation. In a typical ³He MR lung study, the volunteer inhales a liter of ³He from a small plastic bag, and 10 to 15 contiguous 1-cm sections of the lung field are obtained during a 15-22 second breathhold. The first human lung studies with optically polarized ³He reported in 1996 (4,5,6) showed healthy volunteers with a near homogeneous MR signal observed throughout the lung field, except for areas where the gas was displaced by normal structures, such as vessels. In patient studies (5), lesions were apparent as MR signal defects, and obstructive lung disease was depicted with severely inhomogeneous signal intensity. Our study of a patient with chronic obstructive pulmonary disease (COPD) has similar findings (see Figure 4) with significant signal voids that correspond to ventilation defects seen in a nuclear medicine study of the same patient.

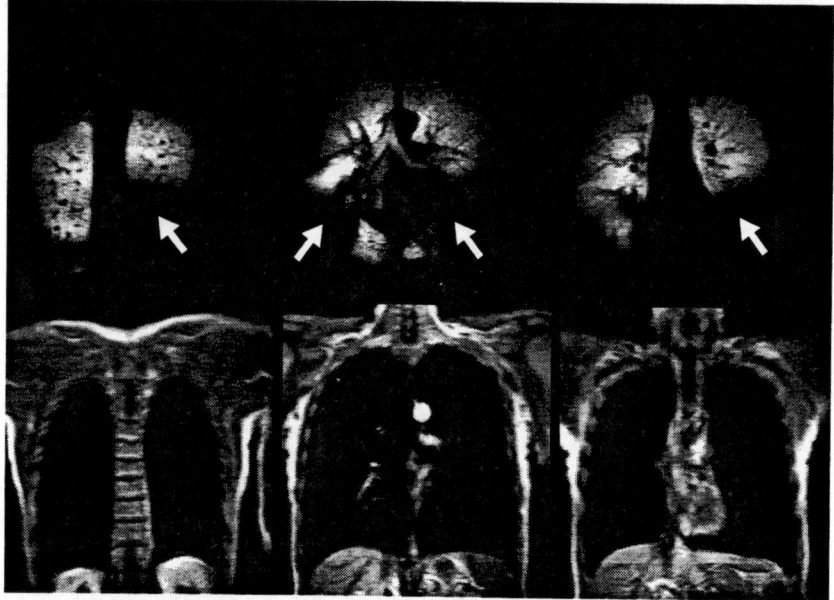

Figure 4: He-3 lung MR images (upper row) and H-1 MR images at same anatomic levels (lower row) obtained from 71-year-old patient with severe chronic obstructive pulmonary disease (COPD). Large areas of low signal intensity (arrows) represent regions with decreased ventilation. The He-3 images were acquired as part of a set of 9 contiguous 1.5-cm thick sections in a 15-sec breathhold with a FLASH acquisition (TR/TE 13/5 ms).

SUMMARY

The precise role that this type of imaging procedure might play in a patient's evaluation and the relative merits of ³He versus ¹²⁹Xe are yet to be determined; but the initial results are sufficiently encouraging to warrant further study. The possibility of imaging the dissolved phases of ¹²⁹Xe in the lung and brain to provide images of tissue perfusion is particularly attractive. This would require higher polarization levels ($\geq 20\%$) and probably some form of echo-train pulse sequence to take advantage of the long T_2 values of these dissolved phases.

Continued improvements in the technology of laser polarized noble gas MRI can be expected to enhance the potential diagnostic value of this new method of imaging, which is readily transferable to many of the MRI centers located worldwide with simple additions to the basic MR platform. It is also possible that very low field MR systems could be developed to reduce the cost of the procedure, since the MR signal strength for this technology does not depend primarily on the value of the applied magnetic field. Furthermore, for some applications, it may be advantageous to introduce polarized dissolved-phase xenon directly into blood or tissue using a compatible solvent carrier as proposed by Bifone et al. (7). This could considerably reduce the burden of xenon in the lung, and also significantly increase the MR signal for regional perfusion studies

ACKNOWLEDGMENTS

The authors thank Joe Camaratta, Wilhelm Dürr, Andreas Potthast, and Phillip Belcher of Siemens Medical Systems for important contributions in bringing the broadband imaging/spectroscopy option and the helium and xenon RF coils into operation, and Stuart Berr, John Christopher, Denise Hinton, and Shella Keilholtz, for valuable assistance with the MR studies. This research was supported by the University of Virginia Pratt Fund, the Dean of the Medical School, Dr. Robert M. Carey, and Siemens Medical Systems.

REFERENCES

1. Anthony P.L. et al., *Phys. Rev. Lett.* **71**, 959-962 (1993), Happer W., Miron E., Schaefer S., Schrieber D., van Wijngaarden W., Zeng X., *Phys. Rev. A.* **29**, 3092-3110 (1984), Driehuys B, Cates G., Miron E., Sauer K., Walter D., and Happer W., *Appl. Phys. Lett.* **69**, 1668-1670 (1996).
2. Albert M., Cates G., Driehuys B., Happer W., Saam B., Springer Jr, C., Wishnia A., *Nature* **370**, 199-201 (1994), Sakai K., Bilek A., Oteiza E., Walsworth R., Balamore D., Jolesz F., and Albert M., *J. Magn. Reson. Series B.* **111**, 300-304 (1996), Wagshul M, Button T., Li H., Liang Z., Springer C., Zhong K., Wishnia A., *Magn. Reson. Med.* **36**, 183-191 (1996).
3. Mugler III, J., Driehuys B., Brookeman J., Cates G., Berr S., Bryant R., Daniel T., de Lange E., Downs III, J., Erickson C., Happer W., Hinton D., Kassel N., Maier T., Phillips C., Saam B.T., Sauer K., Wagshul M., *Mag. Reson. Med.* **37**, 809-815 (1997).
4. MacFall, J., Charles H., Black R., Middleton H., Swartz J., Saam B., Driehuys B., Erickson C., Happer W., Cates G., Johnson G., Ravin C., *Radiology*, **200**, 553-8 (1996).
5. Bachert P., Schad L., Bock M., Knopp M., Ebert M., Großman T., Heil W., Hofmann D., Surkau R., and Otten E., *Magn. Reson. Med.* **36**, 192-196 (1996).
6. Kauczor H., Hofmann D., Krietner K., Nilgens H., Surkau R., Heil W., Potthast A., Knopp M., Otten E., and Thelen M., *Radiology* **201**, 564-8 (1996).
7. Bifone A, Song Y-Q, Seydoux R, Taylor R, Goodson B, Pietraβ T, Budinger T, Navon G, Pines A. *Proc Natl Acad Sci* **93**, 12932-12936 (1996).

Polarized Electrons at Jefferson Laboratory

Charles K. Sinclair

Jefferson Laboratory, 12000 Jefferson Avenue, Newport News, VA 23606

Abstract. The CEBAF accelerator at Jefferson laboratory can deliver CW electron beams to three experimental halls simultaneously. A large fraction of the approved scientific program at the lab requires polarized electron beams. Many of these experiments, both polarized and unpolarized, require high average beam current as well. Since all electrons delivered to the experimental halls originate from the same cathode, delivery of polarized beam to a single hall requires using the polarized source to deliver beam to all experiments in simultaneous operation. The polarized source effort at Jefferson Lab is directed at obtaining very long polarized source operational lifetimes at high average current and beam polarization; at developing the capability to deliver all electrons leaving the polarized source to the experimental halls; and at delivering polarized beam to multiple experimental halls simultaneously. Initial operational experience with the polarized source will be presented.

INTRODUCTION

At Jefferson Lab, polarized electrons are presently delivered by a 100 kV GaAs photoemission electron gun of quite conventional design (1). This gun has no load lock system. Polarization orientation at the injector is accomplished with a Mainz style "zee" spin manipulator (2). An identical gun is used in an off line laboratory for photocathode development studies and polarization measurements. Several smaller ultrahigh vacuum chambers are used to address specific issues, such as photocathode cleaning and activation techniques, reduction of field emission from electrode structures, etc.

Approximately 50% of the approved scientific program at the lab requires polarized electron beams. Many of these experiments require ~ 80% beam polarization, and most of the experiments not explicitly requiring high polarization would use it if available. Many experiments, whether requiring polarized beam or not, require average beam currents of 100 μA or greater. Some experiments which do not require polarized beam desire to operate from a photoemission electron gun, because of the ability to deliver non-standard beam time structures by modulating the laser illuminating the cathode.

In the present CEBAF injector, all electrons must originate from the same cathode. Thus, when delivering polarized beam to one experiment, beam delivery to other simultaneous experiments is an additional load on the polarized source. This point is important, since the general experience with photoemission polarized electron sources is that the operational lifetime of the photocathode is strongly correlated with the total charge delivered, rather than simply the clock hours the cathode is used.

Finally, the specifications on the beam spot size and energy spread delivered to the experimental halls translate into very demanding beam quality specifications in the injector. The injector includes an emittance filter and a sophisticated three-beam chopping and bunching system designed to meet these requirements. The three beam chopping system also provides independent current control for the beams to the three experimental halls. For a typical DC beam from the thermionic electron gun, only a few percent of the electrons leaving the cathode reach the experimental halls. Such large beam losses are very undesirable when polarized beam is being delivered.

As a result of the above realities, the polarized source program at Jefferson laboratory is directed toward achieving long photocathode operational lifetimes from the existing gun; to delivering polarized beam to more than one experimental hall simultaneously; and longer term, to developing a "best technology" photoemission polarized source. In developing this latter source, we will examine all of the issues which are believed to affect photocathode operational lifetime, and incorporate the best practical solutions to these issues.

OPERATIONAL LIFETIME

By operational lifetime, we mean the time during which a photocathode can deliver the required beam conditions, not simply some decay constant. This operational lifetime is the convolution of a number of effects, such as:
- initial cathode quantum efficiency
- static vacuum in the vicinity of the cathode (pressure and composition)
- vacuum degradation during operation with beam
- available laser power
- useful photocathode area
- electron losses from the photocathode to the experimental target

Some of these issues are intimately connected with the requirements for a particular polarized source. For example, if a very small emittance must be delivered, only a small area of the photocathode may be illuminated. Thus to effectively use a large cathode area, it is necessary to either move the laser spot at the cathode and correct the electron beam steering downstream, or move the cathode itself. We address several of these issues in the following sections.

Initial Quantum Efficiency

Cleaning the surface of the semiconductor to be activated as a photoemitter is a very important step in obtaining a high initial quantum efficiency. For the best results, it is necessary to prepare an atomically clean semiconductor surface. Bulk GaAs can be successfully cleaned by wet chemistry and in-vacuum heat treatment. Unfortunately, the very thin layers which provide the highest polarizations cannot be cleaned with wet chemistry, since these processes remove too much material. It is also very difficult to clean semiconductors which contain, for example, aluminum or silicon.

Atomic hydrogen has been demonstrated to remove difficult contaminants on many III-V, II-VI, and elemental semiconductors, such as carbon on GaAs, oxygen on silicon, etc. (3). No chemical cleaning is required prior to the use of atomic hydrogen. The process does not remove material from the bulk semiconductor, so it is very suitable for use with the very thin materials which give high polarization. In the case of GaAs, atomic hydrogen exposure passivates the surface. This allows us to clean a GaAs wafer in one system, and transfer it through air into our electron guns, without loss of the benefit of the cleaning. This is a real advantage when a non-load-locked gun is used. We have been routinely using this process for a year, and have prepared high quantum efficiency photocathodes in three different guns using this process.

Figure 1 shows a view of the system we constructed to evaluate atomic hydrogen cleaning. In this ultrahigh vacuum system, we can clean a semiconductor as well as activate it and measure its lifetime at low average current. We typically operate with a hydrogen pressure of ~30 mtorr in the pyrex dissociation chamber. The RF discharge operates at about 100 MHz, with 40 W of forward power. The sample to be cleaned is held at ~300 C during the cleaning. The time necessary to clean a sample is clearly dependent upon the geometry, and we have not established a minimum time required. For our relatively poor geometry, cleaning times of 30 to 45 minutes are adequate. It is clear that these times could be shortened with improved geometry. During the cleaning, hydrogen in the main chamber is pumped by a combination of a non-evaporable getter and an ion pump. Once the hydrogen flow is stopped, the chamber pressure quickly recovers to $\sim 10^{-10}$ torr.

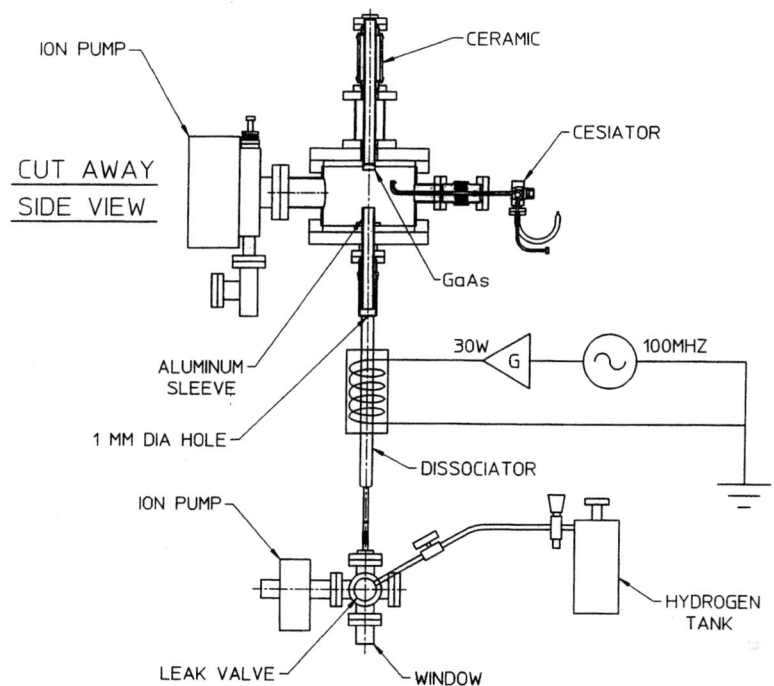

FIGURE 1. Schematic view of the ultrahigh vacuum test system constructed to evaluate atomic hydrogen cleaning and activation of GaAs photocathodes.

Following the cleaning step, the cathode is heated above 450 C to remove the hydrogen.

A typical result for quantum efficiency versus wavelength for an atomic hydrogen cleaned bulk GaAs wafer is shown in Figure 2. The dopant density of this sample was $3.3 \times 10^{18}/cm^3$, and the sample was only degreased prior to introduction into vacuum. We routinely achieve quantum efficiencies above 10% at 780 nm, and above 6% at 862 nm.

We have also constructed a small "roll-around" atomic hydrogen cleaning system. This system is not baked. GaAs wafers are cleaned in this system, and then transferred to one of our electron guns. Even though the samples have been transferred through air, we obtain very high quantum efficiencies on these cathodes as well. Finally, we have adapted an atomic hydrogen source to be

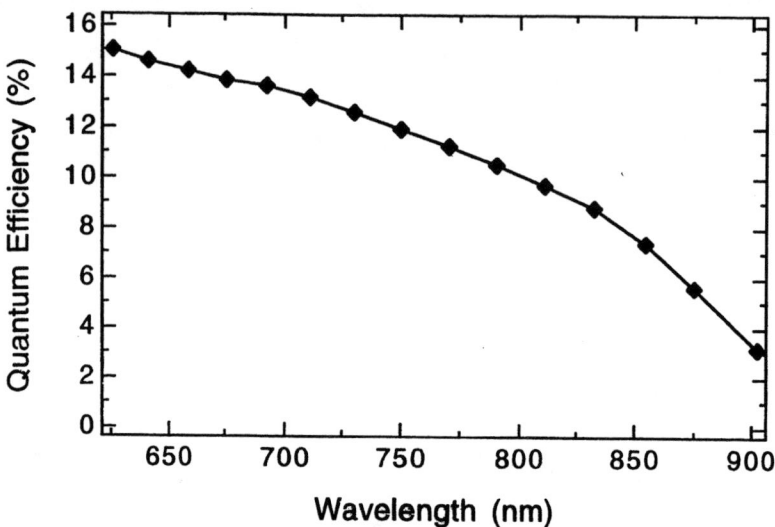

FIGURE 2. Typical Quantum Efficiency of a GaAs photocathode prepared after atomic hydrogen cleaning.

mountable directly on our non-load-locked guns. Given the success we have experienced with this process, we will incorporate it into the "best technology" gun we are developing.

Vacuum Improvements

The vacuum environment in the vicinity of the photocathode is a very important factor in its operational life. Certain gases, such as water and carbon dioxide, degrade the quantum efficiency when present at extremely low partial pressures. Even in vacuum chambers where these chemically active gases are not present in significant amounts, any residual gas in the vicinity of the cathode may be ionized by the emitted electrons, and damage the photocathode by backbombardment. Accordingly, steps to improve the vacuum environment in any region which can communicate, vacuum-wise, with the photocathode can be expected to improve the cathode lifetime.

In the polarized sources developed to date, ion pumps and NEG pumps are employed. These pumps are normally mounted downstream of the anode aperture, and as a result, the effective pumping speed at the cathode is reduced.

We are examining a number of potential pumping schemes for use in our "best technology" gun project. In the meantime, we are making a significant change in the pumping arrangement on our existing gun. We are adding a large diameter cylindrical chamber on the cathode side of the anode chamber. This chamber contains an array of NEG pump modules which will have an initial pumping speed for active gases over 2000 L/sec. A wire mesh cylinder provides a grounded surface radially inside the NEG array. The open fraction of the wire mesh is so large that it does not represent a significant conductance restriction.

Small amounts of electron beam loss in the vicinity of the cathode can result in desorbed gases which reach the cathode. Specular reflections of the incident laser beam from the GaAs surface and the surfaces of the optical input window provide a potential source for such electron beam loss. We have replaced our input window with one AR coated for our operational wavelengths. This change should reduce electron beam losses from this origin by close to an order of magnitude. These two improvements to our present vacuum system are planned for testing in September, 1997.

Improved Laser Power

Our present laser is a semiconductor diode oscillator-amplifier (4). The oscillator is RF gain switched at either 499 Mhz or 1497 Mhz, providing a train of short duration (60 to 80 ps FWHM) optical pulses locked to either the fundamental RF frequency of the accelerator, or the frequency of the bunch train delivered to one of the three halls. The purpose of this gain switching is to provide electron pulses which are short enough to pass through our chopping system without beam loss. In addition to the RF structure provided by gain switching, temporal control of the amplifier current allows us to produce the complex macropulse structure used during accelerator tuning.

Over the past year, the output power of this laser has been increased from 100 mW to 360 mW. These output powers can be obtained at either RF frequency. The ultimate power limitation on a system of this type arises from damage to the output facet of the amplifier. The manufacturer has indicated that we should be able to operate up to about 600 mW from our present amplifier, and we have demonstrated 500 mW in the laboratory. By chosing the proper diodes, the laser is operable over a broad wavelength region of interest. At the present time, we have systems operating at ~780 nm and ~862 nm. With a 1% quantum efficiency photocathode, 500 mW will deliver 3.15 mA at 780 nm, and 3.47 mA at 862 nm.

In January 1998, we will install a laser system which incorporates three separate 499 MHz lasers, each illuminating the same spot on the photocathode. This will allow us to have independent control over the current to each of the three experimental halls, eliminating the beam losses which would otherwise be experienced when the halls operate with different beam current. Much more information on lasers of this type is presented in the talk of Matt Poelker at this workshop.

Electron Losses from Photocathode to Target

Loss free transport of the beam from the photocathode to experimental target is an essential element in achieving the maximum operational lifetime from a photocathode. To obtain loss free transport through a complex injector such as ours, it is necessary to have an accurate model of the system, incorporating all of the physical phenomena present. Such a model allows us to develop a full set of initial settings for the beamline elements, and by iterating between measurement and model, provides guidance toward the final injector setup.

Our model is based on an in-house version of PARMELA developed by Hongxiu Liu (5). His model incorporates measured fields for all magnetic elements, calculated fields for all RF elements, and accurately includes the effects of space

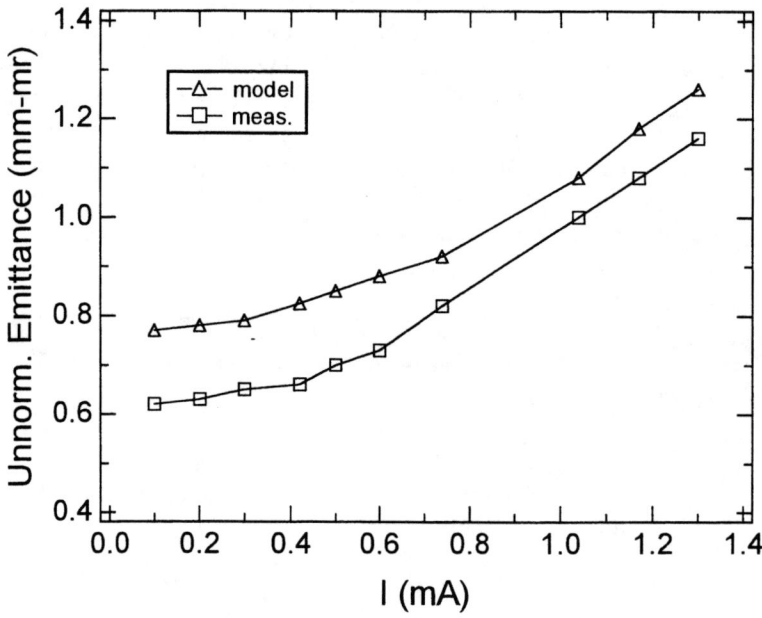

FIGURE 3. Comparison of modelled and measured emittance growth through the "zee" spin manipulator, as a function of the DC beam current.

charge. He has managed to incorporate such elements as double-focussing electrostatic bends and Wien filters, both of which are astigmatic, into his code. An example of the quality of the results he has obtained is shown in Figure 3. Several years ago the emittance of the beam transmitted through the "zee" spin manipulator was measured to grow at unexpectedly low average beam currents. We suspected that this growth was a result of space charge effects at the very small beam waists produced by the short focal lengths (10 cm) of the electrostatic bends, but were not able to calculate this. As demonstrated in Figure 3, the present code has convincingly captured this effect.

Calculations based on this model indicate that although we use a short pulse laser and operate with very low microbunch charge, space charge is sufficient to cause unacceptable bunch lengthening from the cathode to the apertures of the chopping system. A pre-buncher has been added to the system to counteract this bunch lengthing, and code is being applied to determine the best way to set this prebuncher up. The model indicates that we should be able to deliver at least 100 µA to an experimental hall from the polarized source by a combination of:
- careful attention to the transverse optics
- replacement of the "zee" spin manipulator with a Wien filter plus solenoid
- careful optimization of the pre-buncher field

These changes are planned for installation in January 1998.

THE BEST TECHNOLOGY GUN PROJECT

With the changes to our existing polarized source indicated above, we believe that we will have done as well as we can based on the use of the present gun. It is clearly desirable to incorporate a load-lock onto the polarized gun. In planning to do that, our ideas enlarged to attempt to incorporate the best technology which might be applied to all of the identified problem areas of photoemission polarized electron guns. Our reasons for enlarging the scope of the work stemmed from the observation that essentially all of the polarized gun designs built to date have been developed as variants of the original gun built at SLAC (6). That gun was essentially a thermionic gun design modified to incorporate the GaAs photocathode. Variations to this design, often clever, have been made to address one or another of the real or perceived problems in operating a GaAs gun. We have decided to move a step back, and attempt to examine the best possible technologies for achieving the desirable features of a photoemission gun, to the extent that these are known. In particular, we will examine:
- the choice of materials for vacuum system construction
- pumping methods to provide the lowest practical ultimate pressure

- electrode materials and treatments which will minimize field emission
- charge drainage over the inner surface of the primary insulator (7)
- load lock schemes for cathode introduction and activation
- incorporation of high sensitivity, field emission based vacuum diagnostics
- incorporation of atomic hydrogen cleaning

No doubt additional issues will arise as we become more deeply involved with this project. We do not have a rigid time scale for this work. Instead, we anticipate that the changes to the existing source scheduled for January 1998 will provide us enough operational flexibility to give us time to complete the new gun work without short-changing it. It is our intention to test all sufficiently novel aspects of the new gun in a meaningful way before incorporating them into the final design. We anticipate that this project will require about 18 months. Our intention is to install two of these guns at the injector, and operate from each in alternation.

SIMULTANEOUS BEAM POLARIZATION TO MULTIPLE EXPERIMENTAL HALLS

The beam delivered by the CEBAF accelerator is recirculated up to five times through a pair of equal energy linear accelerators, and then deflected through either 0 degrees, or + or - 37.5 degrees to reach the three experimental halls. As a consequence, there is considerable precession between the injector and any experimental hall. Exactly longitudinal polarization is available in all three halls at integral multiples of 2.115 GeV. However, there are over 400 possible energy combinations between 2 and 6 GeV which provide simultaneous longitudinal polarization in any two halls (8). Thus, it is practical to plan for simultaneous operation of two experimental halls with polarized beam.

In general, the delivery of longitudinal polarization to even one experimental hall requires that the polarization be oriented correctly leaving the injector to arrive longitudinal at the experimental hall. To verify this orientation, as well as measure the polarization accurately, we have developed a Mott polarimeter operating at 5 MeV (9). After this location in the accelerator, the only elements which can precess the spin are the beamline dipoles, and their effect is accurately calculable. The maximum analyzing power of gold at this energy is 0.52, and the device is operable with high average beam currents (~10 µA), permitting accurate polarization measurements to be made rapidly.

It appears possible to make an accurate determination of the beam energy by measuring the net precession between the injector and the experimental halls. To do this, the polarization is swept in the plane of the accelerator at the injector.

The 5 MeV Mott polarimeter is used to verify that the polarization is in the plane of the accelerator, and to measure the projection of the polarization in this plane. Mfller polarimeters in the experimental halls measure the projection of the polarization as well. The net precession from the injector to the experimental hall is thus determined. This precession depends only on the geometry of the accelerator and the energy gains through the two linacs. It appears possible to measure the energy with an absolute precision of about 10^{-4} by this technique.

INITIAL OPERATING EXPERIENCE

The polarized source was pressed into operation unexpectedly in February 1997, when a machine protection element of the thermionic gun failed. It was operated for 5 weeks, delivering typical currents of 30 µA CW. The highest current reached briefly was 140 µA. Cathode operational lifetimes during this run were reasonable.

The source was scheduled for operation for physics from mid-July through the first week of August, 1997. Polarized beam was delivered to two experimental halls simultaneously. Hall A conducted polarization transfer measurements from ^{16}O, while Hall C studied helicity correlated effects in preparation for parity violation measurements. Helicity correlated effects were also studied in the Hall A beamline on a parasitic basis. The source delivered 50 to 70 µA CW to Hall A, and 10 to 15 µA CW to Hall C. The transmission from the photocathode to the experimental halls was ~70%, demonstrating that we are already achieving some fraction of the anticipated gain from the use of the pre-buncher. For the duration of these runs, we delivered over 8600 µA-hours to the experimental users, and over 12,500 µA-hours from the photocathode. During all this operation, the photocathode lifetime was exceptionally poor. We had known beam scraping in the early portion of the beam transport system from the gun, but were not able to eliminate it, and we believe that this was the origin of the poor cathode lifetime. Studies will be conducted later this fall in an attempt to diagnose and hopefully eliminate this problem.

Despite these difficulties, the experimenters were able to accomplish some significant measurements. The Hall A experiment employed a focal plane polarimeter to measure recoil proton polarization, and completed a series of measurements on ^{16}O, using a water target. In addition, they were able to demonstrate from measurements on hydrogen that we will be able to do a very good job measuring G_{Ep} using this technique (10). The groups studying helicity correlated effects have concluded that the beam quality is close to good enough for parity violation measurements to proceed. Helicity correlated position asymmetries were at the 20 nm level, and helicity correlated beam energy variations were at the level of 10^{-8}. The helicity correlated intensity asymmetries were typically 10 to 20 ppm, and were strongly correlated with the correct optical alignment of the Pockels cell used to reverse the beam polarization. By imaging the exit of the Pockels cell onto the photocathode (which was not the case for the above measurements) and employing previously developed feedback stabilization, the experimenters believe they will have a system adequate for their planned measurements (11). A full "dress rehearsal" of a parity violation measurement is planned for this coming December, with the complete experiment to follow in April 1998.

ACKNOWLEDGEMENTS

Many people have contributed substantially to the work presented here. Particularly noteworthy contributors include Philip Adderley, Bruce Dunham, Joe Grames, Danny Machie, John Hansknecht, Curt Hovater, Reza Kazimi, Hongxiu Liu, Matt Poelker, and Scott Price and Bill Schneider.

REFERENCES

1. B. M. Dunham, "Investigations of the Physical Properties of Photoemission Polarized Electron Sources for Accelerator Applications", Ph.D. Thesis, University of Illinois, Champaign-Urbana, IL, 1993
2. D. A. Engwall et al., *Nuclear Instruments and Methods* **A324**, 409-420 (1993).
3. There is a very extensive literature on this subject. Recent examples include: Y. Luo et al., Appl. Phys. Lett. **67**, 55-57 (1995); Y. Okada and J. S. Harris, J. Vac. Sci. Technol. **B14**, 1725-1728 (1996); Z. Yu et al., Appl. Phys. Lett. **69**, 82-85 (1996).
4. M. Poelker, Appl. Phys. Lett. **67**, 2762-2764 (1995).
5. H. Liu and B. Dunham, Jefferson Laboratory Technical Note TN-96-067; H. Liu, Jefferson Laboratory Technical Note 96-051.
6. C. K. Sinclair et al., in A.I.P. Conference Proceedings No. 35, American Institute of Physics, 1976, pp. 424-431.
7. F. Liu et al., "A Method of Producing Very High Resistivity Surface Conduction on Ceramic Accelerator Components using Metal Ion Implantation" in *Proceedings of the 1997 Particle Accelerator Conference*, Vancouver, to be published.
8. C. K. Sinclair, Jefferson Laboratory Technical Note TN-97-021.
9. J. S. Price, et al., these proceedings
10. P. Rutt, Private Communication.
11. K. Kumar, Private communication.

Status of the MAMI–Source of Polarized Electrons

B2–Collaboration at MAMI: K. Aulenbacher[†], H. Euteneuer[†], D. v. Harrach[†], P. Hartmann[†], J. Hoffmann[†], P. Jennewein[†], K.-H. Kaiser[†], H. J. Kreidel[†], M. Leberig[†], E. Reichert[‡], M. Schemies[‡], J. Schuler[†], M. Steigerwald[‡], and M. Zalto[†]

[†] *Institut für Kernphysik, Joh. Gutenberg Univ., D–55099 Mainz, Germany*
[‡] *Institut für Physik, Joh. Gutenberg Univ., D–55099 Mainz, Germany*

Abstract. The MAMI source of polarized electrons is based on electron photoemission from III–V–semiconductors. Strained layers of GaAsP are used at present. Typical degree of spinpolarization of emitted electrons is 75 % at a wavelength of light irradiating the cathode of 830 nm. Recent work has been focused on the improvement of source operational reliability and the improvement of the capture efficiency of the interface between source and accelerator. Appreciable gain in stability was obtained by shortening the injection line and setting up a new source right at the injection point of MAMI and by replacing the Ti:Sapphire laser used so far by an infrared AlGaAs–diode-laser as a light source. Installation of a new harmonic prebuncher in the injection line to MAMI increased the capture efficiency by more than a factor of three to 44 %. Beam currents up to 10 μA and spinpolarized to 75 % are available at MAMI for long term experiments now. First tests of driving the source with a 2.45 GHz pulsed laser synchronized to the accelerator–RF were successful already. Implementation of this technique may bring the capture efficiency to a value near unity in not too far future.

INTRODUCTION

The number of experiments asking for a spinpolarized electron beam proposed at the Mainz race track microtron MAMI is steadily increasing [1]. Examples are listed in table 1.

The source of polarized electrons applied in these investigations [2] is based on the photoemission of electrons from III–V–semiconductor–cathodes, that emit highly spinpolarized electron ensembles if irradiated with circularly polarized light of a photon energy right at the gap energy of the semiconductor

TABLE 1. Eperiments with polarized electron beams under way at MAMI

Physics quantity to be determined	Reaction
nucleon formfactors	$^1H(\vec{e},e'\vec{p})$, $^2D(\vec{e},e'\vec{p})$, $^2D(\vec{e},e'\vec{n})$, $^3\vec{H}e(\vec{e},e'n)$
structure functions	$^2D(\vec{e},e'\vec{p})$, $^{16}O(\vec{e},e'\vec{p})$
E2/C2 in N$\to\Delta$	$^1H(\vec{e},e'\vec{p})\pi^0$
parity violation	$^1H(\vec{e},e')$
Gerasimov–Drell–Hearn sum rule	$^1\vec{H}(\vec{\gamma},\gamma')$

[3,4]. The same process is used in the sources installed at Bates[1] [5], at ELSA[2], at Jefferson Lab[3] [6], at KEK[4] [7], at NIKHEF[5] [8] and at SLAC[6] [9,10]. The Mainz source has been applied to experiments for more than 2000 hours already, mainly to measurements of the electric nucleon formfactors in reactions $^1H(\vec{e},e'\vec{p})$ [11], $^2D(\vec{e},e'\vec{p})$ [11], $^2D(\vec{e},e'\vec{n})$ [12], and $^3\vec{H}e(\vec{e},e'n)^7$ [13].

Recent source R&D has been focused on the improvement of operational reliability and the improvement of the capture efficiency of the interface between source and accelerator. Measures taken are: a) Setting up a new source right at the injection point of MAMI to shorten the length of the 100 keV–injection line, which proved to be very sensitive to stray magnetic fields, and to avoid space charge blow up of the beam in pulsed operation b) addition of a harmonic prebuncher to the injection line, c) tests of driving the source with a 2.45 GHz pulsed laser synchronized to the accelerator-RF, and d) tuning the spin to longitudinal orientation at the accelerator exit by fine tuning the end energy of the machine. The paper will discuss the topics in the order mentioned.

R&D OF THE MAMI–SOURCE OF POLARIZED ELECTRONS

$\vec{e}$–Source Mark III

Figure 1 shows another MAMI source of polarized electrons this time installed right at the injection point of the microtron. Its upper part consisting of the gun, the preparation chamber, and the load lock is a copy of the system described in [2]. In figure 2 the triode configuration of the gun is sketched. It has been designed for low electric field strength at the electrode surfaces with help of the wellknown EGUN–code [14]. Cathodes may be exchanged

---

[1]) talk given by M. Farkhondeh at this conference
[2]) contribution to this conference by W. v. Drachenfels and S. Nakamura
[3]) talk given by Ch. Sinclair at this conference
[4]) talk given by T. Nakanishi at this conference
[5]) talk given by M. van den Putte at this conference
[6]) talk given by J. Clendenin at this conference
[7]) talks of P. Becker and of D. Rohe at this conference

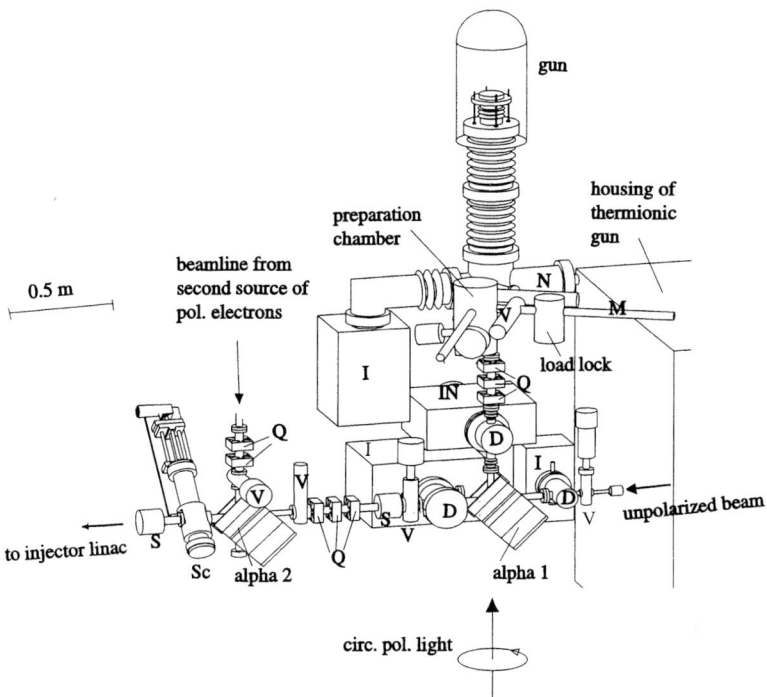

FIGURE 1. MAMI source of polarized electrons. alpha = α–magnet, D = differential pumping stage, I = ion getter pump, N = NEG pump, M = linear movement manipulator, Q = quadrupol. S = solenoid, Sc = wire scanner, V = all metal valve.

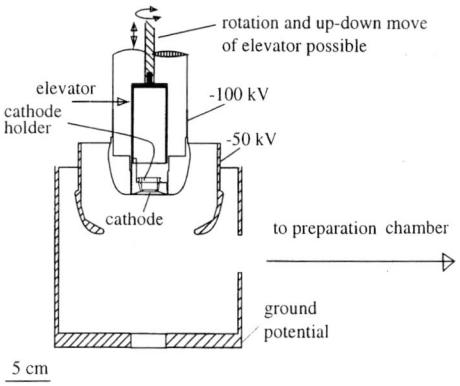

FIGURE 2. Section through the 100 keV gun.

by lowering the *elevator* indicated and passing the cathode crystals held in a molybdenum puck through a side opening to and from respectively the preparation/storage chamber.

Strained layer $GaAs_{.95}P_{.05}$–cathodes [15] are currently installed in the MAMI source purchased from Technical University of St. Petersburg. Figure 3 shows quantum efficiency QE and spinpolarization P of emitted electrons as a function of photon energy of light irradiating a sample of this cathode type. The behaviour seen is typical for this type of cathode. A maximum

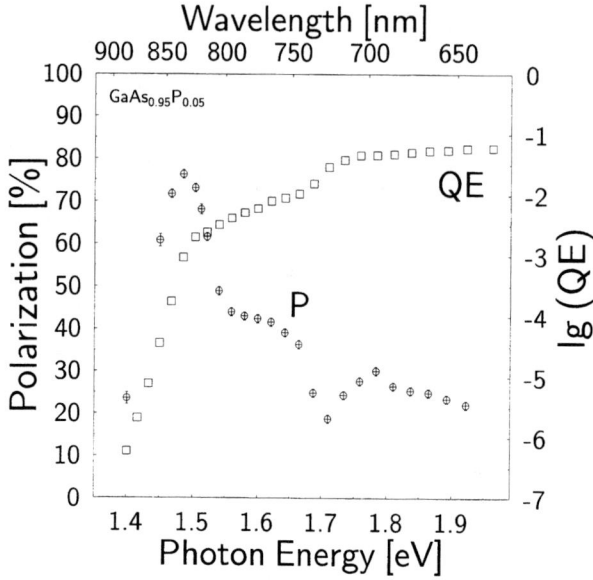

FIGURE 3. Photo electron emission from $GaAs_{.95}P_{.05}$. Quantum efficiency **QE** and polarization **P** of emitted electrons as a function of photon energy of irradiating light.

degree of spinpolarization of 76% is reached at a wavelength of 830 nm for this particular sample. The quantum yield at this point is around 2×10^{-3}.

Linearly polarized 830 nm light is produced by a 200 mW AlGaAs–diode–laser[8], transformed to circular polarization by a pockels cell, transmitted through a beam expander, and finally focused to a 0.2 mm diameter spot at the cathode via a vacuum window in the vacuum chamber of α–magnet labelled **alpha 1** in figure 1.

The 100 keV beam produced by the gun is guided by a beamline that merges one meter downstream at **alpha 1** in figure 1 with the line coming from the thermionic gun of MAMI. The transmission of the beam from cathode to the injection linac of MAMI is better than 95 %. The transverse beam emittance

---

[8] model SDL–5430

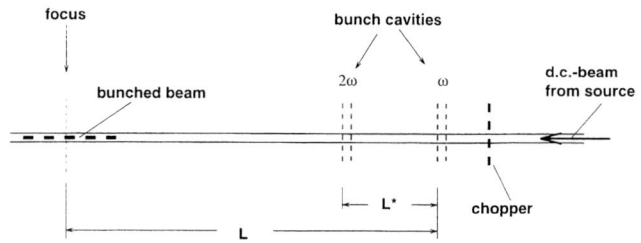

FIGURE 4. Arrangement of harmonic prebuncher cavities.

amounts to $0.3\ \pi \cdot mm \cdot mrad$ measured with help of a wire scanner at the entrance of the injection linac.

Great care has been taken to preserve good UHV-conditions in the gun as well as in the part of the beam line immediately following the gun. Pumping is done with a combination of ion getter and NEG pumps. Hydrogen partial pressure in the gun is $4 \times 10^{-11} mbar$ while all other contributions to the residual gas stay in the low $10^{-12} mbar$ range or below during 20 μA-operation of the gun.

Harmonic prebuncher

The capture efficiency of the RF-chopper-buncher system at the injector linac entrance, that was run until the end of 1996, was in the range 12 % to 15 % typically. So most of the polarized beam was lost there. In 1997 another prebunching cavity working at 4.9 GHz, the first harmonic of the 2.45 GHz-radiofrequency of MAMI, was added to the bunching system and at the same time the distance between the whole prebuncher and the graded-β capture section of the injector was enlarged to 1.18 m. The arrangement is sketched in figure 4. The introduction of 2ω-modulation at a point different from that of the ω-cavity produces also a 3ω-sideband in the velocity modulation of the beam. By proper choice of distance L*, RF-amplitudes, and RF-phases a longitudinal third order focus may be achieved at a point L downstream of the buncher cavities [16]. The installation of this type of prebuncher arrangement enlarged the phase acceptance of the injector to 180^0 and the capture efficiency to 50 %.

2.45 GHz pulsed beam driven by pulsed laser light

The accelerator sections of MAMI are driven by RF-power of 2.45 GHz. Electrons are accepted for acceleration in a fraction of the time of a RF-period only. An appreciable part of a d.c.-beam produced at the source is rejected therefore. This does not matter if the source is a thermionic gun with

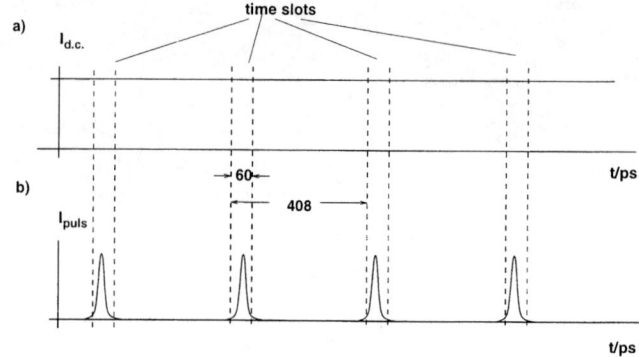

FIGURE 5. Time structure of MAMI–acceptance.

ample current capability but may cause harm if the electrons come from a photo cathode with limited quantum yield. Figure 5a sketches the case for an acceptance time slot that is 15 % of the period, a situation encountered at MAMI before installation of the new prebuncher cavity. A capture efficiency near unity should be obtainable if the source is driven by a pulsed laser with a repetition frequency of 2.45 GHz in synchronism with the MAMI-RF, that produces an already bunched beam as indicated in figure 5b. This scheme already works successfully at Jefferson Lab[9] [17]. Spinpolarization of the beam is not affected by driving the source in pulsed mode[10] [18]. We tested driving the MAMI source of polarized electrons in a joint experiment together with M. Ciarocca and H.Avramopoulos from the University of Athens [19]. The laser used was a harmonically mode locked, external cavity GaAlAs diode laser developed in Athens [20]. Its layout is sketched in figure 6a. It produced a 2.45 GHZ pulse train with 842 nm pulses of a shape seen in figure 6b. An enhancement of capture efficiency from 12 % to 52 % was observed using this laser. The test was run before installation of the above mentioned new prebuncher system. We expect to increase the capture efficiency to a value near 100 % by using both the new prebuncher and a 2.45 GHz laser in near future. Current tests with a mode locked Ti:Sapphire laser [21] are very promising.

SPINADJUSTMENT BY FINE–TUNING THE ACCELERATOR ENERGY

The source of polarized electrons Mark III presented above has been installed as near as feasible to the injection point of MAMI. So the installation did not leave space to implement a spin rotator like that of the hitherto ap-

---

[9] talk of M. Poelker at this conference
[10] talk of P. Hartmann at this conference

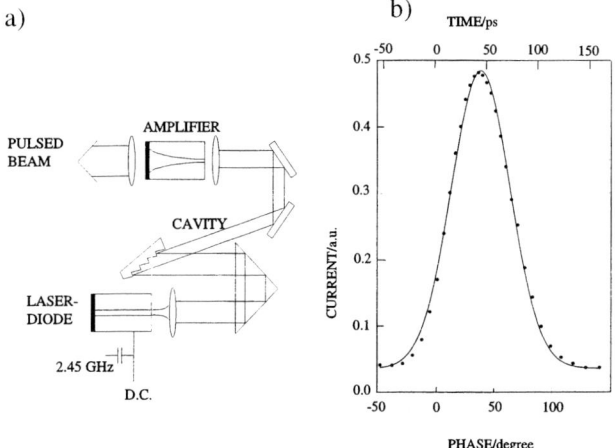

FIGURE 6. a) Harmonically mode locked, external cavity GaAlAs diode laser [20] synchronized to 2.45 GHz MAMI–RF. b) Shape of electron bunces emitted as measured by the method presented in [18].

plied source [2,22]. In order to get the right spin orientation at a target downstream of the accelerator, in most cases longitudinal polarization, the energy of the third stage of the MAMI microtron cascade may be adjusted. Figure 7 sketches the trajectories of the beam in the final race track microtron. The polarization vector precesses faster than momentum in the two dipol magnets of the microtron because of the anomalous magnetic moment of the electron [23]:

$$\Phi = \Phi_0 + 2\pi a(n+1)(\gamma_0 + \frac{n}{2}\delta) \qquad (1)$$

with $\gamma_0 = \frac{E_0}{m_0 c^2}$; $\delta = \frac{\Delta}{m_0 c^2}$; $a = \frac{g-2}{2}$ and g = g–factor of the electron
MAMI–energy may be variied slightly by variation of E_0 as well as Δ by a common factor. One gets a variation of spinorientation:

$$\Delta\Phi = 2\pi a(n+1)(\gamma_0 + \frac{n}{2}\delta) \cdot \frac{\Delta E_n}{E_n} \qquad (2)$$

With n = 90 the final energy is E_{90} = 855 MeV and

$$\frac{\Delta\Phi}{\Delta E} = 45 \frac{deg}{MeV} \qquad (3)$$

Spintuning via fine–adjustment of the MAMI end energy has been tested successfully and applied in production experiments already.

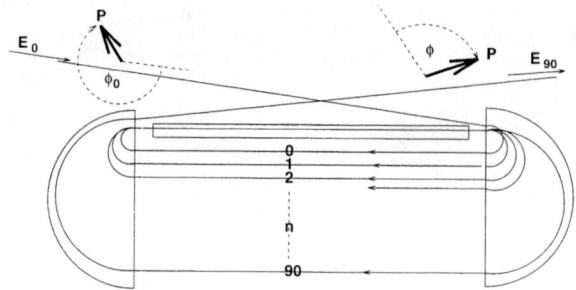

FIGURE 7. Beam trajectories in the third race track microtron of MAMI schematically. Injection Energy $E_0 = 179.7$ MeV, energy gain per turn $\Delta = 7.5$ MeV, number of turns n, end energy $E_{90} = E_0 + n \cdot \Delta = 855$ MeV

TABLE 2. Present Status of the MAMI source of polarized electrons

Source parameter	Measured value	Remarks
Cathode type		strained layer GaAsP
Spinpolarization	75 %	mean over several samples
Current at source	23 μA	
Current at target	10 μA	
Capture efficiency at injection	44 %	
Capture efficiency projected with pulsed mode	$\geq$80 %	
Spintuning		by adjustment of accelerator energy

CONCLUSION

Table 2 summarizes the present status of the source of polarized electrons installed at the Mainz race track microtron. Main achievements obtained are: a) Improved beam emittance and gain in operational stability by installation of another source right at the injection point of the accelerator. b) Threefold enlargement of capture efficiency by addition of a harmonic buncher cavity to the injection line. Another factor of two may be expected in near future by driving the source by a 2.45 GHz pulsed laser that produces a bunched beam already at source level. c) Spin tuning of the accelerated beam may be accomplished via energy tuning.

REFERENCES

1. "Jahresbericht 1994–1995," Institut für Kernphysik der Johannes Gutenberg Universität in Mainz, 1995.
2. K. Aulenbacher, C. Nachtigall, H. G. Andresen, J. Bermuth, T. Dombo, P. Drescher, H. Euteneuer, H. Fischer, D. v. Harrach, P. Hartmann, J. Hoffmann, P. Jennewein, K.-H. Kaiser, S. Köbis, H. J. Kreidel, J. Langbein, M. Petri,

S. Plützer, E. Reichert, M. Schemies, H.-J. Schöpe, K.-H. Steffens, M. Steigerwald, H. Trautner, and T. Weis, "The MAMI–source of polarized elctrons," *Nucl. Instrum. Methods A*, vol. 391, pp. 498 – 506, 1997.

3. D. T. Pierce, "Spin–polarized electron sources," in *Experimental Methods in the Physical Sciences, Vol. 29 A* (R. Celotta and T. Lucatorto, eds.), (New York, London, Toronto, Sydney, San Francisco), p. 1, Academic Press, 1995.

4. E. Reichert, "Sources of polarized electrons.," in *Proceedings of the International Workshop on Polarized Beams and Polarized Gas Targets, Köln, 1995* (H. P. gen. Schick and S. Sydow, eds.), (Singapore, New Jersey, London, Hong Kong), pp. 285–300, World Scientific, 1995.

5. G. D. Cates, V. W. Hughes, R. Michaels, H. R. Schäfer, T. J. Gay, M. S. Lubell, R. Wilson, G. W. Dodson, K. A. Dow, S. B. Kowalski, K. Isakovich, K. S. Kumar, M. E. Schulze, P. A. Souder, and D. H. Kim, "The BATES polarized electron source," *Nucl. Instr. Meth.*, vol. A278, pp. 293–317, 1989.

6. D. A. Engwall, B. M. Dunham, L. S. Cardman, D. P. Heddle, and C. K. Sinclair, "A spin manipulator for electron accelerators," *Nucl. Instr. Meth. A*, vol. 324, p. 409, 1993.

7. T. Nakanishi, S. Okumi, M. Tawada, K. Togawa, C. Takahashi, C. Suzuki, Y. Kurihara, H. Matsumoto, T. Omori, Y. Takeuchi, M. Yoshioka, H. Horinaka, T. Baba, M. Mizuta, T. Kato, T. Saka, and K. Nishitani, "Polarized electron source development in Japan," in *Proceedings of Spin96, Amsterdam 1997* (C. W. de Jager, T. Ketel, P. Mulders, J. Oberski, and M. Oskam-Tamboezer, eds.), (Singapore, New Jersey, London, Hong Kong), pp. 712 – 716, World Scientific, 1996.

8. Y. Bolkhovityanov, A. Gilinsky, C. de Jager, E. Konstantinov, S. Konstantinov, I. Koop, V. Korchagin, V. Kozak, F. Kroes, B. Lazarenko, A. Mamutkin, B. Militsyn, A. Nikiforov, V. Osipov, N. Papadakis, S. Popov, E. Pozdeev, M. van den Putte, G. Serdobintsev, Y. Shatunov, D. Shevelev, T. Sluijk, A. Terekhov, and N. Vodinas, "The polarized electron source at NIKHEF," in *Proceedings of SPIN96, 14–16 September 1996, Amsterdam*, pp. 730–732, World Scientific, 1996.

9. H. Tang, "The SLAC–source of polarized electrons.," in *Proceedings of the International Workshop on Polarized Beams and Polarized Gas Targets, Köln, 1995* (H. P. gen. Schick and S. Sydow, eds.), (Singapore, New Jersey, London, Hong Kong), World Scientific, 1995.

10. R. Alley, H. Aoyagi, J. Clendenin, J. Frisch, C. . Garden, E. Hoyt, R. Kirby, L. Klaisner, A. Kulikov, R. Miller, G. Mulhollan, C. Prescott, P. Sáez, D. Schultz, H. Tang, J. Turner, K. Witte, M. Woods, A. D. Yeremian, and M. Zolotorev, "The Stanford linear accelerator polarized electron source," *Nucl. Instr. and Meth.*, vol. A365, pp. 1 – 27, 1995.

11. D. Eyl, A. Frey, H. Andresen, J. Annand, K. Aulenbacher, J. Becker, J. Blume-Werry, T. Dombo, P. Drescher, H. Fischer, P. Grabmayr, S. Hall, P. Hartmann, T. Hehl, W. Heil, J. Hoffmann, J. D. Kellie, F. Klein, M. Meyerhoff, C. Nachtigall, M. Ostrick, E. W. Otten, R. O. Owens, S. Plützer, E. Reichert, R. Rieger, H. Schmieden, R. Sprengard, K.-H. Steffens, and T. Walcher, "First measure-

ment of the polarization transfer on the proton in the reactions H($\vec{e}$,e'$\vec{p}$) and D($\vec{e}$,e'$\vec{p}$)," *Z. Phys.*, vol. A352, p. 211, 1995.
12. F. Klein, "The electric formfactor of the nucleon," *Prog. Part. Nucl. Phys.*, vol. 36, pp. 53–68, 1996.
13. M. Meyerhoff, D. Eyl, A. Frey, H. G. Andresen, J. R. M. Annand, K. Aulenbacher, J. Becker, J. Blume-Werry, T. Dombo, P. Drescher, J. E. Ducret, H. Fischer, P. Grabmayr, S. Hall, P. Hartmann, T. Hehl, W. Heil, J. Hoffmann, J. D. Kellie, F. Klein, M. Leduc, H. Möller, C. Nachtigall, M. Ostrick, E. W. Otten, R. O. Owens, S. Plützer, E. Reichert, D. Rohe, M. Schäfer, L. D. Schearer, H. Schmieden, K.-H. Steffens, R. Surkau, and T. Walcher, "First measurement of the electric formfactor of the neutron in the exclusive quasielastic scattering of polarized electrons from polarized ^{3}He," *Phys. Lett. B*, vol. 327, pp. 201–207, 1994.
14. W. B. Herrmannsfeldt, *Electron trajectory program, SLAC report 166*. 1979.
15. P. Drescher, H. G. Andresen, K. Aulenbacher, J. Bermuth, T. Dombo, H. Euteneuer, N. N. Faleev, H. Fischer, M. S. Galaktionov, D. v. Harrach, P. Hartmann, J. Hoffmann, P. Jennewein, K.-H. Kaiser, S. Köbis, O. V. Kovalenkov, H. J. Kreidel, J. Langbein, Y. A. Mamaev, C. Nachtigall, M. Petri, S. Plützer, E. Reichert, M. Schemies, K.-H. Steffens, M. Steigerwald, A. V. Subashiev, H. Trautner, D. A. Vinokurov, Y. P. Yashin, and B. S. Yavich, "Photoemission of spinpolarized electrons from strained GaAsP," *Applied Physics A*, vol. 63, pp. 203–206, 1996.
16. V. T. Shvedunov, M. O. Ihm, H. Euteneuer, K.-H. Kaiser, and T. Weis, "Design of a prebuncher for increased longitudinal capture efficiency of MAMI," in *EPAC96, Vol. 2* (S. Myers, A. Pacheco, R. Pascual, C. Petit-Jean-Genaz, and J. Poole, eds.), (Bristol and Philadelphia), pp. 1556 – 1558, Institute of Physics Publishing, 1996.
17. M. Poelker, "High power gain–switched diode laser master oscillator and amplifier," *Appl. Phys. Lett.*, vol. 67, pp. 2762–2764, 1995.
18. P. Hartmann, J. Bermuth, J. Hoffmann, S. Köbis, E. Reichert, H. G. Andresen, K. Aulenbacher, P. Drescher, H. Euteneuer, H. Fischer, P. Jennewein, K. H. Kaiser, H. J. Kreidel, C. Nachtigall, S. Plützer, M. Schemies, K.-H. Steffens, M. Steigerwald, H. Trautner, D. v. Harrach, I. Altarev, R. Geiges, K. Grimm, T. Hammel, E. Heinen-Konschak, H. Hofmann, E.-M. Kabuß, A. Lopes-Ginja, F. Maas, and E. Schilling, "Picosecond polarized electron bunches from a strained layer GaAsP photocathode," *Nucl Instr. Meth. A*, vol. 379, pp. 15–20, 1996.
19. M. Ciarocca, H. Avramopoulos, P. Hartmann, J. Hoffmann, K. Aulenbacher, E. Reichert, J. Schuler, and H. Trautner, "2.45 GHz synchronized polarized electron injection at MAMI," *Nucl. Instr. Meth*, vol. to be published, 1997.
20. M. Ciarocca, H. Avramopoulos, and C. N. Papanicolas, "A modelocked semiconductor laser for a polarized electron source," *Nucl. Instr. Meth. A*, vol. 385, p. 381, 1997.
21. J. Hoffmann, P. Hartmann, C. Zimmermann, D. v. Harrach, E. Reichert, K. Aulenbacher, H. Euteneuer, K.-H. Haiser, S. Köbis, M. Schemies, M. Steiger-

wald, H. Trautner, K. Grimm, T. Hammel, H. Hofmann, E.-M. Kabuß, A. Lopes-Ginja, F. E. Maas, P. Piatosa, and E. Schilling, "Selfstarting modelocked Ti:Sapphire laser at a repetition rate of 1.039 GHz," *Nucl. Instr. Meth. A*, vol. 383, pp. 624–626, 1996.

22. K. H. Steffens, H. G. Andresen, J. Blume-Werry, F. Klein, K. Aulenbacher, and E. Reichert, "A spinrotator for producing a longitudinally polarized electron beam with MAMI.," *Nucl. Instr. Meth.*, vol. A325, pp. 378–383, 1993.

23. V. Bargman, L. Michel, and V. L. Telegdi, "Precession of the polarization of particles moving in a homogeneous electromagnetic field," *Phys. Rev. Lett.*, vol. 2, pp. 435–437, 1959.

Polarized Electrons at MIT-Bates

Manouchehr Farkhondeh, David Barkhuff, George Dodson,
Evgeni Tsentalovich, Bin Yang, Townsend Zwart,
Ernie Ihloff and Christopher Tschalær

MIT-Bates Linear Accelerator Center [1]
Middleton, MA 01949, USA

Abstract. A description of the MIT-Bates polarized electron source will be presented. Improvements to the polarized injector in recent years, including implementation of a multiple gun system, have made delivery of pulsed polarized beams for medium energy experiments routine and trouble–free. The spin orientation from the source is controlled by a Wien filter which has the added benefit of sweeping backstreaming positive ions away from the photocathode surface. The present challenge for the Bates polarized source is an unprecedented set of requirements on the stability and quality of the beam for the SAMPLE parity violation experiment now in progress, including limits on the source lifetime and helicity correlated beam position differences. The current status of the source lifetime is discussed. Recent results in laser beam position differences are presented as is a description of the diagnostics implemented on the laser transport line for fast and accurate measurements of these differences. Finally, we will discuss the plans for using a mode–locked laser system to achieve the higher laser power necessary for the South Hall Ring.

INTRODUCTION

With the pulse structure and the duty cycle at Bates, peak currents of tens of mA with pulses as long as 20 μs are needed from the polarized source. These peak currents, the long pulse length and a 600 Hz repetition rate fall in a very difficult regime for commercially available laser systems. Table 1 lists the injector parameters for both the pulsed and South Hall Ring (SHR) mode of operation.

The parity violating SAMPLE experiment at Bates requires a very high quality beam. The specifications for this beam are listed in Table 2. With the maximum $\sim$4 W peak laser power currently available at the photocathode,

---

[1] Work supported in part by the Department of Energy under contract DE-AC02-76ER03069.

Quantum Efficiencies (QE) of greater than 0.5% are needed to obtain the SAMPLE peak currents of ~12 mA. So far, this requirement has made it impractical to use high polarization photocathodes which have low QE's.

SAMPLE also requires relatively long lifetime from the source so that a minimum of 2 coulombs a day on target may be accumulated for periods of several days between activation cycles. Each activation, which includes heat cleaning, requires about 10-12 hours of downtime. The hydrogen part of the SAMPLE experiment requires 1200 hours of beam on target to achieve the required statistical precision. With the linac capture efficiency of about 1/3, the experiment therefore requires lifetimes of more than three days at an average current of $120\mu A$ in the injector.

The SAMPLE experiment measures a parity violating asymmetry between the yield of positive and negative helicity electrons scattered from a liquid hydrogen target at backward angles. The measured asymmetry is extremely small, $\sim 10^{-6}$. It is therefore crucial to suppress any helicity–correlated differences in the properties of the electron beam which may introduce false asymmetries in the measured yield. This includes differences in position, angle and energy of electrons incident on the hydrogen target for the two helicity states. For example, the experiment requires helicity–correlated beam position differences to be controlled at the level of a few hundred nanometers at the target [1]. This is particularly demanding since the spot size of both the laser beam on the crystal and the electron beam on the target have full width half maxima of a few millimeters. To date, about 1/4 of the data for the SAMPLE experiment have been taken. More progress in improving the source lifetime is essential before the experiment can be finished.

TABLE 1. Polarized Injector Parameters

	Pulsed Mode	SHR Mode Extraction	SHR Mode Storage
Injection Energy	360keV	360keV	360keV
Pulse Width	$16\mu s$	$2.6\mu s^a$	$2.6\mu s$
Intensity[b]: Peak	12 mA	40-160mA[c]	30^d-120mA
Average	$120\mu A$	65-260μA	$\leq 1\mu A$
Duty Factor	1×10^{-2}	$\sim 10^{-3}$	$\sim 10^{-5}$
Repetition Rate	600Hz	600Hz	$\sim$ 1Hz
Linac Capture Fraction	$1/3^e$	1/3	1/3

[a] $1.3\mu s$ for beam loading transient
[b] based on linac capture fraction of 1/3
[c] for extracted average currents of 10-40μA on target
[d] with stacking 4 pulses
[e] with CW laser

INJECTOR SYSTEM

The polarized injector at Bates consists of a commercial high power CW laser system, a 60 kV diode electron gun, a 60 keV Wien filter, a 300 kV acceleration column, a vertical transport line and an achromat with two 45° dipoles. The gun chamber and associated instrumentation are located inside an elevated Faraday cage with 11 sets of corona rings to isolate the high voltage. An isolation transformer supplies utility AC power into the cage for instrumentation controls and ion pumps.

Recently, the injection line has been completely rebuilt. This rebuild included installation of a new acceleration column of larger aperture, removal of other small apertures, installation of additional ion and NEG pumps, improvement of the injector alignment, and replacement of the solenoid lenses with solenoids made of two coils wired back-to-back. These new solenoids allow changes in the beam focusing without affecting the spin orientation. The injection line was successfully commissioned early this year.

Gun Chamber

The design of the MIT diode guns currently in use is based on prints from the University of Illinois and Jefferson Lab with some non-critical modifications. The effusion Cs source was replaced by 4-cell Cs channel dispensers [2]. Remote cesiation of the photocathode is now routine. A few remote cesiations with both high voltage and beam showed no apparent sign of damage to the photocathode.

Until 1995, the polarized injector consisted of a single gun with no spare. A crystal change required a minimum of two weeks downtime to complete a bakeout, during which no polarized beam was available. In 1995, we embarked on and completed a program to build a twin gun system. The basic goal was to have a direct replacement gun available for immediate installation with a ready-to-run photocathode under UHV conditions. When the guns are interchanged, only a small transition section between two valves is baked. A

TABLE 2. Polarized beam requirement for the SAMPLE experiment.

Beam Parameter	Specification
Average current on target	40 μA
Intensity stability	1%
Beam FWHM on target	~2.5 mm
helicity–correlated intensity differences (in 1/2 hour)	$\leq$10 ppm
helicity–correlated position differences (in 1/2 hour)	$\leq$ 250 nm
Time between full activation	$\geq$ 3 days
Total hours at 40 μA	1200 hours

typical gun interchange now requires only about four days of downtime. We now have three identical guns, one on the injector, one as spare ready for installation and one for high polarization photocathode tests.

Laser System

The laser system consists of a 30W CW Ar laser pumping a CW Ti:sapphire laser. The system parameters appear in Table 3. The Ar laser is chopped with a phase–locked electro-mechanical chopper. With 29 W Ar, peak powers as high as 8 W at 750 nm are extracted from the Ti:sapphire. About 4 W is delivered to the photocathode after insertion losses in the remotely adjustable mirror transport system and active feedback systems. The total flight path between the $\lambda/4$ Helicity Pockels Cell (HPC) and the cathode is about 15 m with a point to point imaging. A commercial feedback system [11] is used to control the pulse-to-pulse intensity variation to better than 0.6%. The helicity–correlated position differences in the laser beam are described in detail below.

PERFORMANCE

In our guns, QE's of 3-5% for bulk GaAs at $\sim$750 nm are typically achieved. Photocathodes are heat cleaned to 600°C and activated using the "yo–yo" method of Cs and NF_3 deposition. With bulk GaAs, polarizations of 32-38% are measured with Möller polarimeters. Dark lifetimes typically exceed several hundred hours for each of the three guns. In one measurement at 60 kV, the dark current was at or below 25nA including the corona discharge from a Keithley picoammeter placed on top of the gun at 60 kV. At 100 kV, the dark current could only be measured using the current monitor of the high voltage power supply. No noticeable dark current above 50 nA, the minimum sensitivity of the power supply, was detected. In all three guns, the initial QE uniformity across the crystal surface is within 10-15%. However, after running beam at high average currents, a QE gradient across the surface of

TABLE 3. Laser System Parameters (with Chopping)

	Pump LASER	IR LASER
Type	Argon Ion	Ti:sappire
Model	COHERENT Inova 425	Spectra-Physics 3900S
Peak Power	30W	8W [a]
Wavelength	458-520nm	700-900nm
Stability	$\leq$0.3%	1-2% no feedback
		0.4 -1% feedback

[a] $\sim$4W @ crystal

the photocathode is developed which always shows higher QE at "9 o'clock" than at the center as seen by a CCD camera through a dedicated view port. A possible explanation is that some asymmetric field may exist in the gun which guides damaging ions non-uniformly toward the cathode.

The beam lifetime is a function of the average current. In 1996-1997, delivering 10-15 μA on target for the Out-Of-Plane Spectrometer System (OOPS) experiment, we obtained an effective lifetime of about 400 hours with periodic recesiation. At these relatively low average currents the beam delivery for the polarized source is routine and trouble free.

Source Lifetime for SAMPLE

To deliver 40 μA to the SAMPLE target, peak currents of 12–13 mA and average currents of 120-130 μA in the injector are needed. With 4 W laser power at the crystal, QE's of greater than $\sim$0.5% are needed. To efficiently deliver 1200 hours of beam for SAMPLE, it is essential to run 2-3 days before stopping the experiment for 10-12 hours for heat cleaning and activation. During the past 2 years, we have observed lifetimes as long as 10 days and as short as 24 hours between activations. These unpredictable results have made it difficult to efficiently run the experiment. During two periods of long lifetime in 1995 and 1996, 20-25% of the total data proposed for SAMPLE experiment was collected and the physics results have been published [7].

In October of 1995 during the SAMPLE experiment, the lifetime was as long as 80 hours between activations. In April and May of 1996, lifetimes as long as 190 hours were observed. To date, no polarized electron beam of higher average current has ever been delivered with such long lifetimes between activations. The QE for the period of May 13–22 is shown in in Figure 1. Clearly, most of data was taken at QE's below 0.7%.

Later in May and June of 1996, lifetimes as short as 24 hours were observed. With the effusion source at that time, there was evidence of excess Cs in the gun chamber. Now, with Cs channel dispensers, chambers show the desirable signs of under-cesiation.

In March and June of this year, during two SAMPLE development runs, lifetimes as long as a few days, adequate to meet the requirements of the SAMPLE experiment, were obtained. However, later in June, lifetimes as short as 24 hours were again observed. We do not yet fully understand the conditions under which longer lifetimes have been achieved. With a closer examination of lifetime and optics simulations, we now suspect that the beam scrape–off near the gun or at or above the Wien filter may be responsible for this hit–or–miss behavior.

Recent detailed optics studies show that with the larger laser spot needed to reduce cathode charge limits, the present cathode electrode angle of 39° may cause over–focusing of the beam at the first solenoid lens immediately after

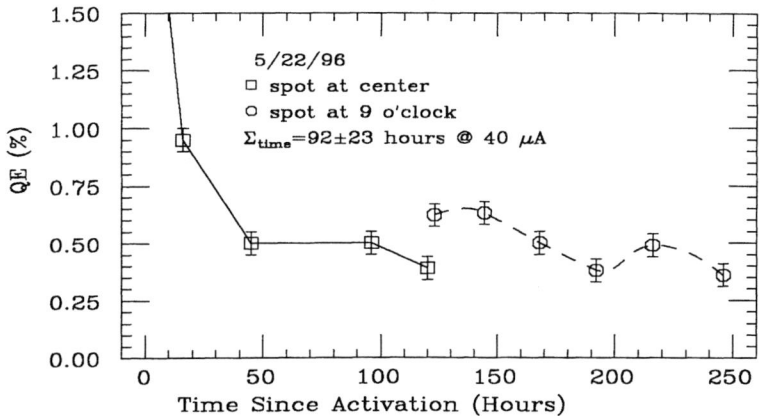

FIGURE 1. QE at 12 mA peak current vs time since activation for the period of May 13-23 1996. The total accumulated beam time at 40 μA equivalent on target (120 μA in the injector) was about 92±23 hours. The period with the laser spot at the center of the crystal is shown by the solid line. The spot was later moved to a different spot (called "9 o'clock" and indicated by the dashed line) near the edge which usually had higher QE than the middle.

the gun. This over-focused beam then becomes large at small apertures above or inside the Wien filter. To address this, a new cathode electrode of 25° is being installed in one of the guns for tests. In the past, backstreaming ions from the acceleration column have long been suspected of contaminating the photocathode and causing QE degradation and variation across the surface. With the Wien filter between the column and the gun, and with improved vacuum in the new injector, the ions from the columns are now less suspect. Ions between the anode and cathode are also suspect. Their possible effects can only be reduced by further improvements to the gun chamber vacuum. A series of lifetime studies are planned for early Fall to address these concerns.

Cathode Charge Limit

In our injector, measurements with beam show a continual reduction of the effective QE with increasing laser power. By adding NF_3 or Cs, QE is only partially recovered. This behavior appears to be consistent with a photocathode charge limit at high laser power which has also been observed at other laboratories [3–5] and was discussed during this and other workshops [6]. This behavior is shown in Figure 2.

One way of reducing the charge limit is to increase the laser beam diame-

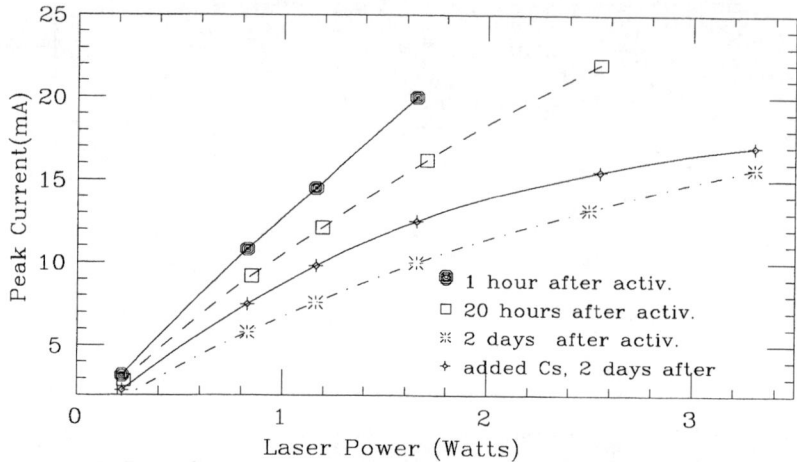

FIGURE 2. Maximum peak current vs. laser peak power at 760 nm shown for several conditions. A continual reduction of QE with time and a partial recovery with recesiation are observed.

ter on the photocathode. When we doubled the laser area, we measured an increase of about a factor of two in the maximum peak current at the same laser power. As discussed in the previous section, to fully utilize larger spot size we may need to reduce the cathode electrode angle to better control the beam envelope immediately below the gun.

HELICITY–CORRELATED DIFFERENCES

As mentioned above, the SAMPLE experiment has very stringent requirements for the helicity–correlated differences of the electron beam. These helicity–correlated differences all have their origin in the laser system where a single state of linearly polarized light is converted to left or right circularly polarized light by the helicity determining $\lambda/4$ Pockels cell[2]. Subsequent elements can amplify or reduce these effects, but an effective approach is to minimize them at their source.

To address this problem we have implemented a set of diagnostics to monitor the properties of the laser beam downstream of the HPC. These diagnostics are built around a set of diodes [12] which are segmented in quadrants. The analog signals from each of the quadrants are integrated and digitized in a CAMAC system and the results are then analyzed on a VAX Station 4000.

---

[2]) Lasermetrics model 1058FW.

By comparing the relative weight of the four quadrants a centroid can be accurately determined. Typically, differences in the centroids of the two helicity states of the laser beam can be measured to 100 nm in ten minutes.

We have found that to consistently achieve good results for helicity correlated position differences, a key step is to establish the proper alignment procedure for the HPC. When we follow this alignment procedure, we consistently measure circular polarization in excess of 99.9 % and position differences of less than 100 nm within 30 cm of the exit of the Pockels cell.

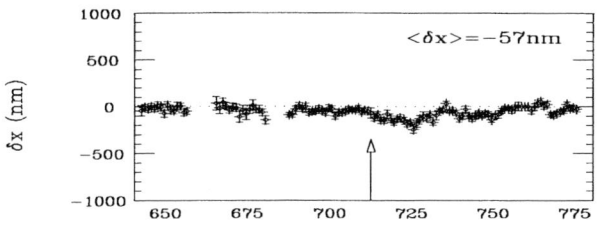

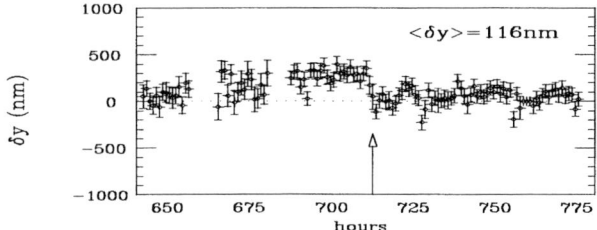

FIGURE 3. Helicity–correlated centroid differences of the laser beam measured at the location of the GaAs crystal. The Pockels cell high voltage was adjusted at the time indicated by the arrow to minimize these differences.

Our situation is complicated by the relatively long distance, 15 m, between the laser source and the GaAs crystal. Traditionally, we have located the HPC within one meter of the laser cavity and transported the two circularly polarized laser states through a balanced system of four mirrors to the crystal. This has the advantage that the HPC is accessible and easily adjusted, but the disadvantage that the mirrors may amplify or introduce helicity-correlated effects. Recently we have tried another approach in which the HPC is located after the mirror transport line but, due to constraints of the electron beam geometry, is still 6 m from the crystal. This avoids any problem with the mirrors, but has the disadvantage that the HPC is located in a radiation area where access is restricted.

Over the past six months we have made an extensive series of measurements of the helicity–correlated laser beam centroids. Figure 3 shows a measurement of the position differences of the two helicity states measured over a period of six days. In this case the HPC was located after the mirror transport system. The averaged asymmetries are -57 nm in X and 116 nm in Y. In the upcoming Fall run we will evaluate the helicity–correlated differences of the electron beam with the HPC installed in this new location.

The helicity–correlated intensity differences must also be maintained below ∼10 ppm in each half hour. This is done with an active feedback system. An old feedback system which involved real-time adjustment of the HPC high voltage to correct helicity–correlated intensity differences introduced helicity-correlated position differences. A new scheme has been developed to overcome this problem. Now, the HPC high voltage for each helicity state is fixed at the point which minimizes position differences. In addition, we have added a new $\lambda/10$ wave plate, an additional "correction" Pockels cell, and a polarizer for the sole purpose of helicity–correlated intensity feedback. This has been installed, fully tested with beam and works very well.

SOUTH HALL RING REQUIREMENTS

One of the design goals for the South Hall Ring (SHR) is to inject polarized beam into the ring with peak currents as high as 40 mA. A universal superconducting spin rotator, or "Siberian Snake", will provide longitudinal polarization at the interaction point in the ring and at the fixed target location in extraction mode. With a 30% injection efficiency, peak currents as high as 120 mA, pulses of 2.6 μs duration at a few Hz may be required from the polarized gun for the stored beam. Stacking four pulses would reduce this peak current to 30 mA. For the extraction mode, the peak injector current will be as high as 100 mA at 1.2 kHz. For the storage mode, high power pulsed lasers available on the market will be adequate to meet the goals. Two laser systems are under development at the Institute of Communication and Computer Systems (ICCS) [8] at the Technical University of Athens. These laser systems are intended for use with all three modes of linac operations, i.e., pulsed, storage and extracted modes.

New Laser Systems

In collaboration with ICCS, a mode-locked external cavity laser system is under construction for Bates. The system uses commercially available components. It consists of a Ti:sapphire laser coupled to a tunable cavity and seeded by diode lasers. The laser output would be locked to the 2856MHz linac RF frequency. A description of this laser system is presented in this and other workshops [8,9]. With this system, peak powers of 1.1 nJ in each 60 ps micro-pulse have been observed. This should be compared with 0.8 nJ, the peak power corresponding to the linac-captured fraction of the existing laser system. However, mode-locking would increase the beam capture fraction to nearly 100% from the current ∼30%. The potential advantage of an increase of a factor of three in capture efficiency would be a reduction of the average current in the injector by a factor of three. This should result in a longer lifetime of the photocathode. The laser system is designed for two central wavelengths of 780±10 nm and 830±10nm.

A second laser system is also under construction at ICCS for meeting the higher power requirements of SHR. This Master Oscillator Power Amplifier Fiber System (MOPA) is based on technologies utilized in the telecommunication industry. The system produces radiation at ∼ 1500 nm. The radiation is frequency doubled to near 780 nm. This pulsed laser system is operated at 2500 MHz, with a bunch-width of 1 μs and at a repetition rate of 1.2 kHz which is suitable for both modes of SHR. To date, 80 nJ peak power at 1.5 nm wavelength has been obtained [10]. Even with 20-30% efficiency of the doubling crystal, we expect peak powers an order of magnitude higher than the existing laser system provides.

REFERENCES

1. Bates experiments 89-06 (R. McKeown and D. Beck, contacts), and 94-11 (M. Pitt and E. Beise, contacts), also see E. Beise *et, al.*,"International Symposium on High-Energy Spin Physics" (SPIN96), Amsterdam, World Scientific, ISBN 981-02-3052-4, 89 (1996).
2. SAES Getters/USA Inc., 118 Charcot Ave., San Jose, CA 95131 478.
3. A. Herrera-Gomez, et. al., SLAC-PUB-95-6966.
4. R. Alley,et.al, Nucl. Instr. and Meth. in Phys. Res. A 365(1995) 1-27.
5. S. Nakamura, *et. al.*, (SPIN96) Amsterdam, World Scientific, ISBN 981-02-3052-4, 708 (1996).
6. D.A. Orlov,*et. al.*, Workshop on Polarized Sources and Low-Energy Polarizations, (SPIN96) Amsterdam, World Scientific, ISBN 981-02-3052-4, 717 (1996).
7. B. Mueller, *et. al.*, Phys. Rev. Lett. **78**, 3824 (1997).
8. H. Avramopoulos, *et. al.*, (SPIN96) Amsterdam, World Scientific, ISBN 981-02-3052-4, 670 (1996).
9. H. Avramopoulos, *et. al.*, "Mode-Locked Lasers at Bates", in these proceedings.
10. H. Avramopoulos, *et. al.*, private communications.
11. Conoptics LASS II, modified for pulse operation.
12. SPOT-9D from UDT Sensors, Hawthorne, CA.

The SLAC Polarized Electron Source[*]

J.E. Clendenin, R. Alley, J. Frisch, T. Kotseroglou, G. Mulhollan,
D. Schultz, H. Tang, J. Turner, and A.D. Yeremian

Stanford Linear Accelerator Center, Stanford University, Stanford, CA 94309

Abstract. Since 1992, the SLAC 3-km linac has operated exclusively with *polarized* electrons. The polarized electron source is highly reliable, remotely operated and monitored, and able to produce a variety of electron bunch profiles for high-energy physics experiments. The source and its operating characteristics are described. Some implications drawn from the operating experience are discussed.

INTRODUCTION

Polarized electrons were first accelerated to high energy in 1974[1] using the SLAC 3-km linac and a Li atomic-beam source.[2] The source had a relatively high polarization, but the peak currents were low and the stability of the beam was marginal. Following a proposal first made in 1974,[3] a polarized electron source[4] based on photoemission from GaAs was developed at SLAC and operated successfully for a high-energy experiment in 1978.

Development of a high peak-current GaAs source matched to the needs of the future SLC was undertaken at SLAC in the early 1980s in parallel with research to find higher-polarization photocathodes. The SLC requires two electron bunches each with a charge in the linac of 5×10^{10} e$^-$ in single S-band buckets, the two bunches separated by 60 ns and repeated at a rate of 120 Hz.

Because of the high charge density, the source was designed to produce 2-ns bunches at >100 kV that would then be rf compressed and accelerated. The rf compression is in two stages utilizing 2 cells at the 16th subharmonic followed by 4 cells at S-band. Compression is completed in the first S-band accelerator section. Because capture plus transmission through this system is limited to about 70%, the gun has to be able to produce up to 10^{11} e$^-$ per 2-ns bunch.

PHYSICS OF GaAs PHOTOCATHODES

Direct band-gap III-V semiconductors, with gaps on the order of 1.5 eV, have long been known as efficient photoemitters into the infrared regime.[5] When highly p-doped, the Fermi level is near the valence band maximum (VBM) while the

---

[*] Work supported by the U.S. Department of Energy contract DE-AC03-76SF00515.

energy bands at the surface are bent downward by as much as half the band gap due to the presence of positive charge at the surface. Using the technique of adding an alkali plus an oxide to the surface, the work function can be lowered from 4 eV to below the conduction band minimum (CBM) in the bulk, creating a negative electron affinity (NEA) surface. Cesium is the most effective alkali, and at SLAC NF_3 is used as the oxide carrier. The process of cleaning the crystal surface and then adding the alkali plus oxide is known as *activating* the surface.

The quantum efficiency (QE) in the red for bulk (i.e., thick unstrained) GaAs is particularly high because the optical absorption length and electron diffusion length for doping concentrations in the mid-range of 10^{18} cm^{-3} are both on the order of 1 µm at room temperature.

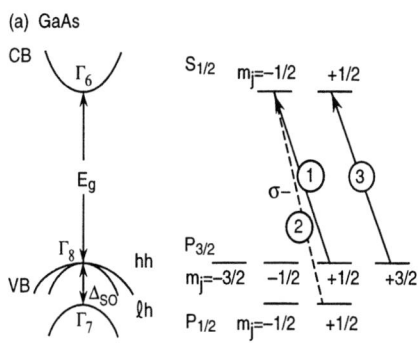

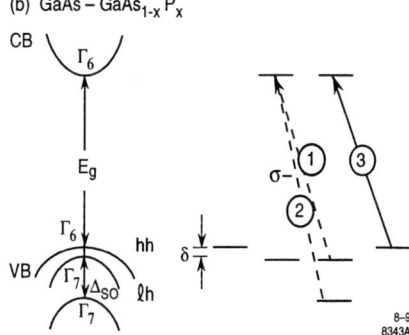

FIGURE 1. Energy level diagram and transition probabilities at the Γ point. Only the transitions for left circularly polarized light are shown. (a) For GaAs, the solid-line transitions are for $E < h\nu < (E_g + \Delta)$. (b) For GaAs-GaAs$_{1-x}P_x$, the solid-line transitions are for $E < h\nu < (E_g + \delta)$.

Although bulk GaAs is an efficient photoemitter, the polarization of the photoemitted electrons is limited to 50%. As shown in Fig. 1(a), there is a two-fold degeneracy in the energy band at the VBM at Γ_8. This band has a $P_{3/2}$ symmetry while the CBM at Γ_6 has an $S_{1/2}$ symmetry. Transitions between these bands are governed by the $\Delta m_j = \pm 1$ selection rule. Because the transition rates from the $m_j = \pm 3/2$ sub-states are 3 times that from the $\pm 1/2$ substates, illumination of the crystal with circularly polarized light will preferentially fill one of the 2 possible conduction band states. The conduction band electrons that then diffuse to a negatively-biased NEA surface will be emitted into vacuum with a theoretical polarization (for the case of left-circularly polarized light) of

$$P_e = \frac{N_+ - N_-}{N_+ - N_-} = \frac{3-1}{3+1} = 0.5.$$

The measured polarizations from bulk GaAs using 100% polarized light vary between 0.25 and very close to 0.5 depending on the experimental conditions.

251

A major achievement in the early 1990s was the discovery that the energy-band degeneracy at Γ_8 could be lifted by directly growing a thin epitaxial layer of the desired III-V material on a substrate with a slightly different lattice constant. This was done first by a SLAC/Wisconsin/Berkeley collaboration[6] using a thin layer of InGaAs MBE-grown on a GaAs substrate, producing a uniaxial compressive strain along the growth direction. However, compared to GaAs, InGaAs has a significantly smaller band gap that is outside the range of high-power tunable laser systems and also results in a lower QE. A better combination was soon found in which a thin layer of GaAs was MOCVD-grown on a GaAsP substrate[7] producing a tensile strain. The cathodes used in the SLAC source since 1993 are of this latter type.[8]

Fig. 1(b) illustrates the allowed transitions for the strained-layer cathodes. If the splitting of the $P_{3/2}$ states at the top of the valence band is larger than the bandwidth (about 25 meV at room temperature), then a narrow-line tunable laser can be adjusted to pump only from the $m_j = \pm 3/2$ substates, resulting in a theoretical polarization of 100%. In practice polarizations of 0.75 to 0.85 are achieved with the SLAC source, depending primarily on the QE. It is interesting that the maximum experimental polarization achieved for strained-lattice GaAs is approximately twice that for thin, unstrained GaAs (both at room temperature).

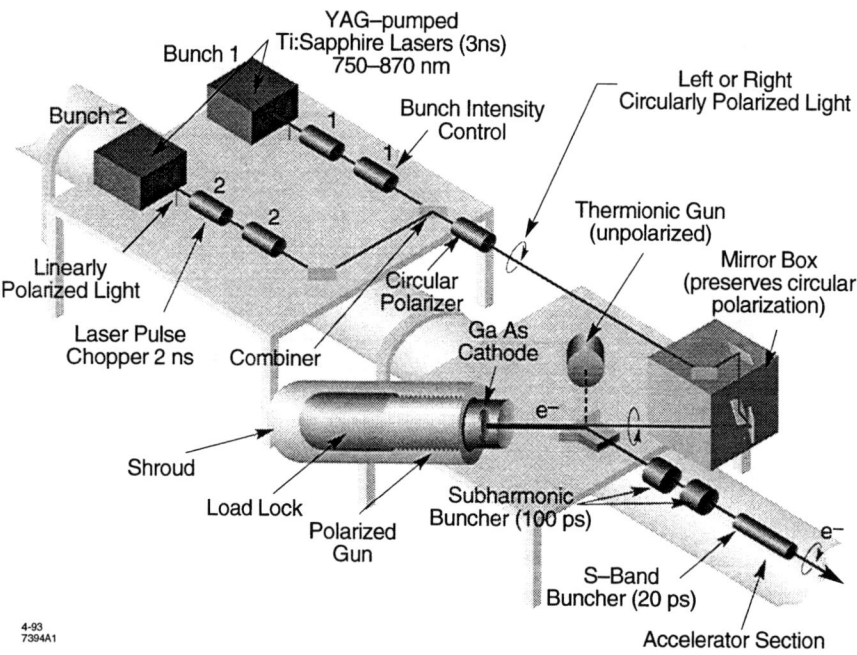

FIGURE 2. The SLAC polarized electron source configured for SLC.

DESCRIPTION OF THE SOURCE

The overall layout of the SLAC polarized electron source[9] is shown in Fig. 2. The gun follows the conventional design for thermionic-cathode guns in which the cathode at high voltage (HV) is supported by a large ceramic insulator which also forms a major part of the vacuum wall. For reliability and ease of operation, the HV is DC. (Pulsed HV would significantly reduce the field emission current.) The Pierce-type electrodes, shown in Fig. 3, were designed[10] using the SLAC Electron Trajectory Program.[11] Special care was taken to minimize the electric field at large radii.[12] The electrodes were fabricated from stainless steel having a very low carbon content and low inclusion density. After machining, they were diamond-paste polished to a 1-µm finish. The gun was assembled in a Class-100 clean room only after a close microscopic inspection of 100% of the electrodes' surfaces showed no remaining contamination or defects >1 µm.

The vacuum for the gun is provided by a 120-l/s ion pump and a 200-l/s non-evaporable getter (NEG) pump. To isolate the gun vacuum system from the 10^{-9} Torr pressure of the rf bunchers, a differential-pumping section follows the gun. It is equipped with a 220-l/s ion pump and a closely-coupled 500-l/s NEG pump. As the RGA spectrum in Fig. 4 shows, the total pressure in the gun is about 10^{-11} Torr, dominated by H_2. The RGA spectrum is not affected by turning either the HV or the beam on or off.

A grounded gas vessel (shroud), which is attached to the rear of the

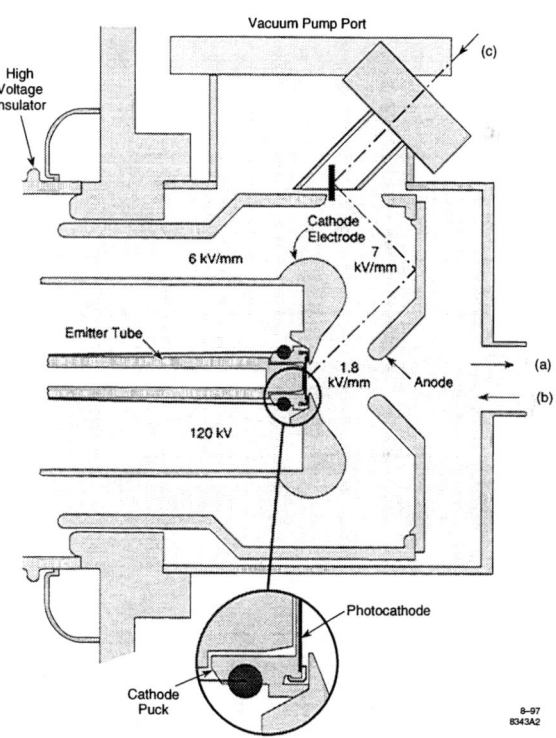

FIGURE 3. Cross section of the gun electrodes. The arrows indicate directions of (a) e- beam, (b) Ti:sapphire laser beam, and (c) beam from diode laser monitoring QE.

gun, prevents discharges from the HV gun terminal to nearby ground level components in the crowded source area. Inside the shroud, a load-lock chamber[13] is attached to the HV end of the gun. It permits the exchange of cathode crystals without backfilling the gun. The cathodes are mounted on Mo pucks allowing them to be activated and then tested at HV in the laboratory for polarization and QE before transferring them to the accelerator gun in a portable vacuum vessel. The operational benefits of a load-lock system will be discussed later.

The cathode temperature and voltage, gun vacuum, leakage current, and dark current, along with the beam current and orbits are monitored using the SLC control system. The beam emittance can be measured using wire scanners at 40 MeV. The energy and energy spread can be measured using a spectrometer at 200 MeV. The principal maintenance activity is to recesiate the cathode every few days. The cesiation is computer controlled, but initiated by the SLC operators. Each cesiation takes about 30 minutes, mostly to run the HV supply down and then back up.

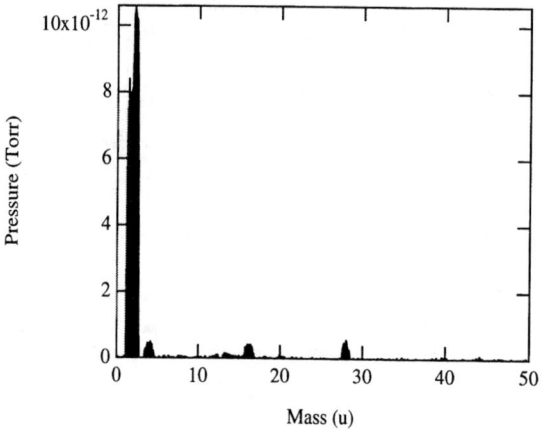

FIGURE 4. RGA spectrum of the gun vacuum during 1997 operation.

A SLAC-built Ti:-sapphire laser system that can be tuned to the band gap energy of most III-V semiconductors was commissioned in 1993 and is still in operation. Two commercial flashlamp-pumped Nb:YAG lasers each operating at 60 Hz are combined to provide 120-Hz pumping. The combined YAG output is split to pump two separate Ti:sapphire cavities, one cavity for each of the 2 required SLC pulses. The Ti:sapphire cavities are each Q switched and cavity dumped, the timing of which is adjusted to provide the desired 60-ns separation. Pulsed Pockels cells with crossed polarizers are used to chop out the desired 2-ns pulse and to control the intensity of the pulses from the first cavity (the *production* cavity[14]) independently of the intensity of the pulses from the second cavity (the *scavenger* cavity[15]). Ten feed-back and 4 feed-forward systems are used to stabilize the laser itself for SLC operation. Over 100 µJ of energy at the cathode can be provided in each laser pulse. The system is remotely monitored and controlled from the SLC main control center via the SLC computer control program with oversight by system physicists. Routine maintenance includes changing the flashlamps every couple months.

For fixed target experiments, a SLAC-built flashlamp-pumped Ti:sapphire laser system has been used.[16] It has provided up to about 75 μJ at the cathode in a 1-2 μs pulse or more recently, with improved reflectors, a similar energy was provided in a 200-300 ns pulse. The flashlamps for this system were changed every 2 weeks to preclude any chance of lamp explosions.

To monitor the QE of the photocathode during accelerator operations, an 833-nm cw diode laser, modulated at 1.5 Hz but gated off during beam time, is rigidly attached to a vacuum window near the cathode as shown in Fig. 3. An optically-powered nanoammeter (at HV) transmits the current drawn from the cathode to a lock-in amplifier via an optical fibre. The modulated signal is detected and continuously recorded by the SLC control system with appropriate conversion constants for the diode laser power. This system is calibrated by a second diode laser whose output can be substituted for the Ti:sapphire's. Finally, the SLC control system periodically runs the Ti:sapphire laser beam through its full energy range using 20 accelerator pulses in succession. For this, the cathode current is measured in the normal manner. The laser beam energy is monitored from pickoff light using calibrated photodiodes. The peak current is recorded online while all the data is saved to an offline disk.

OPERATIONAL CHARACTERISTICS

The gun was originally built without a load-lock chamber. This required the photocathode to be inserted into the gun when it was at atmospheric pressure. The gun then had to be evacuated and baked before the cathode could be activated and tested. Often following a bake the cathode could not be successfully activated. The original design called for the gun to be operated at 150 kV. The gun could be HV processed either before or after the cathode was initially activated, but the processing generally poisoned the cathode requiring reactivation which in turn required addition processing, etc. No matter how much care was taken, this cycle was only rarely broken by production of a fully processed and activated cathode. One such occasion was at the beginning of the 1992 SLC run.

The load-lock, which was built during the 1992 SLC run, completely solved the problem above, but also provided other benefits, including:

a) Experience at SLAC indicates that cathodes which undergo a complete gun bake do not activate as well as cathodes which are not baked.[17] The SLAC load-lock system incorporates a cathode activation system. The load-lock/activation-chamber is itself initially baked after which it is continuously maintained under vacuum. Cathodes are introduced into the load-lock chamber either from air (requiring activation in the load-lock chamber) or from a portable vacuum chamber (if already activated, the cathode may only need to have a little Cs/NF_3 added in the load-lock chamber). When activated cathodes are introduced into a baked gun that has already been HV processed, no additional HV processing is required.[18] With the addition of the load-lock system, activations (but not cesiations) in the gun itself have been eliminated.

b) Load-lock systems permit QE lifetimes to be longer. This is probably due to the constantly-improving gun vacuum. The influence of vacuum on lifetime will be discussed in more detail later. The gun presently in use for SLC has been maintained under continuous UHV conditions since the spring of 1993 and now routinely produces lifetimes in excess of 1000 h.

c) When the gun is not backfilled, it needn't be baked. This virtually eliminates vacuum leaks once the gun is initially evacuated and successfully baked. Consequently the system reliability vastly improves.

d) The benefits of electron scrubbing--from the beam and due to field emission from vacuum components operated at HV--continuously accrue in favor of better vacuum and thus also of longer lifetimes.

e) As a consequence of long lifetimes, the cathodes can more readily be operated with a QE that is close to the minimum needed to produce the charge required by a given experiment. Since the polarization is weakly and inversely dependent on QE, this permits some slight improvement in average polarization.

f) For experiments using polarized electron beams, it is important to know the beam polarization. A constant polarization from the source--which requires a constant QE, i.e., a long lifetime--can reduce the integrated error of polarization measurements.

The cathode is normally operated at 120 kV and 0°C. Immediately after a cesiation the QE (833 nm) is typically 0.005.[19] At the polarization peak, which is typically at ~850 nm, the QE (120 kV) is approximately a factor of 2 lower. The QE decreases with time. It can usually be restored to its original value by adding a small amount of Cs (additional oxide not necessary). For the SLAC source, these cesiations are done discretely since the HV is first lowered to 1 kV, disrupting the beam. If the cesium supply is turned off before (after) the peak in the photoyield is reached, it is called an under- (over-) cesiation.

The time in which the QE decreases by a factor of e is defined as the QE lifetime. Lifetimes computed following each cesiation during the SLC 1994-95 run are shown in Fig. 5. The QE lifetimes are significantly longer when over-cesiation is employed. This is confirmed by the lifetimes associated with the maximum charge.[20]

Although the maximum charge that can be extracted from a semiconductor photocathode appears to be independent of the thickness of the epilayer, for a given cathode it varies linearly with the low-charge QE (measured near the band gap) up to the gun space-charge limit. Lifetimes derived from measurements of maximum charge are also displayed in Fig. 5. The most likely reason the maximum-charge lifetimes appear to be longer than the corresponding QE lifetimes is that at high QE the maximum charge is depressed by the space-charge limit.

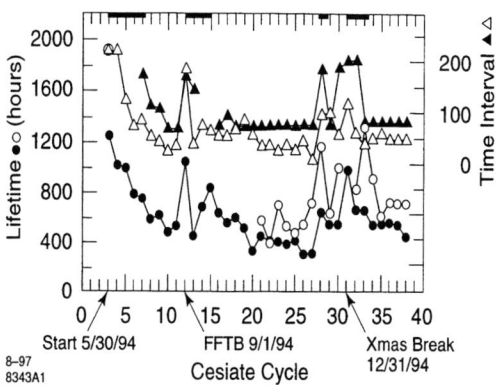

FIGURE 5. Lifetime data for the SLC 1994-95 run as a function of cesiation cycle. The closed (open) circles represent QE (maximum-charge) lifetime, the closed and open triangles represent the cesiation time (in seconds) and days in a cesiation cycle (x10) respectively. The heavy bars at the top represent periods of over-cesiation whereas under-cesiation was otherwise employed.

For SLC, the QE is kept as low as possible since this slightly increases the average beam polarization. During the SLC 1994-95 run, the cathode was typically under-cesiated every 5 days with about 0.1 monolayer of cesium,[21] corresponding to an average Cs disposition rate of 1.4×10^8 atoms cm^{-2} s^{-1}. The minimum partial pressure, P_{min}, of an oxidizing background gas that would be required if the added cesium is continuously oxidized by it alone is given in Table 1. The 1994 RGA pressures are also shown. Clearly O_2 and CO_2 cannot alone be responsible for oxidizing the added Cs. CO_2 is also ruled out because it has been found not to permit complete restoration of the QE,[22] and O_2 is unlikely since the activations at SLAC are made with NF_3 and in a separate chamber. CO has been found to have little or no effect on the QE.[22] Excluding the effect of H*, this leaves water as the most likely oxidizer in the SLAC source causing the QE to decrease. Water molecules are presumably slowly scrubbed from the vacuum walls by a combination of beam interception and field emission from components held at HV.

Table 1. Calculated and Measured Partial Pressures of Oxidizing Gas Species

Gas Species	P_{min} (Torr)	P_{RGA} (Torr)
H_2O	3×10^{-13}	2×10^{-13}
CO	4×10^{-13}	2×10^{-12}
O_2	4×10^{-13}	2×10^{-14}
CO_2	5×10^{-13}	1×10^{-14}

Since the SLAC source is operated above 100 kV, it is possible to minimize beam interception as the beam exits the gun by detecting and minimizing the associated x-rays. Field emission is more difficult to eliminate. Following the introduction in late 1992 of the procedures for preparing and assembling electrodes outlined earlier, it was still found necessary to reduce the gun voltage to about 120 kV in order to consistently achieve a dark current of <50 nA and thus eliminate any gross effect of the HV operation on the QE lifetime. Subsequently the dark current has dropped below 20 nA.[23]

A single strained-layer cathode has been operated in the SLAC source for as long as 0.015 A-h cm^{-2} (equivalent to 1-h of operation of a 10-μA cw beam having a diameter of 0.3 mm) with no obvious sign of permanent cathode damage.[24]

The rms intensity stability of the electron beam out of the source is generally <1% for SLC and about 2% for long pulse operation. This performance depends primarily on the laser stability. The large peak currents required for SLC operation necessitate operating the strained-layer cathodes near their charge saturating limit, thus mitigating the effects of laser jitter. With the laser feed-back/-forward loops operating, the Nb:YAG-pumped Ti:sapphire system has an rms energy stability of 1-2%, while the best achieved from the flashlamp-pumped Ti:sapphire system was about 2%. A highly-stable Nd:YAG pumping laser could probably be designed that would reduce the Ti:sapphire laser jitter to <1%.

Long term stability is ensured not only by additional feed-back loops that keep the electron beam intensity and orbits constant, but also by maintaining the laser-room temperature and humidity to within ±0.2°C and ±5% respectively.

Since 1992, all electron beams for the 3-km linac have been produced by the polarized source. It operates for weeks at a time without any intervention for maintenance or adjustment. The principal operating parameters for the source are shown in Table 2. Note that the total operating time is equivalent to over 3 years of continuous operation. The availability of the polarized source has routinely been ~99%.

Table 2. Source Operating Parameters

Parameter	SLC	Fixed target
Total operating hours	~20,000	~7,000
e$^-$ polarization	0.75 to 0.80	0.80 to 0.85
No. e$^-$/bunch at source	~6x10^{10}	4x10^9 to 10^{11}
Bunch length at source (ns)	2	200 to 2000
Cathode bias (kV)	-120	-120
Cathode temperature (°C)	0	0

CONCLUSION

The stellar performance of the SLAC polarized electron source over the past 5 years invites speculation as to why. Some of the more interesting possibilities are discussed here. Like the SLC itself, this source has proven to be a very important prototype for future linear colliders.

REFERENCES

[1] P.S. Cooper et al., Phys. Rev. Lett. 34, 1589 (1975).
[2] M.J. Alguard et al., Nucl. Instrum. and Meth. 163, 29 (1979).
[3] E.L. Garwin et al., Helv. Phys. Acta 47, 393 (1974).
[4] C.K. Sinclair et al., AIP Conf. Proc. 35, 424 (1976).
[5] R.L. Bell, *Negative Electron Affinity Devices*, Oxford: Clarendon Press, Oxford, 1973.
[6] T. Maruyama et al., Phys. Rev. Lett. 66, 2376 (1991).
[7] T. Nakanishi et al., Phys. Lett. A 158, 345 (1991).
[8] Since 1994, a 100-nm epilayer has been used. The cathode design is similar to sample (3) found in T. Maruyama et al., Phys Rev. B 46, 4261 (1992). The cathodes are cut from 2-inch wafers MOCVD-grown by SPIRE Corporation, One Patriots Park, Bedford, MA 01730.
[9] R. Alley et al., Nucl. Instrum. and Meth. A 365, 1 (1995).
[10] The final design is based on an early design by R. Miller (SLAC). See C.K. Sinclair and R.H. Miller, IEEE Trans. on Nucl. Sci. NS-28, 2649 (1981).
[11] W. Herrmannsfeldt, SLAC Report 331 (1988).
[12] W. Hermannsfeldt (SLAC), private communication.
[13] R.E. Kirby et al., *Procedings of the 1993 Particle Accelerator Conference*, 1993, p. 3039.
[14] The electron bunch it produces will be the electron bunch at the SLC interaction point (IP).
[15] The associated electron bunch goes to the SLC conversion target to generate the positron bunch destined for the IP.
[16] K.H. Witte, *Proceedings of the International Conference on Lasers '93*, 1994, p. 638.
[17] Note that whether or not a given cathode is baked, at SLAC the activation process always includes heating the cathode to about 600°C for about 1 hour.
[18] This may not prove true for guns designed to operate with significantly higher rf fields, such as rf guns.
[19] In cooling from room temperature to 0°C, the QE (833 nm) for these cathodes (close to the band gap energy) at 120 kV decreases by 2×10^{-5} per °C. See P. Sáez, SLAC-R-501 (1997), Fig. 4.7. At low voltage (no Schottky effect) the rate of decrease is about twice this value.
[20] During the commissioning of the SLAC source for the 1992 SLC run, it was discovered that as the laser energy was increased to very high values, the charge that could be extracted saturated at a value well below the space charge limit. For details, see reference(9) and references therein.
[21] A monolayer of Cs is taken to be 6×10^{14} cm^{-2}. A sticking coefficient of 1 was assumed. Calibration of the SAES Cs-dispenser channels was made by R. Kirby (SLAC) using a thermally-stabilized quartz-crystal oscillator.
[22] T. Wada et al., Jpn. J. Appl. Phys. 29, 2087 (1990).
[23] Future polarized sources operating with much higher electric fields may require much more elaborate material selection and preparation. For progress in this area, see H. Matsumoto, *Proceedings of the XVIII International Linear Accelerator Conference*, 1996, p. 626.
[24] Although at the end of use of a given cathode the QE remains fully restorable with the addition of Cs, in some cases a change in the QE profile across the cathode surface has been seen that may be permanent.

The Polarized Electron Source at NIKHEF

M.J.J. van den Putte*, C.W. de Jager*[1], S.G. Konstantinov°,
V.Ya. Korchagin°, F.B. Kroes*, E.P. van Leeuwen*,
B.L. Militsyn*, N.H. Papadakis*[2], S.G. Popov°[†],
G.V. Serdobintsev°, Yu.M. Shatunov°, S.V. Shevelev°,
T.G.B.W. Sluijk*, A.S. Terekhov⋄, Yu.F. Tokarev°

*NIKHEF, P.O.Box 41882, 1009 DB Amsterdam, The Netherlands
°Budker Institute of Nuclear Physics, Novosibirsk, 630090, Russian Federation
⋄Institute of Semiconductor Physics, Novosibirsk, 630090, Russian Federation

Abstract. An overview of the polarized electron source used at the NIKHEF accelerator facility is given. Since its first operation in September 1996 we have obtained currents in the source larger than 150 mA, and polarization degrees up to 88 %. Approximately 350 hours have been dedicated to physics with a polarized ^{3}He internal target. Using one InGaAsP photocathode, 2 μs long pulses were produced with a current of 2 mA in the local linac, with a polarization degree of 82.5 %, at a repetition rate of 1 Hz. Source operations had to be interrupted about every eleventh day for approximately an hour for photocathode reactivation. A photocathode lifetime (1/e) of 180 hours has been achieved.

INTRODUCTION

The accelerator facility at NIKHEF has been upgraded to produce stored beams of longitudinally polarized electrons in the Amsterdam Pulse Stretcher (AmPS [1]) and storage ring. This project has been realized by a collaboration between NIKHEF and the Budker Institute for Nuclear Physics (BINP) together with the Institute of Semiconductor Physics (ISP), both at Novosibirsk. The facility will be used to study scattering of longitudinally polarized electrons on polarized light nuclei ($^1\vec{\mathrm{H}}(\vec{e}, e')$, $^2\vec{\mathrm{H}}(\vec{e}, e'n)$, $^3\vec{\mathrm{He}}(\vec{e}, e'X)$ [2,3]).

The accelerator facility is schematically shown in Figure 1. As the time required for polarization by means of the "Sokolov–Ternov" effect [4] for the

---

[1] Present address: Thomas Jefferson National Accelerator Facility, Newport News, VA 23006, USA
[2] Present address: IASA, P.O.Box 17214, 10024 Athens, Greece

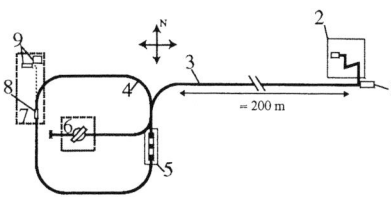

FIGURE 1. A sketch of the upgraded NIKHEF acceleration facility; 1. the thermionic electron source, 2. the polarized electron source (new element), 3. the Medium Energy Accelerator (MEA), 4. the Amsterdam Pulse Stretcher (AmPS) and storage ring, 5. the Siberian Snake (new element), 6. the external target hall, 7. the internal target hall, 8. the internal target, 9. the Compton backscattering polarimeter (new element).

energy range of AmPS is much larger than its beam lifetime, polarized electrons have to be injected into the ring. The 400 keV electrons produced by the Polarized Electron Source (PES) are accelerated to the desired energy of AmPS, by means of a linear accelerator, the Medium Energy Accelerator (MEA). A Siberian Snake is installed in the East straight section of the ring in order to maintain the longitudinal polarization at the internal target (interaction point). The polarization degree of the electron beam at the interaction point is measured by means of a Compton backscattering polarimeter [5].

An overview of the polarized electron source is given in Figure 2. Spin polarized electrons are obtained by means of photo-emission from strained layer III–V-semiconductor crystals (photocathodes). The surface of a photocathode is illuminated with circularly polarized light produced by an optical system based on a tunable Ti-Sapphire laser. A preparation chamber, in which photocathodes are prepared to the Negative Electron Affinity (NEA) surface state, is permanently connected to the photocathode gun. The 100 keV electrons provided by the photocathode gun, are deflected from the laser light path by means of two 45° bending magnets into an electron optical system know as a Z-shape spin manipulator [6,7]. With this manipulator the electron spin can be oriented in an arbitrary angle within 4π. It is used to compensate for the spin precession in MEA and AmPS. Following the spin manipulator, a Mott polarimeter has been installed for the analysis of the polarization degree of the beam in the source. The last element of the source is a post-accelerator, which accelerates the electrons to 400 keV, before the electron beam is deflected into MEA by means of an alpha magnet (270° deflection). The polarized electron source can be operated with repetition rates up to 10 Hz.

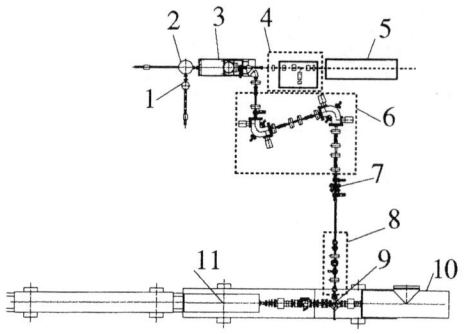

FIGURE 2. An overview of the polarized electron source; 1. the loading chamber, 2. the preparation chamber, 3. the photocathode gun, 4. the optical path, 5. the Ti-Sapphire laser, 6. the Z-shape spin manipulator, 7. the Mott polarimeter, 8. the post-accelerator, 9. the alpha magnet, 10. the thermionic gun, 11. the Medium Energy Accelerator (MEA).

THE PHOTOCATHODE GUN

The photocathode gun (see figure 3) is equipped with a double ceramic high voltage insulator which allows for a permanent connection with the preparation chamber. The gun has two concentric vacuum chambers; the acceleration chamber (the inner vacuum chamber) in which the photocathode is mounted, and the guard vessel (the outer chamber). This design offers several attractive features:

- The permanent connection of the preparation set-up allows for the replacement of a photocathode without opening the vacuum chamber. Thus the number of bake-outs of the gun vessel is minimized, and therefore the source duty factor is high.

- Migration of wall contaminants onto the photocathode surface, during the baking process of the gun vessel, is prevented.

- There is no need for the use of a cesium dispenser in the gun for the activation and reactivation of photocathodes.

- The guard vessel provides, mainly by infrared irradiation, a uniform heating distribution on the high voltage insulators in the baking process of the gun (maximum of 300 °C). This reduces the risk of creating vacuum leaks between the metallic and ceramic parts of the insulators.

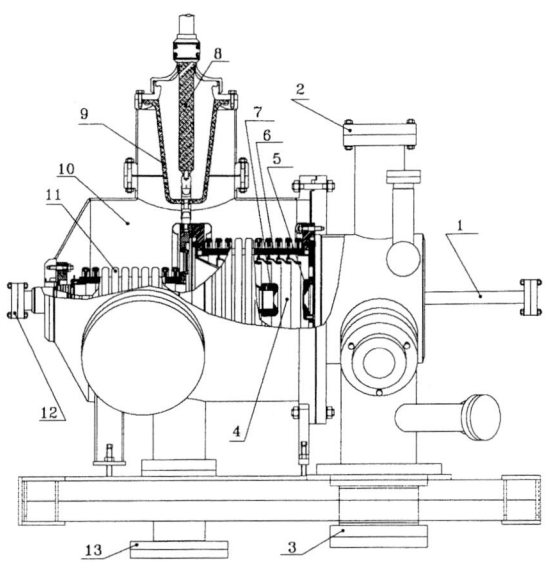

FIGURE 3. The photocathode gun of the polarized electron source; 1. the beam pipe, 2. and 3. pumping ports of the acceleration chamber, 4. the acceleration chamber (inner vacuum chamber), 5. the anode, 6. the high voltage insulator, 7. the cathode electrode, 8. the high voltage cable, 9. the shielding insulator, 10. the guard vessel (outer vacuum chamber), 11. the protection insulator, 12. the port to the preparation chamber, 13. the pumping port of the guard vessel.

- Should nevertheless small leaks occur, as a consequence of the baking process, the guard vessel dramatically reduces the flow of gases, which can contaminate the photocathode surfaces.
- The guard vessel provides an easy access to the preparation chamber, whilst the gun is in operation.

The inner vacuum chamber is pumped by two ion pumps[3] (100 and 250 l/s), and a NEG pump. The vacuum of the inner chamber is better than 10^{-12} mbar. The guard vessel is pumped by an ion pump of 160 l/s. Its pressure is in the lower range of 10^{-7} mbar.

The gun was originally operated with a DC high voltage power supply of 100 kV. This has been replaced by a pulsed high voltage power supply. The latter is operated in "soft high voltage excitation" mode, in view of the high impedance of the gun (5 MΩ). The basic lay-out of the pulsed power supply consists of a charge unit, a control unit, and a high voltage transformer. The

---

[3] All ion pumps are equipped with titanium getters.

charge module loads an internal capacitor to 400 V, and maintains this voltage with a stability better than 10^{-3} until discharged. The start pulse control signal discharges the capacitor onto the high voltage transformer, placed in a tank filled with SF_6 at an absolute pressure of 1.7 bar. The current pulse of the transformer charges the connecting cable to the gun, which has a capacitance of 1500 pF. The relative discharge of the cable by an electron beam pulse of 20 mA is less than $3 \cdot 10^{-4}$. This does not lead to an observable energy spread within the beam pulse. The pulsed high voltage system provides a Gauss like pulse, with a full width of 600 μs, and an amplitude of -100 kV.

THE PREPARATION SET-UP

To obtain for a photocathode the required quantum efficiency of photoemission, its surface needs to be brought to the Negative Electron Affinity (NEA) state by means of an activation (or reactivation) process. This is done in the preparation set-up. The set-up comprises of a glove-box, a transfer vessel, a loading chamber, and a preparation chamber. The preparation chamber is connected to the photocathode gun.

Photocathodes enter the system in a glovebox, where they can be chemically treated (e.g. etching) in a pure nitrogen atmosphere. Subsequently the photocathodes are placed in the transfer vessel, which is mounted on top of the glovebox. It can contain up to three photocathodes. After glovebox treatment the transfer vessel is moved to the loading chamber, whilst the photocathodes are kept under nitrogen atmosphere in the hermetically sealed compartment. After mounting the transfer vessel on top of the loading chamber, which is pressurized with pure nitrogen gas for this operation, the loading chamber is evacuated to a vacuum of 10^{-7} mbar, by means of a turbo pump system (oil free), and an ion pump. The loading chamber is separated from the preparation chamber by a UHV valve. Photocathodes are transported by means of a magnetic manipulator, through the UHV valve, to a carousel in the preparation chamber which can contain up to four photocathodes. A maximum of three photocathodes is kept in the carousel, whilst one photocathode is being used in the gun. A vacuum in the preparation chamber of better than 10^{-12} mbar is maintained by means of an ion pump (250 l/s), and a NEG pump. The preparation chamber contains two heaters (one spare), a cesium and an oxygen dispenser (both electrically operated), and two electron collectors. Of the latter, one is used to monitor the photocurrent during the activation process, whilst the other is used for photocathode quantum efficiency measurements at the He-Ne laser wavelength. The preparation chamber is separated by an UHV valve from the gun. A freshly activated or reactivated photocathode is exchanged with the one mounted in the gun, by means of a second magnetic manipulator, within a typical time of 15 minutes.

THE OPTICAL SYSTEM

The optical system is based on a tunable flashlamp-pumped Ti-Sapphire laser, operating in the wavelength range of 700–850 nm (750–900 nm depending on the mirrors used in the lasing cavity). By means of a polarizer and a remotely controlled electro-optical slicer, both incorporated into the laser system, a linearly polarized light pulse with a length of 0.4–4 μs is produced, from a 25 μs long non-polarized light pulse. The laser light pulse length defines the length of the electron pulses produced.

In the optical path, following the laser system, a lens projects the image of a variable diameter iris diaphragm onto the photocathode surface (magnification of 1.67). The power in a light pulse can be attenuated by means of a remotely controlled rotatable Glan prism. By attenuating the laser light intensity the current of the electron pulses is kept constant, and thus drifts of the electron beam in MEA due to beam loading effects (2.5 MeV/mA) are prevented. By means of a remotely controlled rotatable quarter wavelength plate, left and right handed circularly polarized light is produced. A beam splitter directs part of the laser beam onto a photodiode, which can be used for monitoring the laser power, and light pulse shapes.

The laser and the pulsed high voltage power supply have been synchronized, such that the laser system fires its 2 μs pulse on top of the 100 kV pulse produced by the power supply of the photocathode gun.

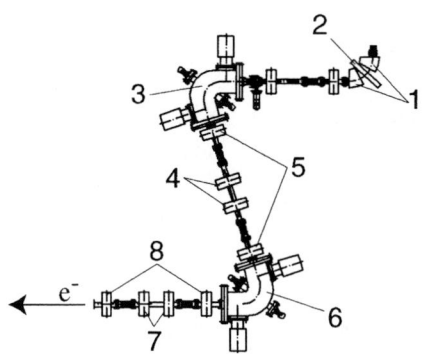

FIGURE 4. The Z-shape spin manipulator; 1. the two 45° magnets, 2. the first pulsed beam current transformer of PES, 3. the first 107.7° degrees electrostatic deflector, 4. the inner pair of solenoids for spin the polar angle rotation, 5. the outer pair of solenoids 6. the second 107.7° degrees electrostatic deflector, 7. the inner pair of solenoids for the spin azimuth angle rotation, 8. the outer pair of solenoids.

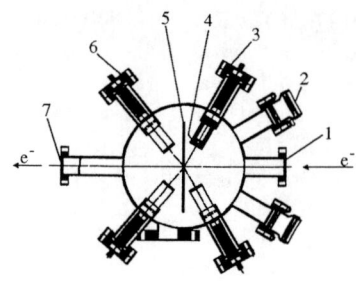

FIGURE 5. The Mott polarimeter; 1. and 7. the beam pipe, 2. a view port, 3. a silicon surface barrier detector at the scattering angle of 120°, 4. the aluminium filter (mounted on all detectors), 5. the rotatable wheel containing foils, a ruby screen, and a hole, 6. a silicon surface barrier detector at a scattering angle of 120°.

THE Z-SHAPE SPIN MANIPULATOR

Depending on the helicity of the light produced by the optical system, the electron spin will be oriented parallel or antiparallel to the velocity of the electrons leaving the photocathode gun. The spin of the electrons can be oriented to an arbitrary angle by means of the Z-shape spin manipulator (see figure 4. The design discussed in ref. [6], with some minor changes made to the Illinois–CEBAF prototype [7], was adopted.

The first electrostatic deflector rotates the electron velocity 90° degrees in the horizontal plane with respect to the polarization vector. The inner pair of solenoids is operated with parallel magnetic fields. With this set the spin is rotated, and its polar angle is set. The outer pair of solenoids is operated with opposite magnetic fields. Their effect on the spin rotation cancels. By placing the electron beam waist between the inner solenoids, the focusing effects of the inner solenoids can be minimized. The outer solenoids are used for second order corrections only. The second electrostatic deflector rotates the electron velocity backwards by 90°, so that it coincides with the projection of the spin in the horizontal plane. With the subsequent set of inner solenoids the azimuth angle of the spin is set.

THE MOTT POLARIMETER

A Mott polarimeter (see figure 5) is installed to measure the polarization degree of the beam in the source. It is sensitive to transverse polarization (90° polar angle). Its vacuum chamber contains a remotely controllable rotatable disk on which four gold foils (gold evaporated onto Mylar foil), 100 nm,

60 nm, 37 nm, and 18 nm thick, are mounted. For the determination of the polarization degree of the electron beam the 100 nm thick foil is used. The other foils are used for the calibration of the polarimeter only. Additionally a ruby screen is installed on the disk, with which the electron beam spot can be observed by means of a camera through a viewport. One position in the disk is kept open for the passage of the electron beam. Four Si(Li) p–i–n Schottky surface barrier detectors, equipped with charge sensitive amplifiers, are used to detect the scattered electrons. Non-elastic low energy electrons are discriminated against by means of aluminium filters installed in front of the silicon detectors. Two detectors, mounted at +120°, and -120°, measure the scattering asymmetry, whilst two silicon detectors at +50°, and -50° can be used to monitor instrumental asymmetries. A Sherman function value of 0.22, for the 100 nm foil, is used.

Detailed information on all beam diagnostic instruments used can be found in ref. [8].

THE POST-ACCELERATOR AND INJECTION

To match the parameters of the polarized electron beam to MEA a two-cavity scheme post-accelerator has been implemented. The first cavity is used for bunching. The second is used for the acceleration to 400 keV. The mode of the RF oscillation in the cavities is TM_{010}. The RF peak power level power in the bunching and the acceleration cavity is of order 50 W and 100 kW, respectively. The electron beam pulse is delayed by 1.2 μs with regard to the RF incident wave front. The measured transmission of the post- accelerator is 50 %.

Leaving the post-accelerator the 400 keV beam is deflected into MEA by means of an alpha magnet. This injection gives the minimal amount of dispersion. A choice can be made between the thermionic electron gun and the polarized electron source by means of the alpha magnet setting. Injection efficiencies into MEA up to 35 % have been achieved (including post-accelerator transmission). The nominal value is approximately 30 %.

SOURCE PERFORMANCE

It is well known that the photocathode lifetime (1/e) is strongly dependent on the quality of the vacuum. With 100 kV DC supplied to the photocathode gun, the vacuum in the gun acceleration chamber deteriorates to typically $2 \cdot 10^{-11}$ mbar. The lifetime of the photocathodes, for continuous operation of the photocathode gun, was limited to approximately 4 hours. Not to be hampered by this small lifetime in operating the source, we switched off the DC high voltage between injections of beam into the ring. In order to improve

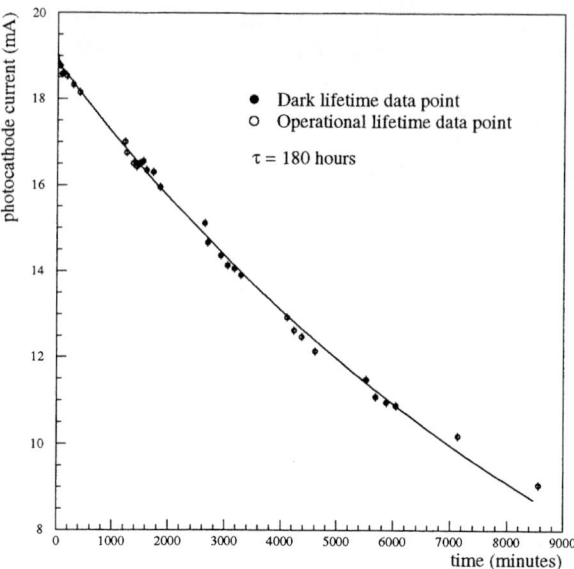

FIGURE 6. The current of an InGaAsP photocathode as a function of time over a period of approximately one week. Dark lifetime data points were taken with the photocathode gun operated for the measurement only. Operational lifetime data points were taken whilst the photocathode gun was operated continuously at a repetition rate of 1 Hz.

the vacuum in the acceleration chamber, whilst operating the photocathode gun, a pulsed high voltage power supply was installed [9].

Photocathode lifetime measurements were then performed on a strained layer InGaAsP photocathode provided by the Institute of Semiconductor Physics at Novosibirsk. The current was measured using the first pulsed beam current transformer (see figure 4). The results of the measurements are given in Figure 6. Both the pulsed high voltage power supply and the laser system were operated at a repetition rate of 1 Hz. The data of the plotted points have been averaged over nine beam pulses. Operating the photocathode gun does not lead to an observable deterioration of the acceleration chamber vacuum, nor to a decrease of photocathode lifetime. The exponential function fitted to all data points has a $1/e$ time of 180 hours, which is a considerable improvement over 4 hours.

The polarized electron source has been used, since September 1996, over several periods of a few weeks. Within the research program on photocathodes the majority of the effort has concentrated on the use of strained layer InGaAsP photocathodes. Pulses of 2 μs length, with averaged currents up to

35 mA, at a repetition rate of 1 Hz, and polarization degrees up to 88 % were obtained. In first test performed on GaAsP photocathodes, we have obtained 2 μs long pulses, with averaged current larger than 150 mA (saturation level of the pulsed beam transformers), at repetition rates of 1 Hz, and polarization degrees up to 50 % (tests are in a very early stage though).

A total of 350 hours of source operation were dedicated to physics with the polarized ^{3}He internal target. For this run we used one InGaAsP photocathode only. For reactivation of this photocathode we had to interrupt the production of polarized electrons by an hour, approximately every eleventh day of operation. The internal transmission up to the post-accelerator was better than 99 %, and a continuous injection efficiency into MEA of around 30 % was obtained. With a laser spot diameter on the photocathode surface of 7 mm we obtained 2 μs beam pulses, with an average current in MEA of 2 mA, and a polarization degree in the source of 82.5 ± 1.5 % (statistical error), at a repetition rate of 1 Hz.

ACKNOWLEGDEMENT

This work was supported in part by the Stichting voor Fundamenteel Onderzoek der Materie (FOM), which is financially supported by the Nederlandse Organisatie voor Wetenschappelijk Onderzoek (NWO), NWO under Grant 713-119 (Novosibirsk), and the EEC Human Capital and Mobility programme under Grants ERBCHBICT-930606, and ERBCHRXCT- 930122.

REFERENCES

1. de Witt Huberts, P.K.A. , *Nucl. Phys.* **A553**, 845C (1993).
2. Ferro–Luzzi, M. , van den Brand, J.F.J. , et al. , *NIKHEF proposal 97–01*.
3. van den Brand, J.F.J. , et al. , *NIKHEF proposal 94–05*.
4. Chao, A.W. , *IEEE Trans Nucl. Sc.* **Vol. NS–30, No. 4**, 2383 (1983).
5. Passchier, I. , *these proceedings*.
6. Reichert E. , *as reported by Sinclair C.K. in Proc. 8th Int. Symp. on High Energy Spin Physics, ed. Heller K.J. , AIP Conf. Proc.* , **187**, 1412 (1989).
7. Engwall, D.A., et al. , *Nucl. Instr. and Meth.* **A324**, 409 (1993).
8. Bolkhovityanov, Y.B. , et al. , *Proceedings of the SPIN96 conference*, ISBN 981-02-3052-4, 700 (1997)
9. van den Putte, M.J.J. , et al. , *submitted as letter to the editor of NIM-A*

A Diode Laser System for Synchronous Photoinjection

Matt Poelker and John Hansknecht

Jefferson Laboratory, 12000 Jefferson Avenue, Newport News, VA 23606

Abstract. A laser system, which is composed of a gain switched diode seed laser and a single-pass diode optical amplifier, is used to drive the polarized electron source at Jefferson Lab. The system emits pulsed laser light synchronized to the accelerating cavity radio frequency (rf) at 1497 MHz or the third subharmonic, 499 MHz. The maximum average output power from the laser system is 500 mW and the optical pulsewidth is 60 to 80 ps. The laser system is compact and very reliable operating remotely for many days without attention.

INTRODUCTION

The first nuclear physics experiment requiring polarized electrons at Jefferson Lab [1] was recently successfully completed. An integral component of the polarized injector is a pulsed diode laser system with a pulse repetition rate synchronized to the accelerating cavity rf (1497 MHz or the third subharmonic, 499 MHz). Electrons are extracted from the photocathode only during the portion of the rf phase when they are accelerated through the machine. In this way, few electrons are lost at the injector chopper. This method minimizes the total charge extracted from the photocathode and prolongs the operating lifetime of the polarized source.

The laser system is composed of a gain switched diode seed laser and a single-pass diode optical amplifier along with various optical components used to deliver the laser beam to the photocathode (Fig. 1). The laser system has been operational for nearly two years and has proven to be very reliable. The system is compact (76 cm x 122 cm x 56 cm) and rests beneath the gun in the injector tunnel. It is completely remotely controlled. Early versions of the laser system have been described in other publications [2].

Pulsed laser light is obtained through the technique referred to as gain switching [3]. Gain switching is a straightforward electrical technique; the pulse repetition rate depends only on the frequency of the applied electrical signal. Consequently, it is a simple matter to change the pulse repetition rate and to lock this frequency to the machine. The electrical drive signal is derived from the

Jefferson Lab master oscillator and as a result, the laser output is very stable with respect to amplitude noise and timing jitter.

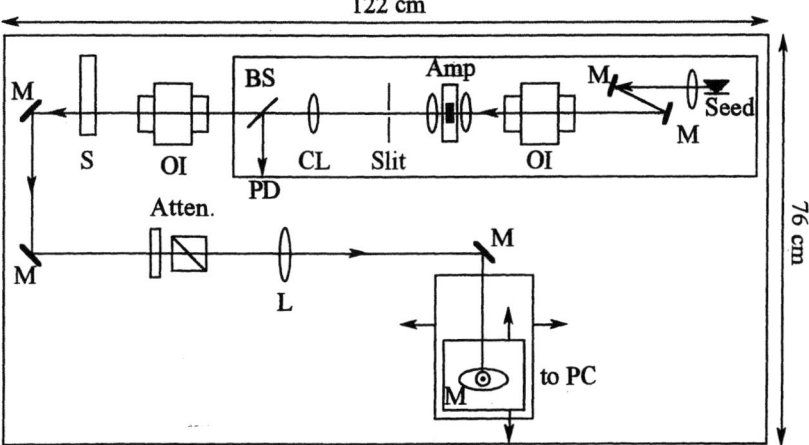

Figure 1. A schematic of the laser system. OI, optical isolator; S, shutter; M, mirror; PC, Pockels cell; CL, cylindrical lens; L, lens; BS, beam splitter.

LASER SYSTEM DESCRIPTION

A seed laser is biased slightly above threshold and driven with approximately 1 Watt of rf at 1497 MHz when delivering beam to three halls, or 499 MHz when delivering beam to one hall. Pulsed light from the seed laser ($\approx$ 5 to 10 mW) is directed through an optical isolator and then focused into a diode optical amplifier (Spectra Diode Labs Model 8630E). Seed laser amplification is a function of diode optical amplifier drive current. Average output power as a function of dc current is plotted for the two different Jefferson Lab pulse repetition rates in Figure 2. The maximum average output power for both repetition rates is 500 mW [4]. The pulsewidth is 60 to 80 ps; pulsewidth increases with amplifier drive current.

The light from the diode optical amplifier passes through a shutter, an attenuator (i.e., mica halfwave plate and stationary linear polarizer) and a focusing lens before passing through a Pockels cell used to obtain left and right circularly polarized light. The Pockels cell is the last optical component the laser beam encounters before passing into the vacuum chamber that houses the photocathode. The focused laser spot can be moved across the photocathode using two stepper motor-controlled translation stages. Light from the diode optical amplifier is nearly diffraction limited; the 3 mm diameter laser beam is focused to a spot on the photocathode with a diameter between 250 to 400 μm (FWHM).

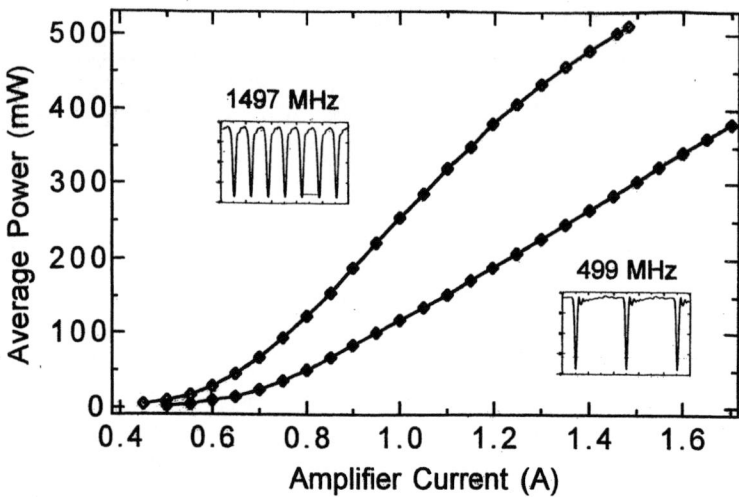

Figure 2. Average output power as a function of diode optical amplifier drive current (dc). The insets show fast photodiode signals for the two different Jefferson Lab repetition rates, 499 and 1497 MHz.

In addition to modulating the seed laser drive current to create rf microstructure on the extracted electron beam, the amplifier drive current can be modulated to create "macropulse" modes of operation for machine tune-up. The amplifier current can be pulsed with a risetime/falltime as short as 500 ns. This allows operators the ability to create specific electron beam time structures similar to those used when operating the thermionic electron gun. The diode amplifier current supply is also interfaced to the machine/personnel safety system to allow rapid termination of electron beam delivery.

ELECTRON BUNCH LENGTH MEASUREMENTS

Electrons travel from the photocathode (at -100 kV) through a Z-style electron spin manipulator, an emittance filter and then the RF chopper which deflects the beam in a circular path across three chopper slits separated 120 degrees in phase. The maximum slit aperture is 60 degrees of rf phase at 1497 MHz which corresponds to a temporal width of 111 ps. The total electron beam path from the photocathode to chopper is approximately 11 m. For a more complete description of the photoinjector, please refer to C. Sinclair's paper in these proceedings.

To determine electron bunch length, two slits were completely closed. The third slit aperture was adjusted to be 10 degrees of rf phase at 1497 MHz which corresponds to a temporal width of 18 ps. The phase of the rf signal applied to the seed laser was adjusted so that the electron bunches at the chopper are swept across the narrow, open slit. The transmitted beam was detected with a Faraday cup downstream of the chopper. Electron bunch length can be determined from a plot of the Faraday cup signal versus laser rf phase. Electron bunch length profiles are shown in Fig. 3 for three different beam currents; 2, 28 and 80 µA. Clearly, electron bunch length increases with increasing beam current. For 2 µA beam current, the electron bunch length is approximately equal to the laser pulsewidth (≈ 50 ps) and beam transmission through the fully-open injector chopper slit is high. At 80 µA however, the electron bunch length has roughly tripled and approximately 60% of the beam is lost on the fully open chopper slit. Based on this observation, a prebuncher was added to the polarized source beamline. For recent operation during the nuclear physics experiment E89033, 75 µA was delivered to the Halls (through three slits) with 100 µA extracted from the gun. It is believed rf settings of various injector components can still be further optimized to provide considerably improved beam transmission. It should be noted that this is work-in-progress. A more detailed account of these measurements and attempts to model the photoinjector will be presented in a future publication.

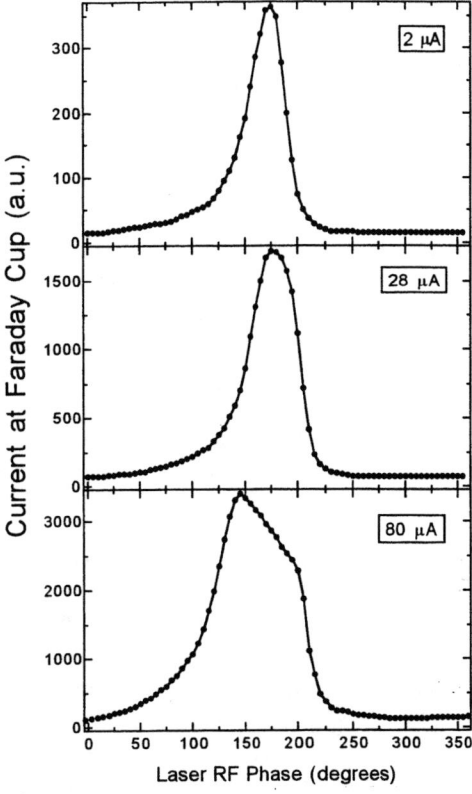

Figure 3. Electron bunch profiles for three different beam currents; 2, 28 and 80 µA.

It is particularly noteworthy that, to a large extent, the laser truly "turns off" between rf pulses. There has been concern expressed that the Jefferson Lab laser system suffers from unwanted amplified spontaneous emission (ASE). That is, the rf pulsed light is contaminated with dc light that results from ASE from within the diode optical amplifier. The bunch length plots in Fig. 3 show that this is not the case. At most, the rf pulsed light is contaminated with 2 to 3 % dc laser light.

FUTURE PLANS

There are a number of improvements planned to further enhance laser system performance. In particular, a remotely controlled mirror mount will be added to the system to provide remote alignment of the seed laser beam into the diode optical amplifier. Despite choosing stable mirror mounts, the seed laser mirrors sometimes require adjustment to obtain maximum laser system output power. (Misalignment may be a result of temperature changes that occur when the laser enclosure is dismantled for system check-out, spring relaxation in various optics mounts, etc.) A remotely controlled seed laser mirror will allow laser power optimization throughout a nuclear physics experiment without requiring access to the injector tunnel.

In the near future, three independent seed laser/optical amplifier laser systems will occupy space on the laser table beneath the gun; one laser system for each hall. The laser power from each of the three laser systems can be adjusted independently to precisely provide the required beam current to the hall, without beam loss at the injector chopper. Three laser systems also means there is more laser power available for each hall (500 mW rather than 500 mW divided by three as is presently the case). This means the gun can operate longer without intervention required to restore the photocathode quantum efficiency. And finally, three independent laser systems may increase machine availability to the nuclear physics users by allowing cw electron beam delivery to one or two halls while providing pulsed beam delivery for machine tune-up from the beam switchyard to the other hall(s).

The present laser systems at Jefferson Lab emit light at 776 nm and 862 nm. To operate at other wavelengths, new laser components must be purchased. Experiments will soon be performed to determine if the present laser systems can be modified to provide 10 to 15 nm of wavelength tunability. Such a modification will be highly useful when operation with high polarization strained-layer photocathode materials begins. A promising technique involves "seeding" the seed laser with light from an external, wavelength tunable source using a partially reflecting mirror. In a recent publication [5], researchers were able to tune the wavelength of a gain switched diode laser over a 40 nm range using a variation of this technique.

ACKNOWLEDGMENTS

The following people made significant contributions to the material in this proceeding; Philip Adderley, Jim Clark, Tony Day, Bruce Dunham, Curt Hovater, Reza Kazimi, Hongxiu Liu, Scott Price, Bill Schneider and Charlie Sinclair. This work was supported by the U. S. Department of Energy Contract No. DE-AC05-84ER40150.

REFERENCES

1. E89033; Polarized electrons on ^{16}O. C. Glasshauser spokesperson.
2. M. Poelker, Appl. Phys. Lett. **67**, 2762-2764 (1995); M. Poelker and J. Hansknecht, Proc. of the 12th International Symposium on High-Energy Spin Physics, edited by K. Jager et.al., (World Scientific, Singapore, 1997).
3. See for example, P.T. Ho, in *Picosecond Optoelectronic Devices*, edited by C. H. Lee (Academic, New York, 1984).
4. The manufacturer of this device, Spectra Diode Labs, recommends that average output power not exceed 500 mW from the output facet of the diode amplifier. The maximum laser power deliverable to the photocathode has been limited to approximately 440 mW as a result of lossy optical components within the laser beam delivery system (eg., optical isolator, Pockel's cell. etc.,).
5. Y. Matsui, et al., IEEE Photon. Technol. Lett. **9**, 1087-1089 (1997).

Strained Semiconductor Structures for Polarized Electrons

Charles W. Tu

Department of Electrical and Computer Engineering
University of California, San Diego
La Jolla, California 92093-0407, USA

Abstract. Limitations of strained semiconductor structures for polarized electrons reported to date are described, in particular, critical layer thickness and partial and unisotropic lattice relaxation. We also proposed a new strain-compensated superlattice to circumvent these problems.

INTRODUCTION

Polarized electron sources made from negative-electron-affinity (NEA) GaAs have many applications in high-energy and condense-matter physics and materials science. Due to the degeneracy of the valence band maximum, the maximum polarization is only 50%. Many researchers have investigated ways to remove the degeneracy in order to achieve the maximum polarization of 100%. One of the promising approach is to use strain to remove the degeneracy, for example, GaInAs on GaAs (1), GaInAs/GaAs strained-layer superlattice (2), GaInP and GaInAsP on GaAs (3), GaAsP on GaAs (4), GaAs on GaAsP/GaAs (5), GaInAsP on GaAs (6), GaAs on GaInP/GaP (7), and AlGaInAs on AlGaAs/GaAs (8). When a thin layer is strained because its lattice constant is different from that of the substrate, its thickness is limited to a critical value before relief of stress in the lattice by generation of misfit dislocations, which are detrimental to material quality. On the other hand, since the strained layer is thin, the quantum efficiency of photo-cathodes suffers.

For the III-V compound semiconductors commonly used, the critical layer thickness of an uncapped layer in angstroms for dislocation generation is roughly one over the lattice mismatch in percent. A 1% mismatch means a critical layer thickness of only 10 nm. This number is just a lower-bound guide since the practical critical layer thickness can be several times that value, depending on the layer structure and growth conditions. In most cases, the thickness of a photocathode is about 0.1 to 0.2 µm to have a reasonable quantum efficiency, but the layer is much thicker than the critical layer thickness. Hence, partial relaxation of the lattice with some residual strain is expected.

In this paper we shall discuss some of the issues in growing strained photo-cathodes and propose a new strain-compensated superlattice for this purpose.

BUFFER LAYER FOR STRAINED EPITAXY

Since the strained InGaAs photo-cathode grown on a GaAs substrate has lower quantum efficiency than that of GaAs (1,2) and the surface preparation of a GaAsP photo-cathode for NEA is probably not as reproducible as that of GaAs (4), a strained GaAs photo-cathode is preferred and is used at SLAC, for example. Then the strained GaAs photo-cathode has to be grown on a thick, strain-relaxed buffer layer, such as GaAsP (5) or GaInP (7), which are grown on a GaP or a GaAs substrate. How to grow a good buffer layer with minimal amount of dislocations is therefore an important issue.

We and others have found that growing a lattice-mismatched ternary layer directly on a binary substrate would result in a large number of dislocations which can propagate to the surface. A compositionally graded buffer layer is required. There are two ways of grading the composition, steps or linear. Although a few groups have reported good results with step grading, which is simple in execution (usually three steps), we and others have found that a linear grading (approximated by, say, 30 steps) is better. Figure 1 compares the cross-section transmission electron microscopy (TEM) images of a linearly graded InGaP layer on GaP and a step-graded InGaP layer on GaP (9). The final indium composition in InGaP is 0.32. It is evident that in the linearly graded case, the dislocations are confined in the graded layer, whereas some dislocations propagate into the top layer in the step-graded case. It should be pointed out that even though the cross-section TEM micrograph of the sample with a linearly graded buffer layer shows no dislocations in the top layer, the field of view is small. In fact, the dislocation density in such top layers is in the range of 10^6 to 10^7 per cm^2, whereas GaAs substrates have about mid 10^4 dislocations per cm^2. This contrast will also be evident later in a broad X-ray diffraction peak for the top layer and a narrow peak for the substrate.

Of course, a small gradient would result in better dislocation confinement in the graded layer, but the graded layer may become impractically thick. The surface morphology of the sample with a linearly graded buffer layer shows a cross-hatch pattern, indicating uniform lattice relaxation, whereas the surface morphology of the sample with step grading does not show a cross-hatch pattern, but a smooth or slightly textured surface. Therefore, we have used the cross-hatch pattern as an indication of a high-quality, strain-relaxed lattice-mismatched layer structure. The cross-hatches are about 20 or 30 nm deep for a 2 μm or 3 μm thick sample, and they may be one source of depolarization since the surface is not atomically smooth.

The strain depends on growth conditions, in particular, growth temperature. We have investigated the growth temperature dependence of strain of a number of Be-doped p-type GaAs layers on a p-type $In_{xo}Ga_{1-xo}P$:Be buffer layer, which is grown on a linearly graded InGaP:Be buffer layer on a GaP substrate (7). Table 1 shows the various parameters for the samples, including strain ($\Delta a/a$ in %), measured residual strain $\varepsilon_\parallel$ from high-resolution X-ray rocking curves, and the substrate temperature (T_g in degree C). We can see that the samples with a large strain all relaxed when the thickness is 100 nm. For samples with a smaller strain, they also relaxed when the growth temperature was too high. Figure 2 shows high-resolution X-ray rocking curves for two samples, grown at (a) 450 °C and (b) 600 °C, respectively. The sharp peaks are from the GaP substrates; the constant-intensity region are from the linearly graded buffer layer; the constant-composition InGaP layers and the top GaAs layers are indicated. We can see that the GaAs peaks are different, corresponding to different strain. In fact, the GaAs peak from the 600 °C sample is in the relaxed position, i.e., no strain, which would not be useful for high-polarization photo-cathodes.

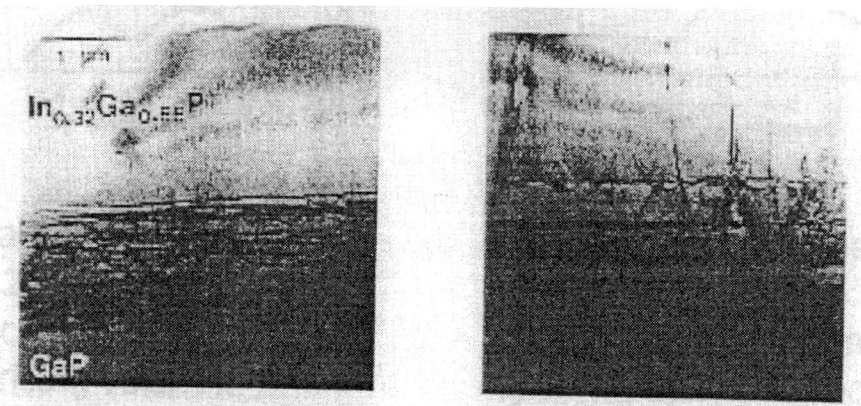

FIGURE 1 Cross-section TEM of an $In_{0.32}Ga_{0.68}P$ layer grown on a linearly graded $In_xGa_{1-x}P$ buffer layer, where x varies from 0 to 0.32, on a GaP substrate (left figure) and one grown on a step-graded buffer layer (right figure).

Another problem of strain relaxation in III-V compound semiconductors is that it can occur asymmetrically, resulting in a tilt of the epilayer with respect to the substrate. Figure 3 shows X-ray rocking curves of two samples of constant-composition $In_{0.53}Ga_{0.47}As$ layers grown (top figure) on a step-graded buffer layer (3 steps with 0.1, 0.2, and 0.3 indium composition) grown on a GaAs(001) substrate and (bottom figure) a linearly graded buffer layer. The asymmetric (224) rocking curves are taken for four different azimuthal angles, 0 to 270 °C. The sharp peaks are those of the GaAs substrates. The three peaks in the left figure are due to the three graded layers. We can see that the sample with a linearly graded buffer layer is relaxed symmetrically, whereas the sample with a step-graded buffer layer does not. This titling of the epilayer can be also a source of electron depolarization.

Sample	xo	thick. (nm)	$\Delta a/a$ (%)	$\varepsilon_{\parallel}$ (%)	Tg (°C)
1533	0.32	100	1.21	0	350
1332	0.34	100	1.06	0	450
1331	0.35	100	0.95	0	600
1531	0.38	50	0.77	61	450
1532	0.39	100	0.69	28	450
1381	0.39	100	0.69	0	600
1382	0.41	100	0.55	0	600

TABLE 1 Parameters for GaAs/In(xo)Ga(1-xo)P samples (7).

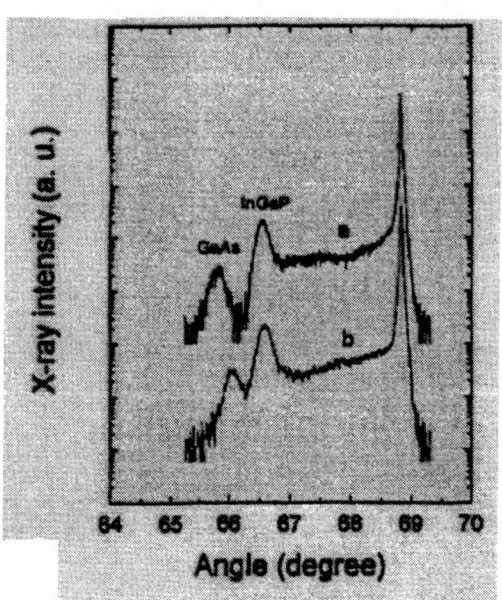

Figure 2 High-resolution double-crystal X-ray rocking curves of two $In_{0.39}Ga_{0.61}P$ layers grown on GaP substrates at a growth temperature of (a) 450 °C and (b) 600 °C (7).

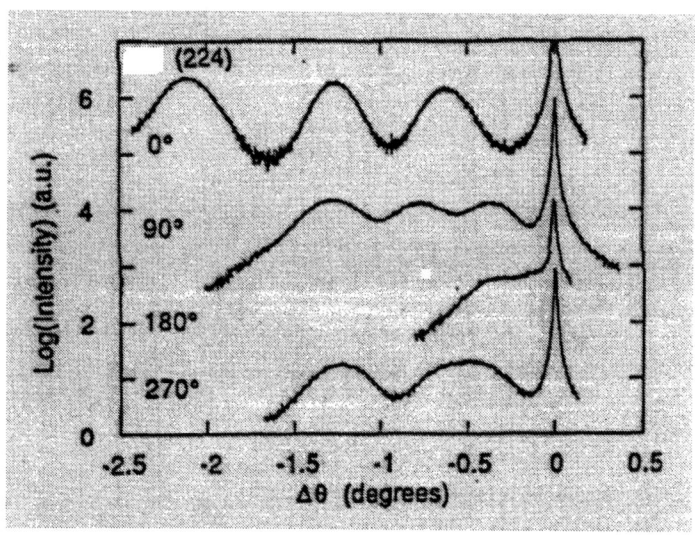

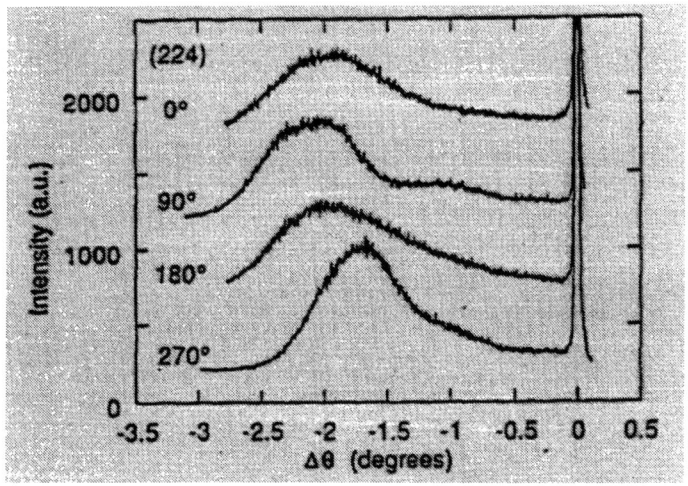

Figure 3 High-resolution double-crystal X-ray rocking curves of (top figure) an $In_{0.3}Ga_{0.7}As$ samples grown on a 3-step-graded InGaAs buffer layer with the indium composition 0.1, 0.2, and 0.3, and (bottom) of an $In_{0.53}Ga_{0.47}As$ layer grown on a linearly graded buffer layer with the indium composition varies from 0 to 0.53 (9).

STRAIN-COMPENSATED SUPERLATTICE

Finally, based on the above considerations we propose a new approach using strain-compensated superlattice to bypass partial and asymmetric lattice relaxation in thick layers and thickness limitation in thin pseudormorphic layers without lattice relaxation. Figure 4 shows X-ray rocking curves (left figure) for strained $InAs_{0.4}P_{0.6}/InP$ (9 nm/13 nm) multiple quantum wells (MQWs) grown on InP substrates for different number of periods, and (right figure) for a strain-compensated $InAs_{0.4}P_{0.6}/Ga_{0.17}In_{0.83}P$ MQW with 30 periods (10). The bottom traces in both figures are simulations based on the dynamical diffraction theory. Here the InAsP layers are quantum wells with a compressive strain of 1.3% with respect to the InP substrate, and GaInP layers are barriers with a compensating tensile strain. We can see that the MQW without strain compensation cannot support many number of periods due to the limitation of the critical layer thickness for dislocation generation. The superlattice satellite peaks broaden as the number of periods increases, and they disappear for the MQW with 11 periods, indicating lattice relaxation. On the other hand, the MQW with strain compensation can support 30 periods (and more). Basically when the net strain is zero, an arbitrarily large number of periods can be supported.

We therefore propose a strain-compensated $Ga_{1-X}In_XP/Ga_{1-Y}In_YP$ superlattice with the average lattice constant equal to that of GaAs so that there is no net strain and an optimal thickness of the well materials can be achieved. The strain and the quantum confinement would remove the degeneracy of the valence band. Such a structure, in principle, bypasses all of the limitations of strained photocathodes reported previously. At present we have grown such samples and are in the process of characterizing them.

SUMMARY

We have shown that in lattice-mismatched epitaxy it is in general preferable to have the constant-composition device layer grown on a linearly graded buffer layer to produce isotropic lattice relaxation with a minimal amount of additional dislocations. Then the top NEA GaAs layer can be strained without too much additional dislocations. The surface morphology, however, is cross-hatched. We propose a strain-compensated GaInP/GaInP superlattice that should have no additional dislocations and can support the optical thickness in terms of polarization and quantum efficiency.

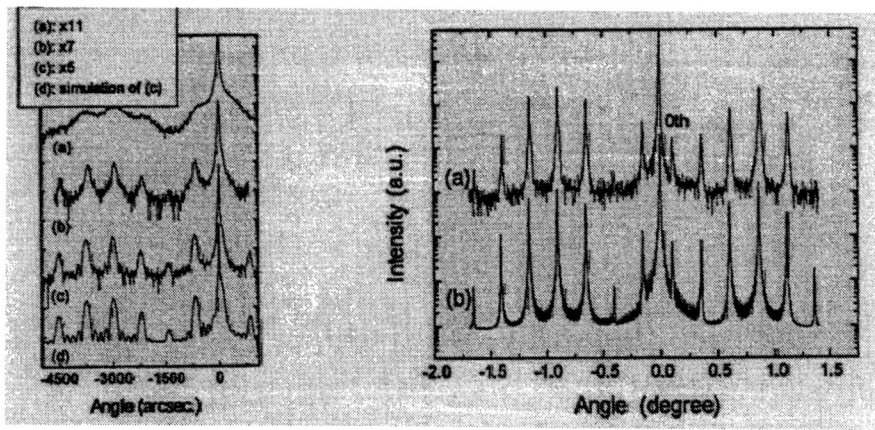

Figure 4 X-ray rocking curves (left figure) for strained $InAs_{0.4}P_{0.6}/InP$ (9 nm/13 nm) MQWs grown on InP substrates for different number of periods, and (right figure) for a strain-compensated

$InAs_{0.4}P_{0.6}/Ga_{0.17}In_{0.83}P$ MQW with 30 periods (10). The bottom traces in both figures are simulations based on the dynamical diffraction theory.

ACKNOWLEDGMENT

I wish to thank Charlie Sinclair for inviting me to give a presentation at this workshop, my former student Wayne G. Bi for the work in Reference 8, and Takashi Maruyama for extensive discussions, collaboration, and suggestions on my presentation.

REFERENCES

1. Maruyama, T., et al., *Phys. Rev. Lett.* **66**, 2376 (1991).
2. Omoti, T., et al., *Jpn. J. Appl. Phys.* **33, Part 1,** 5676 (1994).
3. Drescher, P., et al., *Nucl. Instr. Meth. Phys. Res. A* **381**, 169 1996.
4. Drescher, P., et al. *Appl. Phys. A* **63**, 203 (1996).
5. Bolkhovityanov, Yu. B., et al., *Inst. Phys. Conf. Ser.* **155**, 231 (1997).
6. Nakanishi, T., et al., *Phys. Lett. A* **158**, 345 (1991).
7. Bi, W.G., and Tu, C.W., *J. Vac. Sci. Technol. B***14**, 2282 (1996).
8. Saka, T., et al., *Jpn. J. Appl. Phys.* **32**, L1837 (1993).
9. Chin, T.P., et al., *Mat. Res. Soc. Symp. Proc.* **281**, 227 1993).
10. Chang, J.C.P., et al., *Appl. Phys. Lett.* **60**, 1129 (1992).
11. Mei, X.B., et al., *Appl. Phys. Lett.* **68**, 90 (1996).

Spin Polarization of Photoelectrons from Ordered Semiconductor Alloys

Su-Huai Wei

National Renewable Energy Laboratory, Golden, CO 80401, U.S.A.

Abstract. Recently observed spontaneous CuPt-like ordering of III-V semiconductor alloys provides the opportunity of using them as high quality spin polarized photoelectron source. We show that CuPt ordering of semiconductor alloy causes a splitting at the valence band maximum (VBM) and induces an anisotropy in the intensities of the transitions between these split VBM components and the conduction band minimum. We show that for sufficiently ordered single-subvariant sample, the emitted photoelectrons can have 100% spin-polarization.

INTRODUCTION

High quality spin-polarized electron source with high spin-polarization, high quantum efficiency, high reliability and rapid polarization reversal is very useful in investigating spin-dependent phenomena in high energy, atomic, and solid state physics [1-5]. In all these experiments the quality of the results and the cost of operating accelerators depend significantly on the polarized electron source. Currently almost all of the sources now in use with accelerators are based on photoemission from GaAs or related materials. Several excellent review papers have been written on this subject [1-8]. In the following sections, we will first describe briefly the mechanism of generating polarized electron from zinc-blend semiconductors. This analysis will provide the basis for our proposal of using CuPt ordered semiconductor alloys as candidate of high quality spin-polarized electrons source [9-10]. We will then discuss briefly the optical and spin polarization properties of the ordered semiconductor alloys [9-12].

OPTICAL TRANSITION NEAR THE BAND GAP

The generation of spin-polarized electron from zinc-blende GaAs is based on the symmetry properties of electron wave functions near the band edge and the selection rules governing the optical dipole transition [3]. For direct gap III-V zinc-blende semiconductor, the band edge states at Γ have well defined single particle quantum numbers. The two degenerate conduction states Ψ_c can be described by the two s-like states $(s \uparrow, s \downarrow)$ with $(j = 1/2, m_j = \pm 1/2)$. The valence states Ψ_v can be expanded in terms of the six p-like basis functions $(p_x \uparrow, p_y \uparrow, p_z \uparrow, p_x \downarrow, p_y \downarrow, p_z \downarrow)$ [13], where $\hat{\sigma} = \uparrow$ or $\downarrow$ are the spinors parallel or antiparallel to the z direction. The valence states Ψ_v can be obtained by diagonalizing the corresponding 6x6 Hamiltonian [9]:

$$H_v^{SO} = \frac{1}{3} \begin{pmatrix} 0 & -i\Delta^{SO} & 0 & 0 & 0 & \Delta^{SO} \\ i\Delta^{SO} & 0 & 0 & 0 & 0 & -i\Delta^{SO} \\ 0 & 0 & 0 & -\Delta^{SO} & i\Delta^{SO} & 0 \\ 0 & 0 & -\Delta^{SO} & 0 & i\Delta^{SO} & 0 \\ 0 & 0 & -i\Delta^{SO} & -i\Delta^{SO} & 0 & 0 \\ \Delta^{SO} & i\Delta^{SO} & 0 & 0 & 0 & 0 \end{pmatrix}, \qquad (1)$$

where Δ^{SO} is the spin-orbit splitting at valence band maximum. The six p-like valence eigenstates have $(j = 3/2, m_j = \pm 3/2)$, $(j = 3/2, m_j = \pm 1/2)$, and $(j = 1/2, m_j = \pm 1/2)$ [3]. The transition intensity between Ψ_c and Ψ_v is proportional to the square of the matrix element $I_{c,v} = |<\Psi_c|H_{int}|\Psi_v>|^2$, where H_{int} is the interacting Hamiltonian. For *linearly* polarized light along the [l,m,n] direction we have $H_{int} \propto lx + my + nz$, while for *circularly* polarized light $\sigma^\pm$ with angular momentum parallel and antiparallel to z' we have $H_{int} \propto x' \pm iy'$. The transition matrix elements can be calculated by expanding the wavefunctions and H_{int} in terms of the spherical harmonics Y_{lm} and by noticing that the allowed dipole transitions are for $\Delta m_j = \pm 1$. This leads to the simple selection rule

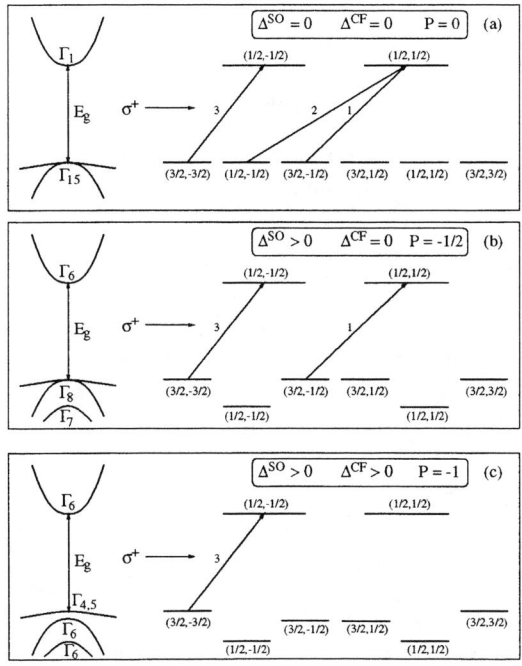

Figure 1. Schematic plot of energy levels at Γ and transition probabilities for zinc-blende system. (a) for $\Delta^{SO} = 0$ and $\Delta^{CF} = 0$, (b) $\Delta^{SO} > 0$ and $\Delta^{CF} = 0$ (e.g., GaAs), and (c) both $\Delta^{SO} > 0$ and $\Delta^{CF} > 0$ (e.g., CuPt ordered $Ga_{0.5}In_{0.5}P$).

$$< s\hat{\sigma} \mid x_\mu \mid p_\nu \hat{\sigma}' > = \lambda\, \delta_{\mu,\nu} \delta_{\hat{\sigma},\hat{\sigma}'} , \qquad (2)$$

where $(\mu, \nu = x, y, z)$ and λ is a normalization parameter.

The transition probability for near gap optical pumping by the circularly polarized light $H_{int} = x' + iy'$ between these band edge states are shown in Fig. 1. Without the spin-orbit interaction ($\Delta^{SO} = 0$) and the symmetry lowering crystal field interaction (Fig. 1a), the six Γ_{15v} valence states are degenerate. In this case the spin polarization P defined as

$$P = \frac{I\uparrow' - I\downarrow'}{I\uparrow' + I\downarrow'} , \qquad (3)$$

will be zero. Here, $I\uparrow'$ and $I\downarrow'$ are the transition intensities for $\uparrow'$ spin and $\downarrow'$ spin, respectively, and the quantization direction is defined by the circularly polarized light. When the spin-orbit interaction is switched on (Fig. 1b), the four $\Gamma_{8v}(j = 3/2)$ valence states are separated from the two $\Gamma_{7v}(j = 1/2)$ valence states by an energy of the spin-orbit splitting Δ^{SO}. In this case, if we can control the optical pumping so that the transition to the conduction band results only from the highest occupied $j = 3/2$ state, the generated conduction electron will have a spin polarization of 50% (Fig. 1b).

The polarized conduction electrons can then be extracted into vacuum by raising the conduction band level of the semiconductor compounds above the vacuum, thus achieving the negative electron affinity (NEA) condition. For GaAs this was done by covering its surface with Cs and O [1-8].

Substantial effort [4-8] has recently be focused on breaking the 50% polarization limit for bulk GaAs. It is clear from Fig. 1 that to achieve higher polarization, the degeneracy at VBM should be removed. Furthermore, consideration of transition probability and band structure near Γ requires the heavy hole (hh) ($j = 3/2, m_j = \pm 3/2$) states having a higher energy than the light hole (lh) ($j = 3/2, m_j = \pm 1/2$) states. Many approaches have been previously utilized to remove the degeneracy by lowering the crystal symmetry. These include (i) using strained semiconductors, e.g., to grow GaAs on GaP_xAs_{1-x} or to grow $In_xGa_{1-x}As$ on GaAs. Polarizations up to 90% have been reported by several groups using this method [4,5]. However, because the thickness of the film which can be grown coherently on a lattice mismatched substrate is reduced when the mismatch increases, the yield of highly polarized photoelectrons is low. (ii) Using multilayer thick semiconductor quantum well such as $GaAs/Al_xGa_{1-x}As$. Attempt in this approach also suffers from low quantum efficiency, possibly due to electron trapping inside the quantum well and depolarization at the well interface. And (iii) using semiconductor compounds that naturally has lower crystal symmetry, e.g., chalcopyrites. However, polarization achieved using these compounds are rather low, partly because of the poor sample quality and partly because of the fact that the crystal field splitting of chalcopyrites are either very small or negative (i.e., the lh state is above the hh state). Table I show our calculated crystal field splitting Δ^{CF} and spin-orbit splitting Δ^{SO} for five chalcopyrite compounds [14]. The reason for the negative crystal field splitting is explained in Ref. 15.

We propose here a different approach of achieving 100% polarization of emitted photoelectrons by using the recently observed spontaneous CuPt ordered III-V semiconductor alloys. We point out that these ordered alloys have

Table I. Calculated crystal field splitting Δ^{CF} and spin-orbit splitting Δ^{SO} for five I-III-VI chalcopyrite compounds. The experimental structural parameters used in the calculation and the experimental band gaps of these compounds are also shown in the table [14].

Compounds	a(Å)	c/a	u	E_g (eV)	Δ^{SO} (eV)	Δ^{CF} (eV)
$CuInS_2$	5.523	1.007	0.214	1.53	−0.02	−0.00
$CuInSe_2$	5.784	1.004	0.224	1.04	0.184	−0.02
$CuAlSe_2$	5.602	0.977	0.259	2.67	0.152	−0.16
$CuGaSe_2$	5.614	0.982	0.250	1.68	0.194	−0.12
$CuInTe_2$	6.161	1.003	0.225	1.01	0.598	−0.00

all the advantages associated with III-V semiconductors. They are bulk-like, have no limit on the thickness in a coherent film growth, and have large positive crystal field splitting. In the next section we will discuss briefly ordering induced effects on band structure and optical transitions. More detailed discussions of the ordering induced effects can be found in Ref. 12.

OPTICAL PROPERTIES OF ORDERED ALLOYS

Spontaneous CuPt-like ordering of $A_xB_{1-x}C$ alloys has been widely observed in vapor phase growth of many III-V systems on (001) substrates [16].

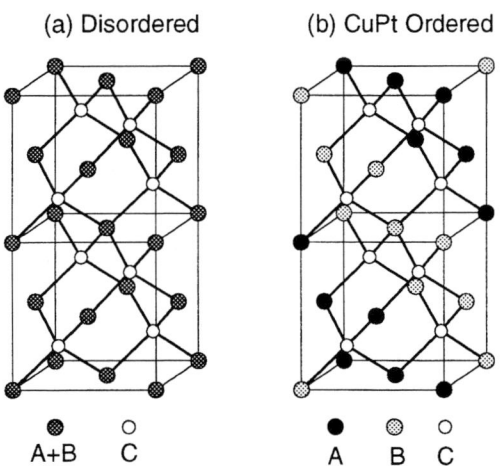

Figure 2. Crystal structures of (a) disordered zinc-blende alloy and (b) fully CuPt ordered alloy.

The ordered phase consists of alternate cation monolayer planes $A_{x+\frac{\eta}{2}}B_{1-x-\frac{\eta}{2}}$ and $A_{x-\frac{\eta}{2}}B_{1-x+\frac{\eta}{2}}$ stacked along the [111] (or equivalent) directions, where $0 \leq \eta \leq 1$ is the long range order parameter. Perfect ordering ($\eta = 1$) corresponds to successive planes of pure A followed by pure B, etc. (Fig. 2).

When a zinc-blende disordered alloy forms a long range ordered (LRO) CuPt superlattice, the Brillouin zone of the ordered structure is reduce to half of the zinc-blende zone. Consequently, two zinc-blende **k**-points (and states associated with them) fold into a single **k**-point in the CuPt Brillouin zone. Those folded states that have the same superlattice symmetry can couple to each other. This coupling lead to energy level shifts and to splitting of those states that were degenerate in the random alloy. For example, the states at $\bar{\Gamma}$ (we use an overbar to denote superlattice states) are constructed from zinc-blende like states at Γ and L^{111} [15]. The coupling between the $\bar{\Gamma}_{3v}(\Gamma_{15v})$ and $\bar{\Gamma}_{3v}(L_{3v})$ states and the coupling between the $\bar{\Gamma}_{1c}(\Gamma_{1c})$ and $\bar{\Gamma}_{1c}(L_{1c})$ states shifts the energy of the valence band maximum (VBM) upwards and the energy of the conduction band minimum (CBM) downwards, thus causing a splitting at the VBM and a lowering of the band gap relative to the random alloy [9,12].

1. Dependence on the degree of LRO

For *spontaneously* ordered semiconductor alloys, the LRO is never perfect [16]. The degree of ordering depends on growth temperature, growth rates, III/V ratio, substrate misorientation and doping. We have derived a general theory to study physical properties of these partially ordered samples [17]. We found that the physical properties P of a partially ordered sample can be well described by

$$P(x,\eta) = P(x,0) + \eta^2 \left[P(X_\sigma, 1) - P(X_\sigma, 0) \right] . \quad (4)$$

This simple equation relates the property P at any degree of LRO η to the corresponding properties in (i) the perfectly random alloy at compositions x and X_σ and (ii) the perfectly ordered structure at composition X_σ.

Using the quasicubic model, the valence band states at Γ for pure (111) CuPt ordering can be described by the 6x6 Hamiltonian [9]:

$$H_v^{SO+O} = \frac{1}{3} \begin{pmatrix} 0 & -\Delta^O - i\Delta^{SO} & -\Delta^O & 0 & 0 & \Delta^{SO} \\ -\Delta^O + i\Delta^{SO} & 0 & -\Delta^O & 0 & 0 & -i\Delta^{SO} \\ -\Delta^O & -\Delta^O & 0 & -\Delta^{SO} & i\Delta^{SO} & 0 \\ 0 & 0 & -\Delta^{SO} & 0 & -\Delta^O + i\Delta^{SO} & -\Delta^O \\ 0 & 0 & -i\Delta^{SO} & -\Delta^O - i\Delta^{SO} & 0 & -\Delta^O \\ \Delta^{SO} & i\Delta^{SO} & 0 & -\Delta^O & -\Delta^O & 0 \end{pmatrix} . \quad (5)$$

As we can see from Eqs. (1) and (5), the effect of (111) ordering is similar to apply a tensile strain along the [111] direction. diagonalizing Eq. (5) gives the eigenstates and the three spin-degenerate energy levels at $\bar{\Gamma}$

$$E_{1,2,3} = \begin{cases} \frac{1}{3}(\Delta^{SO} + \Delta^{O}) \\ -\frac{1}{6}(\Delta^{SO} + \Delta^{O}) \pm \frac{1}{2}[(\Delta^{SO} + \Delta^{O})^2 - \frac{8}{3}\Delta^{SO}\Delta^{O}]^{1/2}, \end{cases} \quad (6)$$

where Δ^O is the ordering-induced crystal field splitting in the absence of spin-orbit coupling. Loosely speaking, levels $|1>$, $|2>$, and $|3>$ are heavy hole ($j = 3/2, m_j = \pm 3/2$; $\Gamma_{4,5v}$), light hole ($j = 3/2, m_j = \pm 1/2$; $\Gamma_{6v}^{(1)}$), and split-off ($j = 1/2, m_j = \pm 1/2$; $\Gamma_{6v}^{(2)}$) states, respectively (Fig. 1c). $\Delta^{SO}(\eta)$ and $\Delta^O(\eta)$ can be calculated using Eq. (4) given the values at the end points $\eta = 0$ and $\eta = 1$. The latter are obtained from self-consistent band structure calculations using the local density functional approximation (LDA) as implemented by the linearized augmented plan wave (LAPW) method. The results for four alloy systems are summarized in Table II [12].

Fig. 3 shows the calculated results (solid lines) of $\Delta E_{12} = E_1 - E_2$, $\Delta E_{13} = E_1 - E_3$, and $\Delta E_g = E_g - E_g(random)$ as function of η for ordered $Ga_{0.5}In_{0.5}P$ alloy. These results are obtained by applying Eq. (4) and the data in Table II to $\Delta^{SO}(\eta)$, $\Delta^O(\eta)$ and $\Delta E_g(\eta)$ and substituting these into Eq. (6). These theoretical results can be used to infer the degree of LRO in a given sample from measured optical data [17]. We have compared in Fig. 3 the calculated results with three sets of experimental data. The solid circles represent the polarized photoluminescence results of Kanata et al. [18], the solid squares are the piezomodulated reflectivity data of Alonso et al. [19] and the solid triangles are the recent photoluminescence data of Ernst et al. [20]. We find that all the fits to our theoretical prediction are reasonably good.

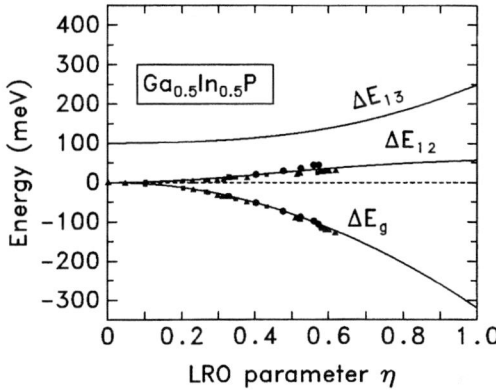

Figure 3. Calculated valence band splitting ΔE_{12}, ΔE_{13} and band gap reduction energies ΔE_g as function of the LRO parameter η for $Ga_{0.5}In_{0.5}P$ (solid lines). The solid circles represent the polarized photoluminescence results of Kanata et al. [18], The solid squares are the piezomodulated reflectivity data of Alonso et al. [19] and the solid triangles are the recent photoluminescence data of Ernst et al. [20].

Table II. Calculated spin orbit splitting (Δ^{SO}) and crystal field splitting (Δ^O) for the perfectly ordered ($\eta = 1$) and perfectly random ($\eta = 0$) phases, and band gap reduction (ΔE_g) of the fully ordered phase relative to the random alloy. Energies are in eV.

	$Ga_{0.5}In_{0.5}P$	$Al_{0.5}In_{0.5}P$	$Ga_{0.5}In_{0.5}As$	$Al_{0.5}In_{0.5}As$
$\Delta^{SO}(0)$	0.10	0.09	0.35	0.34
$\Delta^{SO}(1)$	0.11	0.11	0.35	0.35
$\Delta^O(0)$	0.00	0.00	0.00	0.00
$\Delta^O(1)$	0.20	0.26	0.10	0.24
ΔE_g	-0.32	-0.17	-0.30	-0.14

2. Coupling of ordering with epitaxial strain

Since the ordered samples are grown on a substrate with nominally (001) orientation, strain caused by lattice mismatch between the film and the substrate can exist [16]. We will study here the interesting case of the coexistence of chemical ordering in the direction $\mathbf{G}_{ord} = (111)$ with epitaxial strain in the direction $\mathbf{G}_{substrate} = (001)$. This is different from the case of *artificially* grown strain layer superlattices in which the direction of layer modulation coincides with the direction of strain, so these effects add up co-linearly. We are specifically interested in the effects of the non-co-linear, "vector addition" of ordering and strain on the changes in the valence band splitting ΔE_{12}.

The valence band Hamiltonian corresponding to (111) ordering plus (001) strain at $\mathbf{k} = \mathbf{0}$ is [9]

$$H_v^{SO+O+S} = \frac{1}{3} \begin{pmatrix} \Delta^S & -\Delta^O - i\Delta^{SO} & -\Delta^O & 0 & 0 & \Delta^{SO} \\ -\Delta^O + i\Delta^{SO} & \Delta^S & -\Delta^O & 0 & 0 & -i\Delta^{SO} \\ -\Delta^O & -\Delta^O & -2\Delta^S & -\Delta^{SO} & i\Delta^{SO} & 0 \\ 0 & 0 & -\Delta^{SO} & \Delta^S & -\Delta^O + i\Delta^{SO} & -\Delta^O \\ 0 & 0 & -i\Delta^{SO} & -\Delta^O - i\Delta^{SO} & \Delta^S & -\Delta^O \\ \Delta^{SO} & i\Delta^{SO} & 0 & -\Delta^O & -\Delta^O & -2\Delta^S \end{pmatrix}. \quad (7)$$

Here Δ^{SO} is the spin-orbit splitting at VBM. $\Delta^S(\epsilon) = 3b\frac{C_{11}+2C_{12}}{C_{11}}\epsilon$ is the strain-induced valence band crystal field splitting, where b is the tetragonal deformation potential and C_{ij} are the elastic constants [9]. The epitaxial strain is given by $\epsilon(x) = [a_s - a_f(x)]/a_f(x)$, where $a_f(x)$ and a_s are the lattice constants of the film and the substrate, respectively. $\Delta^O(\eta) = \Delta^O(1)\eta^2$ is the ordering induced valence band crystal field splitting. Solving Eq. (7) gives the valence band levels (in decreasing order) E_1, E_2 and E_3 and their eigenstates as functions of composition x (or strain ϵ) and degree of ordering η.

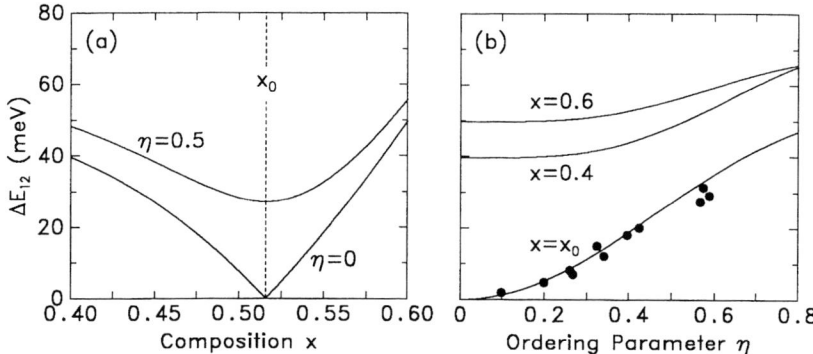

Figure 4. Valence band splitting ΔE_{12} of the $Ga_x In_{1-x} P$ alloy strained on GaAs as a function of (a) composition x at $\eta = 0$ and $\eta = 0.5$ and (b) ordering parameter η at compositions $x = x_0$, $x = 0.4$ and $x = 0.6$. $x_0 = 0.516$ is the lattice matching composition.

Fig. 4 depicts the valence band splitting $\Delta E_{12}[\eta, \epsilon(x)] = E_1[\eta, \epsilon(x)] - E_2[\eta, \epsilon(x)]$ for $Ga_x In_{1-x} P$ alloy grown on GaAs substrate as a function of the film composition x (Fig. 4a) or as a function of the degree of long range order η. The following are the important features: (i) Any non-zero strain leads to a splitting of the VBM (Fig. 4a). (ii) The ΔE_{12} vs x curve have a cusp at $x = x_0$ for random alloy (Fig. 4a), reflecting the change of the VBM from heavy hole when $x < x_0$ to light hole when $x > x_0$. (iii) Chemical order universally leads to a valence band splitting for any $\eta > 0$ (Fig. 4b). For the valence band splitting, chemical ordering is analogous to in-plane *compressive* strain in that both yield a heavy-hole state at the top of the valence band. (iv) The coexistence of ordering ($\eta \neq 0$) with strain ($x \neq x_0$) is predicted (a) to remove the cusp in the ΔE_{12} vs x curves (Fig. 4a). This is so since any amount of chemical order will mix equal amounts of light with heavy hole states at $x = x_0$. (b) Chemical ordering is also predicted to reduce the dependence of ΔE_{12} on x, as illustrated by the flattening of the $\eta = 0.5$ curve near $x = x_0$ in Fig. 4a. (c) (001) strain increases the valence band splitting produced by pure ordering, and vise versa [21]. This reflects the fact that the ordering-induced splitting Δ^O and the strain-induced splitting Δ^S are complementary, so the combined effect is larger than the individual ones.

3. Optical anisotropy in ordered alloys

The splitting of the VBM for the ordered alloy induces an anisotropy in the intensities of the transitions between these split VBM components and the conduction band minimum. Using the calculated eigenstates of Eq. (5) and the selection rule of Eq. (2) we have calculated [9,10] the transition intensities $I_{c,v}$ between the valence states $|1>$ and $|2>$ and the conduction state in (111) ordered $Ga_{0.5} In_{0.5} P$ as a function of η. We consider linearly polarized light with polarization $\hat{e} \| [110]$ (defined as $\Theta = 0°$); $\hat{e} \| [1\bar{1}0]$ (defined as $\Theta = 90°$) and in-between. For random alloy the transition intensities $I_{c,v}$ is independent of Θ.

Figure 5 depicts the calculated normalized intensities as function of the polarization angle Θ. We find that in this case the intensity can be described by

$$I(\Theta) = I_{1\bar{1}0} \sin^2\Theta + I_{110} \cos^2\Theta. \qquad (8)$$

For the $\Gamma_{6c} - \Gamma_{4v,5v}$ transition (Fig. 5a) the intensity is independent of η, since there is no coupling between $\Gamma_{4v,5v}$ and the other two Γ_{6v} valence states. For the $\Gamma_{6c} - \Gamma_{6v}$ transition, however, we see a strong dependence on the ordering parameter η. This is due to the coupling between the two Γ_{6v} states. $I(\Theta)$ can be either an increasing function (at large η) or a decreasing function (at small η) of the polarization angle Θ. Figure 5 compares our calculated results (lines) with the recent polarized electroreflectance data of Kanata et al. [22] (solid dots). We find that the best fit to the measured intensities is obtained using $\eta = 0.58$. This value corresponds to a valence band splitting of $\Delta E_{12} = 34$ meV. The directly measured [18] valence band splitting of this sample is $\Delta E_{12} = 34 \pm 4$ meV, in excellent agreement with the above value. Thus, our intensity analysis [9,10] is consistent with our analysis of the energy levels [17].

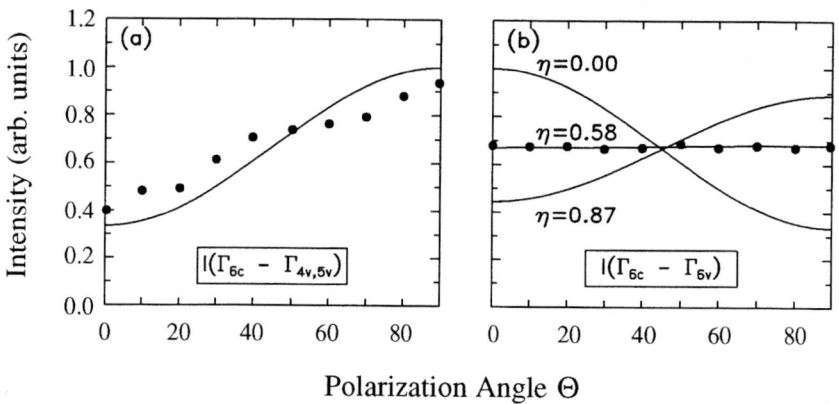

Figure 5. Calculated transition intensity of (111) ordered Ga$_{0.5}$In$_{0.5}$P as a function of polarization angle Θ. (a) The $\Gamma_{6c} - \Gamma_{4v,5v}$ transition. The intensity is independent of η. (b) The $\Gamma_{6c} - \Gamma_{6v}$ transition at $\eta = 0.00$, 0.58, and 0.87. The solid dots in (a) and (b) are the experimental data of Kanata et al. [adjusted to (111) order, see Ref. 22].

4. Spin polarization in ordered alloys

We have studied the ordering-induced changes in spin polarization of emitted photoelectrons [9,10]. We use for this purpose circularly polarized light σ^+ with its angular momentum along the ordering direction $z' = [111]$. We thus have $H_{int} \propto x' + iy'$, where $x' = \frac{1}{\sqrt{2}}(-x+y)$ and $y' = \frac{1}{\sqrt{6}}(x+y-2z)$. The spinors parallel and anti parallel to the [111] ordering direction are given by

$$\uparrow' = \cos\tfrac{\theta}{2} e^{-i\tfrac{\varphi}{2}} \uparrow + \sin\tfrac{\theta}{2} e^{i\tfrac{\varphi}{2}} \downarrow$$
$$\downarrow' = -\sin\tfrac{\theta}{2} e^{-i\tfrac{\varphi}{2}} \uparrow + \cos\tfrac{\theta}{2} e^{i\tfrac{\varphi}{2}} \downarrow ,\qquad(9)$$

where the angles θ and φ are determined by the equation $[\sin\theta\cos\varphi, \sin\theta\sin\varphi, \cos\theta] = \tfrac{1}{\sqrt{3}}[111]$. Figure 6 show the calculated spin intensities $(I\downarrow' - I\uparrow')_{1c}$ and $(I\downarrow' - I\uparrow')_{2c}$ for ordered $Ga_{0.5}In_{0.5}P$ alloys. We find that photoelectrons generated from the $\Gamma_{4v,5v}$ and the Γ_{6v} states are both *fully polarized*. Hence, if the splitting ΔE_{12} is large enough to allow optical pumping only from the highest $\Gamma_{4v,5v}$ state, the generated photoelectrons can be 100% spin polarized.

Note that despite the identical optical response with respect to the linearly polarized light along [110] and [1$\bar{1}$0] of the two $CuPt_A$ [(111) and (11$\bar{1}$)] subvariants, their response to the circularly polarized light is predicted to be different. Using the σ^+ light noted above but for (11$\bar{1}$) ordering, we find that the spin polarization P for the transition from the top $\Gamma_{4v,5v}$ state is only 20% and the total intensity $I\downarrow' + I\uparrow'$ is reduced to 55.56% of the intensity for (111) ordering. This difference can be used to distinguish (111) ordering from (11$\bar{1}$) ordering, which is not possible using the linearly polarized light. This also indicates that in order to obtain the highest efficiency in generating spin polarized electrons, single variant crystals are required.

The coexistence of (001) strain and (111) ordering will mix the hh states with the lh states, thus reduces the spin polarization. In this case the maximum spin-polarization ($< 100\%$) can be obtained in a direction parallel to one of the principle axis defined by the Hamiltonian of Eq. (7).

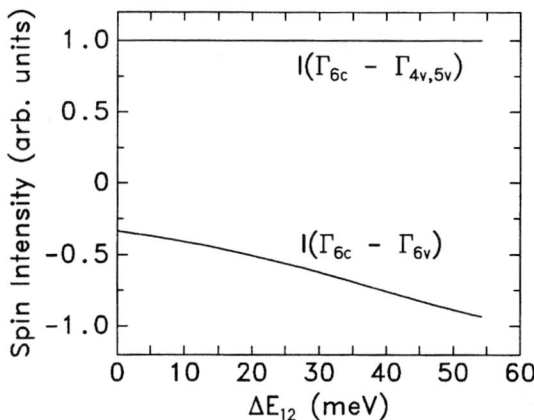

Figure 6. Calculated spin intensity $(I_- - I_+)_{1c}$ and $(I_- - I_+)_{2c}$ (in arbitrary units) of ordered $Ga_{0.5}In_{0.5}P$ as a function of the valence band splitting ΔE_{12}. The angular momentum of the circularly polarized light is in the same direction as the ordering vector. The generated photoelectron from each band is fully polarized.

SUMMARY

In conclusion, we suggest that CuPt-like ordered III-V semiconductor alloys ($Ga_{1-x}In_xP$, $Ga_{1-x}In_xAs$, $Al_{1-x}In_xP$ and $Al_{1-x}In_xAs$) should be good candidates for high quality spin polarized photoelectron source. For sufficiently ordered single-subvariant sample, the emitted photoelectrons can achieve a theoretical 100% spin-polarization. This ultrathin ordered superlattice can be considered as a low symmetry bulk compound. Since it is grown in nominally lattice matched substrate, there is no limit on how thick one can grow high quality coherent samples. Furthermore, due to the strong coupling between the zinc-blende Γ_{15v} and L_{3v} states, the crystal field splitting at VBM is large and positive. The crystal field splitting can be increased further by controlling the residue strain in the sample, thus is capable of achieving high spin polarization at room temperature. Experimental investigation of optical and spin polarization properties of these samples is called for.

ACKNOWLEDGMENT

I would like to thank Dr. A. Zunger for his collaboration in this work and to thank Dr. C. K. Sinclair for discussion on this subject. This work was supported by the U. S. Department of Energy, under contract No. DE-AC36-83CH10093.

REFERENCES

1. J. Kessler, *Polarized Electrons*, 2nd edition (Springer-Verlag, Berlin, 1985).
2. C. K. Sinclair, in *Proc. 8th Int. Symp. on High Energy Spin Physics*, edited by K. J. Heller, (AIP, New York, 1989), p.1412.
3. D. T. Pierce and F. Meier, Phys. Rev. B **13**, 5484 (1976).
4. L. S. Cardman, Nuclear Phys. A **546**, 317c (1992).
5. F. Meier, J. C. Grobli, D. Guarisco, and A. Vaterlaus, Physica Scripta **T49**, 574 (1993).
6. T. Nakanishi, AIP Conf. Proc. **338**, 344 (1995).
7. J. E. Clendenin, *Proc. of the 1995 Particle Accelerator Conf.*, Vol. **2**, 877 (1995).
8. L. Tecchio, *Proc. of the 5th Winter School on Hadronic Physics*, edited by R. Cherubini, P. Dalpiaz, and B. Minetti, (World Scientific, Singapore, 1991), p. 429.
9. S.-H. Wei and A. Zunger, Phys. Rev. B **49**, 14337 (1994).
10. S.-H. Wei and A. Zunger, Appl. Phys. Lett. **64**, 1676 (1994).
11. S.-H. Wei and A. Zunger, Appl. Phys. Lett. **64**, 757 (1994).

12. S.-H. Wei, A. Franceschetti, and A. Zunger, MRS Symp. Proc. **417**, edited by E. D. Jones, A. Mascarenhas, and P. Petroff, (MRS, Pittsburgh, 1996), p.3.

13. We neglect here the small cation d character at the valence band maximum of the III-V semiconductor compounds.

14. S.-H. Wei and A. Zunger, J. Appl. Phys. **78**, 3846 (1995).

15. S.-H. Wei and A. Zunger, Phys. Rev. B **39**, 3279 (1989).

16. A. Zunger and S. Mahajan, in *Handbook of Semiconductors*, 2nd ed., edited by S. Mahajan (Elsevier, Amsterdam, 1994), Vol. 3, p. 1439.

17. S.-H. Wei, D. B. Laks, and A. Zunger, Appl. Phys. Lett. **62**, 1937 (1993).

18. T. Kanata, M. Nishimoto, H. Nakayama, and T. Nishino, Phys. Rev. B **45**, 6637 (1992).

19. R. G. Alonso, et al., Phys. Rev. B **48**, 11833 (1993).

20. P. Ernst, et al., Appl. Phys. Lett. **67**, 2347 (1995).

21. A. Eyal et al., Jpn. J. Appl. Phys. Suppl. **32-3**, 716 (1993).

22. T. Kanata, M. Nishimoto, H. Nakayama, and T. Nishino, Appl. Phys. Lett. **63**, 26 (1993).

Emission of Polarized ps-Electron Bunches from III-V Semiconductor Cathodes

P. Hartmann[1], J. Bermuth[1], J. Hoffmann[1], S. Köbis[1],
H. G. Andresen[1], K. Aulenbacher[1], P. Drescher[2], H. Euteneuer[1],
H. Fischer[2], K. Grimm[1], Th. Hammel[1], D. v. Harrach[1],
H. Hofmann[1], K.–H. Kaiser[1], E.-M. Kabuß[1], H. J. Kreidel[1],
A. Lopes-Ginja[1], F. E. Maas[1], Ch. Nachtigall[2], S. Plützer[2],
E. Reichert[2], M. Schemies[2], E. Schilling[1], J. Schuler[1],
K.–H. Steffens[1], M. Steigerwald[2], H. Trautner[2]

[1] *Institut für Kernphysik, Joh. Gutenberg - Universität, Mainz, Germany*
[2] *Institut für Physik, Joh. Gutenberg - Universität, Mainz, Germany*

Abstract. At the pulsed electron gun testfacility at MAMI we have measured electron bunches from GaAs samples with different layer thicknesses. We observe an increasing dc-polarization with decreasing layer thickness. From the pulse measurements this behaviour may be explained in terms of bunchlength and spin relaxation time.

The principle of operation of the pulsed electron gun testfacility at MAMI is sketched in figure 1. It allows the analysis of short polarized electron bunches generated by illumination of a photocathode with short laser pulses. The generated electron beam is wobbled with a TM_{110} radio frequency resonator over the narrow entrance slit of an electron spectrometer. Since the laser pulse repetition rate is synchronized to the wobble radiofrequency, a stable spatial image of the bunch is generated. This spatial pulse image can be shifted over the slit by varying the phase of the laser pulses relative to the radiofrequency. By measuring the dependence of the transmitted current and the polarization on the phase shift, the bunch profile and the phase resolved polarization are sampled. A detailed description of the setup can be found in Ref. [1].
This article compares pulse measurements of three GaAs samples. Two of them have limited epilayer thicknesses of $0.4\mu m$ and $0.2\mu m$. They were produced at the Ioffe Institute in St. Petersburg. Their doping concentration is $4 \cdot 10^{18}$ cm^{-3}. The third sample is a bulk GaAs sample produced by Siemens in Germany. Its doping concentration is $2 \cdot 10^{19}$ cm^{-3}.
All samples were activated applying caesium and oxygen in a YoYo-procedure. Quantum efficiencies of 13.2 % (bulk), 4.5 % (0.4 μ) and 2.8 % (0.2 μ) were ob-

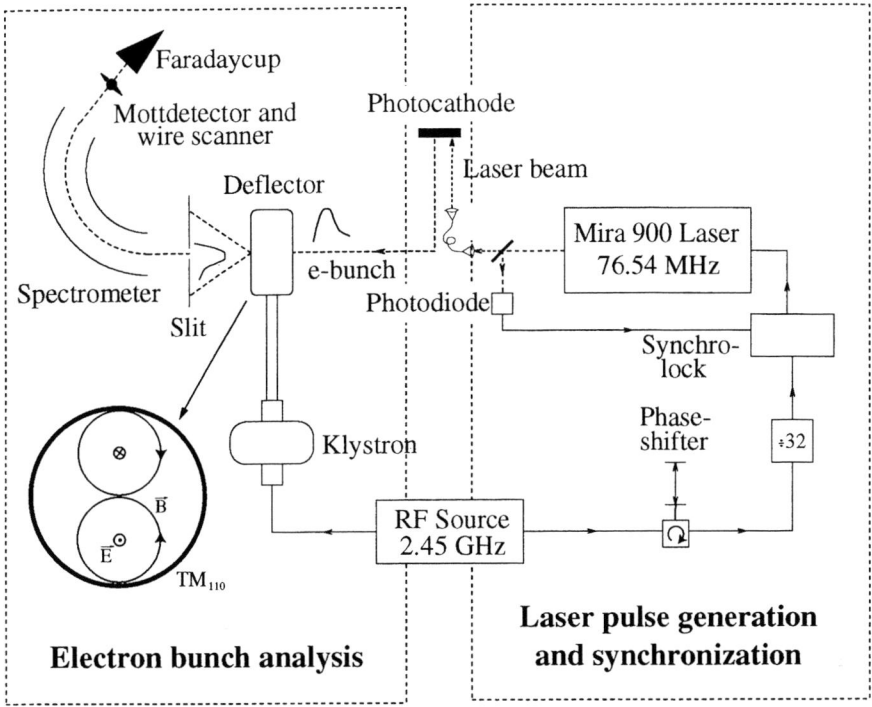

FIGURE 1. Sketch of the principle of the measurement.

tained at 633 nm.

Figure 2 shows a measurement from the bulk sample at a laser wavelength of 840 nm. The bunch shows a steep leading edge and a long tail. The polarization decreases from 43.3±0.6±4.3 % at the beginning of the bunch to zero in the bunch tail. A spin relaxation time of 58.8±1.4 ps was obtained by an exponential fit to the polarization data. The dc-polarization of 26.6±0.5±4.3 % can be reproduced by a weighted mean of the phaseresolved polarization data [2].

The bunch in figure 3(a) was measured with the 0.4 μ-sample. It still shows the typical slowly decreasing trailing pulse edge. Due to the layer thickness this edge decreases faster than the trailing edge of the bunch from the bulk sample. The maximum polarization of 42.1±0.9±4.3 % decreases with a relaxation time of 136^{+11}_{-10} ps. The minimum polarization at the end of the bunch is 25.6±1.6±4.3 %. Due to the long depolarization time and the thin layer the electrons at the bunch end are not completely depolarized. The measured dc-polarization was 38.9±0.9±4.3 %.

Compared to the other two measurements, the bunch of the 0.2 μm-sample in fig. 3(b) shows a completely different pulse profile. This is due to its short decreasing edge, which is dominated by the FWHM laser pulse duration of 5 ps. The maximum polarization of 47.2±2.2±4.3% decreases with a relaxation constant of

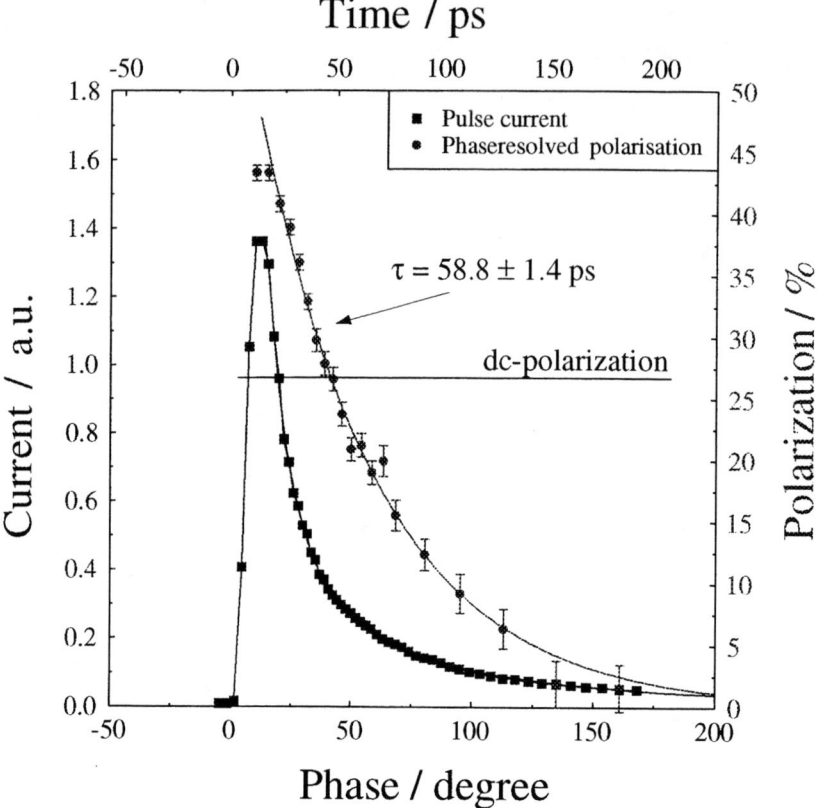

FIGURE 2. Pulse profile and phaseresolved polarization from the bulk sample.

104^{+60}_{-28}ps to 39.6±1.6±4.3%. The dc-polarization was 43.0±1.6±4.3%.

Conclusion

The change of the pulseshape with decreasing layer thickness implies, that electrons in the tail of a bunch were generated deeper inside the crystal than electrons in the leading edge of a bunch. If this was not the case, only the quantum efficiency of the cathode but not the pulseshape would change with decreasing layer thickness.
On their way to the surface the electrons are depolarized. The measured spin relaxation times are comparable to data from measurements of the Hanle-effect [3] and suggest the BAP-process [4] as the dominant depolarization process. Although the data do not agree exactly, the dependence of the spin relaxation time on the doping concentration in Ref. [3] shows the same tendency as the data obtained in this work. The dc-polarization of an unstrained GaAs sample seems to be predictable if one knows the bunch duration and the spin relaxation time from a bunch measurement.

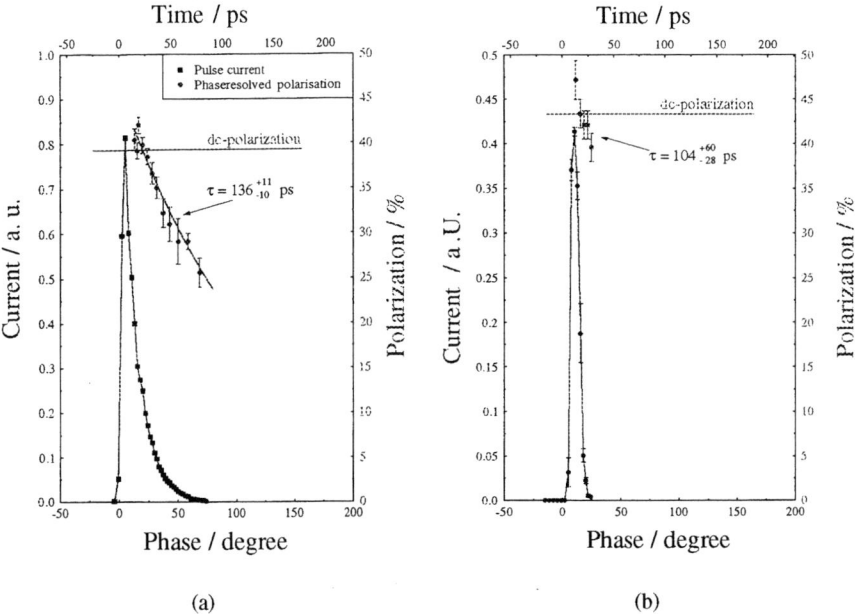

FIGURE 3. (a) Pulse profile and phaseresolved polarization from the 0.4 μm-sample. (b) Pulse profile and phaseresolved polarization from the 0.2 μm-sample.

Spin relaxation itself depends mainly on the doping concentration of the sample and on its temperature, while the bunch duration is determined by the electron diffusion constant, the light absorption constant and the crystal thickness.
Keeping track of all these parameters will lead to a process of "crystal engineering" to obtain the optimized photocathode for photoemission of polarized electrons for accelerator applications.

This project is supported by the Deutsche Forschungsgemeinschaft (SFB 201).

[1] P.Hartmann, J.Bermuth, J.Hoffmann, S.Köbis et al., Nucl. Instr. Meth. A 379 (1996) 15-20

[2] P.Hartmann, J.Bermuth, J.Hoffmann, S.Köbis et al., Proc. 12th Int. Symp. High-Energy Spin Physics (SPIN96), Amsterdam 1996

[3] K. Zerrouati et al., Phys. Rev. B 73 (1988) 1334

[4] G. L. Bir et al., Sov. Phys.-JETP 42 (1976) 705

Highly Polarized Electrons from Superlattice Photocathodes

T. Nakanishi[1], S. Okumi[1], K. Togawa[1], C. Takahashi[1], C. Suzuki[1],
F. Furuta[1], T. Ida[1], K. Wada[1], T. Omori[2], Y. Kurihara[2],
M. Tawada[2], M. Yoshioka[2], H. Horinaka[3], K. Wada[3],
T. Matsuyama[3], T. Baba[4], and M. Mizuta[4],

(1) Dept. of Physics, Nagoya University, Chikusa-ku, Nagoya-464, Japan
(2) High Energy Accelerator Research Organization, Tsukuba-305, Japan
(3) College of Engineering, University of Osaka Prefecture, Sakai-593, Japan
(4) Fundamental Research Laboratory, NEC Corporation, Tsukuba-305, Japan

Abstract. Our collaboration group have continued the research to develop the high performance photocathode and the guns to produce the highly polarized electron beam for high energy accelerator physics. In this report, the experimental results obtained from the photocathodes with superlattice structures, such as, AlGaAs-GaAs, InGaAs-GaAs and InGaAs-AlGaAs are described.

1. Introduction

Nowadays, polarized electron beam is most conventionally produced by using photoemission from the GaAs-type-semiconductor source. This polarized electron source (PES) is based on a combination of two fundamental technologies : laser optical pumping and a semiconductor surface with negative electron affinity (NEA). Followings are considered as the most important performances for PES-photocathodes used at High Energy electron accelerators (1).

1) electron spin polarization (ESP),
2) quantum efficiency (QE),
3) high peak current or high average current of extracted beam
4) cathode life time.

The maximum degree of electron spin polarization (ESP) seems to be governed mainly by two physical mechanisms. One is related to the initial polarization (Pi) of excited electrons in the conduction band, and Pi will be determined by the band-mixing of heavy-hole (hh) and light-hole (lh) bands at the valence band maximum. The other is related to spin relaxation process which will take place both at an inside and a surface of the semiconductor. The depolarization will become significant, if the spin relaxation time (τ_s) is much shorter than the average time (τ_{esc}) required for electrons to escape into vacuum. The spin relaxation time can be measured by photoluminescence method, and is discussed later in section 6.

To overcome the 50% ESP limitation of bulk-GaAs, the fine energy splitting (δ_s) must be given between hh- and lh-bands, and several different types of photocathodes have been developed by our collaboration. They are categorized into two groups. One contains the thin layers of strained GaAs (2), strained GaAsP, and strained GaAs with DBR (distributed Bragg reflector) (3). The other contains unstrained superlattice of AlGaAs-GaAs (4), strained layer superlattice of InGaAs-GaAs (5) and InGaAs-AlGaAs (6). Those superlattice have been fabricated by NEC Corporation using the molecular-beam epitaxy (MBE) method. AlGaAs-GaAs and InGaAs-GaAs are described in sections 2 and 3, and new results of InGaAs-AlGaAs are given in section 4.

The NEA surface plays an indispensable role in electron emission into vacuum, and the quantum efficiency (QE) is mainly governed by the properties of this NEA surface. The QE decay rate (defined as a cathode life time) is the most important performance for stable operation and easy maintenance of the source. The wider band-gap semiconductor cathode is considered as more favorable both for the higher QE, and the longer life time, because it gives the larger NEA. This problem is discussed again in section 4.

The maximum extracted current is determined by space charge limit for a thermionic gun cathode, while there is another limit called "surface charge limit" (SCL) for the NEA cathode of GaAs-PES, which was first observed at SLAC (7), (8). This suppression for emission of conduction-band electrons is considered as caused by electron trapping in the band bending region (BBR) of NEA surface. Therefore, SCL depends on conditions of the NEA state at BBR, as well as, a magnitude of extraction-field-gradient at the surface. The AlGaAs-GaAs superlattice with a highly p-doped NEA surface showed the higher resistance against SCL than that of strained GaAs (9), (10). New experimental data on this SLC effect is given in section 5.

2. AlGaAs-GaAs Superlattice

It was demonstrated in 1991 that electrons with high polarization of ~70% can be extracted from the AlGaAs-GaAs superlattice (4). Since then, various samples of AlGaAs-GaAs superlattice were made and tested, and it turned out that both of ESP and QE are quite sensitive to the amount of Be-dopants. Namely, to avoid the spin depolarization, the medium Be-density ($\sim 5 \times 10^{17}/cc$) inside the superlattice is preferable, while the high Be-density such as $\sim 4 \times 10^{19}/cc$ at the BBR (5 nm thickness) is so effective to increase the QE.

By this modulation doping technique, the QE becomes more than 0.5%, one order of magnitude higher than that of the previous uniformly-doped one, as shown in Fig. 1(a). High QE of this cathode (named as SL#7) was also confirmed at the tests at SLAC, Mainz and Bonn (12). As shown in Fig. 1(b), the thickness of GaAs (AlGaAs) layer is 1.98 nm (3.11 nm), and the total thickness of active layer was chosen to be about 0.1μm (20 pairs) to reduce spin depolarization during the emission process. The calculation of energy band structure by Kronig-Penny-Bastard (11) model shows that an energy splitting (δ_s) defined as the difference between both of the upper levels of hh- and lh-mini-bands is 23 meV.

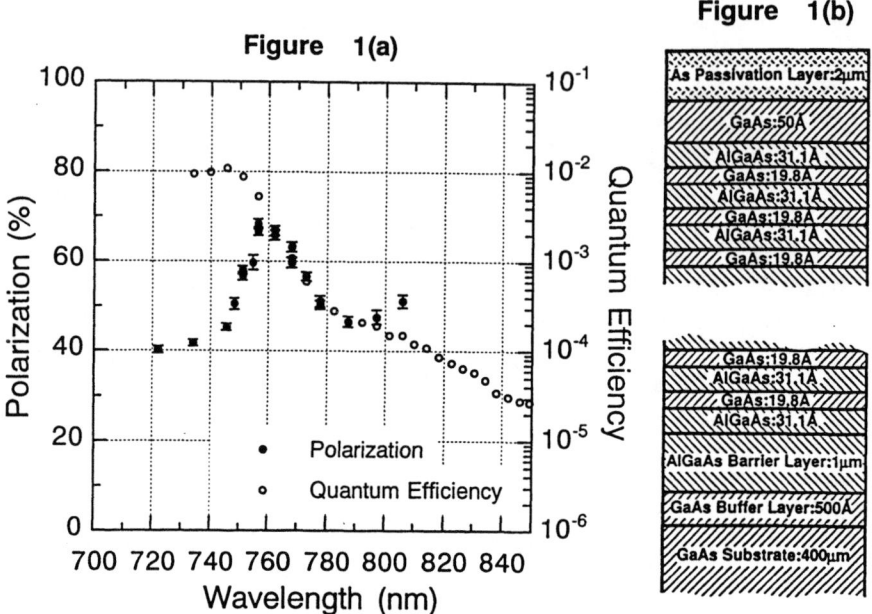

Figure 1(a)

Figure 1(b)

The polarization dependence on the thickness of AlGaAs-GaAs superlattice samples was measured and it turned out that the maximum ESP is limited to be 75% even if the so thin layer of superlattice is used (13). It suggests that the spin depolarization inside the superlattice whose thickness less than $0.1\mu m$ is not so serious, while the initial polarization (the ESP at excitation) itself is insufficient. It is probably due to the band mixing between hh- and lh-bands in the $k\perp$ (wave vector perpendicular to the excitation light) $\neq 0$ states, which is characteristic feature for the superlattice structure.

3. InGaAs-GaAs Strained Layer Superlattice

In order to break the ESP limit of AlGaAs-GaAs superlattice, the strained-layer InGaAs-GaAs superlattice was first studied in 1994 by our collaboration (5).

The band splitting between hh- and lh-bands in the well (InGaAs) layer can be introduced by the strain, in addition of superlattice structure effect. An Indium fraction of 15% was chosen to bring a strain of $\sim 1.0\%$ into the InGaAs layer, and the energy splitting (δ_s) was estimated to be about 30 meV. Four samples with different combinations of surface layer and Be doping density inside the superlattice structure were made and tested. The highest ESP of 91% was obtained by SLS#4 sample, while the better QE was obtained by SLS#2 sample. Specifications and performances for each sample are listed in Table 2 in section 7.

Those results demonstrated that the band mixing between hh- and lh-bands is removed to some extent by the strain, and the ESP higher than 90% can be obtained by this InGaAs-GaAs superlattice. However, the QE is insufficient by more than one order of magnitude compared to that of the AlGaAs-GaAs (SL#7) structure. The reason of this poor QE of InGaAs-GaAs structure is considered as the smaller band-gap than that of AlGaAs-GaAs should bring the smaller magnitude of NEA. This is suggested from several measurements of energy distribution curve of photoelectrons emitted from the NEA surface of GaAs, GaAsP and GaInAsP semiconductors (14).

4. InGaAs-AlGaAs Strained Layer Superlattice

The InGaAs-AlGaAs strained superlattice can have the wider band-gap than that of InGaAs-GaAs, and we started the study of this structure to achieve the QE improvement. Three samples of $In_xGa_{(1-x)}As$-$Al_yGa_{(1-y)}As$ were already made and tested. They have the same fraction parameters of x=0.15 and y=0.35, but different thicknesses of the well (InGaAs) and barrier (AlGaAs) layers.

As the best result among them, the maximum ESP of 80% with the QE of 0.7% was achieved at laser wavelength of 741 nm by the SLSA#2 sample, as shown in Fig. 2(a).

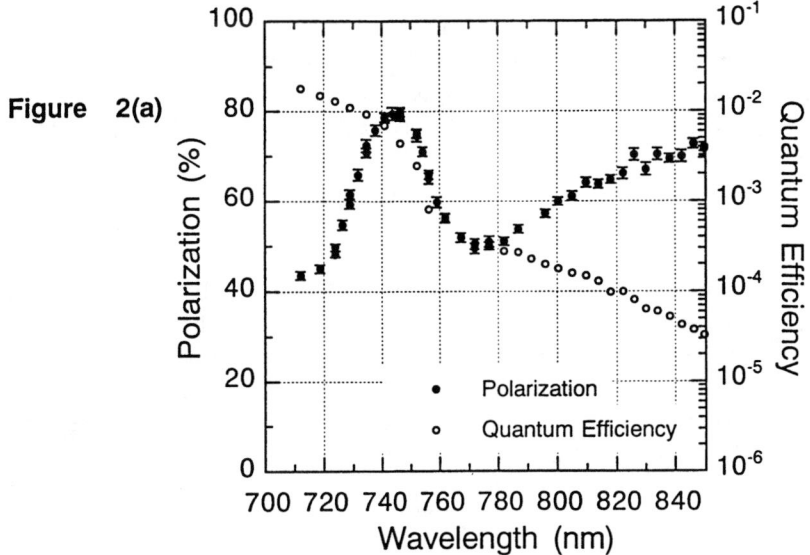

Figure 2(a)

The QE is much improved and becomes nearly equal to that of SL#7. However, as a drawback of this (SLSA#2) sample, the maximum ESP is reduced to ~80% from the best value of ~90% of the InGaAs-GaAs sample. The calculated energy band structure (in meV) is shown in Fig. 2(b). It gives the energy band splitting (δ_s) of 58 meV which is much larger than 30 meV of InGaAs-GaAs, and the reason of lower ESP is not yet understood simply.

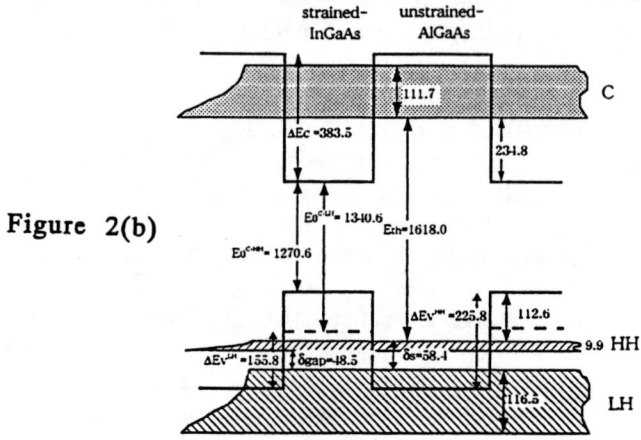

Figure 2(b)

The properties of this InGaAs-AlGaAs structure is not fully probed until now, and further studies to know the upper limits of ESP, QE and SCL will be continued by samples with different structure parameters (6).

5. Surface-Charge-Limit

The "surface charge limit (SCL)" effect is an important problem, especially for an accelerator which requires the high peak current in a short pulse, such as an electron-positron linear colliders (15). Detailed studies of this effect for photocathodes of bulk-GaAs and strained GaAs layer had been done at SLAC (7)(8). The SCL effect for superlattice photocathode was first studied at SLAC using the SL#7 sample because of its best QE performance. The cathode (20ϕ diameter) was biased at -120 kV (~2.7 MV/m field gradient at surface), and fully illuminated over the cathode area by the 757 nm laser lights. A high charge of 37 nC (2.3×10^{11} electrons) in 2.5 ns bunch was extracted, and this value was consistent with a space charge limitation, that is, no SCL effect was observed (9). From this result, the AlGaAs-GaAs superlattice became a promising candidate of polarized photocathodes for future linear colliders, such as JLC and NLC which require the beam with several tens of multi-bunch separated by 1.4 ns or 2.8 ns (15).

However, in this SLAC experiment, the test only for single-bunch generation was performed, and that for multi-bunch generation was not done. The test to produce the multi-bunch beam has been continued at Nagoya Univ.

using a 100 kV polarized gun (named as NPES-2) (10). A Nd:YAG pumped Ti:sapphire pulse laser is used to generate a single bunch pulse of Gaussian shape (6 ns FWHM). The double and quadruple bunch pulse are produced by splitting the original pulse, and the bunch spacing is set to be 15 ns and 25 ns, respectively.

First, a thin layer (0.1 μm thickness) of normal GaAs cathode with a Be-doped ($\sim 5 \times 10^{18}$/cc) surface was tested. Results for extraction of 70 keV electron beam are shown in Fig. 3. The laser wavelength was tuned at 780 nm, where the QE was measured to be 0.48%. The charge saturation curve for the first and the second laser bunch is shown in Fig. 3(a), and the electron bunch shapes for various laser powers are shown in Fig. 3(b).

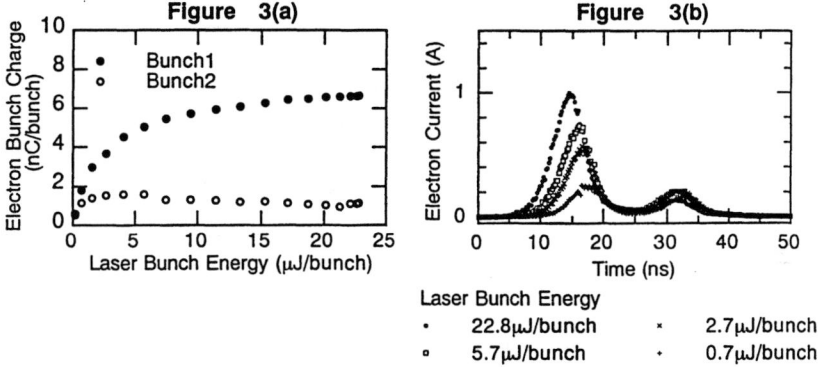

The characteristic features of SCL effect are clearly observed as follows.

(1) The SCL effect is observed in each pulse in Fig. 3(b). As laser power is increased, the electron bunch shape is deformed from original Gaussian shape and the peak position of the pulse is shifted ahead. The saturated charge ($\sim$7 nC/bunch) of the first bunch in Fig. 3(a) is about a half of the estimated space-charge-limitation of 15 nC/bunch.

(2) The SCL effect has a longer memory time than the bunch separation of 15 ns. Most of surface charges produced by the first bunch will be still alive until the second bunch illumination takes place, and suppress the escape of the second bunch electrons into vacuum, as suggested in Fig. 3(b). Fig. 3(a) shows only less than 15% of the first bunch charge can be extracted as the second bunch charge.

It is obvious that such a cathode is not useful for multi-bunch generation, and next the AlGaAs-GaAs (SL#7) superlattice with the highly Be-doped ($\sim 5 \times 10^{19}$/cc) surface was tested. Results are shown in Fig. 4, which were taken under the same gun conditions as thin GaAs test. The laser wavelength was tuned at 748 nm, where the maximum ESP of 69% and the QE of 0.65% was observed. The charge saturation curve for the first and the second laser bunch is shown in Fig. 4(a), and the electron bunch shapes for various laser powers are shown in Fig. 4(b).

As shown in Fig. 4(b), the time shape is not changed both for the first and the second bunch and, the flat-top is observed for higher laser power illumination, which is caused by the space-charge-limitation. It is clearly shown in Fig. 4(a), that there is no suppression for the second electron bunch. The saturated charge level corresponds to ~1.6 A peak current which is consistent with space-charge-limitation.

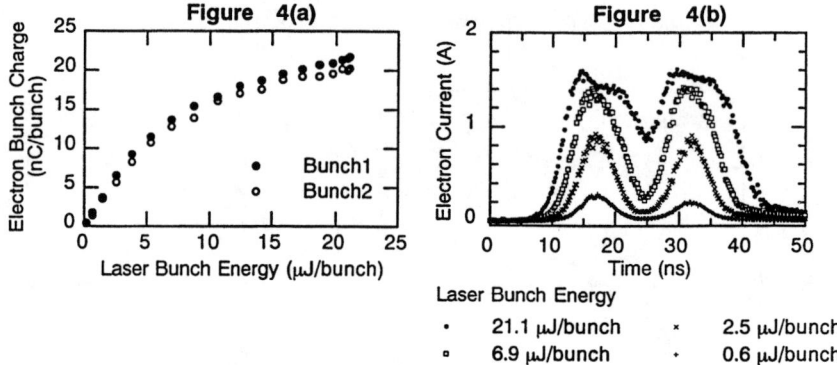

An extension of this experiment to quadruple bunch generation has been performed, and the electron bunch shapes for various laser powers are shown in Fig. 5. This data also shows that the SCL effect is not observed.

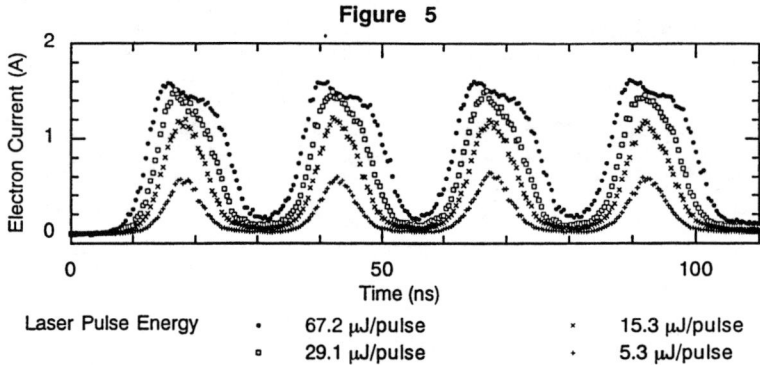

Those data demonstrate the SL#7 cathode is suitable for the multi-bunch generation of high peak current polarized beam. Similar tests are planned for other superlattice to find out which is the essential parameter to remove the SCL effect (10). We expect such a study will lead us to the full understanding of physical mechanism of SCL effect.

6. Spin Relaxation in Superlattice

Spin relaxation time (T_s) and relaxation time (τ) of electrons in conduction band of superlattice can be simultaneously determined by measuring the time evolution of photoluminescence (PL). Such measurements have been done at Osaka Pref. Univ. using a mode-locked Ti:sapphire laser (2 ps pulse width) and a streak camera (± 8 ps time resolution).

The typical data taken for the InGaAs-AlGaAs (SLSA#2) sample is shown in Fig. 6(a), where I^+, I^- denote left- and right-circularly polarized PL components, respectively. The τ and T_s are determined as a decay constant of total PL intensity and PL polarization (shown in Fig. 6(b)), where they are defined as $(I^+ + I^-)$ and $(I^+ - I^-)/(I^+ + I^-)$, respectively. The results were obtained as $\tau \sim 68$ ps and $T_s \sim 85$ ps at room temperature.

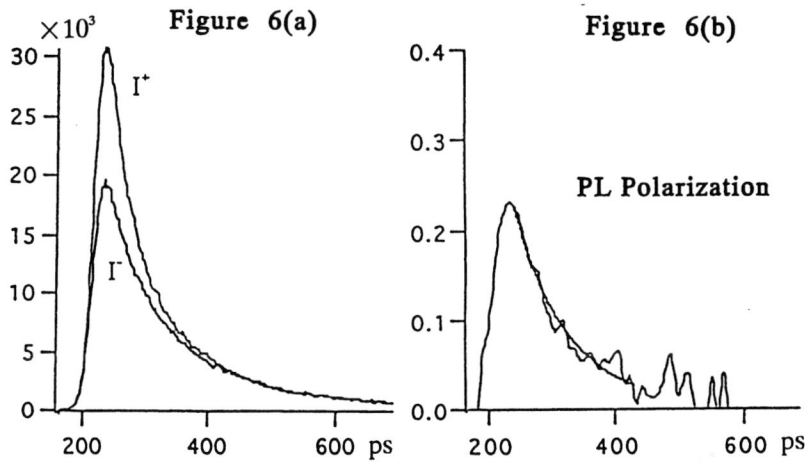

Results of τ and T_s measurements for three different types of superlattice are listed in Table, 1. From those measurements, it becomes clear that T_s is longer than τ for all samples at room temperature and the T_s can be made much longer by cooling the sample at liquid N_2 temperature. It is also obvious that there is no significant difference in the observed T_s values, in spite of the different ESP observed by the extracted electrons.

Besides of the PL method, the time evolution of the intensity and polarization of the extracted picosecond electron bunch was measured by the radiofrequency streak method at Mainz (16). It will be much interesting to compare both results taken for the same cathode, because the former reflects the τ and T_s of the inside electrons, while the latter reflects those of the escaped electrons.

Table 1: Life-time and Spin-relaxation-time of Superlattice

Crystal Name	Temperature (K)	τ (ps)	τs (ps)	λ exc (nm)	λ obs (nm)	Electron Polarization
STP#4	R.T.	45	105	780	800~860	80 % (at 880 nm)
	75	77	165			
SL #7	R.T.	69	95	760	780~860	68 % (at 756 nm)
SLS #4	R.T.	61	71	835	870~940	91 % (at 908 nm)
	77	99	229			
SLSA #2	R.T.	68	85	780	800~870	80 % (at 741 nm)

τ : life time of c-band electrons (1997. 8. 1)
τs : spin relaxation time of c-band electrons
λ exc : wave-length of excitation laser
λ obs : wave-length region of photoluminescence observation
Pol. : the maximum polarization of extracted electrons

7. Discussions

We investigated three different types of superlattice photocathodes. The best achieved performance of these photocathodes in terms of the maximum ESP and the QE at the same laser wavelength are summarized in Table-2, where the various specification parameters are also given.

Making full use of our experimental data, the correlation between the QE and the photon energy which gives the maximum ESP is plotted in Fig. 7.

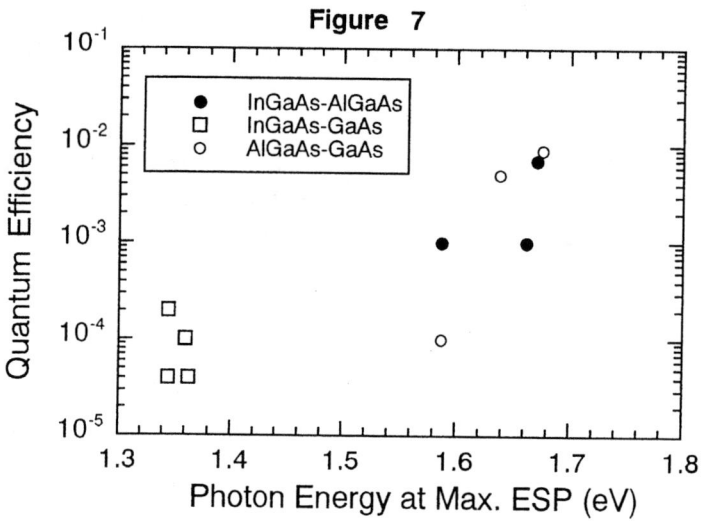

Figure 7

Table 2: Specifications and Performances of $In_xGa_{(1-x)}As$ - $Al_yGa_{(1-y)}As$ Superlattices

(1997.8.1)

Crystal Name	x	y	ε (%)	Lw (ML)	LB (ML)	δs (meV)	Wc (meV)	Surface	Be dope Inside	Be dope Surface	Pol. (%)	Q.E. (%)	λ (nm)
SL #3	0.0	0.35	0.0	7	11	23	89	GaAs	5×10^{18}	5×10^{18}	68	0.01	780
SL #7	0.0	0.35	0.0	7	11	39		GaAs	5×10^{17}	4×10^{19}	68	0.5	756
SL #11	0.0	0.35	0.0	7	25	39	10	GaAs	5×10^{17}	4×10^{18}	68	0.9	739
SLS #1	0.15	0.0	1.1	7	11	30	191	GaAs	8×10^{16}		84	0.01	910
SLS #2	0.15	0.0	1.1	7	11	30	191	InGaAs	5×10^{17}	4×10^{19}	88	0.02	920
SLS #3	0.15	0.0	1.1	7	11	30	191	GaAs	5×10^{17}	4×10^{19}	89	0.004	920
SLS #4	0.15	0.0	1.1	7	11	30	191	InGaAs	8×10^{16}	4×10^{19}	91	0.004	908
SLSA #1	0.15	0.35	1.1	6	25	82	8				73	0.1	745
SLSA #2	0.15	0.35	1.1	5	11	58	112	InGaAs	5×10^{17}	4×10^{19}	80	0.7	741
SLSA #3	0.15	0.35	1.1	7	11	73	73				82	0.1	780

x : fraction of Indium
y : fraction of Alminium
ε : strain of InGaAs layer
Lw: thickness of InGaAs (well) layer
LB: thickness of AlGaAs (barrier) layer

δs : energy splitting between the tops of hh and lh mini-bands
Wc : enegy-broadning of conduction band
Pol. : the maximum polarization
Q.E. : quantum efficiency at the maximum polarzation
λ : laser wavelength which gives maximum polarization

From this plot, we believe we arrive at a conclusion that the wider band-gap superlattice structure is more suitable for the higher QE photocathode.

As a final remark, we consider at moment that the InGaAs-AlGaAs strained layer superlattice is the promising photocathodes due to its potential to provide simultaneously the high ESP, high QE and probably high resistance for SCL-limit. However, the performance limitation is not yet settled rigidly for the superlattice structure, and further studies have enough possibilities to find the best structures for various branches of application.

This work was supported partially by Grants-in-Aids for Scientific Research A07504001, for international Scientific Research 08044068 and for Scientific Research on Priority Areas 09244210.

References

(1) T. Nakanishi, "The New Generation of Photocathodes for Polarized Electron Source" in "Frontier of Accelerator Technology", (Proceedings of the 1994 Joint US-CERN-Japan International School, edited by S.I. Kurokawa et al.), 665-680, (World Scientific, 1996)
(2) T. Nakanishi et al., Phys. Lett. A158 (1991) 345-349
    H. Aoyagi et al., Phys. Lett. A167 (1992) 415-420
(3) T. Saka et al., Jpn, J. Appl. Phys. 32 (1993) L1837-40
(4) T. Omori et al., Phys. Rev. Lett.67 (1991) 3294-97
(5) T. Omori et al., Jpn. J. Appl. Phys. 33 (1994) 5670-80
(6) T. Nakanishi et al., to be published in Journal.
(7) M. Woods et al. J. Appl. Phys. 73 (1993) 8531
(8) R. Alley et al., Nucl. Instr. Meth. A365 (1995) 1-27
(9) Y. Kurihara et al. Jpn. J. Appl. Phys. 34 (1995) 355-358
(10) K. Togawa et al., to be published in Journal.
(11) G. Bastard, Phys. Rev. B24 (1981) 5693-5697
(12) S. Nakamura et al., to be published in Journal.
(13) Y. Kurihara et al., Nucl. Instr. Meth. A313 (1992) 393-397
(14) H.-J. Drouhin et al., Phys. Rev. B31 (1985) 3872
    J. Kirschner et al., Appl. Phys. A30 (1983) 177
    A.S. Terekov and D.A. Orlov, Proc. SPIE 2550 (1995) 157-164
(15) JLC Design Study (April, 1997, KEK-Report)
    J. Clendenin et al., Nucl. Instr. Meth. A340 (1994) 133-138
(16) P. Hartmann et al., Nucl. Instr. Meth. A379 (1995) 15-20

Laser Sources for MIT-Bates and IASA

D. Fraser[*], A. Hatziefremidis, M. Ciarrocca[1], V. Markou,
T. Papakyriakopoulos, T. Houbavlis and H. Avramopoulos

*Department of Electrical and Computer Engineering
National Technical University of Athens
15773 Zographou, Athens, Greece*

[*] *Institute of Accelerating Systems and Applications
PO Box 17214
Athens 10024, Greece*

Abstract: We report the development of a high power mode-locked external cavity diode laser/Ti:sapphire laser system to be used as a source for cw and pulsed electron accelerators. It provides up to 3.4 W average power with a corresponding pulse energy of 1.1 nJ at 2856 MHz repetition rate and tunable between 815-835nm.

There is at present considerable interest in the use of spin-polarized electron beams as a tool for the detailed study of structures in diverse research fields such as atomic, particle and semiconductor physics. The ability to generate and use spin-polarized beams is particularly relevant to rf electron accelerator facilities of either pulsed type such as MIT-Bates Linear Accelerator Center, Boston, USA or continuous wave type such as in Jefferson Lab, Newport News, Virginia, USA, MAMI, Mainz, Germany and IASA-Athens, Greece.

The method of choice for the production of polarized electron beams is by laser illumination of GaAs photocathodes (1). Crucial to the emergence of this experimental tool has been the development of good quality photocathodes and the existence of high power laser systems in the wavelength range between 760-860 nm. High quality photocathodes with quantum yields of up to QE=0.4% and emitted electron beams spin-polarized to greater than 75% have been shown recently for optimum illumination at 830 nm (2). Commercially available lasers in this wavelength range (760-860 nm), are either continuous wave (cw) systems or mode-locked systems with pulse repetition rates in the 76 MHz range and are therefore a poor match to electron accelerators requirements. Specifically the use of cw laser systems results in small capture efficiencies for the produced electrons due to chopper losses (typical transmission is .15-.40 for cw accelerators). This significantly limits the available average beam current into the accelerator and

---

[1] currently with the European Patent Office, Berlin, Germany

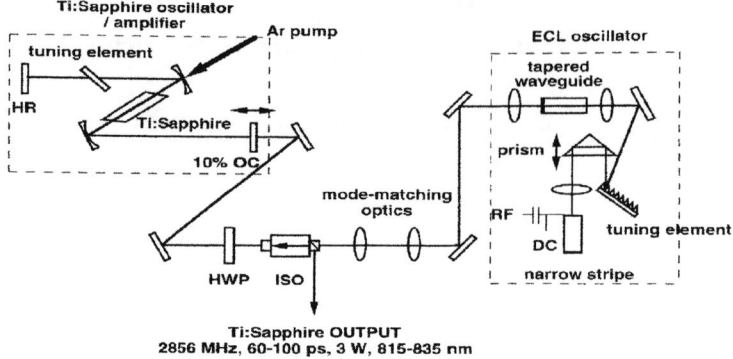

FIGURE 1. Schematic of ECL/Ti:sapphire laser system

makes for inefficient use of the electron beam produced at the photocathode. Furthermore the continuous illumination of the photocathode contributes to a significant reduction of its useable lifetime. Picosecond, mode-locked laser systems operating at the 76 MHz range have also been used to generate polarized electron bunches with very high transmission, but as they operate at a low subharmonic of the accelerating rf field the resulting average beam current has been low (3). The frequency of the accelerating rf field (and the pulse repetition rate of the laser source) is 2856 MHz for MIT-Bates, 1500 MHz for Jefferson Lab, 2450 MHz for MAMI and 2380 MHz for IASA-Athens. Correspondingly the value for the pulsewidth required from the laser source is set by the chopper/buncher employed in each accelerator and typically varies between 50 and 120 ps.

To address the issue of pulse repetition rate from the laser source, a number of research groups have recently developed custom, pulsed laser systems to be used in cw accelerators. These systems have been based on (a) semiconductor diodes that follow the gain switched, master oscillator, power amplifier design (4) and the actively mode-locked, compound external cavity diode design (5) and (b) passively mode-locked Ti:sapphire design (6). These systems have been successfully tested and used in cw accelerators, but as a result of their relatively low average power (up to 500 mW) and pulse energy (less than 150 pJ) they cannot be used in pulsed accelerators. In this paper we report on a laser system capable of delivering a 60 to 110 ps pulse train at 2856 MHz repetition rate, tunable between 815-835 nm, with up to 3.4 W average power which corresponds to a pulse energy of 1.1 nJ. This system has been designed for use in the MIT-Bates Linear Accelerator Center which presently uses a cw Ti:sapphire oscillator.

A schematic of the laser system is shown in Fig. 1. It comprises of a harmonically mode-locked, tunable, external cavity semiconductor diode laser (ECL) which provides the seed signal and an Ar-ion pumped Ti:sapphire oscillator which amplifies it. As the beam profiles from the two oscillators differ,

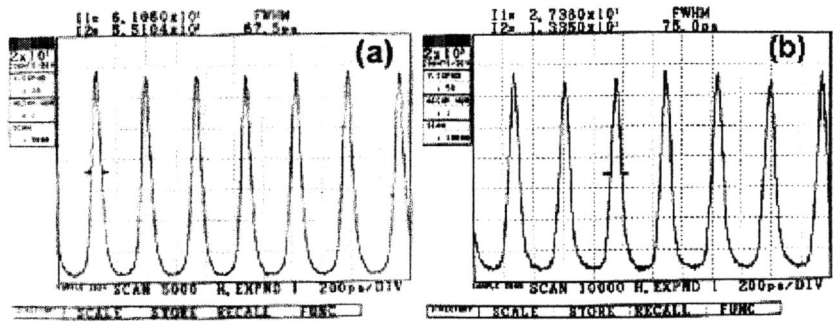

FIGURE 2. Pulse train at 2856MHz from (a) ECL and (b) amplified.

the mode-locked pulse train from the ECL is down collimated with a 5:1 beam expander into the Ti:sapphire oscillator through its output coupler. To avoid mode-locking loss and instabilities arising due to feedback from the Ti:sapphire oscillator, the ECL cavity is isolated using a broad band (40 dB) Faraday isolator which also acts as the output tap of the system on the return beam from the Ti:sapphire oscillator.

The ECL oscillator has been discussed in detail elsewhere (5) and here only a brief description will be given. The compound cavity contains two GaAlAs laser diodes and is formed between the back facet of the narrow stripe diode and the front facet of the tapered amplifier diode. The narrow stripe diode is modulated with a high power rf signal (up to 500 mW) and the tapered amplifier is driven with a dc current of 2 A to provide the mode-locked output pulse train. In the present implementation the diodes used have been coated to allow maximum gain at about 825 nm for use with GaAsP photocathodes and the ECL can routinely provide 60-70 ps long pulses with up to 250 mW of average power with a tuning range between 815 to 830 nm. The repetition rate of a mode-locked laser is set by its cavity length and the ECL oscillator has been configured for use with the 2856 MHz rf master oscillator of the MIT-Bates accelerator. As this frequency corresponds to an impractically short laser cavity of 5.25 cm, the ECL was built with a 31.5 cm long cavity (corresponding to 476 MHz fundamental frequency) and was therefore mode-locked on its 6th harmonic. The choice of this fundamental cavity frequency is fortuitous as its 5th harmonic corresponds to the IASA-Athens master oscillator frequency at 2380 MHz making very easy the testing of the system.

The Ti:sapphire oscillator is a commercial unit with the standard astigmatically compensated cavity which includes a Lyot filter tuning element and which has been modified to allow cavity length adjustment. It employs a 10% output coupler to allow adequate seeding without substantial loss of output power. In the absence of any seeding the Ti:sapphire oscillator provides up to 3.5 W of output at 830 nm for 18 W of pump power. To ensure synchronization with the ECL pulse train, the Ti:sapphire cavity length was set at 63 cm, corresponding to a fundamental frequency of 238 MHz or the 12th subharmonic of the rf frequency.

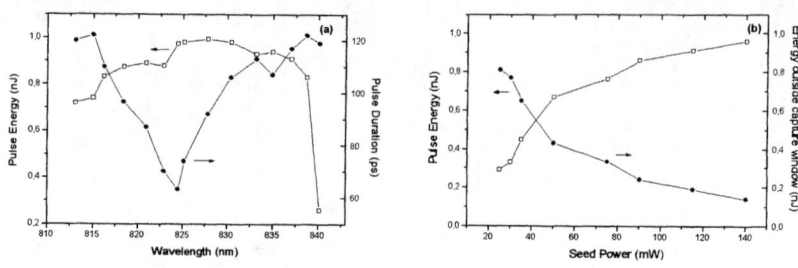

FIGURE 3. (a) Pulse energy and duration against wavelength, (b) pulse energy in and out of capture window (110ps).

Fig. 2a shows the temporal profiles of the pulse trains obtained directly from the ECL operating at 2856 MHz repetition rate with 250 mW average power. The pulses are clean with full width at half maximum (FWHM) is 65 ps and the peak to background ratio is 30:1. Fig. 2b shows the corresponding amplified train from the Ti:sapphire. The pulses are still clean but they are slightly broadened to 75 ps and the peak to background level is now 25:1.

With the present configuration the maximum average power was obtained from the system after optimization of the cavity parameters and was 3.4 W at 2856 MHz, corresponding to 1.1 nJ pulse energy in the 110 ps capture window for the MIT-Bates accelerator. This figure, which can be improved further, must be compared with the obtainable pulse energy of 770 pJ from the present system (cw Ti:sapphire) (7). Apart from the pulse energy increase, the ECL/Ti:sapphire system is expected to lead to a significant improvement in the lifetime of the photocathode because of the threefold reduction in illumination of the photocathode from the mode-locked pulses.

Fig 3a shows the tuning curves of the laser system output when operated at 2856 MHz. These curves have been obtained only by adjusting the tuning elements in the ECL and Ti:sapphire oscillators. The average optical power from the system is determined by the Ti:sapphire oscillator and was approximately constant to 3 W throughout the measured tuning range. The measurements for the pulse energy have been made assuming that there is a 110 ps long, useable capture window. The tuning curve for pulse energy is flat within 20% of the maximum value of about 1 nJ, over a 20 nm range centered at 825 nm and is larger that the tuning range of the ECL oscillator. As the system is tuned further away from the peak wavelength of the ECL, the pulse energy in the output pulse train decreases and a cw background develops. The reason for this tuning range extension is due to a small feedback through the isolator into the ECL. Fig 3a also shows the pulsewidth change against wavelength which varies between 60 and 110 ps. It should be noted however that no optimization of the cavities parameters has been made during these measurements. The key to obtaining high pulse energies from the system is to ensure a high level of seed power from the ECL to the Ti:sapphire. In order to assess this, a variable attenuator was installed after the ECL oscillator and before the beam down collimation optics. Fig 3b shows the pulse energy against the seeding power as measured before the Ti:sapphire output coupler for 2856 MHz operation. For seeding powers lower than 50 mW, more than 50 % of the energy is lost outside the capture window.

We have measured the amplitude stability of the seed pulse train directly from the ECL oscillator and from the Ti:sapphire using a fast photodiode and a 1 GHz analogue oscilloscope. The output stability is better than 1% from the ECL oscillator and about 5% from the Ti:sapphire. The reduction in stability in the amplified signal was traced to the Ti:sapphire oscillator and was not due to the seeding process and we are currently in the process of investigating this.

In summary we have demonstrated a laser system that generates mode-locked pulses with up to 3.4 W average power and pulse energy of 1.1 nJ at 2856 MHz. The system can generate pulse trains with 60-110 ps long pulses and repetition rates at multiples of 476 MHz over a 20 nm tuning range. It is therefore suitable for generating electron bunches for use in continuous wave accelerators but more importantly it can be used in pulsed accelerators where high energy pulses are a prerequisite and no pulsed laser sources are currently available. Scaling to even higher pulse energies should be possible by optimization of the output coupling of the Ti:sapphire oscillator.

ACKNOWLEDGMENTS

The authors would like to thank the MIT-Bates linear accelerator for kindly lending their Ti:Sapphire oscillator and for many fruitful discussions.

REFERENCES

1. T.Maruyama et al, *Phys. Rev. Lett.* **66**, 2376 (1991).

   T. Nakanishi et al, *Phys. Lett.A* **158**, 345-349 (1991).

   F. Meier et al, *Phys. Scripta T* **49B**, 574 (1993).

   J.C. Grobli et al, *Phys. Rev. Lett.* **74**, 2106 (1995).

2. P. Dreschner et al., *Appl. Phys. A* **63**, 203-206 (1996).

3. P. Hartmann et al., *Nucl. Instr. And Meth. A* **379**, 15-20 (1996).

4. M. Poelker, *Appl. Phys. Lett.* **67**, 2762-2764 (1995).

5. M.Ciarrocca et al., *Nucl. Instr. and Meth. A* **385**, 381-384 (1997).

6. J. Hofmann et al., *Nucl. Instr. and Meth. A* **383**, 624-626 (1996).

7. M. Farkhondeh, *private communication*.

A Compton Backscattering Polarimeter for Measuring Longitudinal Electron Polarization

I. Passchier*, D. W. Higinbotham†, N. Vodinas*[1],
N. Papadakis*[1], C. W. de Jager*[2], R. Alarcon*, T. Bauer*,
J. F. J. van den Brand•,*, D. Boersma*, T. Botto*,
M. Bouwhuis*, H. J. Bulten•, L. van Buuren•, R. Ent×,
D. Geurts•, M. Ferro-Luzzi*,°, M. Harvey×, P. Heimberg•,
B. Norum†, H. R. Poolman*, M. van den Putte*, E. Six*,
J. J. M. Steijger*, D. Szczerba•, H. de Vries.*

*NIKHEF, P.O.Box 41882, 1009 DB Amsterdam, Netherlands
†Department of Physics, University of Virginia, Charlottesville, VA 22901, USA
*Department of Physics, Arizona State University, Tempe, AZ 85287, USA
•Department of Physics and Astronomy, Vrije Universiteit, Amsterdam, The Netherlands
×TJNAF, Newport News, VA 23606, and Hampton University, Hampton, VA 23668, USA
°Institut für Teilchenphysik, Eidg. Technische Hochschule, CH-8093 Zürich, Switzerland

Abstract. Compton backscattering polarimetry provides a fast measurement of the polarization of an electron beam in a storage ring. Since the method is non-destructive, the polarization of the electrons can be monitored during internal target experiments. At NIKHEF a Compton polarimeter has been constructed to measure the polarization of the longitudinally polarized electrons stored in the AmPS ring. First results obtained with the polarimeter, the first Compton polarimeter to measure the polarization of a stored longitudinally polarized electron beam, are presented in this paper.

INTRODUCTION

The NIKHEF Compton polarimeter has been constructed to measure the longitudinal polarization of electrons stored in the AmPS ring. The polarized electrons are provided by a recently commissioned polarized electron source (PES) [1]. While Compton backscattering polarimeters are used to

---

[1]) Present address: IASA, P.O.Box 17214, 10024 Athens, Greece
[2]) Present address: TJNAF, Newport News, VA 23606, USA

measure the polarization of transversely polarized stored electron beams [2,3], NIKHEF's detector was the first to measure the polarization of a longitudinally polarized stored beam [4].

In this technique, a circularly polarized photon beam (polarization S_3, energy E_λ) is backscattered from a stored polarized electron beam (polarization P_e, energy E_e).

The cross section for Compton scattering of circularly polarized photons from longitudinally polarized electrons can be written as

$$\frac{d\sigma}{dE_\gamma} = \frac{d\sigma_0}{dE_\gamma}[1 + S_3 P_z \alpha_{3z}(E_\gamma)], \qquad (1)$$

where $\frac{d\sigma_0}{dE_\gamma}$ follows from the energy spectrum for unpolarized electrons and photons and P_z represents the longitudinal component of the electron polarization. For a given E_λ and E_e the asymmetry can be written as,

$$A(E_\gamma) = \frac{N_L(E_\gamma) - N_R(E_\gamma)}{N_L(E_\gamma) + N_R(E_\gamma)} = \Delta S_3 P_z \alpha_{3z}(E_\gamma) \qquad (2)$$

where $N_L(E_\gamma)$ ($N_R(E_\gamma)$) is the number of photons with energy E_γ with incident left (right) handed helicity, and ΔS_3 is the difference between the two polarization states, divided by two. P_e is determined by taking P_z as a free parameter and fitting the measured asymmetry with eq. 2. The relation between P_z and P_e is determined by the lattice of the storage ring.

LAYOUT OF THE POLARIMETER

A schematic layout of the Compton polarimeter is shown in fig. 1. The polarimeter consists of a laser system with its associated optical system and a detector for the detection of backscattered photons.

Laser photons are produced by a 10 W CW Ar-ion laser, operated at 514 nm. Part of the mirrors in the optical path can be controlled remotely, in order to optimize the overlap of the electron and laser beam. A quarter-wave plate is used to convert the initially linearly polarized photons to circularly polarized. A Pockels cell is used to switch the helicity between left and right, while a half-wave plate can be inserted in the optical path to check for false asymmetries by reversing the sign of the Compton asymmetry.

Laser photons interact with stored electrons in the straight section (length $\approx$ 3 m) between the first dipole and second dipole (bending angles 11.25°) after the internal target facility. The backscattered photons leave the interaction region traveling in the same direction as the electrons of the beam and are separated from them after the second dipole. They are detected in a gamma detector, consisting of a block of $100 \times 100 \times 240$ mm^3 pure CsI.

A chopper mounted immediately after the Ar-ion laser is used to block the laser light for 1/3 of the time for background measurements. The chopper is operated at 75 Hz and also generates the driving signal for the Pockels cell.

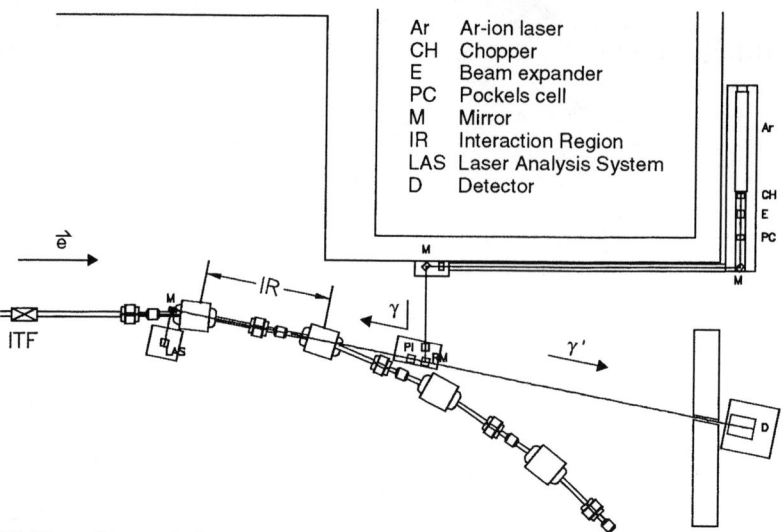

FIGURE 1. Schematic layout of the Compton polarimeter at NIKHEF. Indicated is part of the AmPS ring, the optical system, the gamma-detector and the internal target facility.

RESULTS

The storage ring could only be operated with a 10% partial snake [5]. Therefore, it was necessary to perform all measurements with an electron beam energy of 440 MeV, resulting in a maximum energy for the Compton photons of 7.04 MeV. This energy is lower than that of the design specification (500–900 MeV), resulting in a poor energy resolution. To reduce background at this rather low energy, we performed all measurements with beam currents smaller than 15 mA. The rate of backscattered photons was in the order of 8 kHz/mA at full laser power, in agreement with simulations.

To minimize the effects of false asymmetries (induced by a small steering effect of the Pockels Cell), we performed sets of six independent measurements to determine the electron polarization. Three measurements were done with different electron polarizations injected into the ring (positive helicity, unpolarized and negative helicity). These measurements were repeated with the half-wave plate of the polarimeter inserted in the optical path. The measurements with unpolarized electrons were used to determine and correct for false asymmetries, while the insertion of the half-wave plate was done as a consistency check. Figure 2 shows the asymmetry before and after correction for false asymmetries.

To determine the stability of the polarimeter, one measurement was repeated nine times. To exclude any sensitivity to variations in the polarization of the injected electrons or spin life time, those measurements were performed with unpolarized electrons. The total measurement time was $\approx$ 90 min, while

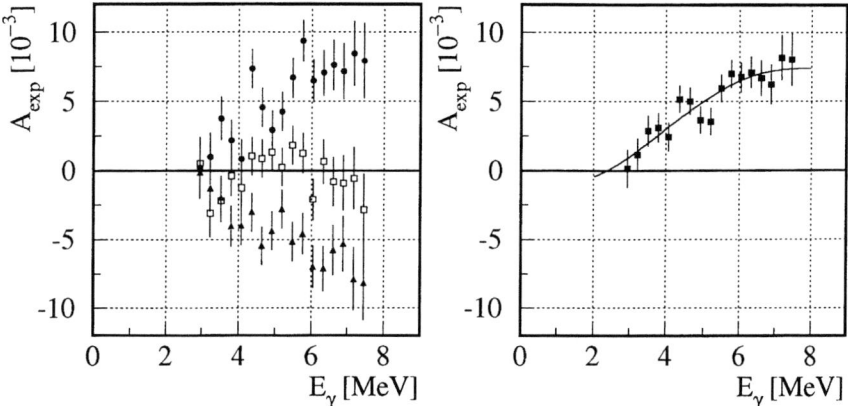

FIGURE 2. Left: Raw asymmetries for electrons with left (●) and right-handed helicity (▲) and for unpolarized electrons (□). Right: the average of the asymmetries for electrons with left and right-handed helicity, taking into account the difference in sign of the asymmetry, after correcting for false asymmetries.

a full set of six measurements normally takes about 60 min. The results are shown in fig. 3 and show good stability on this time scale.

The long-term stability is determined from polarization measurements done typically once a day. These measurements are sensitive not only to variations of the polarimeter, but also to any other time-dependent effect such as a degradation of the cathode used at PES. The results (see fig. 3) show no trend in the polarization of the electrons, indicating a good long-term stability for all components.

The polarimeter has been used successfully to optimize the settings of the Z-manipulator at PES. After the optimization, the spin life time (τ) and initial polarization (P_0) has been determined by combining the data of nine measurements of six minutes each. The combined data have been rebinned as a function of time and the polarization has been determined for each bin separately. We found $P_0 = 61.6 \pm 1.4\%$ (statistical), and $\tau = 4500^{+5900}_{-1600}$ s. The spin life time is in agreement with our calculations. The polarization measured with the Mott polarimeter at PES was $82 \pm 5\%$. The difference between the polarization measured by the Mott polarimeter and by the Compton polarimeter may be caused by depolarization due to the focusing solenoids in the linac or depolarizing resonances during damping of the beam.

CONCLUSIONS

Here, we describe the results of extensive tests done with a Compton backscattering electron polarimeter. The tests have been performed at an electron energy of 440 MeV and a partial snake. The results show that it

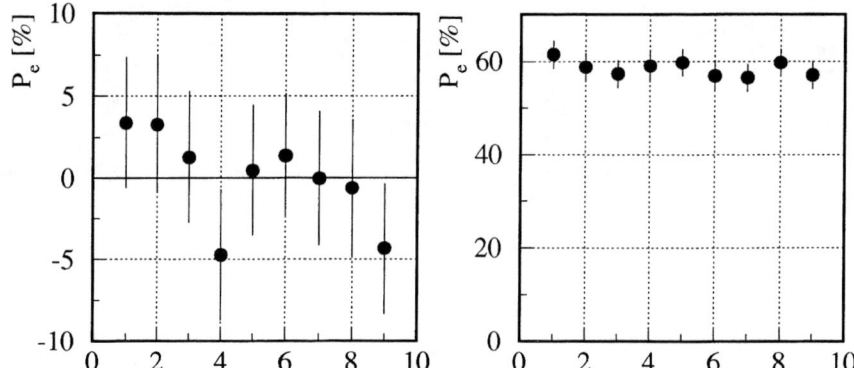

FIGURE 3. Left: short-term stability of the polarimeter. Every data point represent a measurement of the polarization of unpolarized electrons. The time between two measurements is ≈10 min. Right: long-term stability of the polarimeter. Every data point represents a complete set of six polarization measurements. The interval between the measurements is typically one day.

is possible to operate the polarimeter in a reliably manner over a period of weeks. Furthermore, the polarimeter has been used to map out the full dependence of the electron polarization of stored electrons on the settings of the Z-manipulator, and to determine the spin life time and depolarization during acceleration and injection of the electrons.

ACKNOWLEDGMENT

This work was supported in part by the Stichting voor Fundamenteel Onderzoek der Materie (FOM), which is financially supported by the Nederlandse Organisatie voor Wetenschappelijk Onderzoek (NWO), the Swiss National Foundation, the National Science Foundation under Grants No. PHY-9316221 (Wisconsin), PHY-9200435 (Arizona State) and HRD-9154080 (Hampton), Nato Grant No. CRG920219. and HCM Grant Nrs. ERBCHBICT-930606 and ERB4001GT931472.

REFERENCES

1. H. J. Bulten, M. van der Putte *et al.*, contribution to this workshop.
2. M. Placidi and R. Rossmanith, Nucl. Instr. and Meth. **A274** (1989) 79.
3. D. P. Barber *et al.*, Nucl. Instr. and Meth. **A329** (1993) 79.
4. I. Passchier *et al.*, "A Compton backscattering polarimeter for electron beams below 1 GeV," in *Proceedings of the 12th International Symposium on High Energy Spin Physics*, 1996, pp. 807–809.
5. L. V. Alexeeva *et al.*, Phys. Rev. Let. **76** (1996) 15.

The TJNAF Hall A Möller Polarimeter

D.S. Dale, A. Gasparian, B. Doyle, T. Gorringe
W. Korsch, V. Zeps

University of Kentucky Department of Physics, Lexington, KY

A. Glamazdin, V. Gorbenko, R. Pomatsalyuk

Kharkov Institute of Physics and Technology, Kharkov, Ukraine

J.P. Chen, S. Nanda, A. Saha

Thomas Jefferson National Accelerator Facility, Newport News, VA

ABSTRACT

As part of the spin physics program at Jefferson Lab, a Möller polarimeter is being developed to measure the polarization of electron beams of energies 0.8 to 6 GeV. The device uses electron-electron scattering in a set-up in which the polarized beam is scattered from a polarized electron target. A unique signature for Möller scattering is obtained using a series of three quadrupole magnets which provide angular selection, and a dipole magnet for energy analysis. Here, the design and commissioning of this polarimeter will be discussed along with future plans to use its small scattering angle capabilities to investigate physics in the very low q^2 regime.

INTRODUCTION

Within the next couple of years, TJNAF will be the world's premier facility for studying nuclei and nucleons via the $(e, e'x)$ reaction. In Hall A, the combination of two high resolution spectrometers and high quality, high energy 100% duty cycle electron beams will enable separations of electromagnetic structure functions to be made in kinematical ranges and with precision currently unavailable at other facilities. With the added availability of polarized electron beams and focal plane hadron polarimeters, spin physics has become a particularly important component of the TJNAF physics program.

A variety of questions in intermediate energy nuclear physics will be examined including the structure of nucleons and their excited states, properties of few body systems and complex nuclei, and strange quarks and parity violating electron scattering. With the goal of further developing this program of spin transfer measurements at TJNAF, a number of polarimeters exploiting Mott, Möller, and Compton scattering are being constructed.

THE HALL A MÖLLER POLARIMETER

Möller polarimeters exploit the asymmetry in scattering a polarized electron beam off of a polarized electron target [1]. The asymmetry obtained in the scattering for the case in which the target and beam electron spins are aligned *versus* anti-aligned gives the polarization of the incident electron beam. A set of magnets and particle detectors analyze the kinematics of the scattered electrons. The design of the Hall A Möller polarimeter is driven by a number of considerations.

1. Short measurement time: As a polarization measurement requires inserting a solid target into the beam, the physics experiment must be temporarily suspended to measure the beam polarization. As such, high counting rates and large signal-to-background ratios are necessary. In addition, fixed magnetic element positions are required to eliminate overhead associated with changing beam energies over the 0.8 GeV to 6.0 GeV range of operation of the polarimeter.

2. Minimum utilization of space along the beamline: Due to space limitations along the Hall A beamline, the length of the polarimeter was restricted to 7 meters.

3. Minimum perturbations on the beam parameters and trajectory: It was necessary to ensure that the beam was safely steered into the standard Hall A beam dump.

4. High precision: Large acceptance in electron scattering angle, $\Delta\theta/\theta \sim 15\%$, is required to minimize uncertainties in the effective target polarization due to the kinematic broadening from the intra-atomic motion of the target electrons, the so-called Levchuk effect [2].

A top view of the layout of the Hall A Möller polarimeter is shown in figure 1. Polarized electrons from the TJNAF accelerator are incident upon a magnetized iron foil which serves as the polarized electron target. Pairs of electrons which have undergone Möller scattering exit the target and are focussed in a series of three quadrupole magnets. Since the analyzing power for Möller

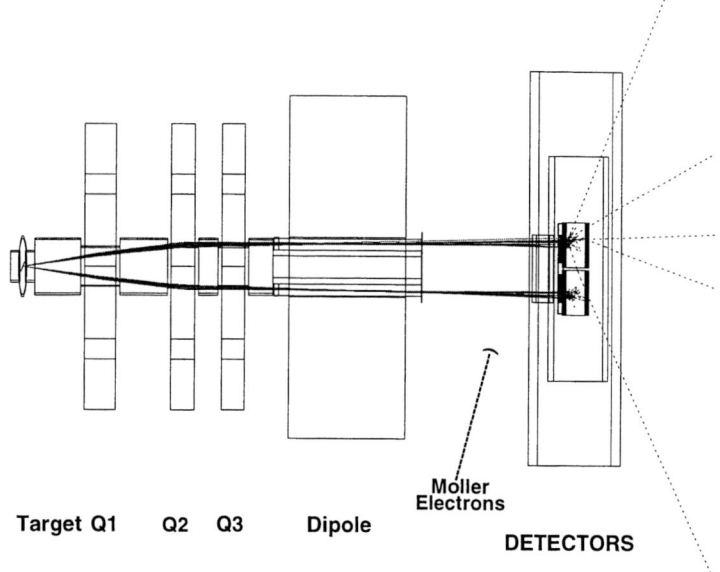

FIGURE 1. Top view of the TJNAF Hall A Möller Polarimeter.

scattering is maximum for scattering at 90 degrees in the center of mass system, the optical arrangement has been chosen to select pairs of electrons, each of which have half the incident beam energy and have symmetric angles with respect to the beam. The quadrupoles transform the trajectories of the Möller electrons such that they are parallel to the incident beam at the exit of the last quadrupole. They are then momentum analyzed in a dipole magnet (bent into the page in the figure) and detected, in coincidence, by a set of lead glass and plastic detectors.

The target has been fabricated using the plasma vapor deposition method in which a 6 μm thick layer of high purity iron (99.9%) is deposited on a 30 μm thick copper substrate. An elliptical area of the copper is etched away in the region of the beam spot, such that when the target is oriented at 20° with respect to the beam, a circular projection of the copper-free region is obtained. The target foil is magnetized in the longitudinal direction by means of Helmholtz coils and is kept at liquid nitrogen temperature. Thus in addition to providing mechanical support, the copper provides good thermal conductivity and thereby mitigates depolarization effects due to beam heating.

The Möller scattering angles are dependent upon the incident beam energy and range from 0.75 to 2.3 degrees. The Möller angle selection is performed in the three quadrupole magnets which provide net focusing in the horizontal plane so that the Möller electrons' trajectories match the acceptance of the dipole.

The dipole magnet serves a dual purpose. It momentum analyzes the scattered electrons and thus physically separates Möller from Mott events, and it also deflects the electrons away from the main beamline and therefore minimizes the length of the polarimeter. In order to not perturb the main electron beam as it passes through the dipole, a piece of iron is placed on the median plane of the dipole with a hole drilled along the length of it to accommodate the beam. This provides magnetic shielding for the electron beam so that it may pass through the dipole without deflection.

The detector package consists of two stacks of four lead glass blocks, with each stack oriented on either side of the dispersive plane of the dipole to detect the Möller electrons in coincidence. The geometrical acceptance of the lead glass blocks is determined by two plastic aperture scintillators, each of which is backed by an array of fourteen hodoscope scintillators, segmented in the dispersive direction. The hodoscope is not part of the normal trigger, and is used for diagnostic purposes.

PRESENT STATUS

The Hall A Möller has been installed in the beamline, and commissioning is presently underway. As part of the commissioning, the Möller asymmetry was measured in Hall A for three different values of the electron spin angle at the injector. The results are shown in figure 2. Based on the beam energy and the spin precession effects in the accelerator, the calculated optimal injection angle for maximum longitudinal polarization in Hall A was -52°. Deviations from this optimal injection angle are clearly correlated with decreased Möller asymmetry.

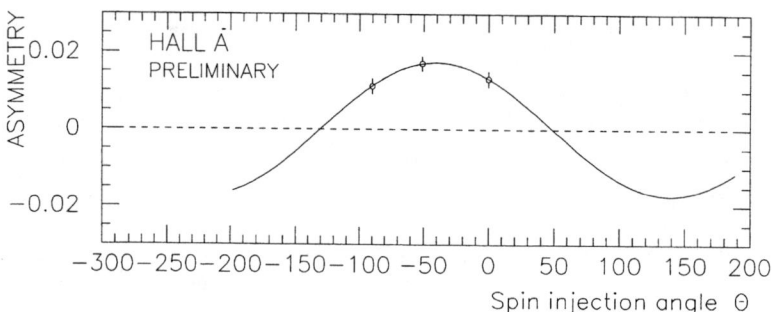

FIGURE 2. Polarization asymmetry as a function of spin injection angle.

THE HALL A MÖLLER POLARIMETER AS A SMALL ANGLE SPECTROMETER

In addition to being a part of the standard beamline instrumentation in Hall A, the small scattering angle capabilities of the Möller polarimeter, coupled with the momentum analyzing capabilities of its dipole, present unique opportunities to do physics in the very low q^2 regime [3]. The QQQD design of the Möller spectrometer will enable one to do electron scattering experiments at electron scattering angles ranging from about three degrees to less than one degree with $\Delta p'/p'$ of about 10^{-3}. As an initial area of investigation, we intend to measure the neutral pion form factor, $F_{\gamma^*\gamma\pi^o}$, at low q^2 via the virtual Primakoff effect [4] [5], i.e. π^o electroproduction in the Coulomb field of a heavy nucleus. The slope of this form factor in the low q^2 range to be measured, 0.005 $(GeV/c)^2$ to 0.04 $(GeV/c)^2$, gives a measure of the mean square $\gamma^*\gamma\pi^o$ interaction radius and is sensitive to the constituent quark mass. Such an experiment can be performed by removing the third quadrupole magnet, installing position sensitive detectors in the focal plane, and placing a series of lead glass photon detectors upstream of the dipole to measure the π^o decay photons from the $Pb(e,e'\pi^o)Pb$ reaction.

CONCLUSIONS

A Möller polarimeter has been built for Hall A of Jefferson Laboratory in support of the extensive planned spin physics program there, and its commissioning is currently underway. The QQQD design of this polarimeter has opened up the possibility of exploiting its small electron scattering angle capabilities to study physics in the very low q^2 regime. Specifically, fundamental studies of the electromagnetic response of the neutral pi meson are planned.

REFERENCES

1. B. Wagner, H.G. Andresen, K.H. Steffens, W. Hartmann, W. Heil, E. Reichert, *Nucl. Inst. Meth.* A294 (1990) 541.
2. L.G. Levchuk, *Nucl. Inst. Meth.* A345 (1994) 496; A. Afanasev, A. Glamazdin, CEBAF preprint PR-96-003.
3. D.S. Dale, A. Gasparian, TJNAF Proposal E97-009, 1997.
4. A. Omelaenko, G.I. Gach, *Ukrainian Journal of Physics*, vol. 18, no, 8, (1973) 1249.
5. E. Hadjimichael, S. Fallieros, *Phys. Rev. C*, vol. 39, no. 4, (1989) 1438.

Mott Scattering of Multi-MeV Electrons from Heavy Nuclei

D. Conti[1], S. Navert[1], K. Bodek[1,2], W. Haeberli[3], S. Kistryn[1], J. Lang[1],
O. Naviliat[1], E. Reichert[4], J. Sromicki[1], M. Steigerwald[4], E. Stephan[1],
J. Zejma[1]

(1) Institut für Teilchenphysik, ETH Hönggerberg, 8093 Zürich, Switzerland
(2) Institute of Physics, Jagellonian University, 30059 Cracow, Poland
(3) Department of Physics, University of Wisconsin, Madison, WI 53706, USA
(4) Institut für Kernphysik, Johannes Gutenberg Universität, D-55099, Mainz, Germany

ABSTRACT

Polarization effects were studied in the elastic scattering of 14 MeV electrons from lead nuclei. The left/right asymmetry was measured at eight scattering angles between 126° and 172°. For each angle the effective analyzing power for scattering foils of various thicknesses was measured. The determined analyzing powers for thin targets are in good agreement with the Mott theory. Effects of the finite size of the lead nuclei are observed. This is the first polarization test of the Mott scattering theory for electrons with energies above 1 MeV.

1. Introduction

Scattering of electrons in the electrostatic field of heavy nuclei, referred commonly as Mott scattering, was one of the first processes calculated in the framework of quantum electrodynamics. Large polarization effects were predicted due to the spin-orbit interaction, which arises as a relativistic correction to the static Coulomb potential [1]. This mechanism became a classic model for spin phenomena in scattering processes.

2. Motivation

Spin dependent Mott scattering is often used to analyze the polarization of electrons in atomic collisions and in solid state studies of magnetic structures at surfaces or in thin films. Precise knowledge of the Mott analyzing power is also important for tests of fundamental symmetries. One spectacular application was the first demonstration [2] four decades ago, that the electrons emitted from unpolarized nuclei in beta decay processes are longitudinaly polarized according to the parity violation hypothesis of Lee and Yang [3]. Since the longitudinal polarization is notoriously difficult to measure directly at low energies, the momentum of the electrons was rotated with respect to their spin axis using an electrostatic deflector, and the resulting transverse polarization was detected in the experiment [2] by means of the Mott technique.

Polarization tests of Mott theory at energies nonaccessible with long-living radioactive sources are rather scarce and of limited precision, in fact above 1 MeV nonexistent. With the renewed interest in the weak interaction, in the context of many proposed extensions of the Standard Model, also a new interest in Mott scattering arose in the last decade. Recent experiments search for *tiny* violations of accepted symmetries, with the expectation to provide a hint of "New Physics" at energy scales which are not accessible directly with the present generation of accelerators [4]. For experiments involving spin observables where the Standard Model predicts a zero value of the polarization (so called "null tests" of the theory), apart from the analyzing power for infinitely thin targets dilution effects arising from multiple scattering in thick scattering foils have to be known precisely. Such dilution effects, decreasing the effective value of the analyzing power, are difficult to estimate, and their experimental determination is essential for calibration of Mott polarimeters.

Motivated by the need of an accurate calibration of the electron polarimeter which was used in our search for time reversal violation in the decay of polarized ^{8}Li [5,6], we performed a new study of the analyzing powers and depolarization effects in Mott scattering of highly relativistic electrons from high Z targets of various thicknesses. In contrast to the many studies at low energies, which were focused on the problem of extrapolation to infinitely thin foils at the standard values of energy and scattering angle (100 keV, 120° [7,8]), the aim of our experiment was to provide a broad scope of data concerning angular dependence of the analyzing power, and to investigate the dilution effects in very thick scattering targets. It was realized that beside the importance for studies of fundamental symmetries, the results of this experiment may be useful for diagnostics of the beam polarization at the intermediate acceleration stages in high energy electron machines. Two such projects are in progress [9,10].

3. Experimental Method

The experiment was performed at the Mainz Microtron (MAMI). The pilot run was performed using a conventional electron source, while in the final data taking a newly mounted strained cathode electron source was used. The beam polarization was measured after acceleration of the beam to 100 keV by a set of Si detectors which detected electrons scattered at 120° from thin gold foils. The mean values of the beam polarization were $P = 0.33 \pm 0.02$ and $P = 0.77 \pm 0.02$ for the two runs, respectively. These values were determined using a special procedure, which was developed using many years of experience in the operation of the polarized electron sources in Mainz [11] and in other laboratories. The quoted error of the polarization reflects combined statistical and systematic uncertainties in the analysis and long time polarization drifts. After polarization analysis, the spin axis of the 100 keV electrons was rotated in a S-shape rotator consisting of four solenoids and two electrostatic deflectors [12], so that a transverse polarization was obtained at the main target of the experiment. The beam was accelerated to 14 MeV by a linac and the first stage of the microtron. Thereafter, the electrons were extracted into the air through a 4 μm thick havar window. The beam spot on the target had a diameter of 5-7 mm. A set of four triple telescope detectors based on plastic stintillators was used to determine the left/right asymmetry in the scattering process. One pair of these detectors was moved to obtain the angular distribution of the analyzing power, while the other was kept at 126° to monitor the polarization of the beam. We used 0.5 mm and 1 mm thick transmission detectors and 10 cm thick stopping counters. The angular acceptance of the detectors defined by a system of tantalum slits was $\pm 1°$ and the solid angle coverage was about 1 msr. Asymmetries were measured for nine lead targets with surface densities varying between 17 mg/cm^2 and 238 mg/cm^2. With a typical beam intensity of 10 nA the event rate was sufficiently large to result in a statistical error in the asymmetry of 0.002 within 10 min. Angular distributions of the analyzing powers were determined at eight scattering angles in the backward hemisphere (126° $-$ 172°) where the polarization effects are large.

Fig. 1 illustrates the improvement in the quality of the energy spectrum measured by the thick scintillators. At the very beginning of the experiment a large low energy tail was observed and the 14 MeV peak of elastically scattered electrons was barely visible. It was found that the main source of this background was backscattering of electrons from the massive Faraday Cup, which was mounted 50 cm downstream the target to monitor the beam intensity. Since the modulations of the beam intensity are very small in this accelerator [13], and since they do not enter our results in the first order, this element of the equipment was therefore removed.

4. Data Analysis

In the first step, the measured spectra were corrected for small electronic deadtime, and "target out" background. The intensity of this background was 2–3 orders of magnitude smaller than the main scattering peak.

Asymmetries ASY were determined using the cross ratio r method [14], calculated from the number of counts N in the left L and right R detector for the $+$ and $-$ spin state of the beam:

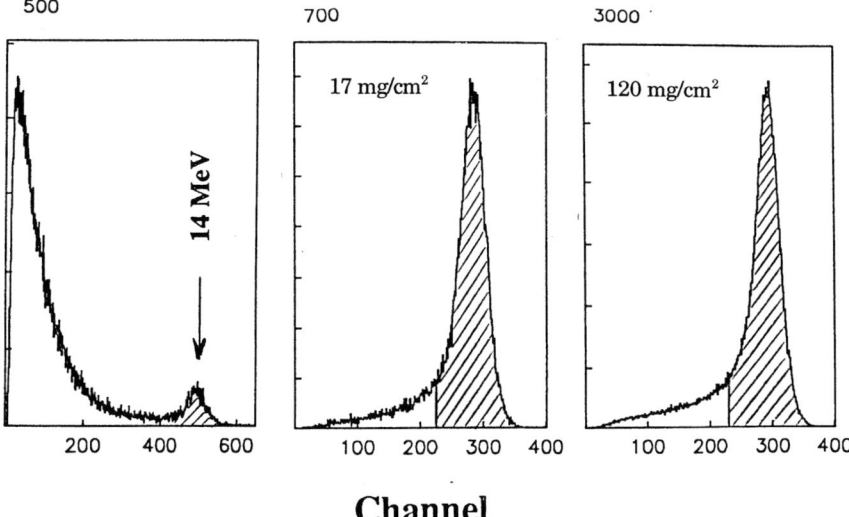

Fig. 1. Energy spectra measured at the beginning of the experiment (left) and after elimination of the backscattering of the electrons in the surrounding of the target during the data collection phase (middle and right) for two target thicknesses. The shaded area shows the integration limits used in the data analysis.

$$r = \sqrt{\frac{N_L^+ N_R^-}{N_L^- N_R^+}} \qquad ASY = \frac{r-1}{r+1}$$

Details of this procedure are well known, therefore here we only recall that solid angles, efficiencies of the detectors and beam intensity variations cancel in the ratio r.

Overall relative positioning of the target, centroid of the beam spot, detectors and defining slits was better than 1 mm, while the distance between the target and the detectors was 60 cm, therefore solid angle and geometry effects were negligible. Relative errors of a single data point, 1–5 % depending on the angle and target thickness, are dominated by counting statistics.

Analyzing powers A at 14 MeV were obtained by dividing asymmetries by the beam polarization, therefore their absolute values are scaled by the theoretical Mott analyzing power at 100 keV and 120°. We note that this theoretical value can be used with confidence: the energy is high enough to make screening of the nucleus by atomic electrons ineffective, and is low enough to neglect effects due to the finite nuclear size.

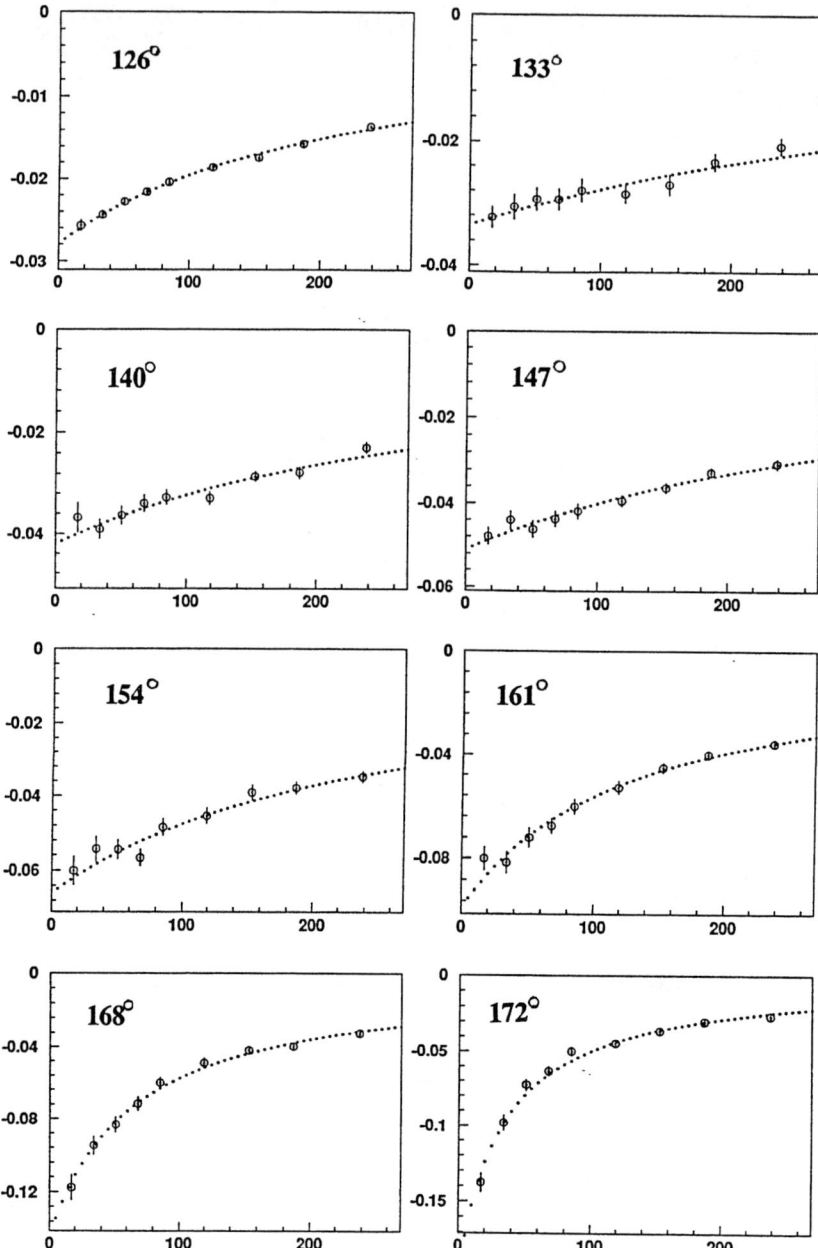

Fig. 2. Asymmetries, $ASY = P \cdot A$, measured as a function of the target thickness t (mg/cm^2) at eight scattering angles. The dilution of the analyzing power at large angles is clearly visible. The dotted line shows a fit with the "inverse" dependence discussed in the text.

5. The Results

The effective analyzing power A diminishes with increasing thickness t of the scattering foil (Fig. 2). At 14 MeV dilution effects of the analyzing power show a flat minimum between 130° and 150°, however they grow rapidly with angles above 160°. Various parametrizations of the dependence $A(t)$ have been considered in the past; most often:

$$A(t) = A_o \cdot (1 - \alpha \cdot t) \qquad \text{linear}$$

$$A(t) = A_o \cdot \exp(-\alpha \cdot t) \qquad \text{exponential}$$

$$A(t) = \frac{A_o}{1 + \alpha \cdot t} \qquad \text{inverse}$$

Here, A_o is the analyzing power for the infinitely thin target and the parameter α represents the strength of the dilution effects. The parametrizations above are chosen so that all the formulas merge in the the first order expansion in the target thickness t. However, notable discrepancies between the proposed parametrizations appear for target thicknesses larger than $1/\alpha$. In addition, a linear dependence is obviously unphysical since it leads to a change of the sign of the asymmetry around $t = 1/\alpha$.

Our data at the angles 161°, 168° and 172°, where the dilution effects are large, select unambiguously the "inverse" parametrization (Fig. 3).

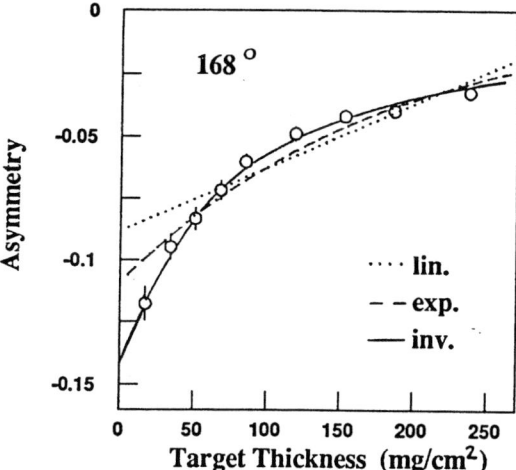

Fig. 3. Example of the fit to the asymmetries measured at 168° as a function of the target thickness t (mg/cm²). A very good description, with $\chi^2/n_D=0.4$ is obtained for the "inverse" parametrization, while the exponential and linear dependence badly fail with $\chi^2/n_D=3.9$ and $\chi^2/n_D=10.4$, respectively. More drastic discrepancies are observed at 172°.

This functional form and a sharp increase of the dilution effects at extremely large scattering angles (Fig. 4) suggest that two subsequent scatterings in the fields of the two different target nuclei are the most important mechanism responsible for the dilution of the analyzing power. This conclusion was confirmed in our microscopic calculations, where the contribution of double scattering was calculated explicitely by integrating over the target volume using cross section data from ref. [15]. Though some approximations were made in this calculation (eg. neglecting polarization effects in the second scattering), the agreement with the experiment is rather good for all scattering angles except 126°. Below 130° many small angle ("multiple") scatterings are probably important in the depolarization of the electrons.

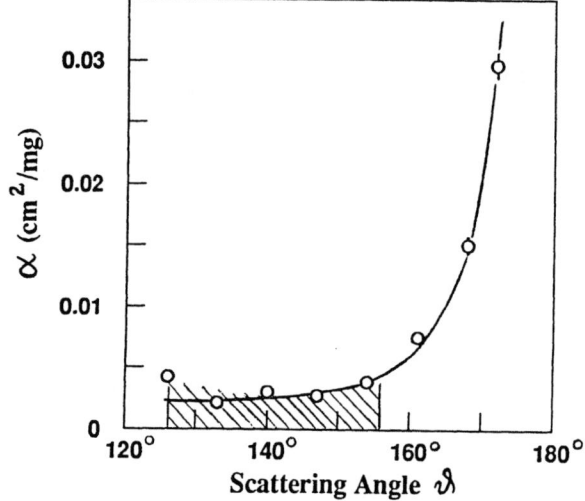

Fig. 4. Depolarization parameter α as the function of the scattering angle. The shaded area shows the angular range used in our time reversal experiment [5,6], which has triggered the present investigation. The curve shows double scattering contribution to the dilution of the analyzing power.

6. Comparison with the Theory.

The measured values of $A(t)$ were extrapolated to infinitely thin foils and the resulting analyzing powers A_o were compared to the theoretical predictions. Our own calculations of the Mott analyzing power at 14 MeV are in very good agreement with the interpolation of the results from the ref. [15]. We therefore use the theoretical values for the point nucleus as calculated by our code and apply the interpolated correction due to finite size of the lead nuclei from the ref. [15]. Since this correction has a smooth energy dependence and should not be very sensitive to the details of the formfactor of the nucleus, the estimated interpolation error is smaller than the experimental accuracy.

Good agreement of the analyzing powers at 14 MeV with the calculations taking into account the finite nuclear size was found. The point nucleus approximation overestimates the measured analyzing powers by roughly 15% in the whole angular range.

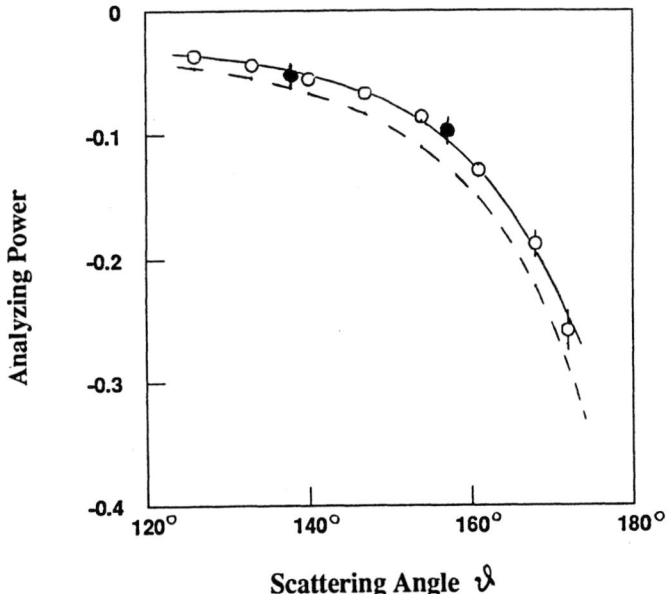

Fig. 5. Comparison of the measured analyzing powers extrapolated to infinitely thin targets and theoretical predictions. Dashed line: point nucleus approximation, solid line: effects due to finite nuclear size included [15]. Open symbols: the results of the final experiment using the strained cathode source providing 77% polarization of the electrons. Filled dots: the results of the test experiment with conventional cathodes and polarization of electrons of 33% only.

7. Conclusions

This first experimental test of the polarization effects in scattering of electrons in a multi MeV energy range from heavy nuclei demonstrates, that the Mott process may be used with confidence in analyses of the polarization of highly relativistic electrons. Beside great importance for studies of fundamental symmetries in weak interactions [5,6], the results of this experiment may be also useful for diagnostics of the beam polarization at the intermediate acceleration stages in high energy electron machines [9,10].

8. Acknowledgements

The authors wish to thank H.K. Kayser and K.H. Steffens from the Johannes Gutenberg Universität Mainz for providing us un unconventional beam for this investigation. J.S. thanks R. Holt, S. Price and C.K. Sinclair for stimulating discussions at this workshop. This work was supported by the Swiss National Foundation.

References

[1] N.F. Mott. Proc. Roy. Soc. A124(1929)425.

[2] H. Frauenfelder, R. Bobone, E. von Goeler, N. Levine, H.R. Lewis, R.N. Peackock, A. Rossi and G. DePasquali, Phys. Rev. 106(1957)386; H. Frauenfelder, A. Hanson, N. Levine, A. Rossi and G. DePasquali, Phys. Rev. 107(1957)643.

[3] T.D. Lee and C.N. Yang, Phys. Rev. 104(1956)254; T.D. Lee and C.N. Yang, Phys. Rev. 105(1957)1671.

[4] for reviews see Advanced Series on Directions in High Energy Physics vol. 14, "Precision Tests of the Standard Electroweak Model", ed. P. Langacker (World Scientific, Singapore, 1995).

[5] M. Allet, W. Hajdas, J. Lang, H. Lüscher, R. Müller, O. Naviliat-Cuncic, J. Sromicki, A. Converse, W. Haeberli, M.A. Miller and P.A. Quin, Phys.Rev.Lett. 68(1992)572.

[6] J. Sromicki, M. Allet, K. Bodek, W. Hajdas, J. Lang, R. Müller, S. Navert, O. Naviliat-Cuncic, J. Zejma and W. Haeberli, Phys.Rev. C53(1996)932.

[7] for a review see eg. T.J. Gay and F.B. Dunning, Rev. Sci. Instrum. 63(1992)1635.

[8] these measurements were perfected by J. Kessler and collaborators in series of experiments, eg. A. Gellrich and J. Kessler, Phys. Rev. A43(1991)204.

[9] S. Price, B.M. Poelker, C.K. Sinclair, P. Adderley, K.A. Assmagan, L.S. Cardman, J. Grames, J. Hansknecht, D.J. Mack and P.Piot, Proceedings of the 12th International Symposium on High–Energy Spin Physics, Sept. 10–14 1996, Amsterdam, The Netherlands, ed. by C.W. de Jager, World Scientific, Singapore, 1997, p. 727; see also contribution of the TJNAF group to this workshop.

[10] E. Reichert, Johannes Gutenberg Universität, communication discussed at this workshop.

[11] M. Steigerwald, Diploma, Johannes Gutenberg Universität, Mainz, 1994.

[12] K. Steffens, H. Andersen, J. Blume-Werry, F. Klein, K. Aulenbacher and E. Reichert, Nucl. Instr. Meth. A325(1993)378.

[13] E. Reichert and H.K. Kayser, personal communication.

[14] S. Hanna, in Symposium on Polarization Phenomena in Nuclear Physics, Karlsruhe, 1965.

[15] P. Ugincius, H. Überall and G.H. Rawitscher, Nucl. Phys. A158(1970)417.

Review of High Intensity Polarized H and D Ion Sources: Recent Progress and Future Projections

Thomas B. Clegg

*Department of Physics and Astronomy, University of North Carolina,
Chapel Hill, NC 27599-3255, USA
and
Triangle Universities Nuclear Laboratory, Duke University,
Durham, NC 27708-0308, USA*

Abstract: Recent development of optically pumped and atomic beam polarized H and D ion sources is reviewed. DC currents available for polarized $H^{\pm}$ beams today range up to 1.8 mA within a normalized emittance of 2π mm-mrad. Areas of investigation needed for possible future improvements of these sources are discussed.

INTRODUCTION

Polarized H and D ion sources providing beams for today's particle accelerators hardly resemble those first developed over three decades ago. Today, available polarized $H^{\pm}$ or $D^{\pm}$ beam intensities of 10μA to 1 mA are a factor of 10^6 to 10^8 higher than the earliest ion source beams used for experiments. Also, today's versatility of beam polarization choices and the reliability of ion source operations are both substantially improved. This makes polarized beams an increasingly frequent choice for nuclear and high-energy physics experimenters in at least fifteen accelerator laboratories worldwide.

Each of these laboratories and accelerators has unique requirements, so polarized source designs and operating parameters reflect this diversity. Polarized ion source techniques and their rapid development have been the subject of frequent conferences and workshops (1,2), so this present review will only report developments since the last in this series held in Cologne in June 1995 (3).

All methods used for producing nuclear polarized $H^{\pm}$ or $D^{\pm}$ beams rely on polarizing first the atomic electron and then transferring the electron's angular momentum to the nucleus using the hyperfine interaction. Three basic types of H or D polarized ion sources are used: Lamb-shift sources (4), optically-pumped polarized sources (OPPIS) (5), and atomic beam sources (ABS) (6). Lamb-shift sources polarize beams of excited $2S_{1/2}$ metastable H or D atoms, while the OPPIS and the ABS polarize beams of $1S_{1/2}$ ground-state atoms. The latter two source types are the only ones being actively developed today, and thus will be the subject of this review.

OPTICALLY PUMPED POLARIZED ION SOURCES

Present Status

In an OPPIS, protons at 3 keV are provided from an electron-cyclotron-resonance (ECR) source which is contained inside a 3 T solenoidal B-field. These protons are then injected into a Rb-vapor neutralizer cell where the Rb atoms are polarized by optical pumping. Entering protons pick up a polarized electron in collisions with the polarized Rb. The resulting atomic hydrogen polarization is then transferred from the polarized electron to the nucleus in a spin-reversal region (8). The nuclear polarized atoms then enter a Na-vapor ionizer cell where they pick up a second electron and emerge as nuclear-polarized, negative ions.

A summary of OPPIS sources performances is included in Table 1. The most active center of optically pumped polarized source development over the past two years has been at TRIUMF in Vancouver (7). At TRIUMF, the OPPIS is used for 40% of their cyclotron operating time.

DC beams

Over the past several years, both output beam intensities and beam polarizations have improved significantly. As the Rb-vapor thickness in the neutralizer cell increases, the beam current increases and typical polarizations decrease slightly. In routine operation, DC polarized H$^-$ ion currents as large as 0.6 mA are available within a normalized emittance $\varepsilon_n = 2\pi$ mm-mrad with a polarization of ~80%. Here ε_n is determined solely by the diameter of the Na-vapor ionizer cell and the magnitude of the ~1200 G solenoidal B-field in which it resides.

Recently larger beam currents were obtained within the same ε_n by enlarging the diameter of the Rb-vapor neutralizer cell. This is then fed by proton beams extracted from the ECR source using a larger diameter, hexagonal close-packed array of 0.95 mm diameter holes. An array of 199 holes with an overall array diameter of 17.5 mm provides DC polarized beam currents as high as 1.8 mA. For this mode of operation, however, the beam's polarization drops to 60% because insufficient laser power is available to promote complete Rb-vapor polar-

Table 1. Performance of optically pumped polarized ion sources.

Institute	Maximum Intensity (mA)				Normalized Emittance 80% π mm-mrad	Polarization % of Max	Duty Factor Fraction	Ionizer Type
	H-	H+	D-	D+				
TRIUMF	0.6 1.2 1.8	- - -	- - -	- - -	2.0	<85 <75 <60	DC	Rb
TRIUMF	30	-	-	-	2.0		2.5x10^{-5}	Rb
KEK/ TRIUMF	-	-	0.38	-	-	70	2x10^{-3}	Rb

ization in the larger diameter neutralizer cell. It is found that 15 W/cm^2 of laser power is necessary to reach 100% Rb-vapor polarization with a Rb-target thickness of 10^{14} atoms/cm^2.

The performance of this DC OPPIS has been driven hard by experiment E497 at TRIUMF which is measuring the parity violating analyzing power A_z for scattering longitudinally polarized 230 MeV protons from hydrogen, with an accuracy of $\pm 10^{-8}$. The target current needed for this experiment after the cyclotron is only 0.2 μA. However, requirements for control of helicity-correlated modulations in the beam on target are quite severe (current $\leq 10^{-5}$; energy ≤ 0.010 eV; position ≤ 20 nm) and have been met.

Pulsed beams

When pulsed H$^-$ polarized beam is required, recent TRIUMF tests show it to be feasible (9) to replace the ECR source in their OPPIS with a bright pulsed proton source which was developed at the Budker Institute in Novosibirsk (10). At TRIUMF, it produced up to 8 A of pulsed H$^+$ beam which was injected into a neutralizer cell containing H$_2$ gas. The emerging fast neutral H atoms then drifted without emittance growth into the strong solenoidal B-field where they were then stripped to provide the H$^+$ beam needed. This source has two advantages: the beam is produced in a region where extensive pumping can be provided and, because all gases are pulsed, the overall pumping and laser power requirements are modest. Output 30 mA, 100 μs beam pulses of H$^-$ at 0.25 Hz were obtained in these initial TRIUMF tests made without optical pumping. Tests with laser pumping to polarize the output beam are planned for early 1998.

It is more difficult with an OPPIS to polarize D$^-$ than H$^-$ (11), but tests using DC lasers at KEK showed that up to 380μA of pulsed ~70% vector-polarized D$^-$ beam can be obtained (12). Since the deuteron has a spin of 1 ℏ, one must attach two polarized electrons to attain large deuteron polarization. Thus, both the neutral–izer and ionizer cells were optically pumped in the KEK tests. Since the cross section for attaching the second electron is smaller than that for attaching the first, the density of polarized Rb-vapor needed in the ionizer cell is higher, requiring larger laser power in the ionizer cell than in the neutralizer cell for its full polarization. Even with this, one obtains output beams with limited choice of polarizations. When the handednesses of the optical pumping lights for the neutralizer and ionizer cells are opposite, and are inverted simultaneously, one obtains beam in one hyperfine state with maximum deuteron polarizations of $P_z=\pm 1$, $P_{zz}=+1$. Alternatively, when the handednesses are the same, and are inverted simultaneously, one obtains beam in two hyperfine states with $P_z=\pm 1$, $P_{zz}= -1/2$.

Future Possibilities

Often experiments benefit considerably from having beams with larger deuteron tensor polarization, or from having deuteron beams which are either purely and maximally vector-polarized ($P_z=\pm 1$, $P_{zz}=0$) or tensor-polarized ($P_z=0$,

$P_{ZZ}=+1$ or -2). Accomplishing this with an OPPIS will require developing radio-frequency transitions to invert the populations of deuterium hyperfine states in the monoenergetic ~6 keV D atomic beam while it moves between the neutralizer and ionizer cells. When coupled with the attractive flexibility afforded by changing the individual handedness of the laser beams used for optical pumping, two transitions would probably suffice to replace the spin-reversal region of a traditional OPPIS.

Since a fast D beam is used in an OPPIS, each beam particle would experience a ~70 ns long rf pulse in a hypothetical rf transition unit 5 cm long. This pulse would be ~1000x shorter than that experienced by a beam traversing a transition unit in an ABS. The associated ~14 MHz pulse bandwidth in this hypothetical OPPIS transition unit is then 1000x larger than for an ABS, but is still comfortably smaller than the ~100 MHz splitting between individual deuterium hyperfine states when $B \geq 100G$. Thus, designing systems to promote transitions between individual pairs of hyperfine states in an OPPIS should be possible.

Adiabatic-fast-passage transitions used in the traditional ABS where the beam velocity spread is large are not necessary for the fast, monoenergetic OPPIS beam. Rather, systems should be designed to promote Breit-Rabi, π-flip transitions for each particle in the beam (11). These designs will, however, require significantly higher input power and/or higher Q than for traditional transition units used in an ABS. Some features of a recently developed large aperture, high-Q transition unit may warrant consideration (13).

ATOMIC BEAM POLARIZED ION SOURCES

Present Status

In an ABS, a thermal beam of H or D atoms emerges as a jet from a dissociator and travels down the axis of a sextupole focusing system. The inhomogeneous sextupole B-field focuses atoms having electron spin projection $m_J=+1/2$ and defocuses those with $m_J=-1/2$. Subsequently, the electron-spin-polarized beam then passes through axial regions where rf and dc electromagnetic fields are applied to cause atomic transitions between selected pairs of hyperfine states; this effectively transfers the electronic polarization of the atom to the nucleus. Finally, the nuclear polarized atomic beam enters a region where ionization occurs, producing either positive or negative polarized ions, as desired. A summary of the techniques used can be found elsewhere (6).

Today, atomic beam sources dominate in overall number and variety of laboratory installations, and new systems are being developed. Steady improvement in performance continues, if more slowly over the past two years than previously. These sources offer the most versatile systems for producing vector- and tensor-polarized deuteron beams. A summary of performances of ABS systems worldwide is presented in Table 2.

Two areas of current investigation promise greater extracted polarized beam intensities: permanent magnet sextupole focusing systems for the polarized atomic beam, and ionizers which store the polarized atoms to enhance the overall ionization efficiency.

Atomic Beam Systems

Permanent magnet sextupoles are now universally used for focusing polarized H or D atoms for jet or storage-cell targets. Several such systems attain output fluxes $\geq 6 \times 10^{16}$ atoms/s into a tube 1 cm diameter and 10 cm long (14,15). This is more than double that available from systems employed in operating polarized ion sources, which have not benefited yet from these gains.

Table 2. Performance of atomic-beam polarized ion sources.

Institute	Intensity (μA)				Normalized Emittance 80% π mm-mrad	Polarization % of Max	Duty Factor Fraction	Ionizer Type [a]
	H⁻	H⁺	D⁻	D⁺				
Saclay	-	700	-	700	8	90	1.5x10⁻³	EB
		2500		2500	6		4x10⁻⁴	
NAC-Faure	-	50	-	40	0.6	60	DC	EB
Saudi Arabia	-	23	-	-	2.2	69(d)	DC	EB
PSI	-	300	-	300	1.2	75(p), 80(d)	DC	ECR
Bonn	-	~30	-	~30	< 0.3	80	DC	ECR
TUNL	8	50	8	70	0.6 (est.)	75(p), 85(d)	DC	ECR
IUCF	-	130	-	90	0.3 (est.)	80(p)	DC	ECR
RIKEN	-	-	-	120	-	80(d)	DC	ECR
RCNP	-	50-100	-	50-100	~0.5	78	DC	ECR
Kyoto	-	50	-	-	-	-	DC	ECR
KFK-Grönigen	-	50	-	100	0.4	70(p), 56(d)	DC	ECR
Uppsala	-	100	-	-	-	40(p)	DC	ECR
Wisconsin	1	-	1	-	-	90	DC	CCB
Brookhaven	40	-	-	-	0.2	75-80	-	CCB
COSY	20	-	-	-	< 0.27	85	10⁻²	CCB
JINR-Dubna	-	-	-	200	-	80(d)	10⁻³	PEI
	-	-	-	500	-	80 (d)	10⁻³	PI
INR/Moscow	-	6000	-	-	1.6 x 2.0	84(p)	2x10⁻³	PI
	1000	-	-	-	1.8 x 1.8	88(p)	2x10⁻³	PI

[a] Ionizer-type: EB, Electron Bombardment; ECR, Electron Cyclotron Resonance; CCB, Colliding Cesium Beam; PEI, Penning Ionizer; PI, Plasma Ionizer

Atomic beam systems for ion sources may be divided into two groups: those operating with dissociator nozzle temperatures ≈35K, and those operating at 80 to 100K. One 35K source, constructed recently at RCNP-Osaka (16), employs permanent magnet sextupoles. No absolute atomic flux measurements are available for that source, but the polarized H$^+$ and D$^+$ currents of 50 to 100μA obtained from their ECR ionizer do not exceed those obtained from similar 35K sources operating at PSI (17) and IUCF (18) which use electromagnetic sextupoles.

Polarized sources having atomic beam systems operating with ≈80K dissociator nozzle temperatures are under construction at Munich (19) and under design at IUCF for the new injector synchrotron for their Cooler (20). Although the Munich atomic beam system provides H fluxes comparable to those available for the HERMES polarized target, it will be early 1998 before we learn whether this leads to polarized H$^\pm$ beams which are greater than for existing sources.

At TUNL, our operating atomic beam source uses electromagnetic sextupoles too, and we are also investigating what gain might be achieved by replacing them at least partially with permanent magnet sextupoles. Calculations predict that if we maintain our present 35K nozzle temperature and 0.5mbar-l/s dissociator gas flow, only modest atomic beam intensity gains (~18%) are possible. Having no comparable test bench with which to measure the atomic beam flux and velocity distribution as the dissociator nozzle temperature and gas flow are raised, we are considering building a hybrid permanent magnet/electromagnetic sextupole system. In so doing, we would seek both to focus the faster beam and to retain tunability to match the sextupoles system's strength to the beam's velocity distribution for a range of dissociator operating conditions.

Ionizer systems

The techniques typically used to enhance the performance of atomic beam systems, higher dissociator nozzle temperatures and gas flows, both enhance the flux and raise the atomic beam velocity for the emerging polarized atomic species. However, traditional ionization schemes require higher beam *density*, not *flux*. To utilize the higher flux effectively, charge-exchange ionizers are needed which store polarized atoms in target cells for passing low-energy ion beams. This idea is attractive because cross sections for charge-exchange with H$^\pm$ or D$^\pm$ are very large and resonant at zero energy (21-23): $\sigma \approx 2.5 \times 10^{-15}$ cm^2 at ~1 keV.

Storage-type ionizers. Systems needed for this are conceptually simple (24): a storage tube, fed with the enhanced flux of arriving polarized D or H atoms, as a charge-exchange target for a high-intensity, unpolarized H$^\pm$ or D$^\pm$ beam. Target thicknesses of $\rho l \approx 8 \times 10^{13}$ polarized atoms/cm^2 have been stored in a 25 cm long tube 1 cm in diameter used in high-energy storage ring applications (25,26). In an ion source, polarized ions produced by charge-exchange inside the tube are extracted along the axis of the ionizing beam and then separated magnetically from that beam after extraction. If polarized atoms of mixed hyperfine state are to be stored in the tube, then the entire tube must reside inside a solenoidal B-field having B≥B$_{critical}$ to maintain the atoms' polarization (27) before ionization.

When compared with the ECR ionizers used today, whose ionization efficiency is ≈ 6 to 8%, the ionization efficiency of such charge-exchange ionizers is theoretically much higher, $I_{in}/I_{out} = \sigma \rho l \approx 40\%$. Selectivity for producing H$^-$ (or D$^-$)

from polarized atoms rather than from unpolarized background molecules is also likely.

Possible disadvantages also deserve mention. As found in polarized atom storage cell targets used in some high-energy storage rings (28), atomic depolarization can occur upon molecular recombination. This occurs on storage cell walls unless care is taken to preserve the cleanliness and integrity of an inert surface which inhibits such recombination. Furthermore, when an unpolarized H^+ beam is used to ionize polarized D atoms, the emerging polarized D^+ beam could be difficult to separate after extraction from any unwanted H_2^+ component in the incident H^+ beam.

Pulsed systems. Tests made at INR-Moscow have shown that storage techniques do enhance the output beam obtained from ionizers operating in a pulsed mode. Already at the 1995 Köln conference, Belov reported (29) that polarized H^- output currents of 600 µA were obtained. Those have recently been increased to 1 mA, and such an ionizer is now being constructed in Moscow for the new Cooler Injector Synchrotron at IUCF (30). At the present meeting, Belov reports on the successful addition of an uncoated aluminum storage cell for polarized atoms inside his D^+ plasma ionizer to obtain similar (1 mA) H^+ currents with very high (94%) polarization for short (1-2 ms) beam pulses.

DC systems. Storage/charge-exchange ionizers which operate DC are much harder to realize. Systems used at Moscow are not capable of DC operation because the gas load from the unpolarized ion source is too great. An alternative proposal (6) has suggested using the intense H^- (or D^-) ion cusp source developed at TRIUMF by Kuo et al. (31) to produce the unpolarized beams needed for charge-exchange. That source produces H^- currents up to 20 mA at 20 keV. To take advantage of the larger charge-exchange cross section at low energy and thus be effective in a charge-exchange ionizer, the beam must be decelerated to ~1 keV before passing through the polarized atom storage cell. To learn whether this might be possible, tests were performed in which a dummy tube of dimensions similar to that needed for a storage cell was installed on the axis of the extracted beam in the TRIUMF source. Measurements showed that the maximum 1 keV beam attainable in a suppressed Faraday cup following this tube was 85 µA for H^- and 35 µA for D^-. Both results are far below the 1 to 2 mA needed to make this competitive for use in a charge-exchange ionizer.

Several factors acted in combination to limit these output currents: 1) the ion optical system used at TRIUMF, optimized for 20 keV beams, was not redesigned for the much lower 1 keV beam energy; 2) radially directed space-charge electric fields increase as $E^{-1/2}$ when the beam energy E decreases, causing significant beam divergence unless space-charge compensation is provided; and, 3) the same charge-exchange cross section for $H^- + D \rightarrow H + D^-$, which is resonant at zero energy and which is so attractive for ionization of polarized D atoms, also strips the low-energy H^- unpolarized negative ion beam efficiently in the background H (or D) gas emerging from the ion source, before the beam ever reaches the storage cell.

Another problem, not present in the tests at TRIUMF which were conducted with the dummy charge-exchange tube in a magnetic-field-free region, is that when a solenoidal B-field is needed in a working ionizer, and when the source of unpolarized beam is placed outside this field, there will be an unavoidable growth of unpolarized beam emittance as the beam enters this B-field. The resulting coherent cycloidal motion of off-axis, low-energy beam particles will then reduce the effective acceptance of the storage cell tube significantly, limiting severely the magnitude of current which could be focused through the tube.

Future Possibilities

A primary conclusion from these TRIUMF tests must be that the most effective ion source for producing low-energy unpolarized ion beam for a charge-exchange ionizer must reside inside the same uniform solenoidal B-field as the polarized-atom storage cell. This will eliminate beam emittance growth when the beam enters the B-field. Furthermore, the B-field will also reduce to some extent the radial growth of beam diameter caused by the beam's internal space-charge electric field.

Any unpolarized ion source used should also have high gas efficiency. This will make it far easier to maintain a good vacuum at low cost and with minimum engineering complication. Good vacuum will, in turn, minimize stripping of the desired negative ions in background gas.

Finally, it is also very clear that carefully designed beam optics for such ionizer systems will be extremely important. This must focus maximum beam through the storage cell for overall ionization efficiency, and minimum beam loss inside the cell to avoid degradation of the cell walls and subsequent molecular recombination.

Ionizing with DC Beams

Negative beams. The unpolarized ion source must operate inside a solenoidal B-field, but the source of Kuo *et al.* at TRIUMF (31), which produces intense H$^-$ or D$^-$ beams, instead requires B=0 on axis. Other DC sources of H$^-$ or D$^-$ beams (32) can operate in a non-zero axial B-field, but the extracted beam intensity is likely to be significantly smaller than 1 mA at ~1 keV, and therefore competitively uninteresting. Also, with an axial B-field imposed, separation of extracted electrons from the desired H$^-$ or D$^-$ beam will be very difficult. Thus, an appropriately intense DC source which produces a primary unpolarized H$^-$ or D$^-$ beam remains to be found.

An alternative scheme worth considering could employ a very intense primary source of unpolarized H$^+$ or D$^+$ beam. Once extracted, this beam could be injected at 0.5 to 1 keV through a cesium vapor charge-exchange cell, much like that used in our ABS at TUNL (33), to convert the H$^+$ or D$^+$ beam with >10% efficiency (34) to H$^-$ or D$^-$ ions. Emerging H$^-$ or D$^-$ beam then could continue along the constant solenoidal B-field into the storage cell to ionize polarized atoms.

Positive beams. Alternatively, reversing this process, the intense H$^+$ or D$^+$ beam might be extracted and injected immediately into the storage cell to produce polarized D$^+$ or H$^+$ ions there by charge-exchange. These could then be extracted and focused at 0.5 to 1 keV through the cesium vapor charge-exchange region to produce polarized negative ions. An attractive feature of this arrangement is that either polarized H$^+$ (D$^+$) or H$^-$ (D$^-$) beam is available, depending on experimental demand. Based on our experience at TUNL (33), one can expect that the positive ion beam intensities will be ~10x those for the negative beams.

For either of the last two arrangements, which utilize an intense H$^+$ (or D$^+$) beam, an ECR source would be a wise choice to provide the positive ion beam needed. Such sources (35) can operate at low internal pressure [we sustain the ECR plasma discharge down to 2×10^{-6} mbar in our TUNL ionizer (33)], have high gas efficiency, and provide beams with excellent brightness and high H$^+$ (or D$^+$) fraction. But, at the ~1 keV beam energies needed, space-charge electric fields in the beam will be very large; the resulting beam blow-up looms as a daunting problem. Another challenge will be the need for careful control of cesium vapor, which if it deposits on the storage cell wall, will likely promote rapid polarized-atom recombination.

Ionization with a DC Plasma

Large space-charge electric fields are eliminated inside plasmas, so one might consider using directly the neutral H or D plasma streaming from a low-pressure ECR-heated discharge as the ionizing medium. This could be guided by the imposed solenoidal B-field to pass from the ECR region down the axis of the polarized atom storage cell. Advantages of this scheme include effective use of the very large low-energy charge-exchange cross section and a small ionization region diameter. Likely disadvantages will be difficulties handling space-charge effects when extracting the polarized beam and when separating it from the remaining unpolarized ion beam downstream of the ionization region.

SUMMARY AND CONCLUSIONS

Since the last of these polarized ion source and target workshops in Köln in June 1995, steady progress continues toward the community's goal of providing enough polarized ion beam that all accelerator experiments use these beams. Progress with OPPIS sources in the last decade is truly impressive. Much of this improvement has come because of rapid laser development. Beam currents from OPPIS sources have improved significantly in the past two years; perhaps more important are the improved beam polarizations which higher power laser systems have enabled.

The best optically pumped and atomic beam sources for H cost roughly the same to construct and maintain, but OPPIS sources for D are more expensive than ABPIS sources because of the additional laser systems required to polarize the ionizer cell. This situation is changing, as low-cost diode laser arrays have been developed and are being used now routinely for optical pumping of Rb. Even as laser systems continue to improve and become less expensive, OPPIS sources will not compete favorably to provide beams for all experimental applications until rf transitions for D are developed and installed.

For ABS systems, charge-exchange ionizers with intense unpolarized H$^\pm$ (or D$^\pm$) beams incident on stored-polarized-atom targets must be developed to make significant output beam intensity gains. This has recently been demonstrated successfully for pulsed systems, but development of DC systems is still problematic. Questions about space-charge effects and about polarized atom recombination on storage cell walls, especially at low temperature and in the presence of metal vapors such as cesium, are still worrisome and need much further investigation.

ACKNOWLEDGMENTS

The author gratefully acknowledges the hospitality of TRIUMF, and specifically the aid of T. Kuo, during measurements using that laboratory's intense negative ion source. He also would like to thank S. Lemaitre for calculations to predict the performance of the TUNL atomic beam source with permanent magnet sextupoles. This work was supported in part by the U.S. Department of Energy, Office of High Energy and Nuclear Physics under grant #DE-FG05-88ER40442.

REFERENCES

1. Mori, Y., ed., *Proceedings of the International Workshop on Polarized Ion Sources and Polarized Gas Jets,* KEK, Tsukuba, February 12-17, 1990, KEK Report 90-15, 1990.
2. Anderson, W., and Haeberli, W., eds., *Proceedings of International Workshop on Polarized Ion Sources and Polarized Gas Targets, Madison, WI, May 23-27, 1993*, AIP Conf. Proc. **293**, 1994.
3. Paetz gen. Schieck, H. and Sydow, L., eds., *Proceedings of International Workshop on Polarized Beams and Polarized Gas Targets, Cologne, June 6-9, 1995*, World Scientific, Singapore, 1996.
4. Clegg, T.B., "Lamb-shift sources -- hopes and realities," *in Proceedings of Workshop on Polarzied Proton Sources, Vancouver,* AIP Conf. Proc. **117**, 63-77 (1983).
5. Zelenski, A., "Optically pumped polarized sources," ref. 3, pp. 111-119.
6. Clegg, T.B., "Atomic beam polarized sources -- recent progress and future possibilities," in ref. 3, pp. 155-166.
7. Levy, C.D.P., "The TRIUMF optically pumped polarized H⁻ ion source," in ref. 3, pp. 120-125.
8. Sona, P.G., *Energie Nucleaire*, **14**, 295 (1967).
9. Zelenski, A.N. and Mori, Y., "Proposal for a pulsed optically pumped polarized H⁻ ion source for high energy accelerators," in ref. 2, pp. 173-178.
10. Davydenko, V.I., *et al., Soviet Physics - Doklady* **28**, 685-687 (1983).
11. Schneider, M.B. and Clegg, T.B., *Nucl. Instrum. Meth.* **A254**, 630-632 (1996).
12. Kinsho, M., *et al., Rev. Sci. Instrum.* **67**, 1362-1364 (1996); also ref. 3, pp. 126-130.
13. Raymond, R. ,"Tests of a prototype large-bore, low-power 2->4 rf transition unit," poster presented at this conference.
14. Wise, T., Roberts, A.D., and Haeberli, W., *Nucl. Instrum. Meth.* **A336,** 410-422 (1993) .
15. Stock, F., *et al.*, Nucl. Instrum. Meth **A343**, 334-342 (1994).
16. Hatanaka, K., *et al., Nucl. Instrum. Meth.* **A384**, 575-581 (1997).
17. Schmelzbach, P.A., "ECR ionizer for intense polarized beams," in ref. 1, pp. 329-338.
18. Derenchuk, V., Brown, R., and Wedekind, M., "Polarized ion source development at IUCF," ref. 2, pp. 72-75.
19. Hertenberger, R., "Beam formation in the Munich atomic beam source," poster presented at this conference.
20. Derenchuk, V.P., "Polarized beam for the IUCF Cooler Injector Synchrotron," paper presented at this conference.

21. Shah, M.B., Elliott, D.S., and Gilbody, H.B., *J. Phys. B: At. Mol. Phys.* **20**, 3501-3514 (1987).
22. Gilbody, H.B. and Ireland, J.V., *Proc. Royal Soc. London* **A277**, 137-141 (1964).
23. Fite, W.L., *et al.*, *Phys. Rev.* **119**, 663-668 (1960).
24. Haeberli, W., *Nucl. Instrum. Meth.* **62**, 355-357 (1968).
25. Ross, M.A., *et al.*, *Nucl. Instrum. Meth.* **A344**, 307-314 (1994).
26. Zapfe, K., *Rev. Sci. Instrum.* **66** (1) 28-31 (1995).
27. Haeberli, W., *Ann. Rev. Nucl. Science* **17**, 373-426 (1967).
28. Stewart, J., "The HERMES H target," presented at this conference.
29. Belov, A.S., "Colliding beam polarized ion sources," in ref. 3, pp. 208-217.
30. Belov, A.S., *et al.*, "Polarized ions from a storage cell," paper presented at this conference.
31. Kuo, T., *et al.*, *Rev. Sci. Instrum.* **67**, 1314-1316 (1996).
32. Collins, L.E. and Gobbett, R.H., *Nucl. Instrum. Meth.* **35**, 277-282 (1965).
33. Clegg, T.B., *et al.*, *Nucl. Instrum. Meth.* **A357**, 212-219 (1995).
34. Morgan, T.J., *et al.*, *J. Phys. Chem. Ref. Data* **14**, 971-1040 (1985).
35. Taylor, T. and Wills, J.S.C., *Nucl. Instrum. Meth.* **A309**, 37-42 (1991).

STATUS OF THE POLARIZED D⁻ ION SOURCE AT KEK

A. Takagi, M. Kinsho*, K. Ikegami, Z. Igarashi and Y. Mori
High Energy Accelerator Research Organization (KEK)
*Japan Atomic Research Institiute (JAERI)

ABSTRACT

A status of the optically pumped polarized negative deutrium ion source at KEK is presented. In order to obtain a large beam intensity with this source, After optimization of an ECR deuteron source, a polarized negative deuterium ion beam current of about 380 µA was obtained. A polarized deuteron beam has been successfully accelerated up to 10.2 GeV at KEK 12 GeV-PS. In this acceleration test, since the OPPIS was operated as a single optically pumped mode.

INTRODUCTION

An idea of the optically pumped polarized ion source (OPPIS) for producing polarized negative hydrogen ion source, which is based on electron-capture reaction of negative hydrogen ions from the optically pumped sodium atoms, was originally proposed by Anderson [1]. The first OPPIS based on this idea was successfully developed at KEK in 1993 [2]. Since then, OPPIS has been developed at various laboratories, LAMPF, TRIUMF and INR [3][4][[5]. Recently, it was foumd at KEK that, with dual-optically pumped scheme, OPPIS was also to generate highly polarized negative deuterium ions which have been thought difficult to make so far [6][7].

Polarized deuteron acceleration tests were performed at KEK-PS [8]. There is one week intrinsic resonance at 8GeV. The polarization was measured by an external polarimeter at 2 GeV following two patterns. The beam was extracted after deccelaration to 2 GeV from 10.2GeV in pattern 1 and just after acceleration up to 2 GeV in pattern 2. respectively. The polarization was almost the same in the two patterns, meening that a polarized deuteron beam has been successfully accelerated up to 10.2GeV at the KEK-PS.

DUAL OPPIS FOR DEUTERONS

Figure 1 shows a schematic diagram of a dual-optically pumped polarized negative deuterium ion source which has been developed at KEK. There are an ECR ion source and a neutralizer in the superconducting magnet which makes a high magnetic field of about 2.7 Tesla. A microwave is fed into ECR source from upstream of it. The superconducting solenoid has three independent coils which allows control of the magnetic field shape. A deuteron beam is extracted from ECR ion source by an electrode system which accelerates the beam to energy of approximately 6 keV. The deuteron beam enters an electron-spin polarized rubidium vapor cell, where a fraction of the deuterons picks up a polarized electron by charge transfer from rubidium and are thus neutralized. We call this cell neutralizer. The electron-spin polarization is induced by optical pumping with circularly a polarized laser tuned to the rubidium D_1 transition at 795 nm. Rubidium is chosen because of its relatively high charge exchange cross section with fast deuteron [9], and the availability of high laser power at the wavelength.

Most of the neutralized deuterium is created in the excited n=2 state. For that reason it is necessary that charge exchange occurs within a high magnetic field which preserves the electron-spin polarization of deuterium atom as the atom decays to the ground state [10]. The ECR cavity extraction electrodes and neutralizer are both contained within the same high magnetic field so as to reduce the effective emittance increase of the deuteron beam as it enters the magnetic field region of the neutralizer. The magnetic field reverses direction in the region between the neutralizer and negative ionizer cell, and as the neutral deuterium beam passes through this region, the nuclear-spin polarization is enhanced by means of a Sona transition [11]. This deuterium atom picks up a polarized electron by charge transfer from rubidium in ionizer. The electron-spin polarization of rubidium atom in the ionizer is induced by optical pumping. A nuclear-spin vector polarized negative deuterium ion beam is created with this scheme. In the downstream of the ionizer, there are einzel lens, Wien filter and a Faraday cup for which the D- beam current is measured. In order to pump the rubidium atoms in the neutralizer and ionizer, two broad band Ti-sapphire lasers are fed into the neutralizer and ionizer from downstream of it, respectively. The thickness and electron-spin polarization of rubidium in the neutralizer and ionizer are measured by Faraday rotation with a linearly polarized laser light tuned to the rubidium D_2 transition at 780 nm. This probe laser fed into the neutralizer and ionizer from upstream of this ion source. Experimental result showed that a negative deuterium ion beam with about 70 % of nuclear-spin vector polarization was obtained with this source. In order to obtain a large beam intensity with this source, we were optimized the ECR deuteron source. Polarized negative deuterium ion beam current of about 380 mA was obtained so far.

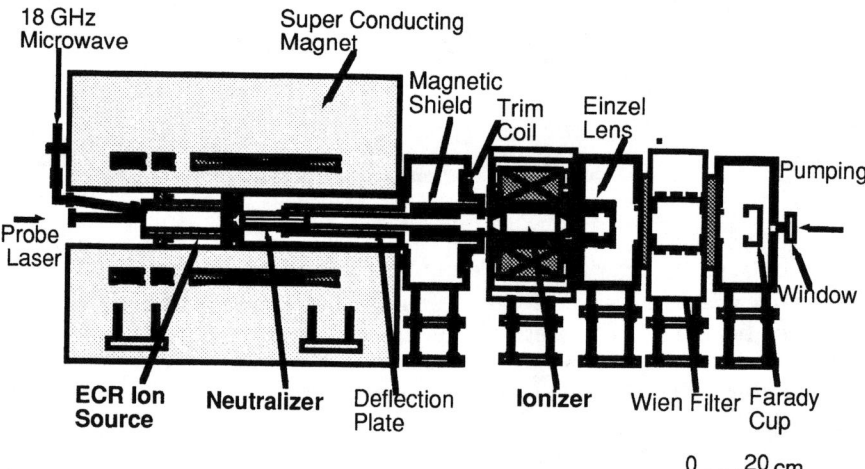

FIG. 1. Schematic diagram of a dual optically pumped polarized negative deuterium ion source at KEK

ECR DEUTERON SOURCE

The ECR deuteron source consists a ECR cavity and a three-electrode extraction system [12]. The ECR cavity is a stainless steel cylinder of 29 cm length and 6 cm i.d. It is mounted to an insulating Teflon™ holder to allow the application of the 6 kV potential for deuteron extraction. The Teflon holder also supports the extraction electrode system. The plasma chamber is made of stainless steel and the inside of it contains a quartz glass tube to evacuate the plasma. External grooves in the stainless steel cylinder hold a Sm-Co hexapole structure to make a good plasma confinement. Microwave is fed into a plasma chamber through a thin vacuum sealed microwave window. Microwave power is generated by a 18 GHz klystron (THOMSON, TH 2463T) which operates with a long pulse duration (up to 1 msec) and high repetition rate (20 Hz). A maximum power of the microwave generated by this klystron is 1 kW.

NEUTRALIZER

Figure 2 shows the experimental data of D⁻ beam current for two types of neutralizer. One has a structure whose entrance and exit holeaperture size were 10 mm dia., respectively. We call it the neutralizer of type A. Another neutralizer has a structure whose entrance hole size was 12 mm dia. and the exit hole size was 16 mm dia. We call it the neutralizer of type B. Figure 2 shows the D⁻ beam current as a function of Rb thickness in ionizer. In this experiment, Rb thickness in neutralizer is kept to be 1.2×10^{14} atoms/cm2. It was found that the D⁻ beam current with he neutralizer of type B increases about 1.5 times larger than that with the type A.

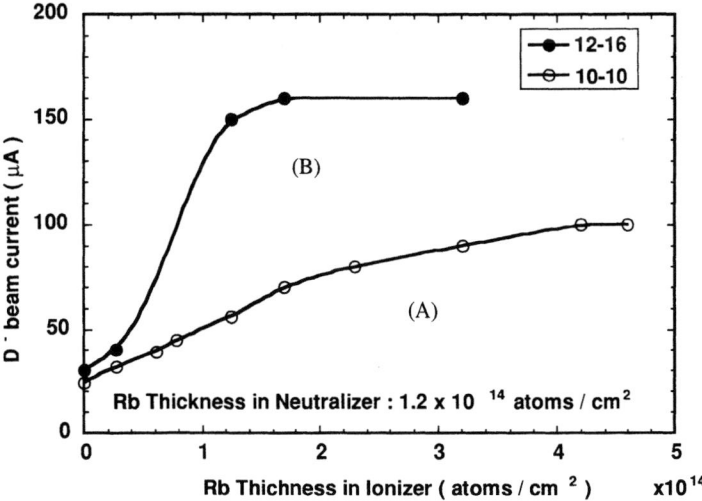

FIG. 2. The experimental result of D⁻ beam current for two types of neutralizer. Rubidium vapor thickness in the neutlarizer is kept at 1.2×10^{14} atoms/cm^2.

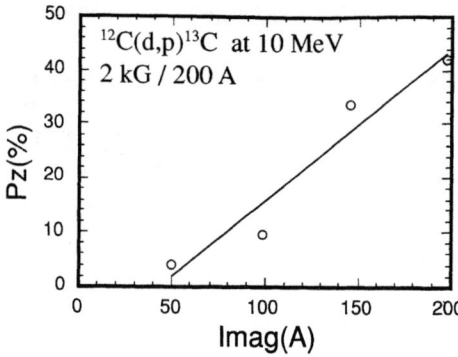

FIG. 3. Measured nuclear-spin vector polarization of a polarized deuteron beam as a function of the ionizer magnet cuurent.

ACCELERATION TEST

Polarized deuteron acceleration tests were performed at KEK-PS[8]. There is one week intrinsic resonance at 8GeV. The polarization was measured by an external polarimeter at 2 GeV following two patterns. The beam was extracted after deccelaration to 2 GeV from 10.2GeV in pattern 1 and just after acceleration up to 2 GeV in pattern 2. respectively. The polarization was almost the same in the two patterns, meening that a polarized deuteron beam has been successfully accelerated up to 10.2GeV at the KEK-PS. In this experiments, since the OPPIS ws operated as a single optically pumped mode, the maximum ideal vector polarization is 2/3.

Thenuclear-spin polarization of a deuteron beam was measured by a low energy polarimeter installed in the low energy beam line of the proton linac (beam energy of 10 MeV) in KEK-PS. Figure 3 shows the measured nuclear-spin vector polarization of a deuteron beam as a function of the ionizer magnet current. Acording the increaseing the ionizer magnet current, the beam polarization was linearly increased whereas the beam intensity of a negative deuterium ions fron the OPPIS was going down.

CONCLUSION

We have been developed the dual optically pumped polarized negative deuterium ion source for producing deuteron with high nuclear-spin vector polarization. With this source, a negative deuterium ion beam with about 70 % of nuclear-spin vector polarization was obtained. In order to obtain a large beam intensity with this source, we are optimizing the ECR deuteron source. Polarized negative deuterium ion beam current of about 380 mA was obtained so far. Polarized deuteron beam acce;eration tests were performed at the KEK-12 GeVPS. A polarized deuteron beam has been successfully accelerated up to 10.2 GeV at KEK. In this acceleration test, since the OPPIS was operated as a single optically pumped mode, the maximum ideal vector polarization is 2/3.

REFERENCES

[1] W.L. Anderson, Nucl. Instr. and Meth. 167, 363(1979).
[2] Y. Mori, K. Ikegami, Z. Igarashi, A. Takagi,and S. Fukumoto, AIP Conf. Proc. No.117,(1983)123.
[3] R.L. York et al., Proc. Int. Workshop on Polarized Ion Sources and Polarized Gas Jets, KEK Report 90-15(1990), p. 142.
[4] L. Buchmann et al., ibid p. 161.
[5] A. Zelenskii et al., ibid p. 154.
[6] M.B. Schneider and T.B. Clegg, Nucl. Instr. and Meth., A254, 630 (1987).
[7] M. Kinsho and Y. Mori, Rev. Sci. Instrum., 65, 1388 (1994).
[8] H. Sato et al., Nucl. Instr. and Meth., A 385, 391 (1997).
[9] H. Tawar, Atomic Data and Nuclear Data Tables, 22, 491 (1978).
[10] E. A. Hinds et al., Nucl. Instrum and Meth., 189, 599 (1981).
[11] P.G. Sona, Energia Nucl., 14, 295 (1967).
[12] M. Kinsho, K. Ikegami, A. Takagi and Y. Mori, Rev. Sci. Instrum., 67, 1362 (1996).

The RCNP Ion Source

K. Hatanaka, K. Takahisa and H. Tamura

Research Center for Nuclear Physics, Osaka University,
10-1 Mihogaoka, Ibaraki, Osaka 567, Japan

Abstract. The RCNP polarized ion source employs cold ($\sim$ 30 K) atomic beam technology and an electron cyclotron resonance ionizer. The source has been intensively operational since fall in 1994 to provide polarized protons to experimental researches at 200 - 400 MeV. Developments have been continued to increase the long term stability of the source and to improve the beam polarization. The source works for two months before cleaning the dissociator. Proton beam intensity from the source is 50 - 100 μ.A and the polarization is 70 % or better after acceleration with the K400 RCNP ring cyclotron.

INTRODUCTION

The Research Center for Nuclear Physics (RCNP), Osaka University, was established in 1971 as the national research center of nuclear physics in Japan. The first polarized ion source was constructed in 1975 [1] and extensive researches have been performed since then with polarized protons and deuterons accelerated by the RCNP K140 AVF cyclotron. Recently the RCNP facility was upgraded by bringing the new ring cyclotron with K = 400 MeV into operation [2]. The existing AVF cyclotron is used as the injector. Frequencies of accelerating voltage at the ring cyclotron are 3 or 5 times of those at the AVF cyclotron depending on the accelerated particle and energy. A flat topping system [3] is introduced to accelerate beams with small energy spreads in order to perform precise and ultra high resolution studies at the intermediate energy region. For this purpose, it is inevitable to inject high quality beams from the injector cyclotron to the ring cyclotron, especially with narrower pulse width than 1 ns for 65 MeV protons which are accelerated to 400 MeV by the ring cyclotron. In general, AVF cyclotrons have a large phase acceptance, and the beam burst width is usually observed as wide as 3 - 5 ns with the RCNP AVF cyclotron. A narrow beam pulse width such as 1 ns can be attained by reducing the RF phase acceptance by slits in the central region of the AVF cyclotron, but it largely sacrifices the beam intensity. The energy spread of the extracted beam from the ring cyclotron becomes about twice of the injected beam energy spread due to a phase compression process which results from the radial distribution of the accelerating RF voltage. This effect requires the energy spreads of the injected beams as small as possible, and we have to cut beams with unacceptable energy spreads by energy defining slits in spite of reducing beam intensity. The aceptance of the ring cyclotron in the transverse phase space is estimated around 3π mm·mrad and is a half of the measured nominal beam emittance from the AVF cyclotron [4]. It is necessary again to reduce the beam spreads in the transverse phase space as well as in the longitudinal phase space to match the injected beam to the acceptance of the ring cyclotron. Consequently, the intensity of the 400 MeV polarized proton beam was restriced to around 10 nA on target with the previous polarized ion source.

A large fraction of the experimental program at the RCNP is concentrated on studies of spin degrees of freedom. More than 60 % of the approved beam time is performed with polarized protons. In order to enhance the opportunities in spin physics research at intermediate energies by the ring cyclotron, the construction of a new high intensity polarized ion source was started in 1993. The new source built at the RCNP is schematically illustrated in Fig. 1. Its design is based on sources in operation at PSI [5], TUNL [6], IUCF [7] and RIKEN [8], which employ the cold ($\sim$ 30K) atomic beam technology and an electron cyclotron resonance (ECR) ionizer. The source was completed in 1994 and has been intensively operated since then to produce polarized protons to perform researches at 200 - 400 MeV.

In this paper, features and performances of the source, developments and results will be described.

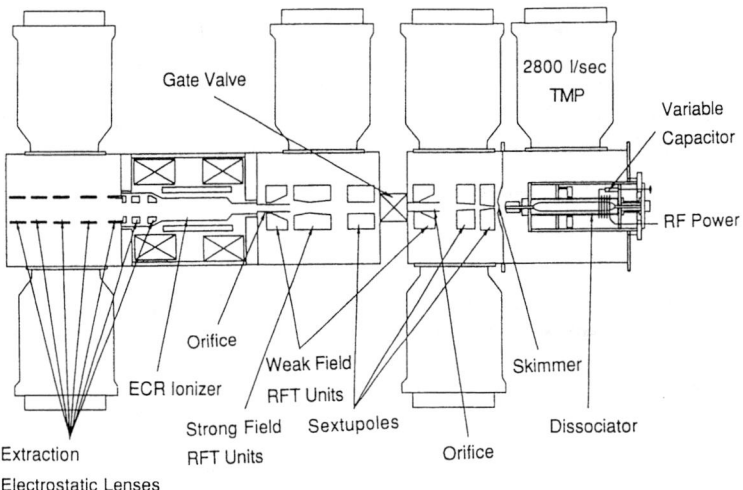

Fig. 1: General layout of RCNP High Intensity Polarized Ion Source.

FEATURES OF THE SOURCE

Dissociator

There are two gas feeding lines for the dissociator, one is for H_2 or D_2 gas and the other for N_2 gas. The gas flow rates are regulated by piezoelectric valves allowing accurate adjustments of the throughputs. The H- or D- atomic beam is produced by 13.6 MHz, 80 $\sim$ 300 W discharge contained in a pyrex/quartz tube of 20 mm inner diameter. The discharge is cooled by water flowing between the dissociator tube and a second, surrounding pyrex tube of a larger inner diameter, 30 mm. The cooling water is supplied with a closed system and the water temperature is controlled around 15°C with an accuracy of 1°C. The RF power is coupled inductively to the discharge with a coil consisting of eighteen turn windings of 2 mm diameter copper wire around the tube. A ceramic vacuum capacitor variable from 3 to 45 pF is connected in series to a terminal electrode of the coil to adjust the coupling condition. A grounded electrode is installed at the end of the dissociator in order to prevent the RF power from transmitting downstream and heating up elements there, especially the cold nozzle. The plasma is observed to be terminated around the grounded electrode. The plasma condition is always monitored on an oscilloscope with a signal picked up by a two turn coil. Atoms emerging from the discharge pass through a Macor section at the end of the tube and flow into an aluminum nozzle. The Macor serves to isolate the nozzle thermally at $\sim$ 30 K from the high temperature upstream. The nozzle is coverd with a copper block of a large volume, 60 cc, working as a heat bath. The outer shape of the nozzle is square to get a good heat conduction with the block, and thin indium foils are inserted between the nozzle and the copper block. The copper block is cooled by conduction through braided cables of oxygen free copper connecting to the cold head of a two stage closed-cycle helium refrigerator with a cooling capacity of 9 W at 20 K. The nozzle temperature is directly measured with a gold-chrome steel thermocouple and is stabilized with accuracy of better than 1 K utilizing a sheath heater. The N_2 gas is fed through 3 mm diameter holes on both a Macor and a dissociator tube. Cooled atoms emerge as a directed jet from a 3 mm diameter nozzle orifice into the first vacuum chamber. An atomic beam is formed at the entrance to the second vacuum chamber when the beam passes through a 4 mm diameter skimmer aperture placed 25 mm from the end of the nozzle which separates the first and the second chamber.

Sextupole magnets

Atomic beam enters the first sextupole magnet 25 mm beyond the skimmer orifice between the first and the second chamber. For the sextupole magnets, permanent magnets were chosen besed on the discussion in ref. [9]. The strong magnetic field is required to improve the atomic beam transportation through sextupole magnets, and a larger pole tip field can be obtained with permanent sextupole magnets than with electromagnets. Wise described the comparison between the loss in beam transmission due to the reduced field strength and the improvement in beam attenuation due to the intra-beam scattering in the magnet bore, when they used electromagnet sextupoles with six radial gaps along which pumping can occur. They estimated the formar loss is larger than the latter gain and prefered permanent magnets to electromagnets [9]. The magnets are based on designs proposed by Halbach [10] and use Nd-B-Fe sintered material. Each magnet is made from 24 wedged-shaped segments. To improve pumping of the volume inside the magnet bore, the magnets were constructed in several short sections as shown in fig. 1. Two short magnets are used for the first separation of the atomic substates, and they are separated by 30 mm. A long sextupole magnet is installed after the first weak-field RF transition unit. The dimensions of the three individual magnets are described in Table 1.

TABLE 1. Permanent sextupole magnets.

	No.1	No.2	No.3
Entrance bore radius (mm)	8	11	15
Exit bore radius (mm)	11	15	15
Outer radius(mm)	47	47	47
Length (mm)	50	70	100
Segments	24	24	24
Type	Tapered	Tapered	Straight

The magnetic field in the bore of the magnet can be calculated from the radius of the bore, r_i, the outer radius of the magnet, r_o, and the remnant field of the magnet material B_0, using formulae derived by Halbach (ref. [10]). For a setupole magnet consisting of M wedge-shaped segments one finds:

$$B(r) = \frac{3}{2} B_0 (\frac{r}{r_i})^2 (1 - (\frac{r_i}{r_o})^2) \cos^3(\frac{\pi}{M}) \times \frac{\sin(\frac{3\pi}{M})}{\frac{3\pi}{M}} \quad (1)$$

The magnets were enclosed in welded stainless steel (SUS316L) cans of wall thickness 1 mm to avoid the rapid deterioration of the magnet material which results from the exposure of the material to atomic hydrogens. The magnet material is NEOMAX-35H with the remnant field $B_0 = 1.245$ T and was supplied by Sumitomo Special Metals Ltd. [11]. Measurements were made of the magnetic field using a calibrated Hall probe. The agreement between measurements and calculation from eq. (1) is satisfacory for all the magnets.

Sizes and positions of sextupole magnets were optimized from calculations using a Monte-Carlo code for atomic beam transport which was written by Roberts at the University of Wisconsin [12]. In the present analysis, the radius of the effective ionizing region was taken to be 10 mm. The separation between the second and third sextupole was determined to 272 mm [13].

The ECR ionizer

The ECR ionizer was designed mainly following Schmelzbach [5]. The entrance of the ionizer is 375 mm downstream from the exit of the last sextupole magnet. The inner diameter of solenoid coils is 180 mm. The current in the entrance and exit solenoid coils can be adjusted independently in order to optimize the shape of the axial mirror field. At the entrance of the ionizer, the solenoid coil has a 50 mm thick iron yoke with a hole of 36 mm in diameter to shield the RF transition units against the stray field from the ionizer. At the exit, an iron yoke of the same thickness but a larger hole, 160 mm in diameter, is equiped in order to increase the magnetic field arround the second coil. Since there are no return yokes between the first and the second coil (Fig. 1), the mirror ratio is relatively small, 2.1 by the first coil and 1.75 by the second coil. The sextupole magnet consists of six NEOMAX-35H bars which are 20 mm wide, 15 mm high and 210 mm long. They are magnetized in the width direction, and are assembled in a sextupole magnet with a 110 mm tip-to-tip diameter. The measured magnetic field is well reproduced by the code Poisson. A 3 mm thick quartz vessel is used as a discharge container. The microwave frequency is 2.45 GHz and the maximum power is 200 W, which is supplied by a magnetron amplifier. The microwave power is fed radially in between the poles of the sextupole magnet. All the microwave components are in the atmosphere and no special window is required. In usual operation, the required microwave power is less than 20 W and the cooling by forced air is enough. The solenoidal coils have the ability to produce B_{min} of 230 mT for the axial mirror field. We have possibilities to increase the microwave frequency higher than 6 GHz. Four electrode accel-decel extraction system is applied to extract the high intensity beam. The plasma potential is defined by the first electrode with an opening of 50 mm and is 15 kV for protons accelerated to 65 MeV by the AVF cyclotron. The plasma potential is determined according to the accelerated energy by the AVF cyclotron with the constant orbit condition. A supporting N_2 gas is put directly into the plasma region through a small pyrex tube, which is important to keep the ECR plasma with the smallest amount of the supporting gas. The lower the pressure is in the plasma region, the smaller background and the higher beam polarization is expected.

DEVELOPMENTS OF THE SOURCE

The source was assembled in February 1994 and connected to the AVF cyclotron in the beginning of September after half a year performance tests off line. The source has been intensively operated since then to provide polarized protons for experimental researches at 200 - 400 MeV. Although the available time was limited, some performance tests and developments have been continued to realize the stable operation of the source.

Dissociator

The material of the nozzle was changed from copper in the original design to aluminum. The aluminum nozzle produces the atomic beam of the same intensity as the copper nozzle. Usually the dissociator works for two months by reducing the N_2 gas flow rate. Even if the nozzle is occluded with frozen ammonia, the dissociator is easily recovered by a heat cycle. The life time of the dissociator seems to be determined by a subtle characteristic of a pyrex tube which depends on the production lot [15]. We also tested quartz tubes in place of pyrex tubes, and found the more stable operation was was obtained with quartz tubes than with pyrex tubes.

Atomic beam intensty was measured with a compression tube which was installed in place of the ionizer. The size of the compression tube is the same as that of the ionizer vessel. Figure 2 shows the pressure difference measured with a hot cathode ion gauge as a function of the microwave power fed to the dissociator.

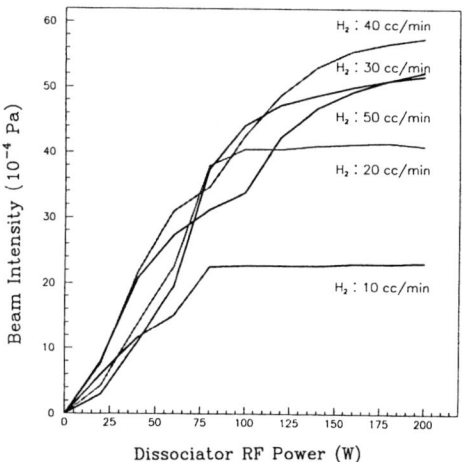

Fig. 2: Measured atomic beam intensities at the ionizer position.

Since the compression tube has not been calibrated, those data are only relative. In the measurements, N_2 gas was fed at the rate of 0.02 std-cc/min near the nozzle through the Macor section. In the usual oprational condition where the microwave power is 100 W, the optimum gas flow rate was 30 std-cc/min. For the gas flow rate of 40 std-cc/min, higher atomic beam intensity was achieved with higher microwave power, but the shorter life time is anticipated for the dissociator tube. When the gas flow rate was increased upto 50 std-cc/min, atomic beam intensities decreased. This fact may come from the effect of intrabeam scatterings in the beam path.

Atomic beam section

When the source began to deliver polarized protons for experiments in November 1994, the beam polarization was much lower than expected. The beam line polarimeter measurements [16] indicated an average polarization of p = 0.55 or less. Extensive investigations have been performed to optimize source parameters in order to improve the polarization. It was found there were two kinds of sources of unpolarized protons; one was due to residual hydrogen molecules and the other recombined molecules. The former is now reduced by an improved pumping of the plasma region. The latter was due to atoms from the dissociator which were not ionized in the ECR ionizer and recombined on the surface of the quartz vessel containing the plasma or on electrodes. In our source, since we use permanent magnet sextupoles to focus the atomic beam, there are no easily adjustable parameters for the atomic beam transport system. Diameters of the nozzle, the skimmer aperture, and the spacing between them were searched to improve proton polarization without sacrificing beam intensities. The optimized diameter of the nozzle orifice was 3 mm which was the same as the original design. The skimmer aperture diameter was reduced from 6 mm to 4 mm. Additional orifices were installed at locations of two weak field RF transition units to reduce the undesired flow of molecules and atoms to the ionizer. Each orifice has a conductance of 6 ℓ/sec for air at room temperature. One of two turbomolecular pumps originally installed at the dissociator chamber was moved to the extraction chamber downstream the ECR ionizer to improve the vacuum at the plasma region. These modifications improved the average polarization to 70 % or better.

The ECR ionizer

The correlation between atomic beam intensities and poron intensities were measured [13]. Atomic beam intensty was measured with a compression tube which was installed in place of the strong field RF transition units after the last sextupole magnet (Fig. 1). Proton beam intensity was also measured after a 90-degree analyzing magnet. The beam intensities for 30 std-cc/min are three times larger than those for 15 std-cc/min. It was observed that beam intensities were neary same when the hydrogen gas flow rate was larger than 25 std-cc/min. It can be seen that atomic beam intensities measured with the compression tube show a saturation effect as a function of the microwave power compared with proton currents. At higher microwave power than 160 W, proton currents begin to saturate.

Figure 3 shows the beam emittance measured at the vertical injection line to the AVF cyclotron. 100 % of the beam were found within an emittance of 300 mm mrad at 15 keV.

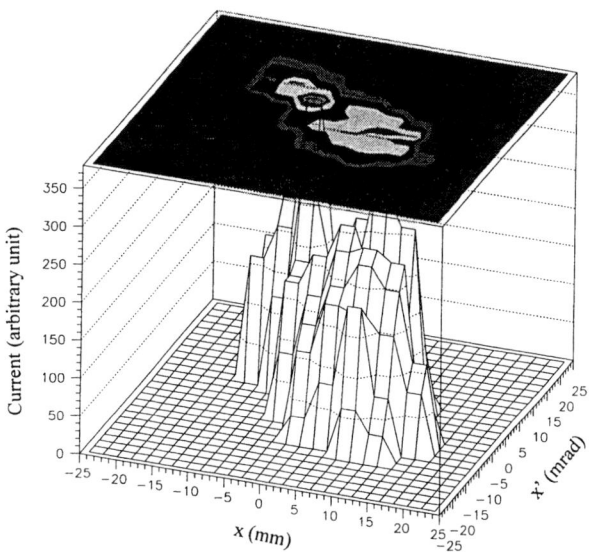

Fig.3 Beam emittance measured at the vertical injection line to the AVF cyclotron.

Even if the Stern-Gerlach state separation and the RF transitions are perfect, the proton polarization is not always 100 %. When the hydrogen atoms are ionized at the magnetic field B mT, the maximum achievable polarization for the beam with substates 1+4 or 2+3 is given by [14]

$$P = \frac{1}{2}(1 + \delta) \qquad (2)$$

where

$$\delta = \frac{x}{\sqrt{1+x^2}}, \text{ and } x = \frac{B}{50.7}.$$

The ECR condition at the frequency 2.45 GHz requires the magnetic field of 87.5 mT and this field strength predicts the maximum beam polarization of 93 % from eq. (2). In this model, proton polarization of 7 % is predicted even if all the RF transitions are turned off. AT the RCNP source, the measured proton polarization was about 5 % without any RF transitions. For this reason, several laboratories have selected higher frequency microwave sources which operate between 3 and 4 GHz [17]. Since our solenoidal coils have the ability to produce B_{min} of 230 mT for the axial mirror field, operational tests were performed at 3.5 GHz. An appreciable increase in the proton polarization was not observed, although the beam current was 1.3 to 2.0 times larger than at 2.45 GHz. Most probable reason is the larger radial extent of the ECR plasma at 3.5 GHz than at 2.45 GHz, because the sextupole field was not increased. Background gas in the ionizer, coming from unpolarized hydrogen atoms lost from the beam and recombined into molecules, can thus more easily re-enter the plasma, become ionized and extracted, and dilute the emerging beam polarization. For the present design of the ECR ionizer, the resonance surface is originally extended rather in the radial direction than in the longitudinal direction [5]. The higher beam intensity at 3.5 GHz shows that the denser ECR plasma is attained when the microwave frequency is increased [18]. In usual operation, the full proton intensity from the source is 50 - 100 μA after the analyzing magnet.

SUMMARY

Features and performances of the newly constructed polarized ion source at the RCNP was described. It employs cold ($\sim$ 30 K) atomic beam technology and an electron cyclotron resonance ionizer. After the commissioning of the source, many developments were performed to increase the long term stability and the beam polarization. With an alminum nozzle, the source works for two months before cleaning the dissociator. Beam polarization was improved by optimizing the geometry of the atomic beam section to preventing unwanted molecules or atoms from entering the ECR ionizer. Higher frequency for the ECR microwave did not improve the beam polarization appreciably, although the beam intensity was increased. It may be necessary to increase the sextupole field strength at the ionizer to get a radially smaller extent of the ECR plasma. The beam emittance was measured at the vertical injection line to the AVF cyclotron, and 100 % of the beam were found within an emittance of 300 mm mrad at 15 keV.

During the past three years the surce has been used to perform experimental researches with 200 - 400 MeV protons from the ring cyclotron. In usual operation, the proton current is 50 - 100 μA from the source and the beam polarization is 70 % or better after acceleration.

ACKNOWLEDGEMENTS

The authors like to thank P. Schmelzbach, T. Clegg, M. Wedekind (IUCF), H. Okamura, Y. Mori (KEK) and their colleagues for providing informations on their sources, W. Haeberli and A. Roberts for making available their computer code to calculate the atomic beam transport. We would like to acknowledge the support by the RCNP members, especially the continuous encouragement of the Director, H. Ejiri throughout the work.

REFERENCES

1. K. Hatanaka, N. Matsuoka, H. Sakai, T. Saito, H. Tamura, K. Hosono, M. Kondo, K. Imai, H. Shimizu and K. Nisimura, Nucl. Instr. and Meth. 217 (1983) 397.
2. I. Miura et al., Proc. 13th Int. Conf. on Cyclotrons and their Applications, Vancouver, 1992, eds. G. Dutto and M.K. Craddock (World Scientific, 1993) p.3.
3. T. Saito et al., Proc. 14th Int. Conf. on Cyclotrons and their Applications, Cape Town, 1995, ed. J.C. Cornell (World Scientific, 1996) p.169.
4. I. Miura et al., RCNP Annual Report, 1976, p.47.
5. P.A. Schmelzbach, Proc. Workshop on Polarized Ion Sources and Polarized Gas Targets, Madison, 1993, eds. L.W. Anderson and W. Haeberli, AIP Conf. Proc. 293 (1994) p.65.
6. T.B. Clegg, in: High Energy Spin Physics, ed. K.J. Heller, AIP Conf. Proc. 187 (1989) p.1227.
7. V. Derenchuk et al., Proc. Workshop on Polarized Ion Sources and Polarized Gas Targets, Madison, 1993, eds. L.W. Anderson and W. Haeberli, AIP Conf. Proc. 293 (1994) p.72.
8. H. Okamura et al., Proc. Workshop on Polarized Ion Sources and Polarized Gas Targets, Madison, 1993, eds. L.W. Anderson and W. Haeberli, AIP Conf. Proc. 293 (1994) p.84.
9. T. Wise, A.D. Roberts and W. Haeberli, Nucl. Instr. and Meth. A336 (1993) 410.
10. K. Halbach, Nucl. Instr. and Meth. 169 (1980) 1.
11. Sumitomo Special Metals Ltd., 4-7-9 Kitahama, Chuo-ku, Osaka, Japan.
12. A.D. Roberts, private communication.
13. K. Hatanaka, K. Takahisa, H. Tamura, M. Sato, I. Miura, Nucl. Instr. and Meth. A384 (1997) 575.
14. R. Beurtey, Proc. 2nd Int. Symp. on Polarization Phenomena of Nucleons, Karlsruhe, 1965, eds. P. Huber and H. Schopper (Birkhäuser, Basel, 1966) p.33.
15. P.A. Schmelzbach, private communication.
16. H. Sakai, H. Okamura, H. Otsu, T. Wakasa, S. Ishida, N. Sakamoto, T. Uesaka, Y. Satou, S. Fujita and K. Hatanaka, Nucl. Instr. and Meth. A369 (1996) 120.
17. T.B. Clegg, Proc. Int. Workshop on Polarized Beams and Polarized Gas Targets, Cologne, 1995, eds. H.P. Schieck and L. Sydow (World Scientific, 1996) p.155.
18. R. Geller, Proc. Int. Conf. on the physics of multiply charged ions and Int. workshop on E.C.R. ion sources, Grenoble, 1988, ed. S. Bliman, Suppl. Journal de Physique C1 (1989) p.C1-887.

Polarized ions from a storage cell

A. S. Belov, S. K. Esin, L. P. Netchaeva, V. S. Klenov,
A. V. Turbabin, and G. A. Vasil'ev

*Institute for Nuclear Research of Russian Academy of Sciences,
Moscow, 117312, Russia*

Abstract. A test of a storage cell in the ionizer of the atomic beam-type polarized ion source is described. The cylindrical aluminum storage cell with uncoated walls and with internal diameter 1.5 cm had been placed in the charge-exchange region of the plasma ionizer. A pulsed polarized hydrogen atomic beam was injected into the storage cell from one side, and deuterium plasma was injected from the other side through an orifice with a diameter of 3 mm. The tests were made with a pulsed deuteron ion beam of 5 mA. Results of intensity and polarization measurements for the polarized proton beam extracted are presented. The intensity gain obtained with the storage cell in comparison with operation without the storage cell was measured to be 9.0. The polarization of the proton beam from the storage cell reaches 0.94±0.03. Possible schemes for production of polarized negative hydrogen ions and polarized ^{3}He^{++} ions with a storage cell are discussed.

INTRODUCTION

Polarized negative hydrogen ions beams produced by pulsed atomic beam-type polarized ion sources have reached intensities up to 1 mA (1). Nevertheless, there is still a need to increase the intensity further, mainly in connection with projects whose aim is the acceleration of polarized beams in the high-energy accelerators such as HERA and RHIC.

In conventional atomic beam-type polarized ion sources polarized ions are produced by ionization of polarized atoms with thermal energy in a free atomic beam. The current of polarized ions extracted from an ionizer is proportional to the density of polarized atoms in the atomic beam. The density of polarized hydrogen atoms in free atomic beams achieved so far does not exceed 10^{12} at/cm^3. Scattering of atoms with thermal energy during the atomic beam formation introduces a severe limitation for a further increase of the density of polarized atoms.

However, there are several ways to increase the atomic density in an ionization region. One possibility is connected with use of high-field sextupole separating magnets with superconducting coils. Calculations (2,3) show that a gain of 2.7 in atomic beam density can be obtained by using sextupole magnets with magnetic pole tip field of 4.5 T in comparison with conventional electromagnet sextupoles.

Another possibility is to extend the storage of polarized hydrogen atoms in an ionization region of polarized ion sources (2-4). This idea is not new; W.Haeberli proposed to use a storage cell for fast polarized hydrogen atom production (5).

Considerable progress has been made in the storage cell technique during the past few years in connection with the development of gaseous polarized targets for storage rings. A target thickness of hydrogen atoms of $0.75 \cdot 10^{14}$ at/cm^2 in dc mode of operation has been achieved (6) in the storage cell target developed for HERMES experiment. This is five times more than the thickness of the target formed by the **pulsed** free atomic hydrogen beam in the resonant charge-exchange plasma ionizer of the polarized ion source at the INR, Moscow (7).

Thus, we can suppose that use of a storage cell in the resonant charge exchange plasma ionizer could lead to increase about one order of magnitude in the intensity of polarized beams produced by the INR polarized source.

However, the first tests of the storage cell made at INR, Moscow with a pyrex glass cell showed that losses of polarized ions in the storage cell with dielectric walls and with an intense plasma inside the cell become large. The efficiency of polarized ions extraction is decreased, probably due to inhomogeneous charging of the dielectric walls by plasma particles which induces instabilities in the plasma. The dielectric walls of the storage cell (Teflon or dry film) also can be damaged by the intense plasma flux.

In this respect an important result had been obtained by W. Haeberli and J.S.Price (8). They found that there is only small depolarization of deuterium atoms in a strong magnetic field in a storage cell made from aluminum alloy and with uncoated walls.

Only direct experimental tests of using such a storage cell in the ionizer of polarized ion source can provide answers to complex questions such as the efficiency of radial confinement of the polarized ions produced, the efficiency of extraction of the polarized ions, the depolarization rate of hydrogen atoms, possible changes of properties of the storage cell walls and so on.

In this paper we present results of the tests of an aluminum storage cell with uncoated walls placed into the resonant charge-exchange plasma ionizer. Polarized protons are produced in this ionizer by using the resonant charge-exchange reaction between polarized hydrogen atoms and unpolarized deuterium ions in the deuterium plasma.

We discuss also the use of a storage cell in the resonant charge-exchange ionizer for production of polarized negative hydrogen ions and polarized ^{3}He^{++} ions.

TEST OF THE STORAGE CELL IN THE PLASMA IONIZER

1. Apparatus

A schematic layout of the apparatus used is shown in Fig.1. The atomic beam-type polarized ion source with the resonant charge-exchange plasma ionizer at INR, Moscow has been used for the test. The source as well as the low energy polarimeter is described in detail in ref. (9).

The uncoated aluminum storage cell was placed inside and on the axis of the ionizer solenoid. The cell had internal diameter 1.5 cm and was 25 cm in length. During the tests described the cell temperature was 300 K. Polarized hydrogen atoms produced by the atomic beam apparatus of the polarized source were injected into the storage cell along the axis of the cell through its one open end. The opposite end of the cell was closed by a diaphragm with a central hole 3 mm in diameter. With this geometry most of the atomic beam is stopped inside the cell near the diaphragm. So it was important to make the diameter of the orifice in the diaphragm small enough to restrict conductance of the cell. The deuterium plasma was injected into the storage cell through this orifice. Polarized protons are formed in the storage cell due to resonant charge-exchange between the polarized hydrogen atoms and the unpolarized deuterium plasma ions and are extracted from the cell in the direction opposite to the atomic beam. The polarized protons were separated from the D^+ ions also extracted by use of a bending magnet.

In order to compare operation of the ionizer with the storage cell relative to its operation with a free atomic beam, the storage cell was replaced by a simple diaphragm with central orifice 3 mm in diameter (the same as in the storage cell).

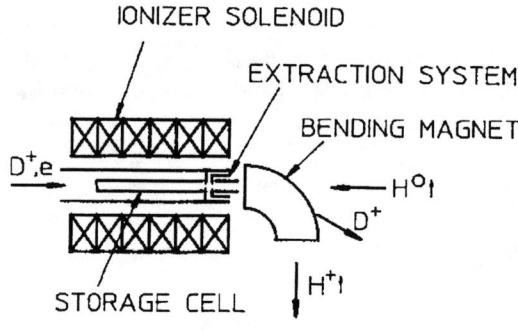

FIGURE 1. Schematic diagram of the plasma ionizer with the storage cell.

This diaphragm was installed in the position of the respective orifice of the storage cell.

The dissociator and the plasma source worked in a pulsed mode of operation. This allowed us to study the dynamics of polarization and intensity of polarized protons extracted from the storage cell. A pulse of the dissociator rf generator and a pulse of the plasma injector together with a typical pulse of the atomic hydrogen beam recorded in the ionizer region are shown schematically in Fig. 2.

Duration of the rf dissociator pulse was 3 ms. Duration of the plasma pulse was 200 μs. A time delay between the beginning of the rf dissociator pulse and the beginning of the atomic beam pulse in the ionizer was measured to be 0.8 ms. This delay is explained by a time-of-flight of hydrogen atoms with most probable velocity $2 \cdot 10^5$ cm/s along the distance (140 cm) between the dissociator and the ionizer plus time for formation of the atomic beam in the dissociator (about 0.1 ms) after the start of the rf discharge.

A typical unpolarized peak D^+ ion current recorded after the bending magnet in these tests was about 5 mA. We believe that it is possible to transport much larger plasma fluxes through the storage cell even with a small input orifice diameter. Probably the best way to do this is to place the plasma source and the storage cell in the same longitudinal magnetic field. This is not so in our case; our plasma source is placed outside the ionizer solenoid. Hence there is a magnetic mirror formed by a fringing field of the solenoid which should reflect plasma ions, reducing efficiency of the plasma injection into the solenoid and the storage cell. It is worth noting that with operation of the polarized source with free atomic beam it was possible to inject into the ionization region plasma fluxes with ion currents of the order of an ampere (9).

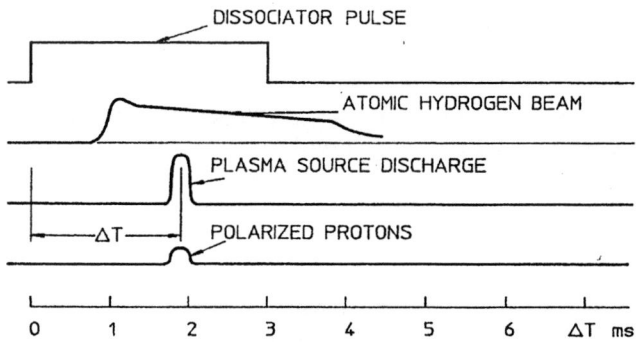

FIGURE 2. Schematic diagram showing the pulse of the dissociator rf generator, the atomic hydrogen beam, the plasma source discharge pulse and polarized proton current. Changing the time delay between the dissociator pulse and the plasma source discharge pulse (ΔT) allowed us to study the dynamics of storage of polarized hydrogen atoms.

2. Intensity Measurements

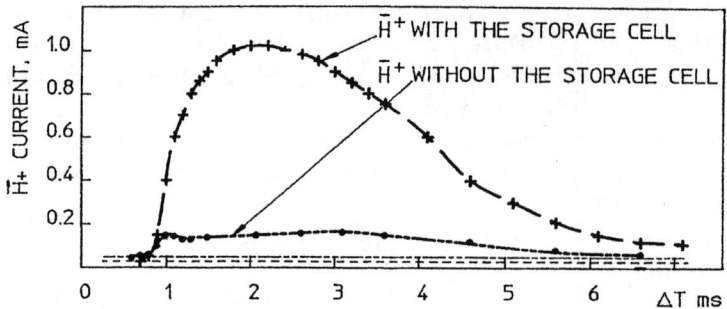

FIGURE 3. Dynamics of polarized H^+ ions current from the ionizer with the storage cell in comparison with the H^+ current obtained without the storage cell. The background ion currents are also shown (dashed line: with the storage cell, the dashed - dotted line: without the storage cell).

The dynamics of the polarized H^+ ion current extracted from the ionizer with the storage cell and recorded after the bending magnet in comparison with the polarized H^+ current obtained without the storage cell is shown in Fig.3.

First of all, we see that the polarized proton current extracted with the storage cell is significantly larger than the current without the storage cell.

For quantitative comparison we have to subtract the background currents (the currents obtained with the rf dissociator off). Also we note that even without the storage cell in our geometry we got some compression effect; the H^+ current is increasing during the period of time between 1.2 ms and 3.1 ms while the atomic beam intensity during the same time is decreasing. This effect is due to the diaphragm installed in the ionization region instead of the storage cell. In this way we got an almost closed volume in the ionization region with internal diameter 6.6 cm and 25 cm in length and with walls made from stainless steel which is filled by the atomic hydrogen beam in about 1 ms time to an atom density comparable with the atomic beam one. Unfortunately this influences the polarization measurements obtained without the storage cell. For comparison of the intensity we take the $H^+\uparrow$ current without the storage cell at beginning of the pulse, which is equal to 110 µA (with the background current subtracted). The maximum $H^+\uparrow$ current obtained with the storage cell is 995 µA. Thus, the intensity gain due to the cell is $K_i = 9$.

The expected increase of the polarized hydrogen target thickness due to the storage cell (or, in other words, the ratio of the target thickness with a storage cell to the target thickness with a free atomic beam) we can estimate for the static case from the formula:

$$K(\infty) = \frac{Sv}{2C_{tot}}, \qquad (1)$$

where S is the cross-area of the atomic beam coming into the storage cell, v is the average velocity of atoms in the beam, and C_{tot} is the total conductance of the storage cell from the point where the atoms are stopped.

For the storage cell used: $C_{tot} \approx 12.9 \; 10^3 \; cm^3/s$, the atomic beam diameter is 1.5 cm, $v = 2 \; 10^5 \, cm/s$. With these parameters the formula (1) yields $K(\infty) = 13.7$.

In comparing the coefficients K_i and $K(\infty)$ we have to take into account the possible difference in the efficiency of the extraction of polarized protons from the ionization region with and without the storage cell.

We can compare the extraction efficiencies from the initial increase of the proton current with the storage cell. We get (assuming that there are small losses of atoms in the storage cell for a small time interval after the beginning of the pulse):

$$K(t) = \frac{t}{L/v}, \quad (t \ll \tau) \qquad (2)$$

where τ is the characteristic time constant for the storage cell, and L/v is the flight time of hydrogen atoms with the velocity $2 \; 10^5 \, cm/s$ along the cell, which is 125 µs in our case.

The formula (2) predicts 2.25 times lower initial rate of increase for the ratio of the proton current with the storage cell to the proton current without the cell than we have with the data shown in Fig. 3. This means that the extraction efficency with the storage cell is higher than without the storage cell. Thus, the total gain in the proton current is presumably due to both factors: the increase of the extraction efficency (factor 2.25) and the increase of the atomic hydrogen target thickness (factor 4) with the storage cell.

3. Results of Polarization Measurements

The results of the polarization measurements are shown in Fig. 4. The proton polarization without the storage cell was strongly affected by dilution of the polarized current by unpolarized background protons. The background current was measured with the rf dissociator off. It was equal to 40 µA without the storage cell in these measurements. Taking into account the dilution, we calculated the polarization of protons arising from the atomic beam (dividing the measured values by the dilution factor). These corrected data also are shown in Fig.4.

The upper theoretical limit of the nuclear polarization of hydrogen atoms in the

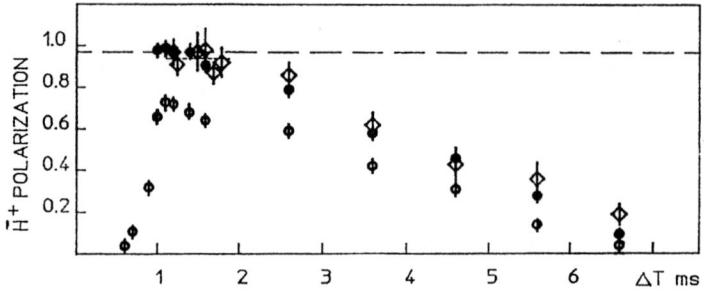

FIGURE 4. The proton polarization versus time. Open circles: without the storage cell; rhombus: with the storage cell; black circles: polarization divided the dilution factor without the storage cell.

atomic beam (0.96), is shown in Fig. 4 by the dashed line. The corrected data for the polarization without the storage cell coincide within the accuracy of the measurements with this theoretical polarization during the first millisecond after the beginning of the atomic beam pulse. The decrease of the polarization without the storage cell for later times is explained by the compression effect without the storage cell discussed in the previous paragraph of the paper.

With the storage cell the measured polarization is high during the time interval 1-2 ms. The averaging of the data for this interval gives a polarization value of 0.94 ± 0.03. Depolarization of atoms inside the storage cell leads then to a decrease of the polarization. The relaxation time observed is about 5 ms. The dilution of the polarized protons in this case is small (only 2% for the time interval 1- 2 ms) due to higher intensity of the polarized beam.

DISCUSSION OF POSSIBLE APPLICATION FOR POLARIZED SOURCES

1.Polarized hydrogen and deuterium ions

The tests described in this paper are encouraging for application to pulsed polarized hydrogen and deuterium ion sources with the resonant charge exchange ionizer because in this case all main parameters of the polarized beams produced can be improved including intensity, polarization (due to elimination of the dilution problem) and emittance (due to a smaller initial ion beam radius).

However, it is necessary to carry out a study of the storage cell with high intensity plasma fluxes passing through the storage cell for more definite forecast because many phenomena involved are very complicated, and it is difficult to make an estimation of the source performance without an experimental test with the plasma fluxes increased one order or two orders of magnitude in comparison with the test described here.

For the transportation of intense plasma fluxes through the storage cell with an internal diameter about 10 mm it will be necessary to place a plasma injector in the same longitudinal magnetic field in which the storage cell resides. This is not a problem for plasma injectors producing plasma with positive deuterium or hydrogen ions because there are many different plasma sources operating with a longitudinal magnetic field of about 0.1 T. But this is certainly a more difficult task for a plasma injector producing plasma enriched by negative deuterium ions which are necessary for the direct production of polarized H^- ions.

If the polarized positive hydrogen ion sources with the storage cell will provide a significantly larger intensity in comparison with existing polarized sources, then the conventional method of double charge-exchange of polarized positive ions in an alkali vapor can be used for production of high-intensity polarized H^- and D^- ion beams.

Important recent achievements in the transport of very low energy ion beams (10) could help to solve the space charge problem arising in conventional double charge exchange which uses very low energy proton beams (500-1000 eV) to get the highest efficiency (up to 50% in a Mg vapor target).

A combination of the resonant charge-exchange ionizer with the storage cell and a laser driven source of polarized hydrogen or deuterium atoms is also promising because the laser-driven source produces fluxes of polarized thermal hydrogen atoms up to 10^{18} at/s (11). The respective polarized gaseous target thickness can be ~10^{15} at/cm^2. Dilution of polarized atomic hydrogen by hydrogen molecules is not important in this scheme due to selectivity of the resonant charge-exchange reaction.

2. Polarized $^3He^{++}$ ions

Use of a storage cell for the polarized ^{3}He ion production would be natural due to the very large relaxation time of polarized helium atoms.

A scheme for a polarized $^3He^{++}$ ion source based on a resonant charge-exchange between polarized ^{3}He atoms and unpolarized $^4He^{++}$ ions in a storage cell has been proposed recently (12). The scheme is shown in Fig. 5.

Polarized thermal ^{3}He atoms in the proposed source could be produced by optical pumping. The polarized atoms are then transported into a storage cell in which the polarized gaseous target is formed. The target thickness with an uncooled storage cell can be about 10^{14} at / cm^2, and nuclear polarization can reach about 50% (13).

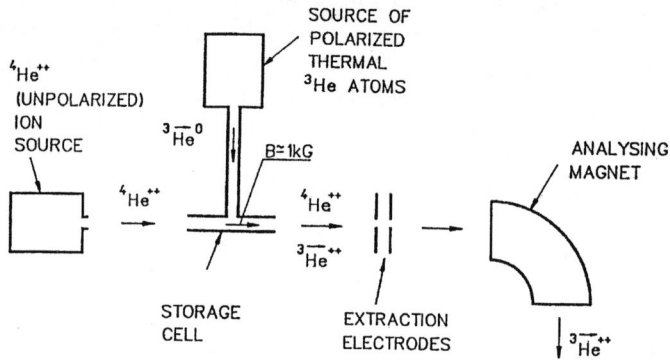

FIGURE 5. Schematic diagram of a ^{3}He^{++} ion source with a resonant charge - exchange ionizer.

If unpolarized ^{4}He^{++} ions are injected into the storage cell then polarized ^{3}He^{++} ions can be produced in the cell by the reaction

$$^3\text{He} + {}^4\text{He}^{++} \Rightarrow {}^3\text{He}^{++} + {}^4\text{He}$$

The cross-section for this reaction at the relative energy of the interacting particles of 50 eV is $\sigma_{ce} = 4 \cdot 10^{-16}$ cm^2 (13). Thus about 4% of unpolarized ^{4}He^{++} ions passing through the target (for the target thickness 10^{14} cm^{-2}) will undergo the double charge - transfer, and respective current of the polarized ^{3}He^{++} ions will be formed in the storage cell. For a pulse α source the storage cell could also operate in a pulsed mode which allows an increase in target thickness to the level of $2 \cdot 10^{15}$ cm^{-2}. In this case the current of polarized ^{3}He^{++} ions will be comparable with the unpolarized ^{4}He^{++} ions current.

A useful feature of this scheme for the direct production of the polarized ^{3}He^{++} ions is that the charge-exchange can be made in a relatively weak magnetic field (~ 1 kG) which is necessary only for a radial confinement of the polarized ^{3}He^{++} ions produced. This means that the emittance of the polarized beam will be relatively small (~ 0.5 π mm mrad).

CONCLUSIONS

The study of a storage cell with uncoated aluminum walls in the ionizer of a pulsed atomic beam type polarized ion source shows that essential improvements of intensity, polarization and emittance of pulsed polarized sources can be obtained with the storage cell. With a small plasma flux through the storage cell an intensity gain 9.0

results from the use of the storage cell. The polarization of the H$^+$ ions produced was measured to be 0.94 ± 0.03 for the time interval 1-2 ms after the beginning of the atomic hydrogen pulse. Essential depolarization of hydrogen atoms has been found in the storage cell with an estimated relaxation time of 5 ms. Perspectives for using the storage cell technique together with the resonant charge exchange ionizer for production of polarized H$^+$, H$^-$, D$^+$, D$^-$ and ^{3}He^{++} ions are encouraging.

ACKNOWLEDGMENTS

We would like to thank V.P. Yakushev and V.G. Dudnikov for useful discussions and A. Krisch for encouragement of this study. The work was supported in part by Russian Foundation for Basic Research and SPIN Collaboration.

REFERENCES

1. Belov A. S., et al., Rev. Sci. Instrum. **67**, 1293-1295 (1996).
2. Belov A. S., et al., " Study of a Polarized Hydrogen Ion Source with Deuterium Plasma Ionizer", in *Proceedings of the Eighth International Symposium on Polarization Phenomena in Nuclear Physics*, Bloomington, IN, USA, 1994, AIP Conf. Proc. **339**, 1995, pp. 643-649.
3. Belov A. S., "Colliding Beam Polarized Ion Sources", in *Proceedings of the International Workshop on Polarized Beams and Polarized Gas Targets*, H. P. gen. Schieck and L. Sydow eds., Cologne, Germany, 1995, pp. 208-217.
4. Clegg T. B., "Atomic beam sources - recent progress and future possibilities", in *Proceedings of the International Workshop on Polarized Beams and Polarized Gas Targets*, H. P. gen. Schieck and L. Sydow eds., Cologne, Germany, 1995, pp. 155-166.
5. Haeberli W., "Sources of Polarized Negative Ions ", in *Proceedings of the Second Intern. Symp. on Polarization Phenomena in Nucleons*, P. Huber and H. Schopper eds., Experientia Supplemement **12**, (Birkhauzer, Basel, Switzerland, 1966), p.64.
6. Stock F. et al., "The HERMES target source for polarized hydrogen and deuterium atoms", in *Proceedings of the International Workshop on Polarized Beams and Polarized Gas Targets*, H. P. gen. Schieck and L. Sydow eds., Cologne, Germany, 1995, pp. 260-264.
7. Belov A. S. et al., *Nucl. Instr. Meth.* A **333**, 256-259 (1993).
8. Price J. S. and Haeberli W., *Nucl. Instr. Meth.* A **326**, 416-423 (1993).
9. Belov A. S. et al., *Nucl. Instr. Meth.* A **255**, 442-459 (1987).
10. Dudnikov V. G., private communication.
11. Gao H. et al., "Design and development of a novel laser-driven polarized H/D source", in *Proceedings of the International Workshop on Polarized Beams and Polarized Gas Targets*, H. P. gen. Schieck and L. Sydow eds., Cologne, Germany, 1995, pp. 67-71.
12. Belov A. S., " A scheme for a polarized ^{3}He ion source with a resonant charge-exchange plasma ionizer", presented at the 7 th RCNP International Workshop on Polarized ^{3}He Beams and Gas Targets and their Application, Kobe, Japan, January 20-24, 1997.
13. Milner R. G., "Polarized ^{3}He internal gas targets", in *Proceedings of the International Workshop on Polarized Ion Sources and Polarized Gas Targets*, L.W.Anderson and W. Haeberli eds., Madison, WI, 1993, AIP Conf. Proc., **293**, pp. 235-238.
14. Schrey H. and Huber B. Z. Physik **A273**, 401-403 (1975).

OPPIS DEVELOPMENT FOR PRECISION EXPERIMENTS AND HIGH ENERGY COLLIDERS

A.N. Zelenski[1,2], V.I. Davydenko[1], G. Dutto[2], A.A. Hamian[3],
V. Klenov[1], C.D.P. Levy[2], I.I. Morozov[1], P.W. Schmor[2],
W.T.H. van Oers[3], G.W. Wight[2]

(1) Institute for Nuclear Research, Russian Academy of Sciences, 117312 Moscow, Russia
(2) TRIUMF, 4004 Wesbrook Mall, Vancouver, BC, Canada V6T 2A3
(3) Dept. of Physics, Univ. of Manitoba, Winnipeg, MB, Canada R3T 2N2

Abstract.
The TRIUMF OPPIS (Optically Pumped Polarized Ion Source) provides a precision quality beam for the experiment on parity non-conservation in proton-proton scattering at 221 MeV beam energy. The pulsed OPPIS applications for the future RHIC and HERA polarization facilities are discussed. The polarization technique of the radioactive nuclide beam is proposed for β-NMR condensed matter studies with a new ISAC facility at TRIUMF.

INTRODUCTION

Collider experiments with polarized proton beams, approved at RHIC [1] and under consideration at HERA [2], will provide fundamental tests of QCD and the electroweak interaction. Polarized beams should allow better identification of new objects produced in proton-proton collisions, and will expand the limits of searches for possible manifestations of New Physics beyond the Standard Model. Such experiments will require the maximum available luminosity, and therefore polarization must be obtained as an extra beam property without sacrificing intensity. Typical currents for unpolarized H^- ion injectors are in the 20-50 mA range. With the lower current polarized sources, the use of multiturn charge exchange injection into a booster ring will partially compensate the intensity loss, but only a 20-30 mA source will completely solve the problem. A 1.64 mA dc polarized H^- ion current was obtained at the TRIUMF OPPIS, with the promise of further increase to the 2-3 mA range [3]. The ECR-type primary proton source used at the TRIUMF OPPIS has a comparatively low emission current density and high beam divergence, which limits further current increase and gives rise to inefficient use of the cw laser power for optical pumping. In pulsed operation, suitable for high energy accelerators, the ECR source shortcomings have been avoided by using an INR-type OPPIS with a high brightness proton source situated outside the magnetic field [4]. Studies performed in collaboration with INR, Moscow and BINP, Novosibirsk have demonstrated the feasibility of producing 20 mA polarized H^- ion currents using this scheme. Proposals for pulsed OPPIS developments for future polarization faciles at RHIC and HERA are considered below.

The TRIUMF OPPIS provides a high quality beam for precision measurements of parity non-conservation in proton-proton scattering at 221 MeV [5,6]. The source operates very reliably and delivers beam for about 40% of the cyclotron operational time.

β-NMR studies of surfaces with polarized radioactive nuclide beams were proposed for the new ISAC (Isotope Separator and Accelerator) facilities at TRIUMF [7]. The application of the optical pumping polarization technique for radioactive nuclides is described for ^{8}Li and ^{17}Ne beams.

POLARIZED BEAM FOR PARITY NON-CONSERVATION STUDIES

At present the TRIUMF OPPIS is heavily used for a parity violation experiment (E497). The goal of this experiment is to measure the parity violating longitudinal analyzing power A_z in proton-proton scattering at 221 MeV to an accuracy of $\pm 0.2 \times 10^{-7}$. This imposes severe constraints on the polarized beam quality, and much of the OPPIS development at TRIUMF has been pushed by this demanding experiment. The initial expectation that spin-reversal-correlated modulations of the beam parameters should be smaller in the OPPIS than in the ABS has been demonstrated experimentally, although it took some time to understand the origins of such modulations, to develop the apparatus to measure them and to find the optimal set of source parameters which minimizes the modulations.

The optimum beam current required at the target for the parity experiment is only 0.20 μA, but to minimize the helicity correlated modulations, most of the beam intensity must be sacrificed for beam quality. Therefore, high brightness source performance is required, and the ongoing high current OPPIS development at TRIUMF is of benefit to the parity experiment.

At present, the polarized beam quality delivered by the OPPIS meets the stringent requirements for the parity experiment. The helicity correlated beam properties as measured by the parity apparatus, and the associated systematic error corrections, are presented in Table 1 for a data set taken in February-March 1997. This data set demonstrates that a statistical accuracy of $\delta A_z =$

TABLE 1. Helicity Correlated Beam Properties and Corrections (ΔA_z) for the February-March 1997 Data Set.

Item	Average Value	$10^7(\Delta A_z)$
P_x	(-0.02 ± 0.01) %	0.01 ± 0.09
P_y	(0.03 ± 0.01) %	0.02 ± 0.20
yP_x	$(1.2 \pm 0.5) \mu$ m	-0.001 ± 0.002
xP_y	$(-3.9 \pm 0.5) \mu$ m	0.00 ± 0.02
$\Delta\sigma$	$(0.05 \pm 0.04) \mu$ m	-0.07 ± 0.20
Δx	(4 ± 20) nm	-0.03 ± 0.15
Δy	(-3 ± 20) nm	0.23 ± 0.16
$\Delta I/I$	$(-2.5 \pm 0.1) \times 10^{-5}$	0.05 ± 0.05
Total Correction		0.2 ± 0.4

$\pm(0.2 \times 10^{-7})$ will be obtained in about 600 hours of data taking. Analysis of the February 1997 data gives the result $A_z = (1.1 \pm 0.6) \times 10^{-7}$, and demonstrates significant progress in tracking down systematic errors. The proposed extension of the parity experiment to 450 MeV has been approved at TRIUMF [8].

PROPOSAL FOR A POLARIZED 800 GEV PROTON BEAM AT THE HERA E-P COLLIDER

Studies of the hadron spin structure functions in collisions of polarized electrons with polarized helium-3 , hydrogen and deuterium internal targets are in progress at DESY (HERMES experiment). The proposal to significantly expand the kinematic range of these studies and measure the gluon contribution to the proton spin in collisions of an 800 GeV polarized proton beam with a 30 GeV polarized electron beam was recently examined [2]. TRIUMF is a part of the SPIN Collaboration, which is working on a proposal for polarized proton acceleration in HERA to 800 GeV and experiments with the polarized beams.

The TRIUMF task is the development of a high intensity polarized H$^-$ ion source. A polarized H$^-$ ion current of 10-20 mA is required to provide sufficient polarized beam luminosity for the above experiments. The feasibility of producing 10 mA of polarized H$^-$ ion current in the INR-type pulsed OPPIS was demonstrated in experiments with an atomic H injector at BINP, Novosibirsk and experiments at TRIUMF on optical pumping of high density Rb vapor in the presence of a high intensity proton beam [9]. A current of 20-30 mA may be feasible in the "combined" polarization scheme, where space charge compensation is easier to achieve [10]. The development of a pulsed OPPIS is in progress at TRIUMF. The atomic H injector is being constructed and tested on a test bench. The primary proton beam is produced in a pulsed source (Fig.1), then focussed by a solenoidal magnetic lens (labelled FS in Fig.4), and converted to a high intensity, converging atomic hydrogen beam in a pulsed gaseous hydrogen cell with neutralization efficiency about 95%. A unique feature of the proton source is the efficient plasma cooling during expansion between plasma generator and extraction system; the plasma temperature drops to about 0.2 eV at a distance of 10 cm from the generator, where the extraction system is situated. A four grid multiwire accel-decel extraction system was used in experiments on high current proton production at low beam energy (less then 5 keV). The grids are made of 0.2 mm molibdenium wire, with 1.0 mm spacing between wires. The wires are positioned on the mounting electrodes in precisely cut grooves, and fastened by point welding. The mutual grid alignment accuracy is better than 0.02 mm. The gap between the first and second grids is 0.8 mm, between the second and third grids, 1.2 mm, and between the third and fourth grids, 2.0 mm. For a beam energy of 4 keV, the optimal extraction is obtained by using an extraction voltage distribution as follows:(+ 4 kV) on the first grid, (+3.2) kV on the second grid, (- 7.0) kV on the third grid, and (+0.1) kV on the fourth grid. The total extracted proton current is about 8 A. The ion extraction system together with the plasma expansion cup are adjustable to provide the required accuracy of the low divergence beam alignment through the long polarizer.

Spatial charge compensation during focussing is an important issue for high current, low energy beams. This compensation was achieved by using a posi-

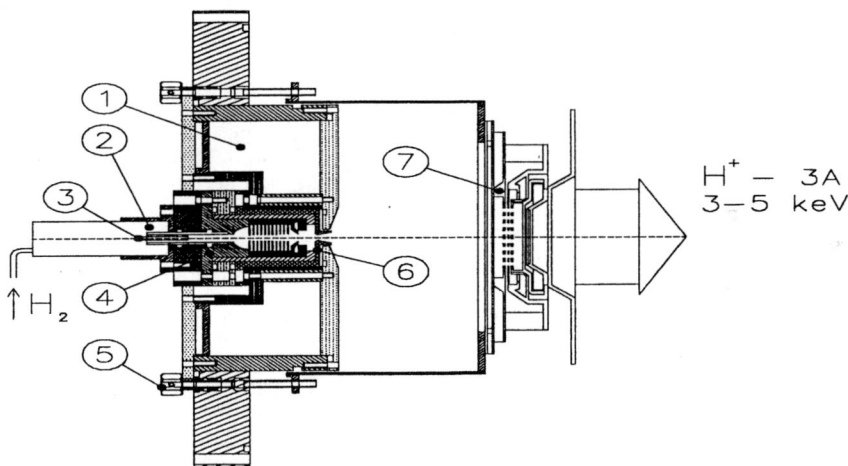

FIGURE 1. The BINP-type high brightness proton source: (1) solenoid; (2) pulsed valve; (3) triggering electrode; (4) cathode; (5) adjusting screw; (6) anode; (7) ion extraction system.

tively biased grid in the middle of the focussing solenoid. Also, the injection of carbon oxide gas into the solenoid improved the compensation, and allowed a significant (about 20%) increase of the atomic beam intensity within the ionizer acceptance. The beam intensity was measured both by calorimetry and from the ionization in the sodium ionizer cell. In the latter case, the well known H$^-$ ion yield in sodium vapor was used to estimate the atomic beam intensity. An oscillogram of the H$^-$ ion pulse is shown in Fig. 2. The pulse duration is about 100 μs, and is completely determined by the pulse duration from the extraction power supply. The results of the two measurement techniques as functions of beam energy are presented in Fig. 3; both methods give quite consistent results. A very high H$^-$ ion current of 28 mA was obtained at 4.0 keV beam energy, which is ideal for use with the spin transfer polarization technique [10], whereby most of this current can be polarized to greater than 80%.

It is planned to install the atomic injector at the TRIUMF OPPIS setup, as shown in Fig. 4. The optical pumping of the high density, large diameter Rb vapor cell will be achieved using a pulsed Ti:sapphire laser. Nearly 100% Rb polarization has already been obtained in an experiment with a pulsed laser under development. The H$^-$ nuclear polarization will be measured and optimized using the powerful OPPIS diagnostic tools.

The high current, low energy polarized H$^-$ ion beam must be accelerated immediately after the ionizer to 20-50 keV to prevent increase of the beam divergence due to space charge effects. This can be done by biasing the whole source to a potential of 20-50 kV. After acceleration in a two gap system, which will also provide the required focusing, the beam will be deflected by a bending magnet through 47.5 degrees to preserve longitudinal polarization. Alternatively, a 15 degree bend plus a solenoidal rotator could be used to align the spin vertically. The beam will then be injected into an RFQ accelerator.

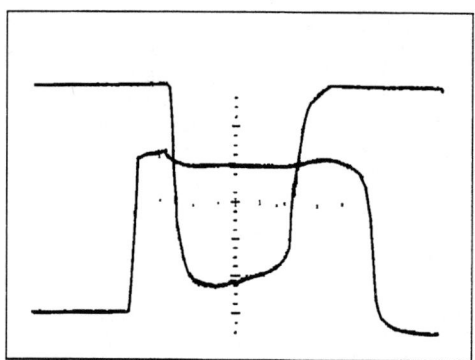

FIGURE 2. H$^-$ ion current pulse at 4.0 keV beam energy. The positive pulse is the extraction voltage.

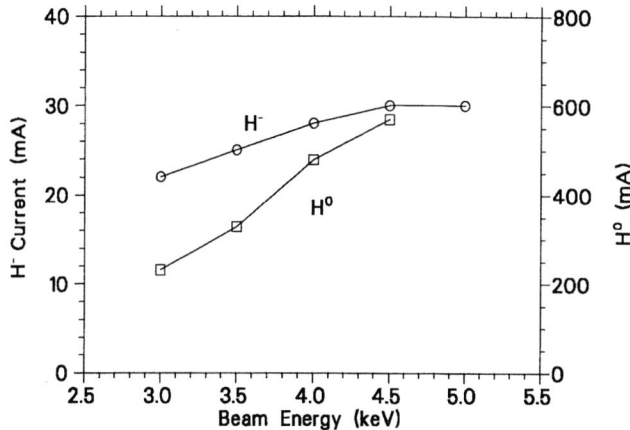

FIGURE 3. H⁻ ion current and H⁰ beam intensity within the ionizer acceptance as functions of the beam energy.

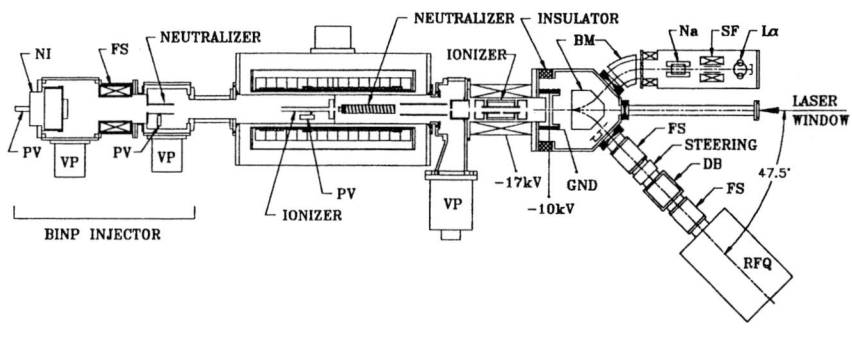

FIGURE 4. Pulsed OPPIS layout.

OPPIS INJECTOR FOR RHIC

The polarization facilities at RHIC will provide 70% polarized proton-proton collisions at energies up to $\sqrt{S} = 500$ GeV and a luminosity of 2×10^{32} cm^{-2} s^{-1} [1]. The polarized injector must produce in excess of 0.5 mA H⁻ ion current during the 300 μs pulse, or current × duration > 150 mA μs, within a normalized emittance of less than 2π mm mrad. This is an ideal application for the TRIUMF-type OPPIS, where a 1.64 mA current was obtained in dc mode. The required pulsed operation will greatly simplify and reduce the cost of the laser system, while providing the best source performance due to ample optical pumping laser power.

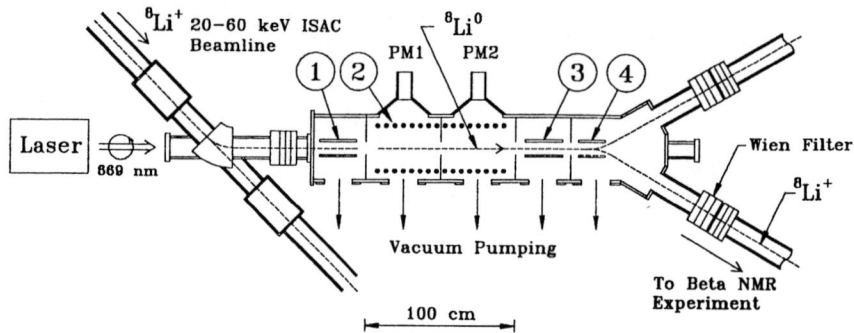

FIGURE 5. Proposed scheme for ^{8}Li beam polarization: 1) Na vapour neutralizer, 2) optical pumping region with magnetic shield and guiding magnetic field, 3) ionizer cell, 4) bender.

The OPPIS is not limited to H$^-$ ion beams. A vector spin polarized D$^-$ ion beam can be produced in the OPPIS using a dual optical pumping scheme, with an efficiency similar to that of the polarized H$^-$ ion beam, as was demonstrated for the KEK OPPIS [11]. A large amount of nuclear spin polarized ^{3}He atoms can be produced by direct optical pumping of helium atoms in metastable 2S states [12]. Hence, polarized ^{3}He^{++} ion beams of 1 mA intensity can be produced via ionization in charge exchange collisions, or in the ECR-ionizer.

There is now an agreement between BNL, KEK and TRIUMF for upgrading the KEK OPPIS for the RHIC polarization facilities. This upgrade will provide a very high intensity polarized injector of H$^-$ and D$^-$ ion beams for high energy spin physics at RHIC.

POLARIZED RADIOACTIVE BEAMS FOR MATERIAL STUDIES AT ISAC

Implantation of several keV energy β-radioactive beams in a material surface (high temperature superconductors, or semiconductors being of greatest interest), and observation of the spin precession due to the local magnetic field can be a useful tool for surface study, similar to the μSR technique for bulk materials [13]. The polarization precession can be detected by measuring the β decay asymmetry. For example, an ^{8}Li$^+$ ion beam intensity in excess of 10^8 s^{-1} will be available from the TRIUMF ISAC facility with a 10 μA proton beam at the production target. The polarization will be produced by direct optical pumping of ^{8}Li atoms in the setup shown in Fig. 2. The 20 keV Li$^+$

beam will be neutralized in a sodium vapor cell and then optically pumped by a colinear 671 nm wavelength dye laser beam ($^2S_{1/2}$ to $^2P_{1/2}$ transition). The optical pumping region must be shielded from external magnetic fields and a homogeneous longitudinal field of a few G will be provided. The laser power density required for pumping both F=3/2, 5/2 states is about 0.5 W cm^{-2} for a multimode laser with a bandwidth of about 500 MHz, the latter being determined by the hyperfine splittings of 382 MHz in the $^2S_{1/2}$ state and 44 MHz in the $^2P_{1/2}$ state. After polarization, the Li beam is ionized to Li$^-$ in a second sodium cell with an efficiency of about 10%, or to Li$^+$ in a gaseous argon cell with about 30% efficiency. The ion beam is then bent and transported to the sample. This bending prevents direct deposition of sodium vapor on the sample surface and provides a convenient entrance for the laser beam. The ion beam can easily be transported a few meters, thus simplifying the obtaining of an ultrahigh vacuum in the analyzing chamber. It is estimated that about 10% of the primary beam can be optically pumped to 80-90% polarization.

Another choice of probe is a ^{17}Ne beam, which can be polarized by optical pumping in the metastable Ne* ($^3P_2$) state. In the sodium neutralization cell, approximately 10-20% of the initial 20 keV Ne$^+$ beam is produced in this metastable state, which can optically pumped using the $^3P_2$-$^3D_3$ transition at 640 nm . Preferential selective ionization of the metastable atoms in charge exchange collisions will permit the obtaining of nuclear polarizations as high as 50% in a Ne$^+$ ion beam suitable for implantation.

CONCLUSIONS

The powerful techniques of optical pumping and polarization transfer collisions are very successfully implemented in the high current OPPIS, which meets the demands of the new generation of high energy accelerators and colliders. The OPPIS also provides high quality beam for precision experiments at TRIUMF. We believe that development of new polarization facilities at RHIC and HERA will benefit from OPPIS technology and will, in turn, boost the further development of polarized sources.

ACKNOWLEDGEMENTS

We would like to thank J. Alessi, D. Barber, V. Davydenko, R. Kiefl, A. Krisch, Y. Mori, S. Page, T. Roser and T. Sakae for useful discussions. We acknowledge the support of the SPIN Collaboration and INR-Moscow in this work.

REFERENCES

1. G.Bunce *et al.*, *Particle World* **v3** (1992) 1-12.
2. *Prospects of the Spin Physics at HERA*, DESY-Zeuten, DESY Report (1995) 95-200.
3. A.N.Zelenski *et al.*, *Proc. of the 1995 IEEE PAC*, Dallas (1995), 864.
4. A.N.Zelenski *et al.*, *Nucl. Instr. Meth.* **A245** (1986) 223-229.
5. A.N.Zelenski *et al.*, *Proc. 12th Int. Symp. on High Energy Spin Physics*, Amsterdam (1996) Ed. C.W. de Jager, World Scientific (1997) 634.
6. A.N.Zelenski *et al.*, "Polarized Beam for the TRIUMF Parity Violation Experiment", W.D.Ramsey *et al.*, "The TRIUMF Parity Violation Experiment", Proc. of 6th Int. Conf. on the Intersections between Particle and Nuclear Physics, Big Sky, Montana, 1997, ed. R.E. Mischke and T.W. Donnelley, to be published in AIP conference proceedings.
7. "A proposal for an intense radioactive beam facility", TRIUMF Report, TRI-95-1 (1995) (private communication).
8. TRIUMF experimental proposal E761, spokespersons J.Birchall, S.A. Page, W.T.H. van Oers.
9. A.N.Zelenski *et al.*, *Rev. Sci. Instr.* **67** (1996) 1359.
10. A.N.Zelenski *et al.*, *Proc. Conf. on Polarized Sources & Targets*, Cologne (1995) Ed. H.P. van Schieck, World Scientific (1996) 111.
11. M.Kinsho *et al.*, *ibid.*, **10** p.126.
12. E.Otten, *ibid.*, p.3.
13. TRIUMF experimental proposals E815, E816 and E817, spokeperson R.Kiefl.

A New Method to Produce Cold Atomic Hydrogen and Deuterium Beams

Victor L. Varentsov*, Alexei A. Ignatiev*, Erhard Steffens[†], Norbert Koch[†]

*St. Petersburg Institute for Informatics and Automation, 39, 14th Line, St. Petersburg, 199178, Russian Federation
[†] Physikalisches Institut, Universität Erlangen-Nürnberg, Erwin-Rommel Str. 1, D-91058 Erlangen, Germany

Abstract. A new method to produce cold atomic hydrogen and deuterium beams is described. This technique can be applied to development of internal polarized targets. The proposed scheme is to combine the low density beam of atoms from a dissociator of a conventional atomic-beam source with a supersonic carrier gas jet formed by a circular converging-diverging Laval nozzle surrounding the central one. A computer code for a realistic description of the formation of the atomic beam inside the supersonic carrier gas jet has been developed and applied for computer simulations of different nozzle geometries, various combinations of gas and carrier gas, and different mixtures of partly dissociated gas, like $H - H_2$ and $D - D_2$. The calculations indicate that with a carrier gas flow of about six times higher than the inner atomic beam, the central beam is confined, which should result in a better forward intensity and much lower beam temperature. An experimental study of this new method is in progress.

INTRODUCTION

The goal of this work is to improve the present technique of cold atomic hydrogen and deuterium beams with the use of a new method of beam formation. Such beams are used in sources of spin-polarized hydrogen and deuterium targets for nuclear and particles physics experiments. The immediate application of improved target sources would be to improve the performance of scattering experiments at high energy storage rings.

The HERMES experiment at DESY (Hamburg) is presently running with a hydrogen target in order to study the spin structure of the proton. Its physics potential is partly limited by the available luminosity. With a substantial

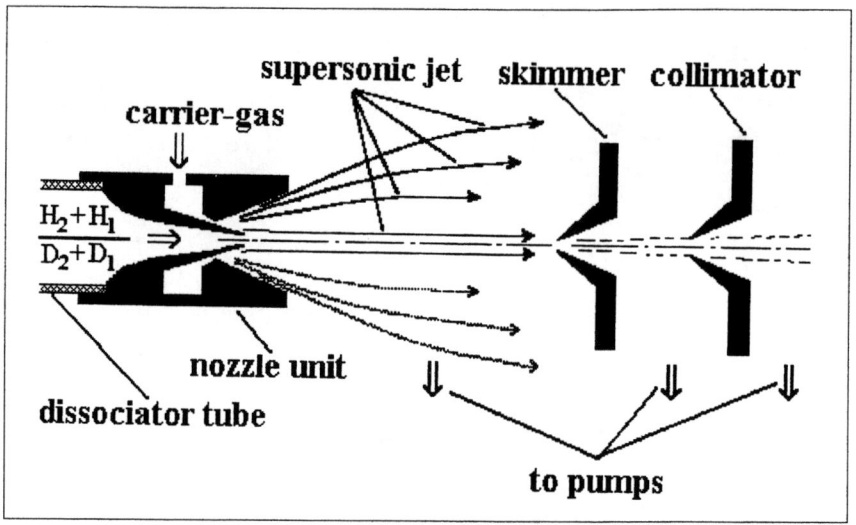

FIGURE 1. Schematic of the method to produce cold hydrogen and deuterium atomic beams.

increase in target thickness, new physics like the spin dependence of open charm production and the question of gluon polarization could be attacked.

DESCRIPTION OF THE METHOD

The proposed method is a result of the development of a new approach to the supersonic internal gas-jet target design [1,2] for cold atomic and deuterium beams production. It combines the low-density beam of atoms from a dissociator of a conventional atomic-beam source (ABS) with a supersonic carrier gas jet formed by a circular converging-diverging Laval nozzle surrounding the central one.

Schematic of the new method is shown in Figure 1.

The conical converging nozzle of the dissociator of the Atomic Beam Source (ABS) (see, e.g., ABS for polarized hydrogen [3] and deuterium [4]) is placed on the axis of the Laval nozzle, leaving a small annular slit between the Laval nozzle throat and the central conical nozzle for the supersonic carrier gas expansion into vacuum.

The supersonic carrier gas flow creates a rarefaction zone behind the inner tube end. In this manner conditions for effective pumping of a binary mixture of molecular and atomic hydrogen $(H + H_2)$ or deuterium $(D + D_2)$ from the dissociator are produced.

In fact, we are dealing here with a complex interaction of two gas streams in the nozzle unit, when the low density gas of $H + H_2$ (or $D + D_2$) from

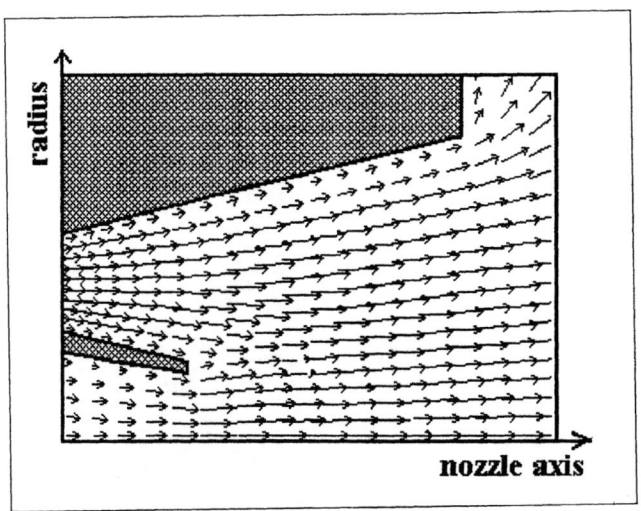

FIGURE 2. The gas velocity field in the vicinity of the inner tube end.

dissociator with Mach number about 1 that is directly injected through the inner cone into the higher density supersonic carrier-gas jet. There is some competition between these two gas streams: on one hand gas from the inner tube tries to expand and on the other hand the supersonic carrier-gas jet tries to squeeze it in the region of the inner tube end. In other words, the carrier-gas jet fulfills in this case a function of some additional "artificial nozzle with very soft and fast moving walls".

As an illustration in Figure 2 we present the calculated gas velocity field in the vicinity of the inner tube end, where the gas from the dissociator is injected into the supersonic carrier-gas jet.

By means of computer simulations, which will be described in the next section, it has been shown that in certain conditions the central atomic hydrogen (or deuterium) beam can be confined, which should result in a better forward intensity and much lower beam temperature. Such a beam should be transported with much higher efficiency by the system of Stern-Gerlach sextupole magnets of a polarized source. As a result, a considerable increase of the output flow rate is expected.

NUMERICAL INVESTIGATIONS

A new version of a computer code for a realistic description of the formation of the atomic beam inside the supersonic carrier gas jet has been developed. The code is based on the full system of Navier-Stokes equations for multi-component gas mixtures.

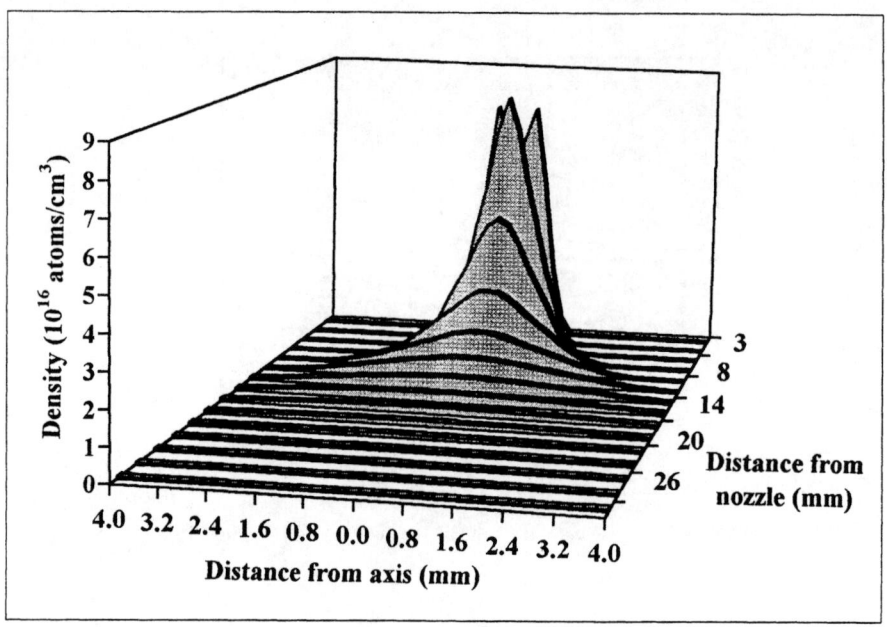

FIGURE 3. The density distribution of atomic hydrogen in the supersonic H_2 carrier gas jet.

The previous versions of this code have been used for simulations of formation of internal molecular-beam targets from medium and heavy nonvolatile elements by means of a carrier gas jet (a binary gas mixture of target-particles and carrier gas), as well as for simulations of conventional supersonic gas-jets of pure gases [5,6].

The new code has been applied for simulations of different nozzle geometries, various combinations of gas and carrier gas, and different mixtures of partly dissociated gas, like $H-H_2$ and $D-D_2$. Some results of these calculations for an improvement of the ABS [3] of the HERMES experiment at HERA/DESY are presented in the following tables and pictures. The flow rate from the nozzle unit required for this new method proposed here is about a factor $30 \div 40$ higher compared with the present atomic beam of the HERMES target resulting in a flow rate of 40 mbar l/s of H_2. Such a high gas flow can only be pumped by roots pumps. It is foreseen to apply roots pumps of 8000 m^3/h in the first pumping stage, which results in an effective pumping speed of 1600 l/s, which has been used in the calculations. The remaining system, starting with the second stage of the HERMES ABS, can be used as before.

Figure 3 and Figure 4 show the density distributions of atomic and molecular hydrogen flows in the supersonic H_2 carrier gas jet, respectively. The confining effect of the supersonic carrier gas jet on the atomic beam formation is visible.

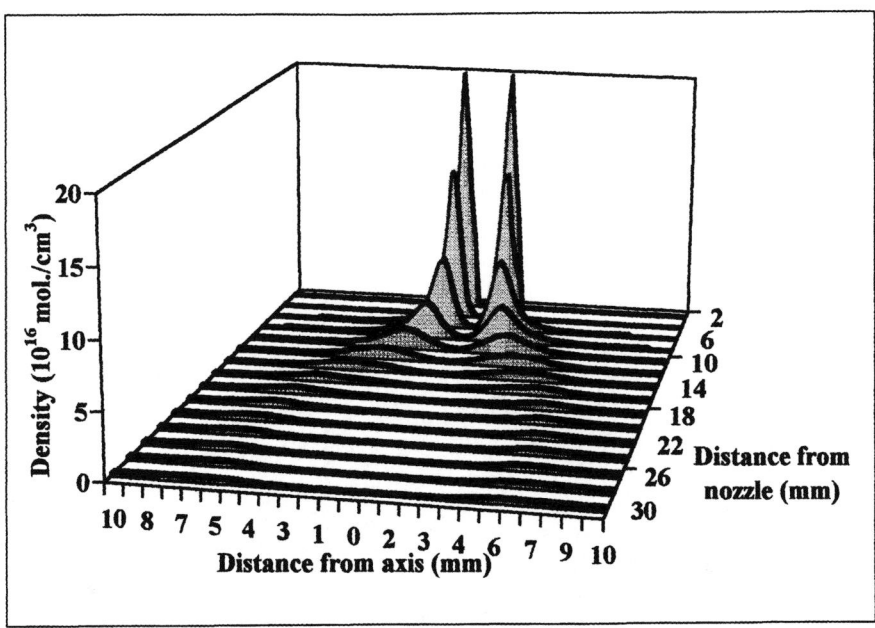

FIGURE 4. The density distribution of molecular hydrogen in the supersonic H_2 carrier gas jet.

In Table 1 the calculated main characteristics of cold atomic hydrogen and deuterium beams formed by the use of H_2 carrier gas jet are listed. The calculations have been made for nozzle temperature 100 K, degree of dissociation of the gas leaving the dissociator 0.5, skimmer diameter 5 mm.

The results collected in Table 1 indicate that with a carrier gas flow of at least six times higher than the inner atomic beam the mixing of carrier molecules into the central jet of atoms is low. So, the atomic fraction in the beam amounts to as much as 64% (for hydrogen) and 80% (for deuterium) of its initial value in the dissociator nozzle. It should be pointed out that atomic beams formed by the supersonic carrier gas jet have very low temperatures and high forward intensities.

For comparison we also show in Table 2 similar results for He carrier gas jet. One can see that these beams parameters are worse than it in case of H_2 carrier gas. We think that the main reason lies in the fact that He atoms in two times heavier than H_2 molecules.

Figure 5 shows the calculated density, temperature and average velocity of hydrogen and deuterium atomic beams on the jet axis as a function of the distance from the nozzle throat. The focusing effect of the supersonic carrier gas jet on the atomic beam formation is very pronounced in the region of the nozzle exit (4 mm distance).

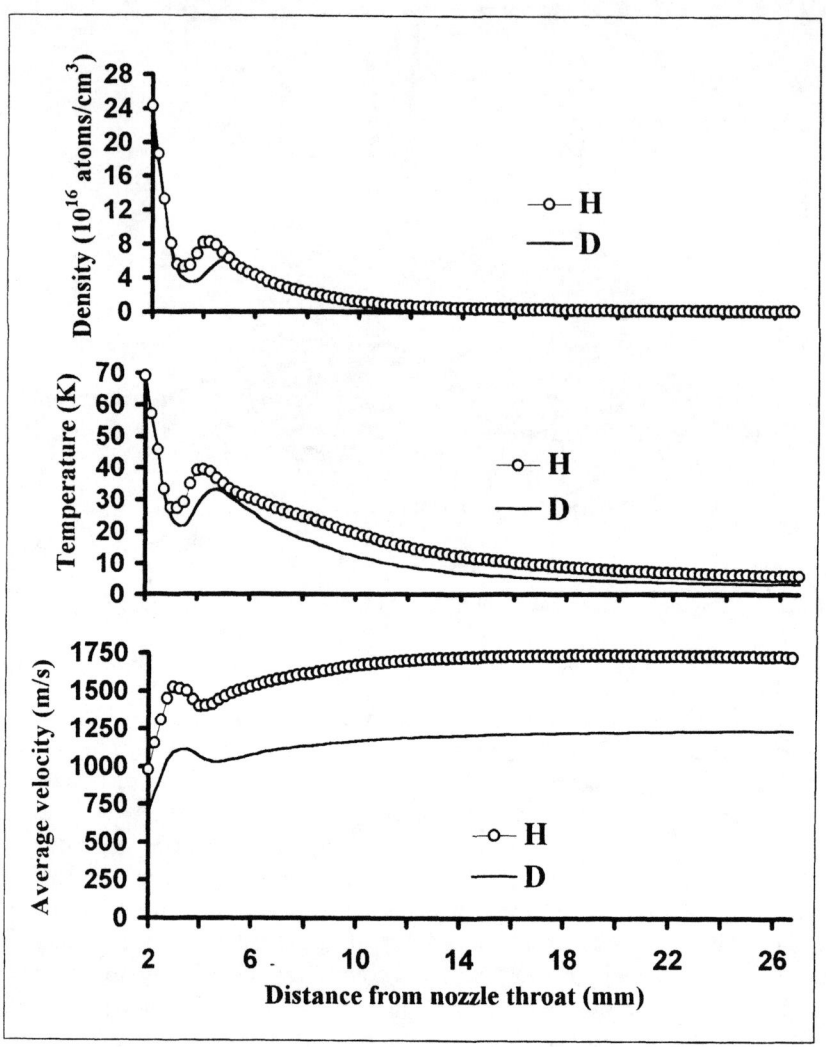

FIGURE 5. Density, temperature and average velocity of hydrogen (H) and deuterium (D) atomic beams on the jet axis as a function of the downstream distance from the nozzle throat.

TABLE 1. Characteristics of cold atomic hydrogen and deuterium beams formed by the use of H_2 carrier gas jet.

	HYDROGEN	DEUTERIUM
Total gas flow through the nozzle unit [mbar l/s]	41.6	27.4
Gas throughput (H_2 or D_2) through the dissociator [mbar l/s]	6.0	4.0
Pressure in the dissociator tube [mbar]	8.45	8.0
Nozzle-skimmer distance [mm]	27	26
Beam temperature [K]	5.7	3.2
Average beam velocity [m/s]	1718	1235
Mach number	7.56	9.39
Atomic fraction in the beam	0.32	0.40
Beam intensity through the skimmer [10^{19} atoms/s]	2.4	2.6

TABLE 2. Characteristics of cold atomic hydrogen and deuterium beams formed by the use of He carrier gas jet.

	HYDROGEN	DEUTERIUM
Total gas flow through the nozzle unit [mbar l/s]	21.7	19.7
Gas throughput (H_2 or D_2) through the dissociator [mbar l/s]	6.0	4.0
Pressure in the dissociator tube [mbar]	8.45	8.0
Nozzle-skimmer distance [mm]	17.5	21
Beam temperature [K]	17.4	7.2
Average beam velocity [m/s]	1604	1203
Mach number	4.65	6.81
Atomic fraction in the beam	0.25	0.32
Beam intensity through the skimmer [10^{19} atoms/s]	2.7	2.0

Summary and Conclusion

In summary, a numerical study of beam formation at the nozzle of a dissociator is presented. Compared with a conventional atomic beam source, it is proposed to confine the inner atomic beam by means of an annular supersonic carrier jet, which is scraped-off after beam formation at the first skimmer. It has been found that the molecular weight of the carrier jet should be as low as possible. Therefore H_2 seems to be the optimum choice. The results show, that the inner beam is slowed down and cooled to very low temperatures not accessible with a free beam. The carrier jet must have an at least six times higher flow rate compared with the central one. The new method presented here may lead to a considerably improved output intensity of atomic beam sources.

REFERENCES

1. Varentsov V.L, Hansevarov D.R., *Nucl. Instr. and Meth.* **A317**, 1 (1992).
2. Varentsov V.L, Hansevarov D.R., Varentsov D.V., *Nucl. Instr. and Meth.* **A352**, 542 (1995).
3. Stock F. et al., *Nucl. Instr. and Meth.* **A343**, 334 (1994).
4. Zhou Z.-L. et al., *Nucl. Instr. and Meth.* **A378**, 40 (1996).
5. Varentsov V.L., Ignatiev A.A., Sokolov E.I., "Computer Simulation of the Supersonic Internal Molecular-Beam Target generation", in *Book of Abstracts of STORI 96, 3rd Int. Conf. on Nuclear Physics at Storage Rings, Berncastel-Kues, Germany*, 1996, p. P32.
6. Varentsov V.L., Okunev I.S., Ignatiev A.A., "A novel technique for the polarized atomic-beam target production", in *Book of Abstracts of STORI 96, 3rd Int. Conf. on Nuclear Physics at Storage Rings, Berncastel-Kues, Germany*, 1996, p. P32, to be published in the STORI 96 Proceedings in Nucl. Phys. A.

Polarization of Deuterium Molecules

J. F. J. van den Brand,[1,2] H. J. Bulten,[1] M. Ferro-Luzzi,[2,3] Z.-L. Zhou,[4] R. Alarcon,[5] T. Botto,[2] M. Bouwhuis,[2] R. Ent,[6] P. Heimberg,[1] D. Higinbotham,[7] C. W. de Jager,[2] J. Lang,[3] D. J. de Lange,[2] I. Passchier,[2] H. R. Poolman,[2] J. J. M. Steijger,[2] O. Unal,[4] H. de Vries.[2]

[1] *Department of Physics and Astronomy, Vrije Universiteit, Amsterdam, The Netherlands*
[2] *NIKHEF, P.O. Box 41882, 1009 DB Amsterdam, The Netherlands*
[3] *Institut für Teilchenphysik, Eidg. Technische Hochschule, CH-8093 Zürich, Switzerland*
[4] *Department of Physics, University of Wisconsin, Madison, WI 53706, USA*
[5] *Department of Physics, Arizona State University, Tempe, AZ 85287, USA*
[6] *TJNAF, Newport News, VA 23606, and Hampton University, Hampton, VA 23668, USA*
[7] *Department of Physics, University of Virginia, Charlottesville, VA 22901, USA*

Abstract. For *molecular* systems, spin relaxation is expected to be suppressed compared to the case of atoms, since the paired electrons in a hydrogen or deuterium molecule are chemically stable, and only weakly interact with the spin of the nucleus. Such systems would be largely insensitive to polarization losses due to spin-exchange collisions, to the interaction of the electron spins with external fields (e.g. the RF-field of a bunched charged-particle beam), and/or to the presence of container walls. Here, we discuss the results of a recent experiment [1] where we obtained evidence that nuclear polarization is maintained, when polarized atoms recombine to molecules on a copper surface (in a magnetic field of 23 mT and at a density of about 10^{12} molecules·cm^{-3}).

INTRODUCTION

Nuclear polarized ^{1}H and ^{2}H atoms are used in various important nuclear and particle physics experiments:

- Scattering of 28 GeV polarized positrons from polarized ^{1}H and ^{2}H is carried out in HERMES [2] at DESY (Hamburg, Germany). Here the goal is to understand how the nucleon spin originates from the dynamics of quarks and gluons. Other laboratories such as CERN and SLAC (Stanford, California, USA) are pursuing the same goal.

- The neutron is a composite object and is expected to possess an inhomogeneous electric charge distribution [3]. This distribution can be studied by scattering medium-energy electrons from a 'quasi-free' polarized neutron inside a polarized deuteron. Such data provide stringent constrains for models of the nucleon.

- An important subject of study is the spin structure of the deuteron. The action of the tensor force results in a non-spherical density distribution, or, in other words, in a small D-wave admixture to the ground-state wave function. Our knowledge of this characteristic S/D-wave structure of the deuteron is still incomplete [4].

Thus, spin-dependent scattering from polarized hydrogen and deuterium nuclei can provide important information on the spin structure of the nucleon and deuteron. These measurements can be optimally performed by scattering particles from pure and highly polarized gas injected into particle storage rings. Recently, several such experiments were carried out at Indiana University (Bloomington, USA) [5], NIKHEF [4], Budger Institute for Nuclear Physics (Novosibirsk, Russia) [6], and DESY [2], while new experiments are being prepared at MIT (Cambridge, USA) [7] and NIKHEF [8]. Although this novel technique has many advantages, in that spin-dependent scattering from pure atomic species of high polarization can be realized, it provides a challenge for attaining sufficient luminosities, especially when polarized hydrogen or deuterium is used. Furthermore, there exist many mechanisms that can depolarize the target atoms, mainly through the interaction of the electron spin with external fields associated with the stored particle beam and/or the presence of container walls. This has prompted a significant effort in the past few years, devoted to the production of intense polarized atomic beams of hydrogen and deuterium, the development of suitable coatings to preserve the nuclear polarization of these atoms, and the realization of polarimeters that can precisely measure the nuclear polarization of the target species.

Until recently, the general belief was that the hydrogen or deuterium *molecules* originating from recombination would not carry polarization. This was based on an incorrect interpretation of previous experimental results [9]. We addressed the issue experimentally [1] and found that our data are consistent with deuterium molecules retaining the tensor polarization of the initial atoms. If correct (it concerns a 2.4σ result), then these data have important implications. Firstly, a significant amount of molecules is always present in an internal target and any uncertainty about its polarization immediately leads to an increase of the systematic error of such an experiment. Secondly, recombination of polarized atoms to molecules has the potential to allow the creation of ensembles of polarized nuclei of unprecedented performance.

Next we will describe our setup. Then we will discuss the experiment that provided the first evidence of tensor polarization in low-density ($\approx 10^{12}$

molecules·cm^{-3}) systems of deuterium molecules. Finally, future plans are described.

EVIDENCE FOR NUCLEAR POLARIZATION IN LOW-DENSITY D$_2$ GAS

For our experiments, we used an atomic beam source (ABS) to prepare deuterium atoms in specific combinations of hyperfine states [10] (the source can be readily modified for use with hydrogen). For a detailed explanation we refer to Ref. [11].

Fig. 1 shows that a medium-field transition unit (MFT) provides a $1-4$ Zeeman transition, while a strong-field transition unit (SFT) is used to induce either a $2-6$ or a $3-5$ transition [12]. Each transition involves a collective change of the nuclear- and electron-spin orientation. Therefore, a decrease of $1/3$ in the amount of atoms detected by an atomic-beam detector (BRP) with a high-frequency transition unit on, indicates a 100% efficiency of the transition. Consequently, deuterium atoms in well-controlled mixtures of hyperfine states can be injected into a storage cell. The substate population was alternated every 10 s, changing the tensor polarization of the deuterium atoms between -2 and $+1$, while keeping the vector polarization at zero. Note that for the used combinations of hyperfine states, the electron polarization of the ensemble of deuterium atoms is kept constant ($P_e = 0$). This choice eliminates uncertainties due to P_e-dependent processes (e.g. recombination).

Two different storage cells, with 15 mm diameter and 400 mm length, were used in our experiment: an uncoated copper cell, and an ultrapure aluminum cell coated with a solution of PTFE3170 liquid Teflon diluted with water. The copper cell was constructed from 10 μm thick copper foil and cleaned with trichloro-ethane before manufacturing. No precaution was taken to avoid natural oxidation of the surface. The PTFE-coated aluminum cell was cooled to approximately 180 K. The copper cell was kept at room temperature. The atoms (molecules) spend about 3 (5) ms in the storage cell, while undergoing about 300 wall bounces. On wall contact, the polarized deuterium atoms will largely recombine to molecules on a copper surface [9], whereas on a PTFE-coated cell surface recombination is strongly suppressed.

The relative amount of atoms and molecules in the two cells was determined by analyzing the fraction of the gas, ionized by the beam of the Amsterdam pulse-stretcher (AmPS) electron storage ring at NIKHEF. We determined the atomic fraction, $\chi = n_D/(n_D + 2n_{D_2})$, where n_i is the areal density of the species. The atomic fraction was corrected for a 2 ± 1 % contribution from dissociative ionization of the molecules by the electrons. We found for the PTFE-coated aluminum (bare copper) cell $\chi = 0.71 \pm 0.02$ (0.26 ± 0.03).

The molecules, $n_{D_2^{unp}}$, coming from background gas or from the nozzle will be unpolarized, and only the molecules, $n_{D_2^{rec}}$, originating from recombination

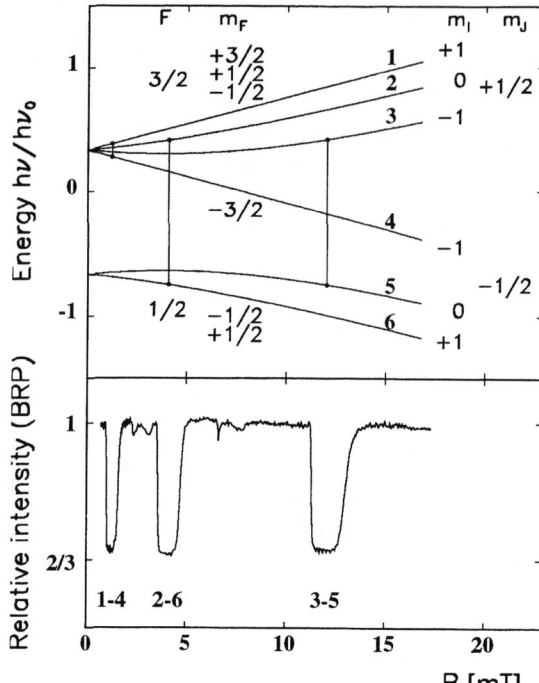

FIGURE 1. *Top panel: hyperfine structure of deuterium as a function of the static field B. I (J) represents the deuteron (electron) spin, $F = I + J$ is the total spin. The lines indicate the Zeeman transitions used in the present measurement. Bottom panel: response of the BRP as a function of the central magnetic field in the MFT and SFT.*

are potentially polarized. Their contribution was determined by turning on and off the sextupole electromagnets as well as by flowing background gas. For the PTFE-coated aluminum (uncoated copper) cell, we found that about 85 (35)% of the molecules in the cell are due to the molecular beam, residual gas in the vacuum chamber and diffused flow from the ABS into the feed-tube, while about 15 (65)% of the molecules in the storage cell originated from recombination of atoms on the walls. Fig. 2 shows that the PTFE-coated cell contains mostly atoms, whereas a substantial molecular contribution from recombined atoms is realized in the uncoated copper cell. The hatched area represents the contribution of atoms and recombined molecules.

The nuclear tensor polarization of the deuterium atoms, $P_{zz}(D)$, has been determined by accelerating the atomic ions, produced by the circulating electrons, to 60 keV, which are then used to bombard a tritiated foil [13]. The angular distribution from neutrons in the reaction $^3H(d,n)^4He$ was used to measure the tensor polarization of the deuterium atoms.

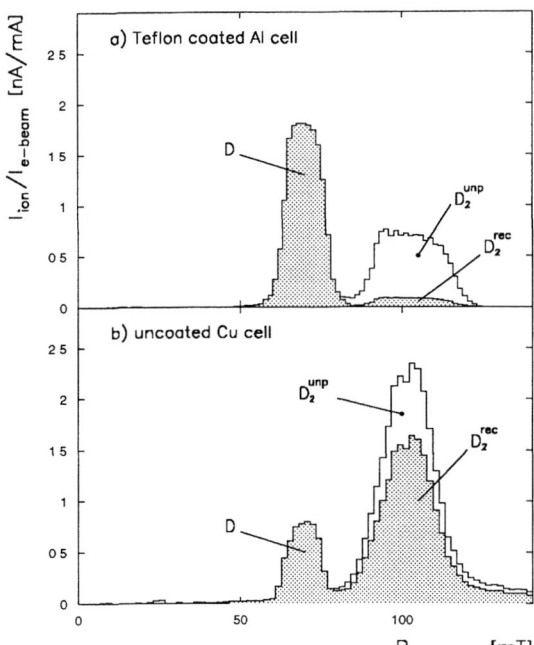

FIGURE 2. Normalized ion current as a function of the magnitude of the Wien filter magnetic field. The peaks at lower and higher magnetic field correspond to D^+ and D_2^+, respectively. Results are shown for PTFE-coated aluminum (top) and uncoated copper (bottom) cells. The hatched area represents the contribution of atoms, and molecules which recombined in the storage cells. The unhatched contribution D_2^{unp} represents unpolarized molecules originating from residual D_2 gas and from undissociated molecular beam.

We determined the total tensor polarization, i.e. of the atoms and molecules, by measuring the asymmetry, $A = \frac{N^+ - N^-}{2N^+ + N^-}$, for elastic electron-deuteron scattering at 704 MeV incident energy. Here, N^+ (N^-) are the yields of scattered electrons for deuterium nuclei with tensor polarization P_{zz}^+ (P_{zz}^-). We found for the elastic electron scattering asymmetries $A^{PTFE} = -0.232 \pm 0.014$ and $A^{cu} = -0.183 \pm 0.043$, where the superscripts PTFE and cu denote that the measurements were carried out with the PTFE-coated aluminum cell and the copper cell. As a check on false asymmetries, we measured the asymmetry for unpolarized target gas and obtained $A_{unp} = 0.000 \pm 0.014$.

Fig. 3 shows the tensor polarization, $\Delta P_{zz}(D) \equiv P_{zz}^+(D) - P_{zz}^-(D)$, of the atoms and the absolute value of the elastic electron-deuteron scattering asymmetry for both the PTFE-coated aluminum and the copper cell. All previous internal-target experiments with polarized hydrogen or deuterium (e.g. Refs. [4-6]) have used storage cells with special surfaces, consisting of drifilm or

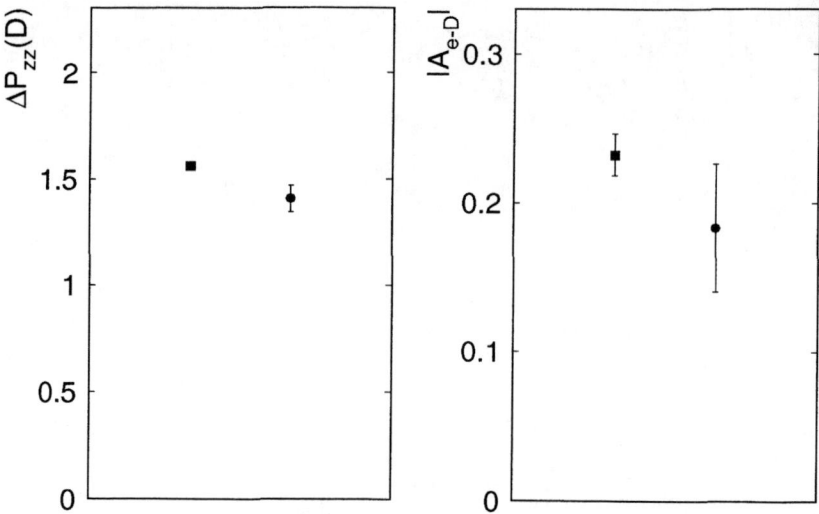

FIGURE 3. *Left panel: tensor polarization, $\Delta P_{zz}(D)$, of the atoms. Right panel: absolute value of the elastic electron-deuteron scattering asymmetry. Data are given for both the PTFE-coated aluminum (squares) and the uncoated copper cell (circles).*

PTFE. However, we observe a significant asymmetry with an uncoated copper cell. It is concluded that the molecules retain most of the tensor polarization of the parent atoms. Compared to a PTFE-coated aluminum surface, atoms exhibit a 10 % polarization loss on copper. Since molecules originate from recombination of these atoms, we expect a similar polarization loss for molecules, and thus for A^{cu}. Assuming that the ratio of molecular over atomic tensor polarization is the same for the copper and the PTFE-coated aluminum cell, we obtain for the nuclear tensor polarization of the molecules from recombined atoms $\Delta P_{zz}(D_2^{rec}) = (0.81 \pm 0.32) \times \Delta P_{zz}(D)$. Such a high polarization indicates the possibility of developing a dense ensemble of polarized nuclei through recombination of polarized atoms.

STUDIES OF FORMATION AND RELAXATION IN NUCLEAR POLARIZED MOLECULES

Several dedicated experiments were proposed or are in progress to study spin-dependent molecular formation and subsequent relaxation:

- At DESY the HERMES collaboration [14] measured the polarization of H_2 molecules by using deep-inelastic scattering of 28 GeV polarized positrons in the HERA ring. Polarized atoms were produced in their

ABS and injected in a drifilm coated aluminum storage cell. The measurements were performed at a magnetic holding field strength of 0.35 T. The atomic fraction as measured by their target gas analyzer (TGA), was varied from 0.97 to 0.46 by changing the cell temperature from 110 K to 260 K, respectively. The molecular polarization can be obtained from a comparison of the results for A_1^p for both conditions. In case the TGA measurements would be representative of the actual atomic fraction in their storage cell, then molecules would carry less polarization then their parent atoms. However, sizable uncertainties are introduced due to the sampling factors that relate the measured atomic fraction to that in the storage cell. Their analysis is in progress.

- The PINTEX collaboration [15] intends to measure H_2 and D_2 polarization for longitudinal magnetic holding fields up to 0.35 T by using a superconducting magnet. An internal target fed by the Wisconsin ABS will be employed in the 'A'-region of the Indiana University Cooler Facility (IUCF). With longitudinal field they will investigate the spin correlation coefficient for pp elastic scattering at 200 MeV beam energy. Their unique feature is that there will be a valve in the cell. With the valve opened, the atomic beam will first strike a recombination zone of copper, or some other surface, where several hundred bounces occur. The atoms then drift into a Teflon-coated beam tube where the elastic scattering takes place. With the valve closed the atoms will only see Teflon. The temperature of the recombination zone will be varied systematically from above room temperature down to about 40 K. The temperature of the beam tube and the recombination zone can be independently varied. They intend to systematically study the dependence on guide field as well.

For the D_2 measurements they will to use an α beam at 125 MeV and take advantage of the tensor analyzing power.

- Measurements with a laser-driven source will be carried out at IUCF by the Illinois group [16]. Hydrogen and deuterium atoms will be recombined in their storage cell and the magnetic field dependence of the polarization will be studied for field strengths up to about 100 mT.

We will continue our studies of molecular polarization with experiments both in AmPS and by using a stand-alone set-up with a 1-2 keV beam from an electron gun. The lower current of the electron gun ($\sim$ 100 μA compared to 100 mA of the high-energy beam) is balanced by the larger ionization cross section at these energies with respect to a 500 - 900 MeV beam.

The hamiltonian of a diatomic molecule in a magnetic field [17,18] can be written as

$$\hbar\mathcal{H} = \omega_I(I_z^1 + I_z^2) + \omega_L L_z + \omega'(\mathbf{I}^1 + \mathbf{I}^2)\cdot\mathbf{L} + \omega''[\mathbf{I}^1\cdot\mathbf{I}^2 - 3(\mathbf{I}^1\cdot\mathbf{n})(\mathbf{I}^2\cdot\mathbf{n})], \quad (1)$$

where **n** is a unit vector in the direction of the molecular axis. The first two terms represent the precession of the nuclear spins, $\omega_I I_z^1$ and $\omega_I I_z^2$, and the precession, $\omega_L L_z$, of the rotational magnetic moment of the molecule. The Larmor frequencies are proportional to the strength of the external magnetic field and are given by $\omega_I = -\gamma_I H_0$ and $\omega_L = -\gamma_L H_0$. The next term represents the *spin-rotation interaction* and one can write $\omega' = \gamma_I H'$. Here, the nuclear spin of one nucleus in the molecule precesses around the magnetic field setup by the orbital motion of the other nucleus. For hydrogen (deuterium) we have that $H' \simeq 27$ gauss ($H' \simeq 13$ gauss). For deuterium H' is about half the value for hydrogen since the molecule is about twice as heavy. The last term in the hamiltonian shows the *dipolar coupling* and for hydrogen molecules the corresponding precession frequency can be written as $\omega'' = 2\gamma_I H'' = \gamma_I^2 \hbar / <r^3>$, where r is the internuclear distance. The corresponding effective magnetic field strength amounts to $H'' \simeq 34$ gauss. For deuterium one finds 19 gauss, but matters are more complicated since this includes a contribution from the electric quadrupole moment (which provides an additional coupling $+eQ \frac{\partial^2 V^e}{\partial z_0^2}$). From these considerations it follows that only in the presence of a strong external magnetic field ($H_0 \gg H', H''$) nuclear polarization may be conserved in molecules. Without such a field any alignment will be lost on a time scale of μs. Note that for $L = 0$ both the spin-rotation and dipolar relaxation mechanisms vanish. Clearly, the magnetic field strength is an important parameter in the relaxation process and its influence should be investigated.

Models of nuclear spin relaxation in diatomic gases can be found in e.g. Ref [18]. However, all previous work focused on systems with densities in the order of 10^{19} molecules·cm^{-3}, or more. For hydrogen molecular gas (deuterium will be discussed later) the main spin-relaxation processes, in that regime, are associated with the spin-rotation coupling and dipolar coupling. In the first case, the total proton spin $I = I_p^1 + I_p^2$ couples to the rotational angular momentum L of the two protons. The projection m_L on the polarization axis undergoes frequent transitions due to the numerous collisions of the molecules. In analogy to a solid state system, it is said that L acts as a 'lattice', i.e. 'spin heat' is transferred from the nuclear spin I to L, until equilibrium is reached. Note that transitions which change L are much less likely than transitions within the same L-multiplet. The second relaxation process is due to the tensor force between the two proton spins I_p^1 and I_p^2. This interaction depends on the relative orientation of the two spins and on the distance between them. Therefore, the relaxation process also depends on the rotational angular momentum L, which defines the angular distribution of the proton point-density in the molecule.

Typical relaxation times in the order of milliseconds have been measured for H_2 gas at densities larger than 10^{19} molecules·cm^{-3}, in good agreement with theory. As explained above, the dominant 'spin thermalizing' process stems from the frequent collisions between the molecules. However, in the case of our storage cell measurements, at pressures below 10^{-4} mbar (densities smaller than 10^{13} molecules·cm^{-3}), the molecules undergo many more collisions with the cell walls than with each other. To our knowledge this regime has not been studied yet. It is known, for example, that tiny admixtures of paramagnetic centers in gases or liquids strongly affect the spin relaxation times. In analogy, direct interaction between the nuclear spins and the cell-wall material might play a crucial role in the spin-relaxation process. For this reason the relaxation processes should be investigated for various wall materials (Al, Cu, Ti, PTFE, Drifilm, etc.).

Measurements [19] indicate that the effective interaction for spin relaxation is about six times weaker than that expected from the hard-sphere model. Furthermore, for HD spin relaxation times three orders of magnitude larger than in H_2 have been observed. This can be explained by the fact that the polarized nuclei of a HD molecule can be in a rotational angular momentum state $L = 0$, which forbids the main relaxation mechanism of spin-rotation and dipolar coupling, while the lowest possible state for polarized protons in a H_2 molecule is the *ortho* state $I = 1$, $L = 1$ (the energetically lower *para* state with $L = I = 0$ obviously cannot carry polarization). For D_2 molecules, the situation is somewhat different as for H_2. On the one hand, in addition to the previously mentioned relaxation mechanisms, the interaction of the deuteron *quadrupole moment* with electric field gradients in the molecule introduces a new spin-relaxation process. On the other hand, because the deuterons obey Bose-Einstein statistics, the lowest states of D_2 molecules are the *para* states $L = 0$ and $I = 0$ or 2. This means that deuterons can carry a polarization up to 5/6. Polarization in solid D_2 at 14.5 T has been realized [20] (up to 0.13) by cooling the molecules in a Pomeranchuk cell to temperatures of about 35 mK. Given the strong influence of the symmetry of the molecular wave function on the relaxation processes, it is of interest to study nuclear spin relaxation in H_2, HD and D_2 molecules.

Finally, we would like to point out the importance of measuring the temperature dependence of the spin relaxation mechanisms. This follows from a consideration of the energies of rotational excitations, $E_r = BL(L+1)$. For H_2, HD and D_2 molecules we find that B equals 7.6 meV (86 K), 5.7 meV (65 K) and 3.8 meV (43 K), respectively. If we are interested in nuclear polarized molecules (e.g. we consider ortho states for H_2) and we neglect ortho-para transitions, then 10B or 860 K is needed for the first rotational transition ($L = 1 \to 3$) in H_2. For D_2 this requires 6B or 258 K, needed for the transition $L = 0 \to 2$. Especially, in D_2 the presence of molecules in $L = 2$ (n_2) compared to $L = 0$ (n_0) states are expected to induce significant spin relaxation mechanisms. The relative occupation numbers are a strong function of the temperature of the cell wall (assuming a large accommodation coefficient), and are given by $\frac{n_a}{n_b} = \frac{N_a}{N_b}\exp^{-(E_a-E_b)/kT}$. For H_2 we find $n_3/n_1 = 0.13$, 4×10^{-4} and 0 for wall temperatures of 300, 100 and 20 K, respectively. For

D_2 we obtain $n_2/n_0 = 2.1$, 0.38 and 1.2×10^{-5}, respectively. Consequently, we expect that the spin relaxation mechanisms will be strongly temperature dependent.

In summary, after having obtained evidence that nuclear-polarized deuterium atoms maintain most of their nuclear polarization when recombining to molecules on a copper surface, we argue that it is important to study the formation process of nuclear polarized H_2, D_2 and HD molecules, and their spin-relaxation mechanisms. It is expected that temperature, strength of the external magnetic field, and type of surface material are important parameters in the spin-relaxation process.

This work was supported in part by the Stichting voor Fundamenteel Onderzoek der Materie (FOM), which is financially supported by the Nederlandse Organisatie voor Wetenschappelijk Onderzoek (NWO), the Swiss National Foundation, the National Science Foundation under Grants No. PHY-9316221 (Wisconsin), PHY-9200435 (Arizona State) and HRD-9154080 (Hampton).

REFERENCES

1. J.F.J. van den Brand et al., Phys. Rev. Lett. **78**, 1235 (1997).
2. K. Ackerstaff et al., Phys. Lett. **B404**, 383 (1997); Science **267**, (1995) 1767.
3. "Inside the Neutron", J.F.J. van den Brand and P.K.A. de Witt Huberts, Physics World, February (1996).
4. M. Ferro-Luzzi et al., Phys. Rev. Lett. **77**, 2630 (1996).
5. W. Haeberli et al., Phys. Rev. **C 55**, vol. 2 (1997).
6. R. Gilman et al., Phys. Rev. Lett. **65**, 1733 (1990).
7. BLAST proposal, MIT-Bates, 1991 (unpublished).
8. Exp. 9701 at NIKHEF, Spokespersons: M. Ferro-Luzzi and J.F.J. van den Brand (1997).
9. S. Price et al., Nucl. Instrum. Meth. **A326**, 416 (1993) and **A349**, 321 (1994).
10. Z.-L. Zhou et al., Nucl. Instrum. Meth. **A378**, 40 (1996).
11. M. Ferro-Luzzi et al., contribution to this workshop.
12. M. Ferro-Luzzi et al., Nucl. Instrum. Meth. **A364**, 44 (1995).
13. Z.-L. Zhou et al., Nucl. Instrum. Meth. **A379**, 212 (1996).
14. B. Braun, private communication (1997).
15. Experiment CE70 at IUCF, T. Wise spokesperson for the PINTEX collaboration: UW-Madison, IUCF, Western Michigan Univ., and Univ. of Pittsburg.
16. M.A. Miller, private communication (1997).
17. N.F. Ramsey, Phys. Rev. **85**, 60 (1952).
18. "The Principles of Nuclear Magnetism", A. Abragam; W. C. Marshall and D. H. Wilkinson eds, Oxford University Press (1961).
19. M. Bloom, Physica **23**, 237 (1957).
20. E. Ter Haar et al., Physica **B 194-196**, 945 (1994).

Has the Atomic Beam Source Reached a Hard Intensity Limit?

Erhard Steffens

*Physikalisches Institut, Universität Erlangen-Nürnberg,
Erwin-Rommel-Str.1, D-91058 Erlangen, Germany*

Abstract. The intensity of Atomic Beam Sources for Storage Cell targets is reviewed. The limitations of present sources are analyzed, in particular losses by residual gas scattering and intra–beam scattering Simple quantitative estimates are given in order to illustrate their relative importance. More general calculations covering the full range of flow regimes are required to extract basic information from the many empirical studies performed up to now. Several new ideas as the Microwave dissociator and the Carrier Jet may substantially improve ABS targets for storage rings.

INTRODUCTION

The so–called Atomic Beam Source (ABS) is based on Stern–Gerlach separation of a thermal hydrogen beam. It consists of a dissociator with nozzle and a beam–forming differential pumping system, sextupole magnets as state selectors and RF transitions for switching nuclear and electronic polarization between high values of opposite sign. The beam of polarized atoms can be ionized to produce polarized H^+ or H^- ions. More recently, such beams have been used directly as a Jet target or for feeding a Storage Cell target [1]. Such "thin" polarized gas targets are exclusively used in storage rings, where in combination with high circulating currents of up to 100 mA very useful reaction rates can be obtained. "Thin" means that the blow-up of the stored beam caused by the target is slow or even compensated for by a suitable cooling mechanism. In practise targets up to areal densities of about $10^{15} H/cm^2$ can be used without affecting the beam life time. For polarized targets of such high densities a storage cell is required. ABS target sources for Jet targets or Storage Cell targets have quite different design criteria. Jet targets require high volume density at the IP. Therefore very slow beams are favorable as produced by the Ultracold Source [2]. Also focusing of a beam with large opening angle might result in a dense target spot [3]. For a storage cell target, the flow of atoms into the narrow feed tube of typically 10 mm diameter and

100 mm length has to be as high as possible, independent of their temperature [4]. Therefore, the following discussion is restricted to sources for Storage Cell targets.

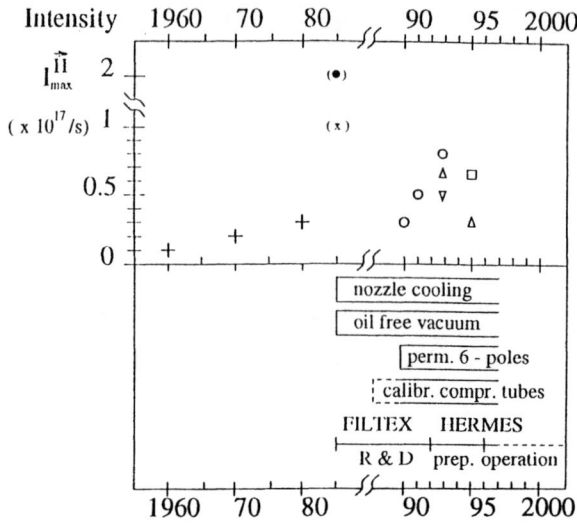

FIGURE 1. History of the intensity of atomic beam stages of ion or target sources.
Legend: + Review Grübler Ann Arbor (1981); (x) ETH source, unspec. acceptance; (•) INR Moscow, pulsed; ∘ FILTEX, CT 10/100, with HFTs; △ Wisconsin, CT 10/130, with HFTs, two (93) and single substate (95); ▽ HIPIOS, CT 22 dia., □ HERMES, CT 10/100, with HFTs. In the lower part the advent of certain source technologies is indicated, together with the FILTEX–HERMES development steps.

It is very instructive to review the performance of atomic beam stages of ion sources or target sources over the years. In Figure 1 selected intensities reported in the literature are given as function of the year, with some information on the measurement condition. It should be noted that for a precise absolute measurement of a flow rate a calibrated compression tube (CT) is required [5]. Such CTs are now in use with systematic errors of less than 10% [6]. Therefore earlier intensity figures quoted in the literature have to be considered with caution. - The diagram illustrates the rather slow progress in intensity over the years. It took 20 years from 1960 to 1980 to increase the intensity by a factor of three and 13 more years to nearly achieve another factor of three. Within the same period, the ion currents delivered by AB sources have gone up by four to five orders of magnitude, which was to a large extent due to improved ionizers. From 1990 on ABS target sources give the highest output flow rate. Those sources designed for high guide field need two RF transitions behind the last sextupole magnet, like in ion sources. The reduction in intensity due to the larger distance to the target cell can be seen by comparing FILTEX-93 with HERMES-95. In the bottom of Figure 1 some new techniques introduced into the source technology are shown. - The general trend exhibited in the diagram suggests the existence of an intensity limit at about $7\text{-}8 \bullet 10^{16}$ H/s of useful beam. In the following, we will attempt to discuss our present understanding of intensity limitations and possible ways out.

ESTIMATE OF THE ABS INTENSITY

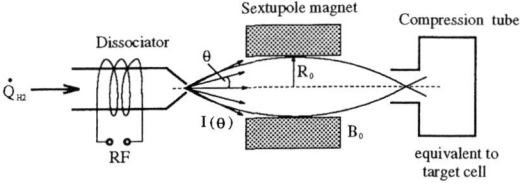

FIGURE 2. Sketch of the ABS components relevant for the atomic beam intensity.

A simple model of an ABS is shown in Figure 2. For the optimum velocity, the beam from the nozzle is focused into the compression tube, which mimics the target cell. The intensity of the nozzle beam $I(\theta)$ is given by $I(\theta) = I_o \cdot cos^n(\theta)$ with I_o being the forward intensity:

$$I_o = \frac{\dot{Q}}{\pi [sr]} \cdot \frac{n+1}{2} \quad (1)$$

The first factor corresponds to effusive flow, while the second factor $f_p = (n+1)/2$ is the peaking factor. The intensity of the focused H_1 beam can be described as follows [5]:

$$i(acc.H_1) = \alpha \cdot I_o \cdot \pi \theta_{max}^2 \cdot T \cdot (1-A) \quad (2)$$

with α = atomic fraction, θ_{max} = maximum initial slope of accepted trajectories, (1-A) = attenuation factor due to residual gas scattering, and T = transmission factor accounting for the velocity dependent focusing. For the HERMES source we have approximately an atomic fraction of $\alpha = 0.5$ and (1-A) = 0.7 which corresponds to 30% scattering losses. The maximum slope can be estimated from the pole tip field B_o. For the wave length λ of the sine–like trajectory $r = r_o \cdot sin(kz)$ we have

$$\lambda = 2\pi/k = 2\pi \cdot [mv_z^2 R_o^2/(2\mu_B B_o)]^{1/2} \quad (3)$$

With v_z = 1840 m/s, R_o = 12.5 mm and B_o = 1.5 T we obtain λ = 1.12 m. The initial slope θ_{max} is given by:

$$\theta_{max} = r'(0) = 2\pi \cdot R_o/\lambda \quad (4)$$

For the present assumptions, we get θmax = 70 mrad, resulting in a solid angle of $\Delta\Omega$ = 15.4 msr.

The forward intensity I_o can be measured using a set-up shown in Figure 3. Many experimental studies have been done to measure and optimize I_o ([7], [5],

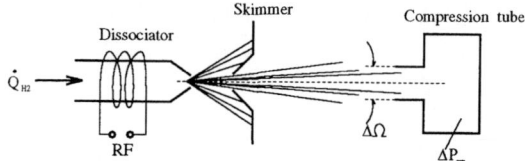

FIGURE 3. Basic arrangement for measuring the Forward Intensity I_o.

[8], [6], [9], [10]). By correcting for attenuation, peaking factors have been extracted, which turned out to be low. Values found were $f_p = 1.1$ [9] and 1.6 [6] for H_1, and 2.0 for He [10]. With $f_p = 1.6$ and $Q_{H_2} = 1.25$ mbar l/s we expect $I_o = 1.7 \cdot 10^{19}\ H_2/sr\cdot s$. Using the parameters mentioned so far and taking T = 0.85 [9] for the Wisconsin source, we calculate from Equ. 2: $i(H_1) = 7.8 \cdot 10^{16}\ H/s$. This is close to the best values reported (see Figure 1). It shows that in the present range of flow rates there are no dramatic losses by attenuation. This leads to the conclusion that a substantial improvement can only come from beam formation.

BEAM FORMATION

Unlike focusing of atoms by the multipole magnets, there is at present no means to perform realistic simulations on beam formation in order to design a beam-forming system consisting of nozzle, skimmer and a vacuum chamber with a certain pumping speed. Usually hydrogen is dissociated by a RF discharge (e.g. 13.56 MHz) at a pressure around 1 mbar. The gas flows out of the cooled nozzle into the first pumping stage towards a conical skimmer. The task is to model a jet of a binary mixture of H_1 and H_2 in the "Viscous Flow" regime. During expansion the transition to "Knudsen Flow" occurs which is finally followed by "Molecular Flow", i.e. independent motion of particles, governed only by the geometry of the walls. Calculations in the viscous flow regime have been performed using Navier–Stokes equations for multicomponent gas mixtures [16]. Molecular flow can be reliably treated by the Monte–Carlo method [11]. Another attempt using the "Direct Simulation Monte Carlo Method" from aerospace simulations [12] has been presented at the Madison workshop. This method has very high demands on computing time which will limit its applicability.

RESIDUAL GAS SCATTERING: Due to the lack of simulations, many empirical studies have been performed by systematic variation of geometry, flow rate and nozzle temperature ([7], [5], [8], [6], [9], [10], [13]). The basic set-up is shown in Figure 3. In some cases magnet dummies were placed on axis for simulating attenuation by the pressure bump in the bore of sextupole magnets. For residual gas "single–scattering losses" characterized by a total loss cross section σ one can derive a simple dependence of the flow rate $\dot{q}$ accepted by the compression tube on the nozzle throughput $\dot{Q}$ (Figure 4, S =

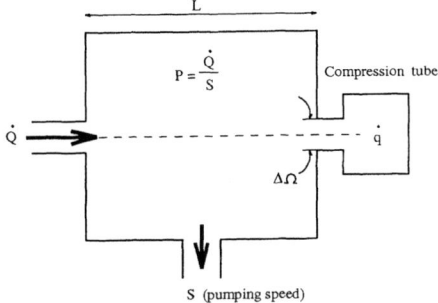

FIGURE 4. Simple model for Residual Gas scattering losses.

pumping speed):

$$\dot{q} = \frac{\Delta\Omega}{\Omega} \cdot \dot{Q} \cdot exp[-C_o \cdot \frac{\dot{Q}}{S} \cdot L \cdot \sigma] \quad (5)$$

$$\dot{q}(\dot{Q}) = a \cdot \dot{Q} \cdot exp(-b \cdot \dot{Q}) \quad (6)$$

Equ. 6 contains two constants a and b. As b is proportional to σ, b can be regarded as attenuation coefficient. Stock [14] showed that the dependence of the beam signals for H_1 and H_2 for various geometries and source conditions could always be fitted by Equ. 2.

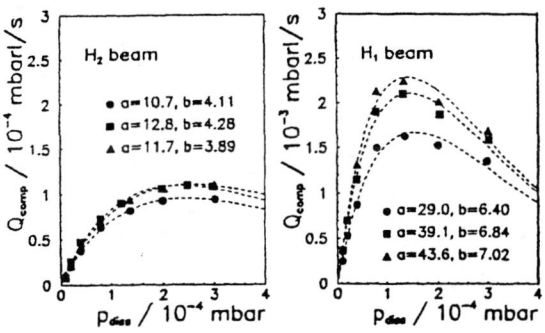

FIGURE 5. Intensity measurements (FILTEX) for 2 mm nozzle with H_2 (left) and H_1 (right) [14]. The fits according to Equ. 6 and the parameters are shown.

In Figure 5 various intensity scans for the (ballistic) H_2 and (focused) H_1 beams are shown together with the fit parameters. In Figure 6 the temperature dependence of the attenuation coefficient b obtained from another study is shown. A strong increase of b by a factor of 2–3 is observed if the nozzle temperature is lowered from 300 to 100 K.

INTRA–BEAM SCATTERING: Apart from residual gas scattering, there might be another important loss mechanism known from ion storage rings, the so-called Intra–Beam Scattering (IBS). Due to the different velocities, particles may scatter on each other, as shown in Figure 7. A simple

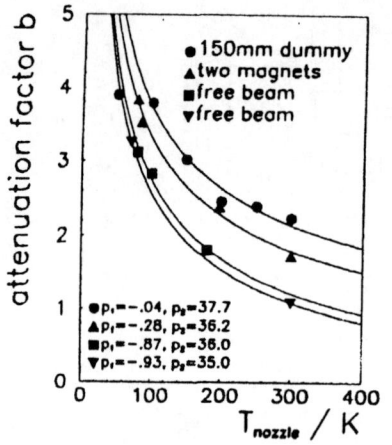

FIGURE 6. Dependence of the attenuation parameter b on nozzle temperature T. The fit was done with $b = p_1 + p_2/T^{1/2}$. A Fit with p_2/T dependence gave similar results [14].

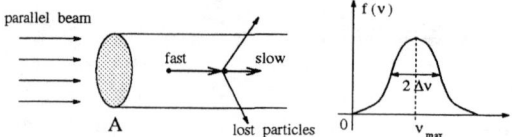

FIGURE 7. Losses due to Intra–Beam scattering caused by the velocity spread.

calculation with parallel beam and Δv being the velocity spread yields the following result. The flow density $\Phi(z)$ is governed by the differential equation

$$d\Phi/dz = -\alpha \cdot \Phi^2 \qquad (7)$$

with $\alpha = \frac{\Delta v \cdot \sigma}{v_{max}^2}$ and σ being the loss cross section for IBS. The solution

$$\Phi(z) = \Phi_o \cdot (1 + z/z^*) \qquad (8)$$

with $z^* = 1/\alpha\Phi_o$ is shown graphically in Figure 8 for $z \leq 30$ cm and three different initial flow densities Φ_o.

Here a loss cross section measured for residual gas scattering of $1.5 \; 10^{-14} cm^2$ [15] has been used, resulting in $\alpha = 2 \cdot 10^{-22}$ m s. Clearly a better σ value for the low relative velocities and the different systems $H_1 - H_1$ and $H_1 - H_2$ is needed. The present result shows that the ratio $\Phi(z)/\Phi_o$ is strongly density-dependent. For a typical ABS flight path of 0.5 and 1.0 m, respectively, the relative intensities are:

Φ_o [$cm^{-2}s^{-1}$]	Φ/Φ_o (z = 0.5 m)	Φ/Φ_o (z = 1.0 m)
10^{17}	91%	83%
10^{18}	50%	33%
10^{19}	9.1%	4.8%

The absolute flow density at $z = 1.0$ m increases only by a factor of 5.8, if Φ_o is increased 100 times. It should be noted that at the present flow

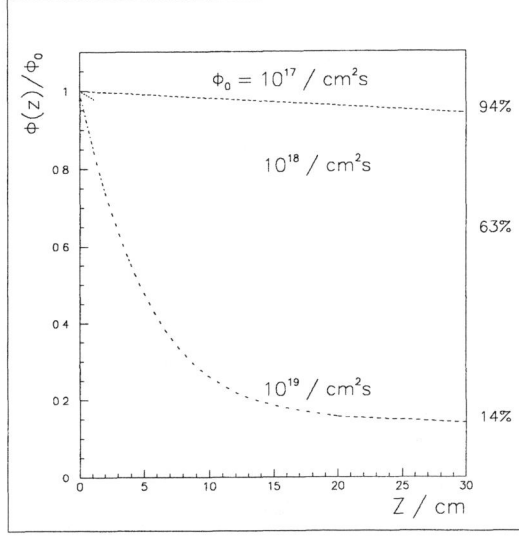

FIGURE 8. The ratio $\Phi(z)/\Phi_o$ calculated with Equ. 8 for various initial densities. For parameters see text.

densities of about $10^{17}/cm^2s$ only moderate IBS losses are expected. Due to the Φ^2 dependence in Equ. 7, a rapid onset of IBS losses is predicted at higher densities, which will probably pose a final limit to ABS intensities. The considerations presented here are clearly too simple to allow for a quantitative prediction and need to be refined.

NEW IDEAS

A weakness of the present concept of beam formation is the conflict between the requirement of *LOW* density in the dissociator for high atomic fraction, and *HIGH* density for obtaining a supersonic cold jet. There are two ideas to improve on this:

CARRIER JET: It has been proposed [16] to immerse the low-density atomic beam from the dissociator into a cold high-density carrier jet of H_2 in order to confine and cool the inner beam. The predicted flow rates of the carrier jet, which is formed by a coaxial annular nozzle, are one to two orders of magnitude higher than the present ones. A vacuum system capable of handling throughputs up to 50 mbar l/s is required for the first stage. According to the calculations, the carrier jet will be removed with high efficiency by the skimmer. The divergence of the free atomic beam after the skimmer is reduced due to the low temperature induced by the carrier jet. An experimental study of this new method is in preparation for the HERMES target.

MICROWAVE DISSOCIATOR (MWD): This new type of dissociator is based on a surface wave discharge at 2.45 GHz and was first developed at CERN [17]. A slit nozzle of 1×14 mm^2 opening at a temperature of 80 K has been used. Atomic fractions of about 80% were observed with 600–800 W of microwave power, a throughput of up to 4 mbar l/s and 1% oxygen admixture. The basic MWD components are a glass or quartz discharge tube with cooled nozzle, like in conventional dissociators, and a coaxial wave guide with a coupling slit, the so–called launcher. Using another rectangular tunable wave guide, effective impedance matching is provided.

A similar system for the HERMES ABS has been built and tested [18]. The tests are performed in a 4–stage pumping system with 14 m^3/s pumping speed. For improved beam formation the nozzle diameter is kept small, e.g. 2 mm, which results in higher pressure at the same flow rate. In contrast to the CERN source, the HERMES MWD runs with 10 times smaller O2 admixture, which increases the period, after which the nozzle needs to be warmed up, to several days. More studies of the long–term stability are underway. It is planned to install the new MWD in the coming shutdown. Compared with the old dissociator at the same H_2 flow rate of 1.25 mbar l/s, the MWD delivers 80% more atoms. As attenuation is the same, a similar increase of the H_1 beam is expected. Increasing the flow rate does not change α and may result in a further increase. Due to the property of the MWD to yield high dissociation at high throughput, it is the optimum dissociator to be operated with a dense carrier jet.

CONCLUSIONS

Due to the low electromagnetic cross sections, there is a strong demand from polarized target experiments at electron rings to increase the target density. Presently, there is no alternative to ABS target sources, which therefore are at the moment in an exciting innovative phase. Several new ideas have been discussed at this workshop. The MWD seems to be very promising and its full potential needs to be exploited. Theoretical methods to calculate the beam parameters after formation and the propagation through the magnet system including scattering losses are under development and need to be improved. The final goal of such simulations is to use them as a tool for the design of more powerful sources. The effect of a dense carrier jet on beam formation has to be studied experimentally and compared with the predictions.

All these developments appear quite promising. In combination they should provide a very significant improvement factor for the beam intensity, resulting in higher target density for experiments at storage rings.

Acknowledgement: I would like to thank R.Hertenberger and N.Koch for useful discussions and A.Golendukhin for critical reading and help in the preparation of the manuscript.

REFERENCES

1. W. Haeberli, AIP Conf. Proc. 128, 251 (1985)
2. V. Luppov, this Workshop
3. W. Kubischta, Proc. WS Pol. Gas Targets for Storage Rings, Heidelberg 1991, p.34
4. E. Steffens, Proc. Symp. Pol. in Nucl. Physics, Paris 1990, Colloque de Physique C6, 221 (1990)
5. W. Korsch, Proc. Symp. HE Physics, Bonn 1990, Vol.2, p.168.
6. F. Stock, Proc. WS on Pol. Ion Sources and Pol. Gas Targets, Madison 1993, AIP Conf. Proc. 293, 22 (1994)
7. D. Singy et al, Nucl. Instr. Meth. A278, 349 (1989).
8. A.V. Sukhanov and D. Toporkov, Proc. Symp. HE Physics, Bonn 1990, Vol.2, p.173.
9. A.D. Roberts et al, Proc. WS on Pol. Ion Sources and Pol. Gas Targets, Madison 1993, AIP Conf. Proc. 293, 10 (1994).
10. G. Cinque and W. Kubischta, Proc. WS on Pol. Beams and Pol. Gas Targets, Cologne 1995, 275 (1996).
11. H. Kolster et al, ibid., p.265.
12. I.D. Boyd and W. Kubischta, Proc. WS on Pol. Ion Sources and Pol. Gas Targets, Madison 1993, AIP Conf. Proc. 293, 44 (1994).
13. R. Hertenberger, this Workshop
14. F. Stock, Dipl.Thesis, University of Heidelberg (1992)
15. D. Toporkov, priv. comm. (1991).
16. V.L. Varentsov et al, this Workshop.
17. G. Boero et al, Proc. WS on Pol. Ion Sources and Pol. Gas Targets, Cologne 1995, 270 (1996).
18. N. Koch, this Workshop.

CONTRIBUTED TALKS

The Polarization Technology at The High Energy Laboratory of The Joint Institute for Nuclear Research

V.P.Ershov, V.V.Fimushkin, G.I.Gai, M.V.Kulikov, L.V.Kutuzova, M.Yu.Liburg, Yu.K.Pilipenko, A.I.Valevich, A.S.Belov[1]

Joint Institute for Nuclear Research, Dubna 141980, Moscow region, Russia
(1)Institute for Nuclear Research of Russian Academy of Sciences, Moscow, 117312, Russia

A spin physics is an important branch of the LHE JINR scientific program. The 200 μA cryogenic polarized deuteron source POLARIS is working at the JINR 10 Gev accelerator long time. There is a plan to use it at the new superconducting ring NUCLOTRON now. To increase the intensity of the polarized deuteron beam a pulsed charge exchange hydrogen plasma ionizer is developed. It will replace electron Penning ionizer. The new ionizer have been fabricated and it is tested now. To extend a physics internal target program at the NUCLOTRON ring a new pulsed polarized cryojet target is developed. Atomic beam stage have been constructed and installed on a test bench. A beam intensities measured by thermal detectors at 4.7 K are $5 - 8 * 10^{16}$ at/s. Looking for new ways of the increasing luminosity of internal target experiments a study of the ultracold atomic beam production is continue. Using the ATOM-H test bench a beam of 10^{17} at/pulse electron polarized cold hydrogen atoms has been got. To increase an input atomic hydrogen flow a new powerful dilution refrigerator is constructed.

Introduction

Over ten years the continuous growing interest to high energy spin physics is demonstrated. It is explained by several exciting results in spin physics experiments. In 1981 the polarized deuteron beam from the cryogenic source POLARIS has been accelerated at JINR synchrophasotron to 4.5 GeV/nucleon energy. (Fig.1). A spin physics program was started at LHE and successfully continues and essential to day. Commission of the new superconducting accelerator NUCLOTRON gives a good chance to extend spin physics ex-

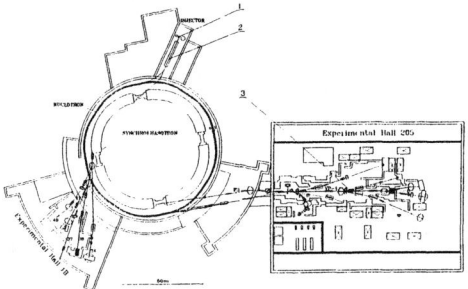

Figure 1: A layout of experimental spectrometers at the synchrophasotron and NUCLOTRON rings. 1 - POLARIS, 2 - 5 MeV/nucleon linac, 3 - two-arm polarimeter ALPHA.

periments. A short review of developments of the polarizing apparatus (jet and ion sources), at the LHE JINR are presented.

Cryogenic source of polarized deuterons POLARIS.

Many years the source runs at JINR 10 GeV synchrophasotron. There is a plan to use it at a new superconducting accelerator NUCLOTRON now, Fig.2 (more detail see [1,2,3,4,5,6]. The source is based on the Stern-Gerlach atomic beam method and has the following features:

- magnetic fields are set up by superconducting magnets operating in a persistent current state,
- vacuum in the source is obtained due to cryopumping of gas,
- cooling of the dissociator, nozzle and skimmer is produced by a thermal contact with the cryostats,
- the source is very compact and requires power only for RF and control systems,
- it is installed on a 700 kV terminal, an information exchange is performed by a fiber glass lines.

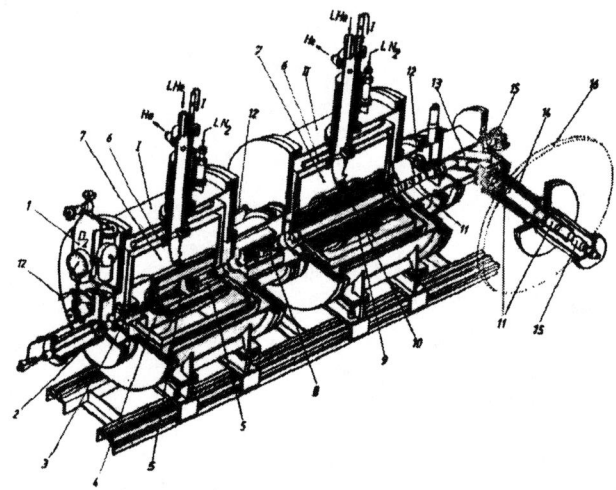

Figure 2: A general view of the polarized deuteron source POLARIS. I - polarized atomic source, II - ionizer. 1 - deuterium volume, 2 - electromagnetic gas valve, 3 - dissociator, 4 - nozzle chamber, 5 - SC sextupole magnets, 6 - nitrogen shield, 7 - helium cryostat, 8 - RF cell, 9- SC solenoid, 10 - electron optics, 11 - ion optics, 12 - vacuum gate, 13 - electrostatic mirror, 14 - solenoid of the spin-precessor, 15 - Faraday cup, 16 - position of the preaccelerator flange.

The polarization of the high energy deuteron beam are the following: P_z^+ = 0.54± 0.01, P_z^- =−0.57±0.01, P_{zz}^+ = 0.76± 0.02, P_{zz}^- =−0.79± 0.02 [7].

The acceleration of the polarized deuterons is usually produced once a year.

The accelerator runs have a 3-5 weeks duration. The source can stable run about a month without sublimation of the condensed gas. A week need of the cryogenic liquids is 200 l of the LHe and 300 l of the LN_2.

To increase the intensity of the polarized deuteron beam from POLARIS a short pulse charge exchange ionizer is developed [6,8] ,(Fig.3). The fabrication of the new ionizer is over and test bench experiments are started.

Polarized deuterons are produced by change exchange between polarized deuterium atoms and ions of hydrogen plasma $D_\uparrow^0 + H^+ = D_\uparrow^+ + H^0$ and ionization of $D_\uparrow^0$ by plasma electrons. The energy of the deuteron beam will be 15 keV at the emittance 0.2πcm*mrad.

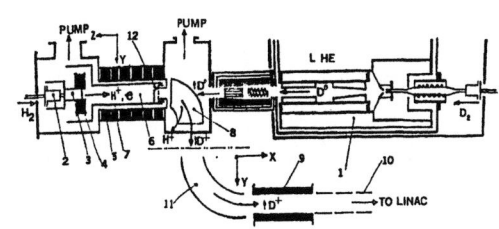

Figure 3: Schematic view of the warm POLARIS charge exchange ionizer. 1 - cryosource, 2 - H_2 gas valve, 3 - hydrogen plasma source, 4 - magnet, 5 - solenoid, 6 - charge-exchange volume, 7 - 15 kV shield, 8 - deflecting magnet, 9 - spin-rotator, 10 - beam lens, 11 - spherical mirror, 12 - extracting grids and forming electrode.

Varying parameters of the arc plasma source (3) $0.3 - 1$ A H^+ beam has been measured at the electrode after extracting grids. Typical signals at Faraday cap after analyzing magnet (8) are $40 - 70$ mA H^+ and $3 - 4$ mA H_2^+. To measure efficiencies of the ionizer at different gas target thicknesses a special pulsed dissociator (simulator of ABS) has been installed before magnet (8) (it is not shown). When D_2 is pulsed into charge exchange volume H^+ signal (mass 1) at Faraday cap is decreased on 10-30 mA. The same time D^+ current (mass 2) is increased up to 3 mA depend on the D_2 gas quantity . There is no difference has been observed if the dissociator is switched on. The efficiencies of the ionizer is estimated as: $\varepsilon = N_D^+ / N_D^0$ where, $N_D^+ = 6.25*10^{18} * I_D * \Delta t$ – is number of D^+ ions after deflecting magnet, $N_D^0 = n_0 * l * s$ – is number of atoms crossed by the plasma beam. The efficiencies of the ionizer at the target thickness $n_0 l = 0.4 - 1 * 10^{14}$ atoms/cm^2 is $3 - 5$%. Using POLARIS atomic beam stage about 0.5 mA polarized D^+ beam at the output of deflecting magnet of the ionizer has been obtained.

To get full advantages of the cryosource the ionizer should be cryogenic too. So we developed a cryoversion of the charge exchange ionizer and did some preliminary tests. It is a combination of the POLARIS LHe cryostat with 60 mm cold bore superconducting solenoid and H^+ plasma source of the warm charge exchange ionizer. The 1.2 A H^+ plasma current is observed after extracting grids. A vacuum in the cryostat due to cryopumping of hydrogen just before a plasma pulse is $2 - 4 * 10^{-6}$ mbar. To improve the vacuum cryosorbshen of hydrogen should be released. The study of the setup will be continued.

Pulsed cryogenic polarized atomic deuterium (hydrogen) jet target.

For polarization experiments at the JINR superconducting accelerator NUCLOTRON the pulsed cryogenic polarized atomic beam hydrogen jet target has been designed and fabricated [9,10]. The target is designed as a cryogenic apparatus with a vertical cryostat with three superconducting sextupole magnets inside (Fig.4). The cryostat has a complicated shape to supply the best evacuation of a scattered gas from the beam region. There are three functions of the source cryostat:
- cryopumping of the gas,
- cooling of the sextupole magnets,
- cooling of a dissociator nozzle.

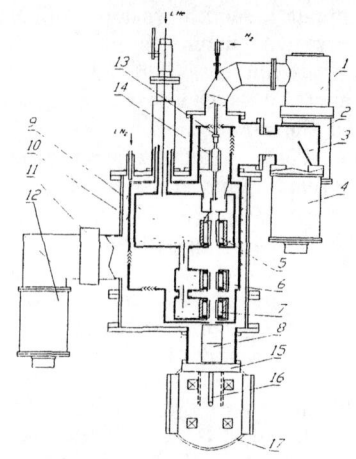

The cryogenic has many advantages, but there is a specific problem. A vapor pressure of solid hydrogen (compare to deuterium) at 4.2K cryosurface is relatively high ($2*10^{-6}$ mbar). To get better vacuum in this case the cryosurfaces and turbomolecular pumps are used (Fig.4).

Figure 4: Schematic of the jet. 1,3,11,15 - valves, 2,4,12 - TMP (500 l/s), 5,6,7 - SC 6p magnets, 8 - RF cells, 9 - cryostat, 10 - nitrogen shield, 13 - gas valve, 14 - dissociator, 16 - storage cell, 17 - interaction box.

Atomic hydrogen is produced by an RF discharge of 80 MHz in a 7 cm^3 pyrex volume. The dissociator tube is closely fitted into a teflon block having a thermal contact with a shield at 90 K. There is a photodiode with a red light filter to check the dissociation discharge. After the dissociator the gas flow runs out through a teflon pipe (l=50 mm, ϕ 6 mm) and a conical aluminum nozzle (l=10 mm, ϕ5/2.8 mm). The nozzle temperature can be varied in a range of 16−80 K by a thermal switch. The atomic beam is formed by a skimmer (ϕ 4.7 mm) and a collimator (ϕ 10 mm).

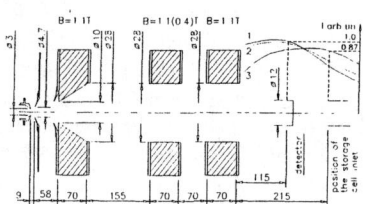

Figure 5: Magnet configuration and the calculated beam intensity into a ϕ 12 mm aperture vs it's position along the beam axis (1,2,3). 1 - B1=B2=B3=1.1T; 2 - B1=B2=1.1T, B3=0.7T; 3 - B1=B3=1.1T, B2=0.4T

Spatial spin separation of the atomic beam takes place in the gradient field of three superconducting sextupole magnets. Maximum dimensions of the magnets are rather small (yoke diameter ϕ 170 mm, lengths 104 mm and 68 mm). The coils and yoke of each SC magnet are placed inside the cryostat volume and separated from vacuum by a stainless steel pipe (ϕ 62/60 mm) inserted into the magnet bore. Pole tips made of FeCoV alloy are combined in a block and inserted into the pipe. The maximum magnetic field at pole tips is 1.1 T. The pole tip configuration is presented in Fig.5.

To produce a nuclear polarization the target will be equipped with two RF cells. The RF cells will be placed behind the magnets. It allows to operate with the maximum intensity. A choice of the RF cells depends on a particular physical task. The transitions 1-3 and 2-4 will be realized to produce both signs of polarization for hydrogen without a magnetic field reversion in the interaction region. The transitions 1-4 and 3-6 will produce both signs of vector polarization for deuterium without the magnetic field reversion in the interaction region. The transitions 3-5 and 2-6 will produce both signs of a tensor polarization for deuterium. It is possible to have another set of cells when a medium field transition sell can be used. There is an intent to use a storage cell for polarized atoms in the NUCLOTRON ring.

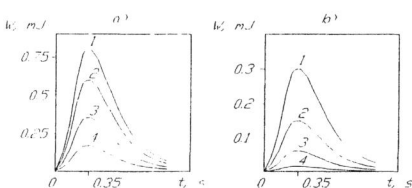

Figure 6: The detector heat loads. Gas and RF pulses are 150 ms. Gas throughput is 0.045 mbar*l/pulse. Nozzle temperature is 27 K. a - deuterium states 1,2,3 , b - hydrogen states 1 and 2. 1 - all magnets are turned on, 2 - the first and second magnets are turned on, 3 - the first magnet is turned on, 4 - all magnet are turned off.

Some preliminary tests of the target are performed. The measurements of atomic beam intensity are performed by a thermal detector under registration of the atomic recombination energy [11]. To match the focusing properties of the magnets and the velocity distribution of the atomic beam the nozzle temperature was scanned over a range of 16 – 80 K. The optimum nozzle temperature is about 27K. The pulse duration is varied from 0.02 to 1 s. The optimum feed rate at 0.15 s pulse is 0.045 mbar*l/pulse or about 0.3 mbar*l/s. The detector signals due to recombination energy of deuterium and hydrogen beams are shown in Fig.6. An incomplete accommodation of the atomic recombination energy released on the detector surfaces [12,13,14] has required a separated calibration of the detector.

It was found a fraction ε of the beam atom recombination energy registered by the detector is 18% for deuterium and 11% for hydrogen [11]. These values are used for intensity measurements. The maximum recombination heat loads on the detector at the optimum source operation are W_D=0.78 mJ/pulse, W_H=0.30 mJ/pulse for deuterium and hydrogen respectively. The averaged beam intensities are estimated as:

$$I_D = W_D/(\varepsilon_D * \Delta t * 0.5 * E_{recomb}) = \quad 8*10^{16} at/s \quad , \quad I_H = 5*10^{16} at/s \quad ,$$

where, $\Delta t-$ is the pulse duration, E_{recomb} – recombination energy of a pair of deuterium (hydrogen) atoms (4.48 eV). Some beam focusing simulations were performed to predict an intensity of the beam at the position of the storage cell inlet (Fig.5). According to simulations the optimum beam focusing into the storage cell inlet is reached for the magnetic field of the second magnet at 0.4 T. Under this a loss of intensity is 13%.

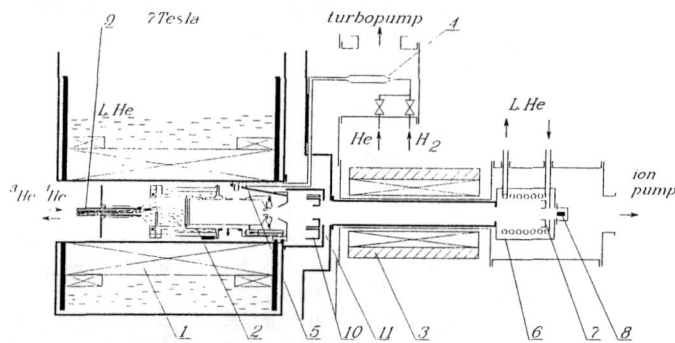

Figure 7: The schematic view of the ATOM-H setup. 1 - SC solenoid, 2 - mixing chamber and cell, 3 - sextupole magnet, 4 - dissociator, 5 - teflon pipe and accommodator, 6 - LN$_2$ shield, 7 - thermal detector, 8 - ion gauge, 9 - final ^{3}He heat exchanger, 10 - peripheral detector, 11 - central detector.

Ultracold source of polarized atomic hydrogen beam.

Ideas of an ultracold source with microwave extraction are discussed in [15,16,17]. Proposal and studies of an ultracold electron polarized atomic beam extraction by high gradient magnetic field of superconducting solenoid instead of microwave technique are described in [18,19,20,21]. We continue a study of the ultracold method using upgrade setup ATOM-H developed for stabilization of atomic hydrogen [22].

Atomic hydrogen is produced by RF discharge in a pyrex dissociator, cooled down in a teflon pipe and accommodator to 8-10 K (Fig.7). A cold atomic hydrogen jet is injected into the low temperature cell (0.25 K) through a vacuum gap. After thermalization ultracold $H\uparrow$ atoms are pushed out of the cell, accelerated by the back side gradient magnetic field of the solenoid up to velocities corresponding to ~ 3 K forming a polarized beam.

The low temperature cell is cooled down by a 10 mW ^{3}He-^{4}He dilution refrigerator. A mixing chamber of the refrigerator and the low temperature cell are designed as a single block. The cell has a high thermoconductivity and its temperature is $0.15 - 0.3$ K. The cell is located in a gradient magnetic field of the superconducting solenoid so the magnetic fields at the positions of entrance and exit holes are about 80% and 50% of the maximum field (7T) respectively.

To study properties of the beam and an efficiency of atomic separation an additional sextupole, thermal and ion detectors were installed.

To prevent recombination of the hydrogen atoms the cell surface is covered by a superfluid helium film. The film thickness depends on a setup vacuum pressure. At the typical pressure of $1.5*10^{-7}$ mbar the film thickness measured by the bolometer is about $3 nm$.

Due to insufficient refrigerator cooling capacity the setup operates pulsewise. Each cycle starts every ten seconds with an injection of helium ($4*10^{17}$ mol/pulse) to restore the cells film. Hydrogen and RF discharge pulses (t = 300 ms) follow one time in four seconds. The hydrogen flow rate is $4.5*10^{17}$ mol/pulse and the flow of atoms is estimated as $6*10^{17}$ at/pulse. A part of atoms entering the cell is recombined and their heat can be measured by the calorimetric method. The same idea is used to measure the beam flux on the thermal detectors. There are two ring thermal detectors which are placed near the cell (central detector $\phi 28/15$ mm, peripheral detector one $\phi 64/30$ mm) and one final detector after the sextupole ($\phi 42/12$ mm).

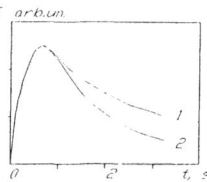

Figure 8: Signals on the peripheral detector. 1 - atomic beam pulse, 2 - electrical amplitude calibrating pulse.

A typical atomic beam pulse and an electrical calibrating signal are shown in Fig.8. The intensity of atomic beam intensity is estimated as $1*10^{17}$ cold at/pulse.

To increase an input atomic hydrogen flow and prevent evaporation of the superfluid film the low thermal resistance cell and new powerful 100 mW dilution refrigerator have been developed. The dilution refrigerator consists of a cryostat with two LHe (4K,1.3K) bathes, still and some heat exchangers. Circulation of ^{3}He-^{4}He mixture is provided by a sealed 35 mmol/s three stage pumping station (4700, 700 m^3/hour Roots pumps and 72 m^3/hour mechanical pump). To prevent a blockage of the refrigerator heat exchangers by frozen impurities the ^{3}He-^{4}He mixture passes through a purification unit.

Hydrogen and deuterium atomic beam polarized DC target [23] for internal experiments at UNK, Protvino was designed and mainly fabricated at JINR . The time when the accelerator will start to work is delayed so the target project temporary is stopped.

[1] A.A. Belushkina et al,- In: High Energy Physics with Polarized Beams and Polarized Targets, Argonne in 1978, AIP Conf.Proc. **51** (1979) 351.

[2] A.A. Belushkina et al, -In: High Energy Physics with Polarized Beams and Polarized Targets, Lausanne in 1980, Birkhauser **EXS-38**, Basel (1981) 429.

[3] N.G. Anischenko et al, -In: The 5th Int. Symp. on High Energy Spin Physics, Brookhaven in 1982 AIP Conf. Proc. **95** (1983) 445.

[4] N.G. Anischenko et al, -In: The 6th Int. Symp. on High Energy Spin Physics, Marseille in 1984, Jorn.De Phys., Colloque C2, Supplement an n° 2, **46** (1985) C2-703.

[5] V.P. Ershov et al, JINR Communication, Dubna, **E13-90-331** (1990).

[6] V.P. Ershov et al. - In: Int. Workshop on Polarized Beams and Polarized Gas Targets, Cologne in 1995, World Scientific, Singapur, (1996) 193.

[7] V.G. Ableev et al., Nucl.Instr. and Meth., **A306**, (1991) 73.

[8] A.S. Belov et al., Nucl. Instr. and Meth., **A255**, (1987) 442.

[9] V.P. Ershov et al. -In: Int. Workshop on Polarized Beams and Polarized Gas Targets, Cologne in 1995, World Scientific, Singapur, (1996) 189.

[10] V.P. Ershov et al. -In: The 12th Int. Symp. on High Energy Spin Physics Amsterdam in 1996, World Scientific, Singapur (1997) 407.

[11] V.P. Ershov et al, Nucl. Instr. and Meth., **A385** (1997) 435.

[12] R.T. Brackmann, W.L.Fite J. Chem. Phys., **34**, N5 (1961) 1572.

[13] G. Marenco et al, J. Vac. Sci. and Techn., **9** (1972) 824.

[14] A. Schutte et al, J. Chem. Phys., **64** (1976) 4135.

[15] T.O. Ninikoski -In: Int.Symp. on High Energy Physics with Polarized Beams and Polarized Targets, Lausanne in 1980, Birkhauser **EXS-38**, Basel (1981) 191.

[16] D. Kleppner and T.J.Greytak -In: The 5th Int. Symp. on High Energy Spin Physics, Brookhaven in 1982, AIP Conf. Proc. **95** (1983) 546.

[17] T. Roser et al., Nucl.Instr. and Meth., **A301** (1991) 42.

[18] M. Mertig, et al, -In: The 9th Int. Symp. on High Energy Spin Physics, Bonn in 1990, Springer-Verlag, **2** (1991) 164.

[19] V.P. Ershov et al.-In: Workshop on Polarized Gas Targets for Storage Rings, Heidelberg, Max-Planck-Institut (1991) 68.

[20] M. Mertig et al.-In: Workshop on Polarized Gas Targets for Storage Rings, Heidelberg, Max-Planck-Institute (1991) 87.

[21] V.P. Ershov et al. -In: Int. Workshop on Polarized Sources and Polarized Gas Targets, Cologne in 1995, World Scientific, Singapur (1996) 185.

[22] V.G. Luppov et al.- In: JINR Rapid Communications, Dubna, **5[31]-88** (1988) 21.

[23] V.P. Ershov et al.- In: Int. Workshop on Polarized Sources and Polarized Gas Targets, Cologne in 1995, World Scientific, Singapur (1996) 256.

A Polarized Atomic-Beam Target for COSY-Jülich

P.D. Eversheim*, M. Altmeier*, O. Felden*, M. Glende*,
M. Walker*, A. Hiemer* and R. Gebel[†]

*Institut für Strahlen- und Kernphysik[1] Universität Bonn, D-53115 Bonn, Germany
[†]Institut für Kernphysik, Forschungszentrum Jülich, D-52425, Germany

Abstract. An atomic-beam target (ABT) for the EDDA experiment has been built in Bonn and was tested for the very first time at the cooler synchrotron COSY. The ABT differs from the polarized colliding-beams ion source for COSY in the DC-operation of the dissociator and the use of permanent 6-pole magnets. At present the beam optics of the ABT is set-up for maximum density in the interaction zone, but for target-cell operation it can be modified to give maximum intensity. The modular concept of this atomic ground-state target allows to provide all vector- (and tensor) polarizations for protons and deuterons, repectively. Up to now the polarization of the atomic-beam could be verified by the EDDA experiment to be $\gtrsim 80\%$ with a density in the interaction zone of $\gtrsim 10^{11}$ atoms/cm^2.

THE ATOMIC-BEAM TARGET

Two experiments are presently planned with an atomic-beam target (ABT) at the cooler synchrotron COSY at Jülich: The EDDA experiment, which measures $\vec{p}$-$\vec{p}$ excitation functions of polarization observables, and a time-reversal invariance (TRI) test [1]. It has been demonstrated that due to its circular design around the beam-pipe the EDDA detector also serves as an ideal internal polarimeter that allows to monitor simultaneously the luminosity too. In view of the space, the EDDA detector left, the design of the ABT had to compromise on the distance of the last 6-pole magnet to the interaction point in the beam-pipe. According to ray-tracking calculations this costs about 30% in density (cf. Fig. 1).

The atomic-beam of the ABT is produced in an atomic-beam source which is essentially identical to the colliding-beams ion source (CBS) of COSY as

---

[1]) This work was supported by the BMWF and the Forschungszentrum Jülich, Germany

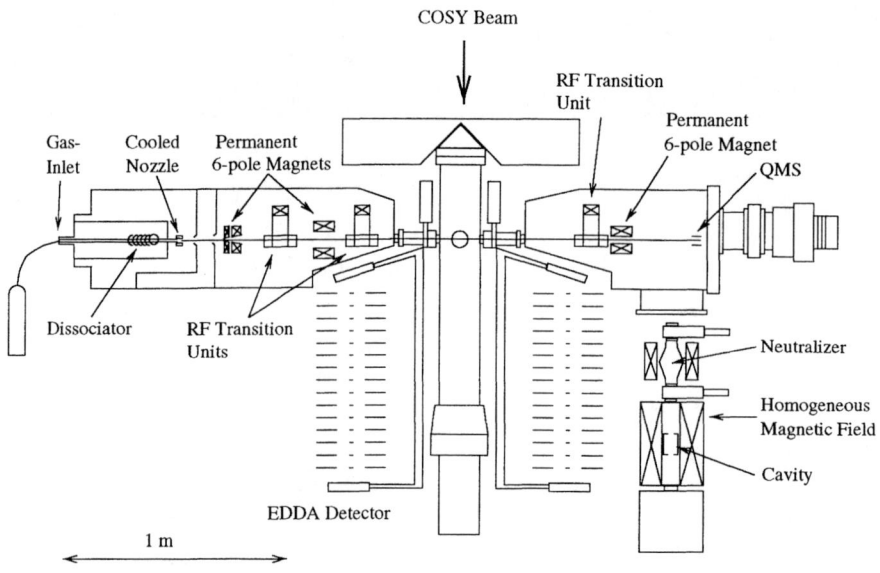

FIGURE 1. The atomic-beam target with the EDDA detector.

described in [2]. The H2/D2 molecules are dissociated in an inductively coupled 350W RF-discharge and pass subsequently through an aluminium nozzle cooled to about 30K, a skimmer and two groups of permanent 6-pole magnets. Thus, the atoms are slowed down with the consequences that the dwell time of the atoms in the interaction region and the solid angle of acceptance of the magnets increases. These beneficial effects are in part offset by gas scattering in- and outside the nozzle. In order to reduce chromatic aberrations the atomic-beam transport system comprises the second (group) of "compressor" 6-pole magnets. Between these groups of magnets or behind the compressor magnet(s) up to three Abragam-Winter RF-transition units can induce transitions between the hyperfine states of hydrogen or deuterium. Therefore, any reasonable population of the hyperfine states and thus all vector- (and tensor) polarizations for protons and deuterons can be achieved, respectively [3].

If the source has to feed a target-cell, the atomic-beam must be intensity optimized. In this case the nozzle temperature will be increased to about 50K with the consequence that the focusing power of the compressor 6-pole magnet has to be increased by an additional magnet.

Differences between the atomic-beam sources of the ABT and the COSY-CBS exist in the front part of the ABT and in the 6-pole magnet design. Starting from the successful design of the electrically driven COSY-CBS magnets, which corresponds to the magnet design of the sources of the ETH-Zürich and PSI [4], the 6-pole magnets for the ABT in COSY are now made out of permanent magnets (Nd-Fe-B alloy [5]).

The final 6-pole magnet design of the ABT was derived from a modified program code [6], which calculates the beam transport through skimmers and magnets by a fast linear matrix approach. The predictive power of the code has been successfully verified by the performance of the polarized ion source of the Bonn isochronous cyclotron and the atomic-beam source of the COSY-CBS. Meanwhile the code comprises an optimization algorithm that tunes the magnet design with respect to a given figure of merit.

A comparison with several atomic-beam transport designs [3], as described in literature, shows that the calculated intensities and densities of the program code exceed the measured ones by consistently $\sim 25\%$. Since no additional loss mechanisms except hitting of obstacles during the beam transport is taken into account, these 25% also give an upper limit for additional losses and nonlinearities. For instance intra-beam scattering in the vicinity of the nozzle is discussed in this respect. On the other hand, we know from dynamic measurements of the dissociator efficiency by means of a quadrupole mass-spectrometer that hydrogen atoms do not necessarily recombine to 100% at ordinary stainless steel surfaces. A possible consequence is that the 25% discrepancy between calculated and true intensity may be even smaller.

The atomic-beam target is usually operated at 0.5 mbl/s hydrogen flow. It could be shown that though the degree of dissociation remains 80% even beyond a flow of 1.5 mbl/s the atomic-beam density drops. On the other hand, in case the turbo molecular pumps are switched off for a while until the pressure rises by a factor three, results in a 5 – 10% reduction in atomic-beam density only. This shows that the pumping speed should be sufficient up to a flow of 1.5 mbl/s. Since our ray-tracking code gives reliable results all through the molecular flow regime, the problem seems to be located in the nozzle or in the immediate vicinity thereof.

At present the measurements of the EDDA detector give for the atomic-beam polarization $\gtrsim 80\%$ and a density of $\gtrsim 10^{11}$ atoms/cm^2. These results agree within the error bars with our expectations, derived from calculations as well as measurements with compression tube and quadrupole mass-spectrometer.

REFERENCES

1. cf. separate contributions in this volume
2. R. Gebel, Ph.D. Thesis, Inst. für Strahlen- und Kernphys., Univ. Bonn (1994)
3. P.D. Eversheim et al., Proc. 3rd Int. Conf. Nucl. Phys. at Storage Rings (STORI 96), Bernkastel-Kues, Germany, Nucl. Phys. A, special issue, accepted for publ.
4. D. Sing et al., Nucl Instr. Meth. A306 (1991) 36
5. P. Schiemenz, A. Ross and G. Graw, Nucl Instr. Meth. A305 (1991) 15
6. C.R. Meitzler, Code STRAHL, Brookhaven National Lab. (1989)

Polarized Beam for the IUCF Cooler Injector Synchrotron

Vladimir P. Derenchuk* and Alexander S. Belov[†]

*Indiana University Cyclotron Facility,[1]
Bloomington, IN 47408, USA
[†]Institute for Nuclear Research of the Russian Academy of Sciences,
Moscow, 117312, Russia

Abstract. The Indiana University Cyclotron Facility (IUCF) is in the process of commissioning a new synchrotron that is designed to rapidly accumulate, accelerate and inject beam into the IUCF Cooler ring. [1] The Cooler Injector Synchrotron (CIS) is filled by the strip injection of 7 MeV negative ions that were pre-accelerated by an RFQ-DTL linear accelerator. This pre-accelerator has an acceptance of 1.0 π mm mrad normalized with an injection energy of 25 keV. The injected beam should be pulsed with a maximum repetition rate of 5 Hz, have a maximum pulse width of 300 μs and a peak beam current that exceeds 300 μA. Beam commissioning will be complete by the fall of 1998 at which time CIS will be used exclusively to fill the Cooler ring with both unpolarized and polarized beams. A pulsed source of polarized negative hydrogen and deuterium ions is under construction at IUCF. The source will consist of a pulsed atomic beam which is ionized by the resonant charge exchange method [2]. Progress in the design and construction of the ABS will be described.

INTRODUCTION

The Cooler Injector Synchrotron (CIS) will replace the IUCF cyclotrons as the beam injector for the Cooler ring. CIS is filled by strip injection of H$^-$ or D$^-$ beam, pre-accelerated by an RFQ/DTL to 7 MeV and 4 MeV, respectively on a 1 Hz to 5 Hz repetition rate. A new pulsed polarized ion source is being designed and built at IUCF to fill the polarized beam requirements for operation of the Cooler with CIS. The Cooler Injector Polarized Ion Source (CIPIOS) must be completed and ready for operation by the end of 1998. A 25 keV H$^-$ beam current of about 300 μA within an emittance of 1.0 π-mm-mrad is required to meet luminosity requirements of experiments in the Cooler

---

[1] This work is supported by a grant from the National Science Foundation, NSF PHY-9724216.

synchrotron. The source is pulsed at a maximum of 2 Hz initially and with a minimum pulse length of 150µs. By comparison, the polarized *positive* ion source at IUCF, HIPIOS [3] normally produces over 150 µA of DC beam in an emittance of 0.5 π-mm-mrad.

The decision was made to build an atomic beam type ion source with a sextupole focusing system using permanent magnets. Matching the beam types to the proposed experiments that require both vector polarized proton and pure tensor deuteron beams, and minimizing the cost and construction time by utilizing existing equipment and expertise were the primary criteria for this choice. The ionizer design is the resonant charge-exchange ionizer similar to one developed by A. Belov at INR-Moscow [2].

IUCF is fortunate to be able to use parts of the University of Washington polarized H$^-$ ion source [4]. Work required to design the atomic beam focusing system and pulsed dissociator testing has begun and will continue until the sextupole assemblies are constructed, scheduled for the spring of 1998. Measurements of the atomic beam characteristics, velocity distribution, flux and molecular fraction, will be used as input for the design of the permanent magnet sextupoles. The sextupoles will be designed for the characteristics of the pulsed dissociator and optimized to focus the atomic beam into a 1 cm diameter at the ionizer.

A combination of an optimally designed pulsed ABS plus resonant charge exchange ionizer will allow us to meet or exceed the CIS and Cooler experiment requirements. Acceleration of unpolarized H$^-$ beams in the RFQ has already been tested; an additional set of RFQ vanes is required to accelerate D$^-$ beams. No additional modification to CIPIOS or the DTL is necessary for accelerating polarized D$^-$ beam except for a change of gas in the ABS and ionizer and replacing some of the transition units.

ATOMIC BEAM SECTION

This source of polarized atomic beam will operate with a nozzle cooled by a He refrigerator. The final operating temperature of the nozzle must be determined as part of the design effort of the sextupole system. The design goal is to transmit a beam of 1 cm diameter with an intensity of about 2×10^{17} atoms/s of peak flux (pulsed) into the ionizer optimized for maximum intensity. For pure vector and tensor states of deuterium, one transition unit after the first sextupole system is required followed by two transition units upstream of the second sextupole. Table I displays the polarization states available for hydrogen and deuterium beams. The transition scheme is patterned after the source at TUNL [5].

A pulsed gas valve similar to the one developed for the CIS unpolarized source will be used to inject hydrogen gas into the dissociator. A 6 kW pulsed RF power supply will be used to excite the dissociator discharge. The velocity

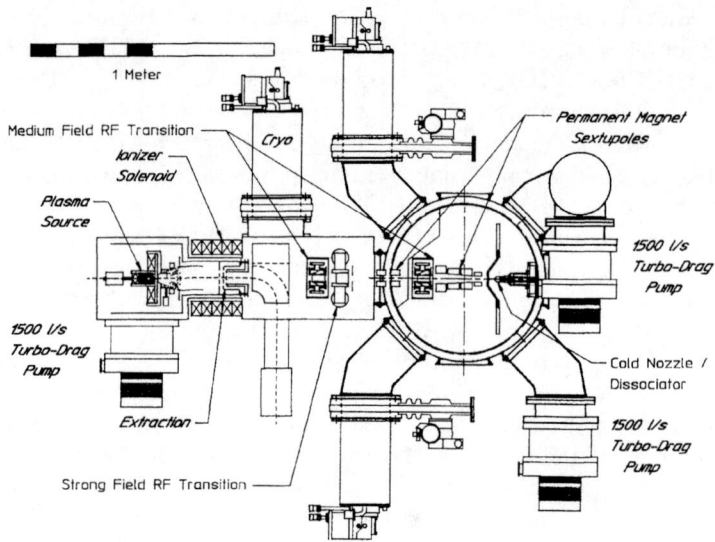

Figure 1. CIPIOS atomic beam section with ionizer. The second set of sextupoles will be electromagnetic.

distribution and the dissociation fraction and flux of the pulsed atomic beam from the dissociator will be measured with a 400 Hz chopper, a fast ion gauge [6], and a time-of-flight mass spectrometer [8].

	Transition Units On			Maximum Pol.		
	MFT1	Sext	Strong	MFT2	P_z	P_{zz}	Relative Intensity
PROTONS:		ON	$2 \to 4$		+1		1
		ON		$1 \to 3$	-1		1
DEUTERONS:		ON		$1 \to 4$	-2/3	0	1
		ON	$3 \to 6$		+2/3	0	1
	$1 \to 4$	ON	$3 \to 5$		0	-2	2/3
	$1 \to 4$	ON	$2 \to 6$		0	+1	2/3

Table 1. Polarization states available for hydrogen and deuterium beams patterned after the TUNL [5] scheme. Note that the use of RF transitions before the second sextupole will change the beam intensity.

Results of the velocity distribution and transmission property measurements of the pulsed atomic beam will be used to determine the dissociator operating parameters and to design the sextupole magnets that will optimally match the atomic beam to the CIPIOS ionizer. Much of the design will rely on experience gained by the polarized target group in Wisconsin [7] and from the work done by Belov [8]. Features of the sextupole geometry will include a large inner diameter for a long focal length and maximum transmission to the ionizer, the possibility of using a less expensive electromagnet in place of the second sextupole set and space for location of the transition units required for tensor polarized deuterium.

Strong field transition units will be constructed from designs already in use at IUCF. The $3 \to 6$ transition unit will be constructed using a HERMES [10] design. New medium field transitions will be designed after existing Wisconsin and TUNL units. This combination of transition units will give the states listed in Table I for polarized protons and deuterons.

The vacuum chamber has been built from existing equipment. Since the dissociator is pulsed, the hydrogen gas load is considerably reduced when compared to DC operation. Initial operation has been with two low compression ratio 1500 l/s turbo-pumps on the dissociator and 1500 l/s cryo pumps on the remaining chambers. Operation with these pumps limits the repetition rate to about a 1 Hz. Sextupoles and transition units will be mounted in a differentially pumped series of 3 vacuum chambers.

The pulsed atomic beam is finally focused between the pole pieces of an extraction magnet into the resonant charge-exchange ionizer by a large bore sextupole magnet. The negative ions are accelerated through a 25 kV extraction system and deflected out of the source by the extraction magnet. A schematic of the proposed atomic beam section with ionizer is shown in Figure 1.

THE CIPIOS IONIZER

In CIPIOS, polarized hydrogen atoms pass through the inhomogeneous magnetic field of the sextupole magnets and are focused into the ionizer. This beam of polarized neutral hydrogen atoms and a jet of deuterium negative ion plasma are injected in opposite directions into a region with a longitudinal magnetic field of about 1.3 kG. Polarized **H**$^-$ ions are formed by the resonant charge-exchange reaction. The use of this technique to ionize polarized atoms was first suggested by W. Haeberli [9]. The ionizer uses the large cross-section, 10^{-14} cm^2 for resonant charge-exchange between low energy ($\sim$10 eV) D$^-$ (H$^-$) ions and polarized **H°(D°)** atoms. The deuterium plasma is produced by a pulsed arc-discharge plasma injector. D$^-$ ions are formed due to a plasma-surface interaction of D$^+$ ions with a Cs coated converter installed at the entrance of the charge-exchange region. The converter (Figure 2) consists of conical parts with ring slits that permit the plasma to penetrate to the inner surface of the cones where D$^-$ ions are formed. The efficiency of D$^-$ ion production depends on the converter surface work function and is improved by the introduction of Cs vapor into the converter. A transverse magnetic field of about 250 G is applied to the converter region in order to decrease the density of the negative-ion-depleting plasma electrons in the charge-exchange

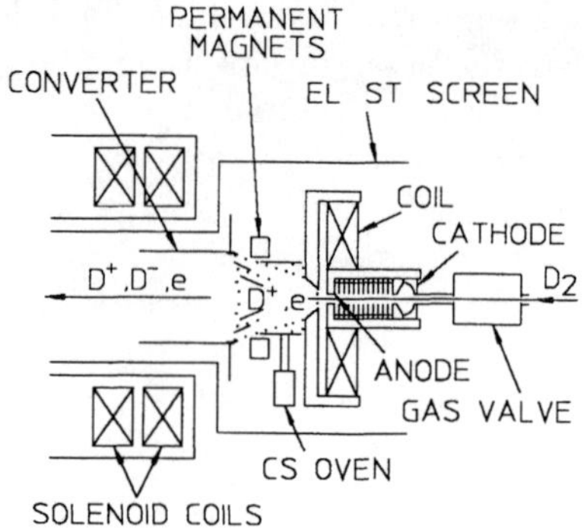

Figure 2. Schematic of the plasma injector and Cs coated converter at the entrance to the charge-exchange region.

region.

Recently, A. Belov *et.al.* [2] have reported that the INR source has produced a pulsed, polarized **H**$^-$ beam, $p_z = 0.87 \pm 0.02$, with a peak current of 1 mA in 1.8 π mm mrad with a pulse duration of 180 μs at a repetition rate of 10 Hz. The IUCF source will be constructed to have a peak beam current of 1 mA during a pulse of duration 300 μs at a repetition rate of up to 5 Hz. The emittance of the extracted beam is expected to be somewhat smaller than the INR source since the emittance is dependent on the size of the atomic beam from the atomic beam source. Switching between polarized **H**$^-$ operation and **D**$^-$ operation will be accomplished by switching the D_2 and H_2 gases in the dissociator and ionizer and by installing a different set of strong field transition units. The extraction magnet setting will vary according to the mass of the polarized ion species.

The ionizer will be constructed by IUCF and INR as part of an agreement of cooperation. Special thanks to Tom Clegg and Willi Haeberli for useful input during the decision stages of the source design. Also the efforts of R. Brown, R. Kupper and B. Lozowski have been greatly appreciated.

REFERENCES

1. D.L.Friesel and S.Y.Lee, 'Status of the IUCF Cooler Injector Synchrotron', 1997 Particle Accelerator Conference, to be published.
2. A.S.Belov, *et.al.*, Rev. Sci. Instrum. **67**(3), March 1996.
3. V. Derenchuk *et.al.*, AIP Conf. Proc. 339, 662 (1995).
4. C.A.Gossett, Proc. Intern. Workshop on Polarized Ion Sources and Polarized Gas Targets, KEK-Tskuba, (ed. Y.Mori, KEK Report 90-15, 1990), p101.
5. D.C. Dinge *et.al.*, Nucl. Instrum. Methods **A357**, 195 (1995).
6. W.R.Gentry and C.F.Giese, Rev. Sci. Instrum. **49**, 595(1978).
7. T. Wise *et.al.*, AIP Conf. Proc. 339, 680 (1995).
8. A.S. Belov *et.al.*, Nucl. Instrum. Methods **A239**, 443 (1985).
9. W. Haeberli, Nucl. Instrum. Methods **62**, 335 (1968).
10. F. Stock for the HERMES target group, AIP Conf. Proc. 339, 674 (1995).

A Microwave based Dissociator for the HERMES–ABS

Norbert Koch

Physikalisches Institut der Universität Erlangen–Nürnberg, 91058 Erlangen, Germany

Abstract. A new type of dissociator for the production of a cooled, highly dissociated hydrogen beam for the application in the HERMES–ABS is presented. In order to increase the dissociation efficiency in the discharge region, a dissociator based on a modified version of a 2.45 GHz Waveguide Surfatron, which sustains the hydrogen discharge via a surface wave, has been built. First measurements show that the degree of dissociation could be improved by a factor of about 1.8 compared to the presently applied rf–dissociator under similar operating conditions.

INTRODUCTION

Although the HERMES–ABS yields a high intensity of 6.5×10^{16} polarised H–atoms/s [1], there is still some improvement potential due to the relatively low dissociation effiency of the currently applied 13.56 MHz radio frequency dissociator (RFD). With this device the achieved degree of dissociation is about 45% at the ABS's nominal throughput of 1 to 1.5 mbarl/s [2]. In order to further increase the intensity of this source, a new dissociator based on a Waveguide Surfatron [3,4] plasma source (WGSD) has been built.

The two main differences between the present RFD and the WGSD are the change from RF to microwave(MW) frequency providing a higher free electron density in the plasma [5], and the coupling of the applied high frequency field to the discharge via a surface wave, which uses the plasma column as its sole propagating medium [6]. The higher electron density leads to an enhancement of the electron impact driven processes in the discharge and especially in the case of a hydrogen plasma to a higher rate of dissociative processes compensating for an incomplete dissociation as well as recombination effects. The surface wave coupling results in a high power transfer efficiency to the discharge as well as stable impedance matching properties due to non–resonant coupling.

Encouraging results with a WGS based dissociator have been obtained in the framework of the target development at CERN, from where atomic fractions of more than 80% up to throughputs of 8 mbarl/s have been reported [7].

DESIGN OF THE DISSOCIATOR

The design of the HERMES–compatible WGSD is shown in fig. 1. With

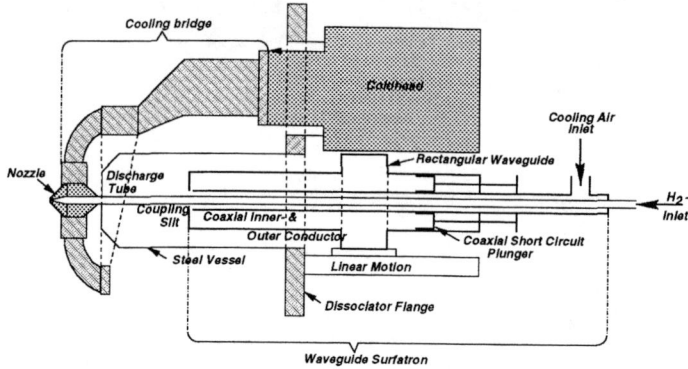

FIGURE 1. Schematic picture of the Waveguide Surfatron based dissociator for the HERMES–ABS

respect to the original WGS, a coaxial transmission line has been added to the waveguide section in order to adapt it to the ABS's geometry. The WGS can be slidden over the discharge tube via a linear motion in order to adjust the optimum distance between the end of the plasma column and the nozzle entrance independent of the microwave input power. Moveable short circuits for the waveguide and the coaxial section, a three stub tuner and the possibility to adjust the width of the coupling slit serve as impedance matching elements. These have made it always possible to tune the reflected power to zero.

The design of the nozzle cooling is similar to that of the present RFD [8] but its temperature regulating system has been improved in order to obtain a better temperature stability of $\pm\,0.5$ K and a shorter nozzle regeneration time of about 2 h, the minimum time between two fills of the HERA machine.

FIRST RESULTS AND CONCLUSIONS

Measurements of the degree of dissociation α have been carried out with a crossed beam quadrupole mass spectrometer placed in the last chamber of a four stage differentially pumped vacuum system with a total pumping speed of 14,000 l/s for hydrogen, yielding high signal–to–noise ratios of better than 5:1 for molecular and 10:1 for atomic hydrogen. Optimizing the WGS geometry as well as the discharge tube cooling and surface preparation resulted in a low level of needed MW–power and a low recombination rate in the discharge tube: about a 350 W of MW–power and an O_2–admixture of 0.1 vol.% are sufficient to obtain the maximum atomic fraction at the ABS's nominal throughput. A comparision of the α–values obtained with the RFD [9] and the WGSD under similar conditions using the same nozzle material (Al 99.5), geometry (2 mm i.d.) and temperature (80 K) is given in fig. 2. It is shown that the degree of dissociation obtained with the new dissociator is a factor of about 1.8 higher compared to the RFD at the HERMES–ABS's nominal throughput.

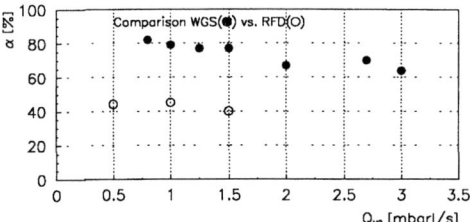

FIGURE 2. Comparison of the throughput dependence of the degree of dissociation measured with the presently applied RFD (circles) and the WGS based dissociator (full dots)

The first results indicate that the WGS based dissociator yields a higher dissociation efficiency compared to the presenly applied RFD under similar operating conditions. Future efforts will be directed towards a long term stability and the installation into the HERMES–ABS.

ACKNOWLEDGEMENTS

I would like to thank DESY–Hamburg, MPI–Heidelberg and the Universität Marburg for technical and financial support as well as W. Kubischta for fruitful discussions.

REFERENCES

1. Stock, F., *et al*, Proc.s Int. Workshop on Pol. Beams and Pol. Gas Targets, Köln 1995, World Scientific, Singapore, 260 (1996).
2. Lorentz, B., Diploma Thesis, Universität Heidelberg (1993).
3. Chaker, M., *et al*, J. Physique Lett. **43**, L71 (1982).
4. Moisan, M., *et al*, J. Phys. E: Sci. Instr. **20**, 1356 (1987).
5. Moisan, M., *et al*, J. Vac. Sci. Technol. **B 9(1)**, 8 (1991).
6. Zakrzewski, Z., Moisan, M., *Microwave Discharges: Fundamentals and Applications*, Plenum Press, New York (1993).
7. Boero, G., Kubischta, W., Leprince, P., *accepted for publication in* Nucl. Instr. and Methods, NIM A, (1997).
8. Funk, U., Diploma Thesis, Universität Heidelberg (1995).
9. Stock, F., PhD Thesis, Universität Heidelberg (1994).

^{3}He Neutron Spin Filter at ILL.

H. Humblot, W. Heil, F. Tasset and D. Hoffmann
Institut Laue-Langevin, Grenoble, F. 38042

Optical pumping of metastable ^{3}He atoms in a plasma at 1 mbar is a very efficient method to produce large quantities of spin polarized ^{3}He gas. Subsequent polarization preserving compression by two stage titanium compressor system enables to prepare neutron spin filter (NSF) cells of about 300 cm3 volume with 4 bar of polarized ^{3}He (Phe = 55-60%) within 4 hours. The NSF may then be closed off by a valve and removed for some remote operation. To keep the ^{3}He polarization close to 50% on the average the NSF is refilled with fresh gas every day. The Cesium coated target cells have a decay time of polarized ^{3}He around 100 hours. First experiences at ILL with NSF will be presented.

A novel approach for a spin–exchange high density ^{3}He target.

Robert C. Welsh, John N. Zerger, Kevin P. Coulter, Todd B. Smith, Timothy E. Chupp

Physics Department, Randall Laboratory, The University of Michigan Ann Arbor, Michigan 48109

Abstract. In the past decade high density polarized ^{3}He fixed targets based on spin–exchange with optically pumped Rb have been used successfully in electron scattering experiments at MIT/Bates and SLAC, as a neutron polarizer, and with proton and pion beams at TRIUMF. In the near future similar targets will be used in physics programs at TJNAF and as neutron polarizers at Los Alamos. Until now, the typical spin exchange target was a sealed design, however there are several advantages to target cells that can be evacuated and filled with a variable pressure of polarized ^{3}He. We have begun development of such an actively filled target, based on technology recently developed for NMR/MRI with hyperpolarized noble gases. The benefits and difficulties of constructing such a "flowing" high density ^{3}He target system are discussed.

INTRODUCTION

For a neutron spin filter, the optimum ^{3}He thickness depends on both the neutron wavelength and the ^{3}He polarization. Neutron polarization is given by the following equation :

$$P_n = \tanh(n_3 \sigma_\alpha l P_3). \tag{1}$$

The ^{3}He density is indicated by n_3, the velocity dependent neutron absorption cross–section by $\sigma_\alpha = \frac{v_o}{v} 5327$ barn, $v_o = 2200 \frac{m}{s}$, the column depth of the spin filter by l, and the ^{3}He polarization by P_3. For experiments at neutron spallation sources, a broad spectrum of wavelengths is available. The variable ^{3}He pressure allows for optimization of the polarizer to specific conditions and experimental parameters such as neutron wavelength. Another advantage of the actively filled cells is that leaks can be tolerated. Target materials containing strong neutron absorbers (e.g. ^{11}B in most conventional glasses) strongly attenuate long wavelength neutrons due to $(1/v)$ cross–sections. Materials such as quartz have been shown to have favorable properties for ^{3}He polarization in addition to neutron transmission. However, quartz has a high leak rate for helium. The leaking gas can easily be replenished in the actively filled cells. Also, leaky seals such as those between dissimilar window and target cell materials would no longer be a problem.

VARIABLE PRESSURE CELL SET-UP

A variable pressure ^{3}He polarization cell was constructed. The cell was manufactured with Corning 7056 glass, a borosilicate. The cell has a 2.5 cm diameter and is approximately 10 cm long. During operation the cell is held in a cylindrically shaped Pyrex oven. The oven has a diameter of 4 cm and a length of 14 cm. Hot air is used to raise the temperature of the oven; for these tests the cell was operated at 170° C, controlled with a RTD and Omron temperature controller device. A metallic valve is used to control the pressure of the cell; the valve is connected to the main body of the cell with a Pyrex capillary tube. The capillary has an inner diameter is 0.05 cm and a length of approximately 10 cm. Assuming that the valve is 100% depolarizing and that no polarization gradient exists transverse to the axis of the capillary, the effective relaxation time, T_1, is given by :

$$\frac{1}{T_1} = \frac{1}{T_1'} + \frac{SD}{LV}. \qquad (2)$$

The intrinsic cell relaxation time is T_1'. The capillary cross–sectional area is given by S and the length by L. The ^{3}He diffusion constant in the capillary is given by D (at the appropriate temperature and density). The volume of the cell is given by V. Figure 1 illustrates the optical pumping cell.

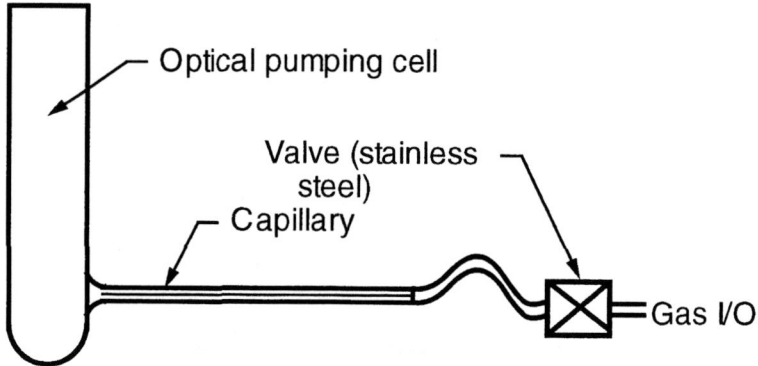

FIGURE 1. Variable pressure optical pumping cell.

RESULTS

We have tested the cell under two conditions : 1) a fill pressure of approximately 1 atmosphere (atm) and 2) a fill pressure of approximately 2.5 atm. In addition to the helium, a small partial pressure of nitrogen was included, about 60 torr. Each cell was optically pumped with approximately 2.3 watts of 795 nm circularly polarized light provided by an argon ion pumped Ti:sapphire laser. Relaxation measurements were made over the course of many hours using the method of *adiabatic fast passage*. Polarization data along with exponential fits are shown in Figure 2. Plot **A** shows the spin–down for the 1.0 atm. fill. The 2.5 atm. fill spin–down is shown in plot **B**. The capillary serves

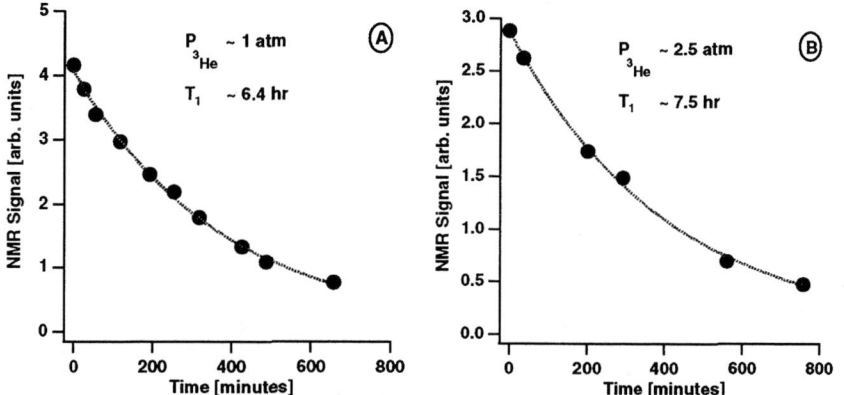

FIGURE 2. Spin-down curves for variable pressure cell. (A) Approximately 1 atm. pressure of ^{3}He. (B) 2.5 atm. pressure of ^{3}He. Lines indicate the exponential fits.

the purpose of isolating the bulk polarized ^{3}He from the depolarizing effects of the stainless steel valve. However, the relaxation times are not sufficiently long for use in neutron spin filter applications. We expect to run further tests to diagnose the performance of the cell.

CONCLUSIONS

A variable pressure cell has been successfully tested with promising results. We expect to conduct further tests with this cell to verify the performance. We have plans for constructing quartz cells and cells with glass valves instead of metal. In the near future we have plans to test two chamber cells, in which the optical pumping cell can accomodate high ^{3}He density and can thus take advantage of pumping with laser diode arrays due to pressure broadening of the Rb line. The secondary chamber would then be filled with polarized gas from the first chamber. We expect to make large quartz secondary cells applicable for use as neutron polarizers.

Development of Optical-pumping Polarized Deuteron Target

Tadaaki Tamae[*], Tamio Yokokawa[*], Itaru Nishikawa[*],
Kazuhiro Abe[*], Osamu Konno[*], Haruhisa Miyase[†],
Itaru Nakagawa[*], Masumi Sugawara[*], Eiji Tanaka[*],
Hiroaki Tsubota[†], Nobuo Yamaguchi[*] and Hirohito Yamazaki[*]

[*] *Laboratory of Nuclear Science, Tohoku University, Mikamine, Taihaku-ku, Sendai 982, Japan*
[†] *Faculty of Science, Tohoku University, Aramaki, Aoba-ku, Sendai 980, Japan*

Abstract. An optical-pumping system of rubidium atoms for a laser-driven polarized deuteron target was constructed. The density and polarization of the rubidium atoms were measured using Faraday rotation. The rotation angle was determined within an error of $0.01°$. Our preliminary result showed a polarization of 0.4 at a gas thickness of 4×10^{13} atoms/cm^2.

INTRODUCTION

The Stretcher-Booster Ring (STB) is under construction in Tohoku University. The ring has three operation modes; the pulse-stretcher mode, the storage mode and the booster mode for injection to a proposed synchrotron-light facility. In the storage mode, the energy is ramped up to 1.2 GeV with rf acceleration in 1 sec after injection. Internal-target experiments and high-energy tagged-photon experiments will be performed using the stored high-intensity electron beam.

We are developing an optical-pumping polarized deuteron target, which will be used in the STB ring. As the first stage of development of the polarized target, an optical-pumping system of rubidium atoms was constructed, and the density and polarization of the rubidium atoms were measured using Faraday rotation.

OPTICAL-PUMPING SYSTEM AND FARADAY-ROTATION MONITOR

A 12 cm long spin-exchange cell of Pyrex glass was coated with a drifilm from dichloro-dimethyl-silane plus water. The cell and the rubidium container were heated with separate Nichrome heaters. The temperature was controlled within 1°. The rubidium atoms were optically-pumped in the magnetic field of a solenoid coil with a circularly-polarized light from a Ti-sapphire laser.

The density and polarization of the rubidium atoms were measured using Faraday rotation. A light from a diode-laser was divided into two beams with a polarizing beamsplitter cube. The wavelength of the light was monitored with a wavelength-meter using one of the beams. Another beam was injected into the spin-exchange cell, and the linear-polarization direction of the beam through the cell was measured with a polarizer on a rotation stage. The photon intensity through the polarizer was monitored as a function of the polarizer angle using the CAMAC system, which recorded the number of pulses to rotate the stage and an ADC out-put of a photo-diode. The data acquisition system was triggered at 50 Hz. The stage rotates 342 degrees in 68sec, that makes 3,420 data points. The intensity curve was fitted by the square of a sinusoidal function, and the phase of the curve was determined within an error of 0.01° from the least square analysis.

EXPERIMENTAL RESULTS

At first we measused a rotation angle of the Pyrex cell using our Faraday-rotation monitor. The rotation angle of glass is proportional to the product of the thickness l of the glass and the magnetic field H as,

$$\theta = VHl, \qquad (1)$$

where H is the Verdet constant. A rotation angle of 0.56 degrees was observed for vacant threefold cells of total 1.4 cm thickness in 2 k gauss at the wavelength of 780.8 nm. The Verdet constant of 2×10^{-4} degrees/gauss cm was obtained from the result. The Verdet constant of glass little depends on the wavelength in this wavelength region.

The thickness and polarization of the rubidium vapor were also evaluated from the Faraday rotation. The rotation angle is given by

$$\Phi = \Phi_0 + P\Phi_p, \qquad (2)$$

where P is the polarization of the rubidium atoms. Φ_0 is the rotation angle for unpolarized atoms, which depends on the thickness of the material, the magnetic field and the wavelength of the probe light. The thickness is determined from the rotaion angle without the pumping-laser, using formulae described in [1]. As Φ_p can be also calculated using formulae described in [1], the polarization of the rubidium atoms is obtained from the additional rotation angle with the pumping-laser.

The measurement was made at temperatures between 150 and 165 °C. The intensity of the pumping-laser was about 0.5 W. The Faraday rotation angle was measured at wavelengths between 780.2 and 780.5 nm, which was typically several degrees in a magnetic field of 2 k gauss. When the wavelength of the pumping-laser was adjusted to the D1 line (794.76 nm), the polarization is almost symmetrical for opposite polarities of the magnetic field, and 0.2 at most. Figure 1 is the result when the wavelength of the pumping-laser was

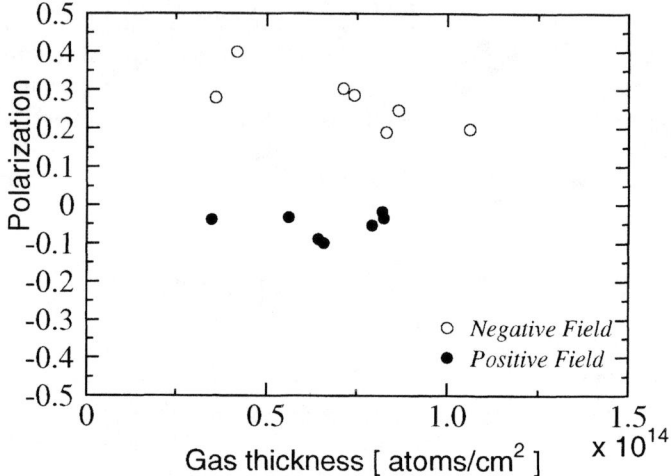

FIGURE 1. The polarization of the rubidium atoms as a function of the gas thickness,when the wavelength of the pumping-laser was adjusted to the σ_+ transition.

adjusted to the σ_+ transition taking account of the Zeeman splitting. In this case, the polarization showed asymmetry from reversal of the magnetic field, and it increased to 0.4 for the σ_+ transition. The reason why the polarization was not very high is assumed that the atomic density of this measurement was rather high out of the power of the pumping-laser.

REFERENCES

1. Mori, Y., Ikegami, K., Takagi, A., Fukumoto, S., and Cornerius, W. D., Nucl. Instr. and Meth. **220**, 264 (1984).

Performance of the Laser Driven Polarized Hydrogen Source at IUCF

R. V. Cadman[1a], K. Bailey[b], J. Brack[c], W. J. Cummings[b],
J. Fedchak[b], B. Fox[c], H. Gao[2b], C. Grosshauser[d], R. J. Holt[a],
C. E. Jones[e], E. Kinney[c], W. Kirwin[c], R. S. Kowalczyk[b],
Z.-T. Lu[b], M. A. Miller[a], W. Nagengast[d], B. R. Owen[a],
K. Rith[d], F. Schmidt[d], E. C. Schulte[a], J. Sowinski[f],
F. Sperisen[f], J. Stenger[d], E. Thorsland[a], D. K. Toporkov[g]

[a] *University of Illinois, Urbana, Illinois 61801*
[b] *Physics Division, Argonne National Laboratory, Argonne, Illinois 60439*
[c] *University of Colorado, Boulder, Colorado 80309-0446*
[d] *Universität Erlangen-Nürnberg, 91058 Erlangen, Germany*
[e] *California Institute of Technology, Pasadena, California 91125*
[f] *Indiana University Cyclotron Facility, Bloomington, Indiana 47408*
[g] *Budker Institute for Nuclear Physics, 630 090 Novosibirsk, Russia*

Abstract. A laser driven source of polarized hydrogen and deuterium has been installed in the Cooler Ring at the Indiana University Cyclotron Facility. Polarized nuclei from the source are injected into a storage tube, and the resulting target has been used in a scattering experiment with 200 MeV proton beam. This paper discusses the performance of the source, including measurements of atomic fraction and electron polarization.

For several years there has been interest in using spin exchange with an optically pumped polarized alkali vapor as a way of obtaining a source of highly polarized hydrogen or deuterium nuclei with a higher flux than can be achieved by other methods [1]. Development of this technique on the bench led to a target which, assuming spin–temperature equilibrium, had a higher figure of merit than conventional targets currently in use [2]. The source was then installed in the Cooler Ring at the Indiana University Cyclotron Facility (IUCF) and has now been tested with the proton beam. The analysis of

---

[1]) Corresponding author, cadman@uiuc.edu
[2]) Present address: Laboratory for Nuclear Science, Massachusetts Institute of Technology, Cambridge, Massachusetts 02139

nuclear polarization data is not complete and is discussed in the paper by M. Miller et al. in this volume. Here we present data on the performance of the source in terms of electron polarization and atomic fraction.

The hydrogen nuclei are polarized in the glassware shown in figure 1. Molecules enter from the right and are dissociated by an rf plasma discharge. The atoms then enter the spin exchange cell. At the left a reservoir of potassium is heated and the vapor flows into the spin exchange cell. Circularly polarized light from a Ti:Al$_2$O$_3$ laser enters the spin exchange cell from the top. Its polarization is transferred to the hydrogen nuclei in three steps: first to the potassium valence electrons through photoabsorption, then to the hydrogen electrons through K–H spin exchange collisions, and finally to the hydrogen nuclei through H–H spin exchange collisions. The polarized hydrogen exits the spin exchange cell and enters an aluminum storage tube through which the proton beam passes. Some atoms pass directly from the spin exchange cell through a hole in the bottom of the storage tube and into the atomic polarimeter [1] which was used for the measurements reported in this paper.

The source used at IUCF differs from other laser driven sources in two ways. First, we have eliminated the transport tube, which decreases the number of wall bounces and thus improves the atomic fraction and polarization. Second, we no longer use a uniform magnetic field. In a typical magnet setting the field in the lower half of the spin exchange cell will vary between 110 mT and 120 mT, but decreases to about 20 mT at the top of the cell. This leads to a region near the bottom of the cell where potassium can be optically pumped without radiation trapping, and a region at the top where polarization is transferred from hydrogen electrons to hydrogen nuclei more rapidly.

Figure 2 shows the performance of the source during a test run of the target. These data were taken with a flow of 1×10^{18} atoms/sec, with a 6 mm diameter hole between the spin exchange cell and the storage tube, and with 1.5 W of laser power at the source. The figure shows that increasing potassium temperature led to an increase in the polarization and a decrease in the atomic fraction. Eventually the high potassium density in the spin exchange cell caused nearly complete molecular recombination.

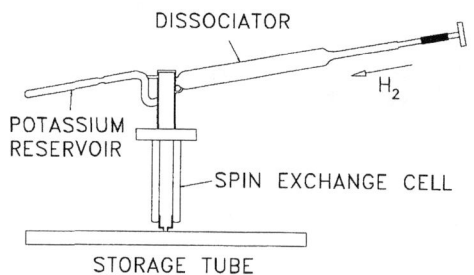

FIGURE 1. Glassware used for the Laser Driven Source

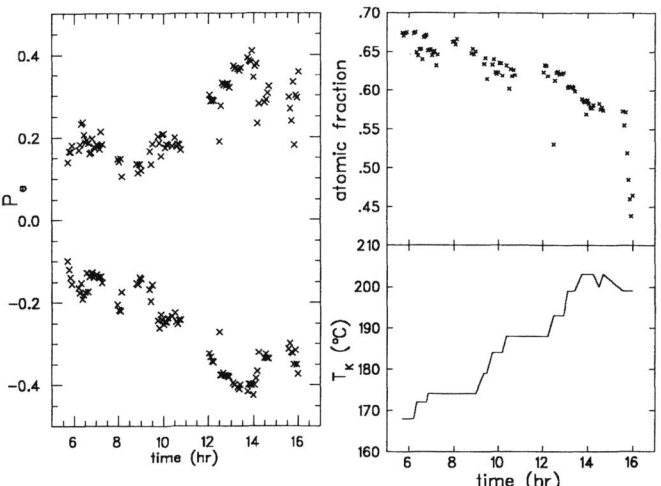

FIGURE 2. Performance of the hydrogen source during tests with the proton beam. Clockwise from left: electron polarization; atomic fraction; and temperature of the potassium reservoir over a 10 hour period.

More recently tests were made without the proton beam in preparation for future beam time. A 4 mm hole was used between the spin exchange cell and the storage tube, and the glass was coated using the afterwash technique described in ref. [3]. As expected from the smaller hole size, we found a lower atomic fraction, about 40%. We also found higher polarizations, and observed that doubling our laser power by adding a second laser made a significant improvement. The best result was a polarization of 63% with an atomic fraction of 35%, again at 1×10^{18} atoms/sec.

We thank D. Tupa for the loan of the two Ti:Al$_2$O$_3$ lasers, our glass blower W. Lawrence, A. Kenyon for helping with assembly and repair of the source, J. Doskow for maintaining our vacuum system, and the rest of the operators and staff at IUCF for all their assistance. This work was supported in part by the NSF under grants PHY94-20787 and PHY94-20470 and by the DOE under contract No. W-31-109-ENG-38 and grant DE-FG03-95ER40913.

REFERENCES

1. Poelker, M., *et al.*, Phys. Rev. A **50**, 2450 (1994).
2. Owen, B., *et al.*, in *Proceedings of the 12th International Symposium on High-Energy Spin Physics, Amsterdam, 1996,* edited by C. W. de Jager, *et al.* (World Scientific, Singapore, 1997), p. 490.
3. Fedchak, J. A., *et al.*, Nucl. Instrum. Methods Phys. Res., Sect. A **391**, 405 (1997).

Status of the Helium Afterglow Injector for MAMI

J. Arianer, S. Cohen, S. Essabaa, R. Frascaria, O. Zerhouni,
F. Szeremeta[1], R. Wurzinger [1] and K. Aulenbacher[†]

Institut de Physique Nucléaire, 91406 Orsay Cedex, France
[†]*Institut für Kernphysik, 55099 Mainz, Germany*

Abstract. For the parity-violation experiment at the Mami cw electron accelerator facility at the University of Mainz [1], we need a stable polarized electron source able to deliver a current $> 150 \mu A$ with polarization $> 60\%$. We develop the Orsay polarized electron source SELPO as an injector for MAMI. SELPO source based on a flowing helium afterglow is now operating in a high voltage platform. The terminal will be tested completely at Orsay and is foreseen to be installed at MAMI before the end of this year. Recent results obtained with an array of capillaries to generate the helium flow and a longitudinal spin direction are described.

I THE PRINCIPLE AND THE LATEST RESULTS

The operating principle of SELPO is based on the Penning ionization involving optically aligned $He(2^3S_1)$ metastable atoms and a singlet ground state target of CO_2 molecules [2]. Due to its spin preserving nature, this reaction generates polarized electrons. The metastable atoms are produced in a flowing helium afterglow and are optically pumped on the D_0 ($^3S_1 \rightarrow ^3P_0$)resonance transition at 1083 nm. At present, we have demonstrated that the SELPO device is able to deliver current (I) and polarization (P) with a figure of merit (IP^2) up to 40 μA. The measured optical properties of the beam have shown that the highest normalized emittance at low current ($1\mu A$) is 0.6 π mm mrad for 80 % of the total current [2]. The upper limit of the beam energy spread is 0.25 eV, corresponding to the experimental resolution. The routine qualities of the prototype design confirm that this kind of source is suitable for injection into an accelerator. The universities of Mainz and Orsay have initiated a collaboration with the goal to install SELPO at the MAMI Accelerator as a very promising alternative to the existing sources based on GaAs photocathodes.

---

[1]) Financial support from IPN Orsay and IKP Mainz

II SELPO PLATFORM

The arrangement of SELPO is composed by 3 main components: the electron source in the terminal, the beam transport and the diagnostics, the laser system. We will comment each of these components separetely.

The terminal consists of a parallelepipedic electrode floated at 100 kV (fig. 1). It is powered by 25 kVA-100 kV transformer. The structure of the terminal is composed by the configuration of SELPO including the electronic and the vacuum system which is managed by a programmable logical computer (PLC). The support is made of two separated platforms. The Roots blower platform stands on four vibrations insulators to damp the mechanical vibrations of the Roots system. The whole source including He purifier, electronics, PLC, laser optics and optical links is supported by the second platform.

To start the preliminary tests at Orsay, we prefered to restrain the optic beamline. The beam is extracted through a 2 mm diameter nozzle (disk shaped electrode) and immediately accelerated up to 3.5 kV by a conical extractor followed by two lenses and two deviators. This first optical part is located inside the terminal. The second part, separated from the first one by a ceramic insulator, is composed of a Wien filter to flip the spin from the longitudinal to the transverse direction. The beam enters then through a diaphragm into the Mott Polarimeter which is ended by a Faraday cup.

In the calculated beamline, it is planned to focus the beam by double solenoids and quadrupoles and also to install monitors for the optical diagnostics of the beam shape and emittance measurements.

III PRELIMINARY RESULTS:

The objective is to improve the electron production by limiting the losses of metastable atoms. For this we need to decrease the radial extension of the atomic beam in order to reduce desexcitations of metastable atoms striking the walls of the container. The classical convergent-divergent Pyrex tube for He injection is now replaced by an array of 200 μm inner diameter capillaries. The experimentally observed profile and the exponential decrease of the density of the metastable atoms (fig. 2) confirm a very significant gain (a factor of 10). Presently, for a He pressure between 0.06 and 0.14 mbar, the metastable density in the chemiionization area varies from 4×10^9 to 3.5×10^{11} cm^{-3}.

The electron production rate is roughly proportional to the square of the density of the metastable atoms. It is therefore very sensitive to the metastable production rate. However, the extracted electron beam is not proportional to the produced one due to the plasma formation following the Penning ionization and the non-optimized extraction geometry. Actually the extraction efficiency is about 16 %. In figure 3 we represent the evolution of the extracted current versus He pressure. At the pressure of 6×10^{-2} mbar where the electron

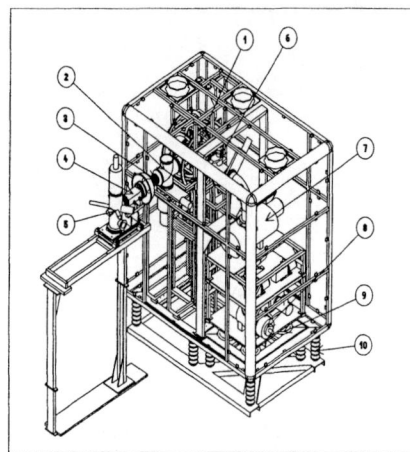

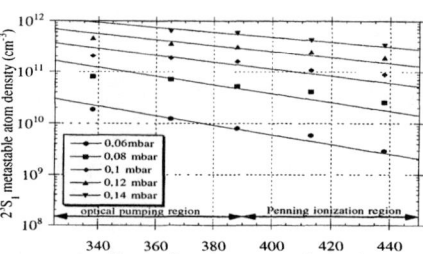

FIGURE 2. The average He(2^3S_1) density vs. distance from the μ-wave cavity

(1) interaction chamber & laser optics
(2) differential pumping stage
(3) H.V. insulator
(4) Wien filter
(5) Mott polarimeter
(6) turbo pump
(7) roots pump
(8) air compressor
(9) vibration insulator
(10) H.V. insulator

FIGURE 1. SELPO terminal

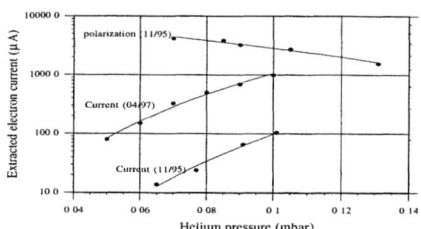

FIGURE 3. The evolution of the extracted electron current vs. He pressure

polarization was 80 % for 1 μA, the electron current reaches now 300 μA. The fluctuation on the beam current is less than 1 %.

Presently, the polarization measurements are in progress. The Mott polarimeter is tested and optimized using a ^{109}Cd radioactive source. The measured energy resolution of the silicon surface barrier detectors (100 μm thick) is of 15 keV at 80 keV. The next steps of the project are the study of the beam characteristics (normalized emittance, beam profile) and the long term stability during 200 hours runs.

Acknowledgments

The authors thank E. Guinault, S. Ducourtieux, N. Duc for their help and G. Roger, J. Baudet, P. Julou for their technical support.

REFERENCES

1. Parity Violation A4 experiment (PVA4). See e.g. E. Heinen-Konschak, Ph. D. thesis, Mainz 1994).
2. J. Arianer, S. Cohen, S. Essabaa R. Frascaria and O Zerhouni, *Nucl. Instr. and Meth.* **A 382** (1996) 371.

Photocathode research at SLAC[1]

G. Mulhollan[†], J. Clendenin[†], E. Garwin[†], R. Kirby[†],
T. Maruyama[†], R. Prepost[‡] and H. Tang[†]

[†]*SLAC, Stanford, CA 94309, USA*
[‡]*University of Wisconsin, Madison, WI 53706, USA*

Abstract. GaAs based photocathode research at SLAC will be described. Recent efforts have focused on both immediate applications and fundamental photocathode properties. This includes revisiting some old measurements with state-of-the-art instrumentation.

Metal coating the surface of GaAs has the potential for reducing the influence of charge trapped in the near surface region. Trapped charge has an adverse effect on the short pulse output properties of the NEA surface by rapidly producing a surface photovoltage which counteracts the NEA condition. A potential pitfall for use of a metal on a polarized electron emitter is depolarization upon transmission though the overlayer [1]. The effects of Pd metal overlayers on the photoemissive properties of GaAs have been probed with XPS, SEM and spin-polarized photoemission. The XPS and SEM measurements were carried out in the deposition system while the polarization measurements were conducted on the sample after transfer through a nitrogen atmosphere. Pd reacts strongly with the GaAs surface making it is necessary to heat the surface to over 500°C after Pd deposition for NEA activation to be possible. Heating to higher temperatures can diminish the XPS Pd signal, but it never disappears which is a signature of the

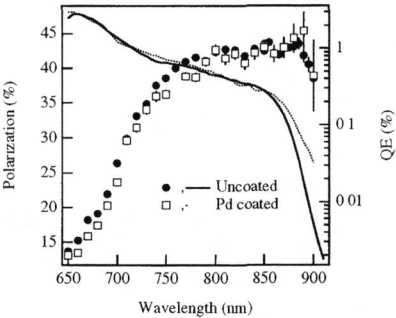

FIGURE 1 Polarization and quantum efficiency from Pd coated and uncoated 100 nm GaAs/AlGaAs as a function of wavelength. The uncoated GaAs was permitted to age for one month making the yield comparable to that of the Pd coated material

[1] Work supported by the U S Department of Energy contract DE-AC03-76SF00515

strong reactivity of the Pd/GaAs interface. While the overall effect of the Pd layer is to reduce the quantum efficiency by a factor of ~10 compared to a freshly cesiated surface, the addition of Pd does not affect the polarization beyond the effects attributable to the lowering of the yield as may be seen in Figure 1. The effect on nanosecond pulse emission is expected to be quantified soon.

The polarization of electrons emitted from strained GaAs has been tracked as a function of the quantum yield over two decades. Beyond the initial rise in polarization due to attenuation of the low energy tail which contains depolarized electrons [2], there is no discernible change in polarization up to a yield drop by more than a factor of 100. Therefore, unpolarized electrons from the $5\times10^{18}/cm^3$ acceptor states contribute less than a factor of 10^{-3} to the near band gap photoemission current and the long wavelength rollover of the polarization seen in strained GaAs (see Figure 4) cannot be attributed to the acceptor state electrons.

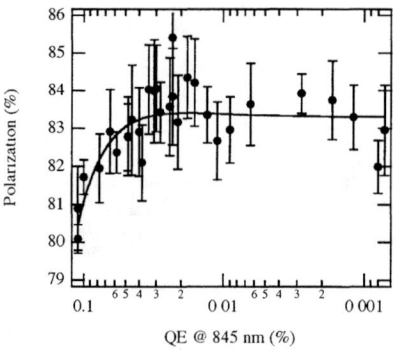

FIGURE 2 Polarization as a function of quantum efficiency. The NEA state was permitted to decay over a period of several weeks

The properties of the NEA surface created by the use of differing activation gases, notably NF_3, O_2 and N_2O, have been quantified for 1μm thick GaAs. While there is little difference in the total yield, activations with O_2 are considerably more troublesome with regards to control. There are stages in the activation process where the partial pressure of O_2 must be actively varied even while using the codeposition technique. Lifetime measurements in a test vacuum chamber demonstrated that by a factor of two the shortest lived photocathodes were obtained with the N_2O activations; this effect is attributable to differences in the sticking coefficients of the gases and in the pumping efficiency of the system.

First results from carbon doped NEA GaAs have been obtained. It was once thought the carbon dopant might diffuse to or segregate onto the surface, lowering the yield. However, there does not seem to be a segregation temperature below the 600°C cleaning value. By all measures, carbon doped GaAs behaves identically to NEA surfaces using Zn or Be for the p-type doping. Carbon is particularly attractive as it has a much lower bulk diffusion constant than either Zn or Be and can readily be used for graded doping profiles.

A precision systematic study is underway to determine the ultimate limit of polarization available from unstrained NEA GaAs. This is being done by measuring the polarization as a function of thickness while varying this parameter on a single sample of unstrained GaAs. The limit seen by extrapolation to zero thickness sets an upper bound for the depolarizing effects in $5 \times 10^{18}/\text{cm}^3$ carbon doped GaAs. The results for this technique applied to MBE grown GaAs are shown in Figure 3. The polarization did not reach the value measured for 100 nm, unstrained GaAs (about 44%). This is partially explained by surface roughening during the sample thinning process. AFM measurements of the surface show structures 100 nm high. The spread in thickness has smeared out the polarization response of the thinnest GaAs. Part of the lower polarization may be attributed to material-specific short spin lifetime.

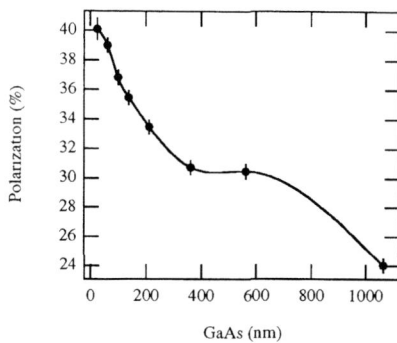

FIGURE 3 Polarization as a function of thickness as measured from carbon doped GaAs. The thickness was reduced via chemical etching

Studies of alternate techniques for growing single strained layer GaAs are underway. Among these is the substitution of a GaAsP pseudo-substrate for the single unit growth of strain inducing and strained layers. We have obtained LPE-grown Te-doped n-type GaAsP pseudo-substrates with a GaAsP thickness of 25 μm. Strained GaAs was grown with standard MBE techniques. Polarization and quantum efficiency values from such material can be seen in Figure 4. This method may ultimately eliminate the need for the costly and occasionally nonconforming growth of the single strained GaAs layer structure as a unit. In addition, the quality of the pseudo-substrate is quite high and its use may reduce the defect density in the final layer. High defect densities are one limiting factor in achieving the highest polarizations. While the best strain obtained in the pseudo-substrate samples is not yet equal to that of the single unit growth material, a number of growth parameters are still under optimization.

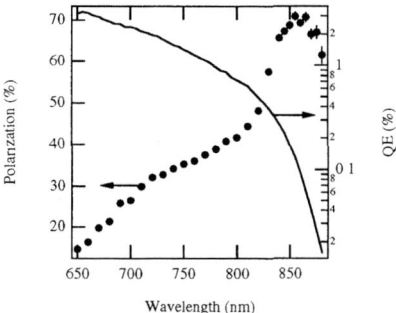

FIGURE 4 Polarization and quantum efficiency as a function of wavelength for GaAs epitaxially grown on a GaAsP pseudo-substrate at 520°C.

REFERENCES

1. Mulhollan, G., Zhang, X., Dunning, F. B. and Walters, G. K., *Phys. Rev. B* **41**, 8122 (1990).
2. Drouhin, H.-J., Hermann, C. and Lampel, G., *Phys. Rev. B* **31**, 3872 (1985).

5 MeV Mott Polarimeter Development at Jefferson Lab

J.S. Price*, B.M. Poelker*, C.K. Sinclair*, K.A. Assamagan[†],
L.S. Cardman*, J. Grames[‡], J. Hansknecht*, D.J. Mack*, and
P. Piot*

*Jefferson Lab,[1]
12000 Jefferson Avenue, Newport News, VA USA
[†]Hampton University, Hampton, VA USA
[‡]University of Illinois at Urbana-Champaign, IL USA

Abstract. Low energy (E_k=100 keV) Mott scattering polarimeters are ill-suited to support operations foreseen for the polarized electron injector at Jefferson Lab. One solution is to measure the polarization at 5 MeV where multiple and plural scattering are unimportant and precision beam monitoring is straightforward. The higher injector beam current offsets the lower cross-sections. Recent improvements in the CEBAF injector polarimeter scattering chamber have improved signal to noise.

INTRODUCTION

Low-energy ($E_k \leq 100$keV) polarimeters using Mott scattering from high-Z foils have been used to measure electron beam polarization in a variety of applications. However, the large cross-section produces significant plural and multiple scattering in the thinnest of free-standing pure metal foils (40 nm thick) which reduces the effective analyzing power [1]. Mott scattering at higher energy has lower cross-section so μA beam currents can be tolerated which facilitates rapid real-time monitoring of beam polarization. Dilution of the analyzing power due to plural and multiple scattering from 0.1 μm thick gold foils is of order a few percent; this reduces sensitivity to target foil thickness. RF structure on μA beam currents in the CEBAF injector make beam position, angle, current, and spot size easy to monitor simultaneously with polarization thus allowing control of systematic uncertainties.

[1)] This work was supported by the USDOE under contract DE-AC05-84ER40150.

PROTOTYPE POLARIMETER

We have installed a prototype polarimeter in the 5 MeV section of the Jefferson Lab Injector. There is little uncertainty in the calculated single scattering analyzing powers upon which the polarimeter design is based [2]. Foils of different Z (e.g. Cu, Ag and Au) and identical Molière scattering distribution were used to make preliminary measurements of target foil analyzing power independent of beam polarization (Figure 1).

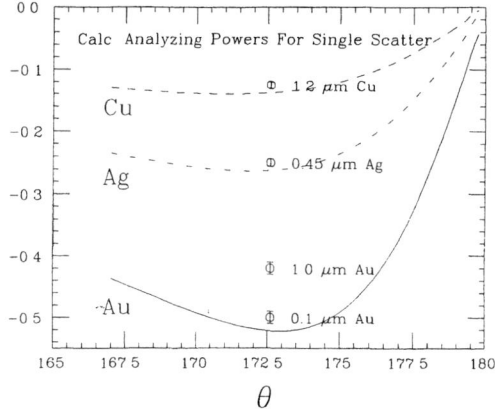

Figure 1. Calculated single-scatter analyzing power for Au, Ag and Cu as a function of scattering angle at 5 MeV (curves) and preliminary measured values (circles).

The 0.1 μm Au foil shows approximately 5% difference from the calculated maximum analyzing power (-0.52) at $\theta = 172.6$ °, however a 1.0 μm Au foil was -0.42. Nuclear size effects at this energy are expected to be of order a percent [3].

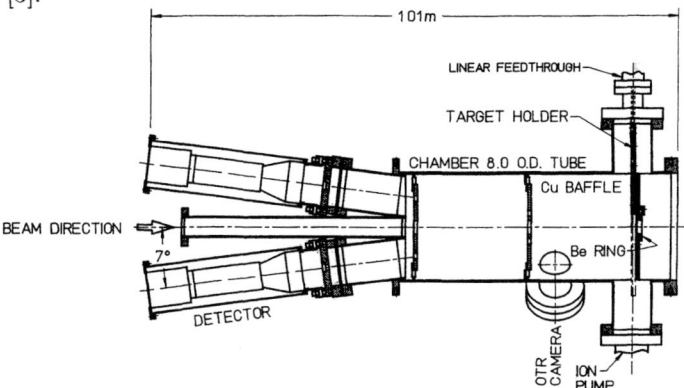

Figure 2. Prototype 5 MeV Mott polarimeter scattering chamber.

Figure 2 shows a side view drawing of the scattering chamber including placement of two of the four scintillation detectors at 173° to the beam line. These four detectors allow simultaneous measurement of the two components of polarization transverse to the beam momentum direction.

Copper aperture plates 1 cm thick have been installed between the target foil and detectors. Four holes in each plate provide line of sight view to the target foil. The downstream portion not shown includes a long Al pipe which serves as a beam dump. A solenoid magnet focuses the scattered beam into a dipole field which deflects the electrons into a short Al dump angled at 40°. Chamber surfaces downstream of the foil are lined with high density polyethylene. The performance of the instrument as a polarimeter is directly related to the signal to noise. Recent efforts have reduced the number of inelastically scattered electrons; compare pulse height spectra in Figure 3. Encouraged by this progress, work continues on reducing background.

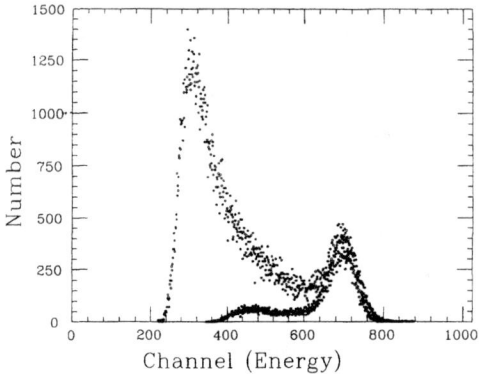

Figure 3. Typical PMT pulse height spectra for 1 μA beam on 0.1 μm Au foil shown before and after scattering chamber improvements.

The NE102a plastic scintillator detectors are equipped with 1 cm diameter apertures cut in 12 mm thick Al which subtend 1° in θ and about 5° in ϕ. Rate measurements with these 0.22 msr apertures give 130 Hz count rate with a 10 μA CW polarized beam on a 0.1 μm Au foil. This agrees with calculated expected rate allowing for the 0.20 mm thick Al vacuum window. Measurements with statistical uncertainty of order 1% can be made rapidly.

RF time structure is present so precision beam position, angle and current monitoring is straightforward. For example, a beam current monitor (BCM) is located 2 m upstream of the polarimeter; measurements of integrated charge can be made for different helicity states. A CCD camera focussed on the target foil can detect the visible radiation emitted as the electron beam passes through the foil. This optical transition radiation (OTR) monitor can easily resolve the profile of a mm diameter beam (see Figure 4). Helicity-correlated changes in spot size and position can be directly and simultaneously measured with polarization. This allows control of and correction for false asymmetries in the measurement due to beam spot changes. It is also a valuable diagnostic for polarized source performance.

Preliminary measurements of beam polarization as a function of beam current over a range 2 to 12 μA show insignificant rate dependence. Background and asymmetry vary little as a function of beam spot location within the central 0.4 diameter on the target foil.

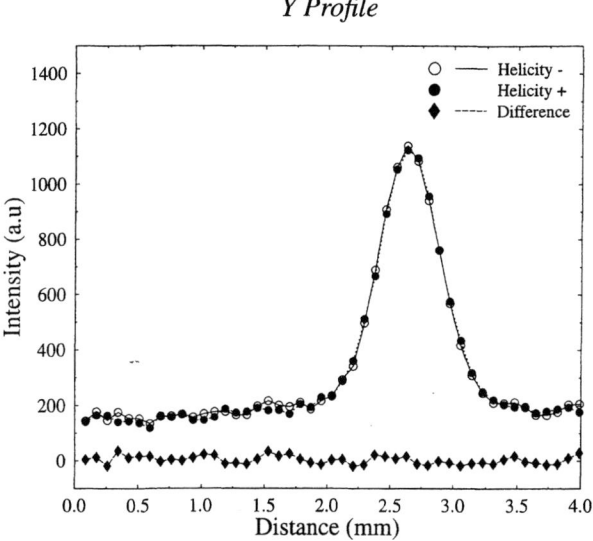

Figure 4. Helicity-correlated OTR beam profiles (summed channels vs y, 1 mm=16 ch) for 2.5 μA beam.

Polarimeter operation is controlled from the Machine Control Center via EPICS, CEBAF's accelerator control software. Accelerator operations personnel measured beam polarization during a recent five week period that required polarized electron delivery to more than one experimental hall. A video monitor presents operators with a real-time image of the beam spot on the target foil. Raw scaler rate, asymmetry, beam current and dead time can be monitored at one second intervals. Helicity-correlated pulse-height spectra for all four detectors are displayed automatically after each measurement period so that operators can evaluate data quality.

SUMMARY

Progress has been made on improving aspects of the prototype 5 MeV Mott polarimeter at Jefferson Lab. Although systematic studies are not complete, recent calibration efforts and operational experience indicate the instrument is performing well. Further chamber modifications and calibration runs are planned.

REFERENCES

1. T.J. Gay and F. B. Dunning, *Rev. Sci. Instr.* **63** (1992) 1635-1651 and T. J. Gay, M. A. Khakoo, J. A. Brand, *et al*, *Rev. Sci. Instr.* **63** (1992) 114-130. T.J. Gay, these proceedings.
2. N. Sherman, *Phys. Rev.* **103** (1956) 1601-1607.
3. P. Ugincius, H. Uberall and G. H. Rawitscher, *Nucl. Phys.* **A158** (1970) 418-432.

Polarization Transport at TJNAF: Simulations and Measurements

J. M. Grames[*,1], D. H. Beck[*], L.S. Cardman[†], J. H. Mitchell[†], B. M. Poelker[†], J. S. Price[†], C. K. Sinclair[†], B. Zihlmann[†]

[*]*Department of Physics, University of Illinois at Urbana-Champaign, Urbana, IL 61801, USA*
[†]*Thomas Jefferson National Accelerator Facility, Newport News, VA 23606, USA*

Abstract. Polarized beam commissioning has been initiated at TJNAF. Measuring the polarization properties of the accelerator, which consists of two superconducting linacs and more than 2200 magnetic elements at its maximum energy, provide useful information about the beam energy and beam transport through the accelerator. Stability of the output beam polarization depends upon time-dependent orbit and energy oscillations. Modeling indicates that the spin transport dynamics are sensitive ($\sim 10^{-2}$) to vertical betatron oscillations in the recirculation arcs. This sensitivity is calculated to be enhanced in the arcs where the spin tune is most nearly matched to the vertical betatron tune. Modeling also suggests that depolarization phenomena associated with the particle trajectory differences for electrons within the accelerator's nominal transverse phase volume are too small to be measured between the injector and an experimental hall. Preliminary results from measurements obtained during polarization development runs will be presented.

INTRODUCTION

The objective of this talk is to describe studies performed at TJNAF that include modeling and simulation of both a finite phase space depolarization mechanism and of a beam polarization sensitivity to the orbit in the recirculation arcs. Preliminary measurements using the accelerator to study these effects are also presented.

A 100 keV photoemission polarized electron gun, driven by a laser system matched to the 1497 MHz fundamental frequency of two superconducting linacs, provide CW beam to the accelerator. The accelerator is racetrack style; linacs, connected by recirculation arcs at either end of the machine, allow for five pass operation. Five simultaneous beams of distinct energy can be

---

[1)] Work supported by the US NSF under contract PHY 94-20787.

separated to traverse the recirculation arcs which connect the linac segments by a series of magnets located both at the beginning and end of each arc. Up to three beams can be extracted from the accelerator and directed to the experimental end stations following the second linac.

SIMULATIONS & MEASUREMENTS

The transported polarization of a beam, to first approximation, can be described by applying the BMT [1] equation to the design momentum particles. In this case, the dipole fields which guide the beam through the machine then determine the bend–plane precession, Θ_{spin}. For a relativistic electron in a pure dipole field it is related to the bending angle, Θ_{bend}, by $\Theta_{spin} = \frac{g-2}{2} \gamma \Theta_{bend}$, where the bend–plane precession depends only upon the beam energy, physical constants, and geometrical factors of the machine. The nominal beam energy and spin tune per arc are shown in Table 1.

TABLE 1. Spin tune in the nine recirculation arcs.

Arc	1	2	3	4	5	6	7	8	9
Energy (MeV)	445	845	1245	1645	2045	2445	2845	3245	3645
Precession/2π	.505	.959	1.41	1.87	2.32	2.77	3.23	3.68	4.14

Simulations were performed to estimate the contribution to beam depolarization due to the finite transverse emittance of the beam. To describe this effect the spin dynamics of particles within the nominal transverse emittance are tracked [2] through the recirculation arcs. The results of these representative particles are then combined to determine the depolarization of the beam. Simulations indicate that the machine emittance ($\epsilon_x = \epsilon_y = 2.0 \times 10^{-9}$ m rad at 1 GeV) incurs less than 1.5 ppm depolarization for all nine recirculation arcs. This reduction is smaller than can be absolutely measured between the injector and end stations.

Another series of simulations were performed to study the sensitivity of the beam polarization orientation to vertical betatron orbits in the arc proper (180° bending section). Such orbits, due to magnet misalignments or fabrication errors, betatron mismatches, or beam energy error, introduce terms to the BMT equation representing the quadrupole fields of the arc optics. These off-axis fields cummulatively couple the bend–plane and vertical components of the beam polarization. The simulations indicate there exists a spin-orbit sensitivity, qualitatively similar to that of resonance depolarization, which appear to be enhanced in the arcs where the spin tune is most nearly matched to the vertical betatron tune. For example, the coupling between the bend–plane and vertical components of the beam polarization for arc 1 ($\nu_s = 0.505, \nu_y = 3$) is two orders of magnitude weaker than for arc 7 ($\nu_s = 3.23, \nu_y = 3$).

To investigate the predicted spin-orbit sensitivity an experiment is planned. Using a series of air core magnets located in the recirculation arcs a pair of vertical betatron oscillations will be alternately excited at the entrance, and then removed at the exit, of arc 7.

In preparation for this experiment a development run measured the polarization at both the injector and then simultaneously in two end stations for various injector polarization orientations. The beam polarizations measured at both the beginning and end of the machine are consistent with one another within statistical precision. The phase advance to both end stations indicated that the average linac energy might have been lower than assumed by as much as 0.25%.

A demonstration for inducing and removing a 3 mm vertical betatron oscillation in arc 7 successfully showed the RMS beam motion varied by $\Delta x \leq 100 \mu m$ and $\Delta y \leq 65 \mu m$ at the end station polarimeter target foil. A plot of the final polarization for a sequence of measurements using two different orbit errors is shown. The accelerator schedule limited the polarization to the bend–plane,

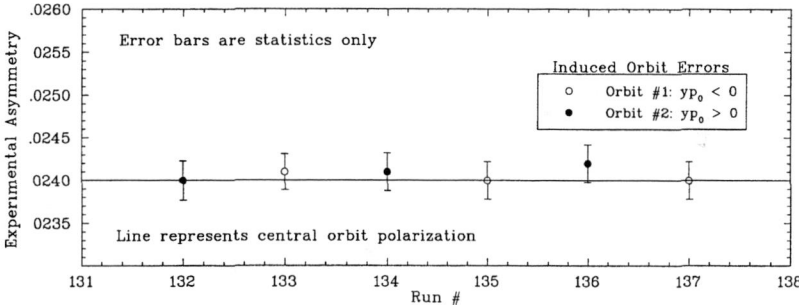

where theory predicts that the polarization is insensitive to vertical betatron oscillations. The measurements statistically agree with this prediction.

CONCLUSIONS

Measurements at TJNAF indicate that the beam polarization is conserved when transported through the machine and that the first tests of the spin–orbit sensitivity agree with prediction. An experiment to investigate the $\sim 10^{-2}$ effects predicted and associated with out–of–plane polarization orientations is planned.

REFERENCES

1. J.D. Jackson, *Classical Electrodynamics*, Wiley, New York, 556–60 (1962).
2. T.H. Fieguth, private communication.

POSTER SESSION

Fiber Optic Applications for Laser Polarized Targets

William J. Cummings and Robert S. Kowalczyk

Physics Division, Argonne National Laboratory, Argonne IL 60439

Abstract. For the past two years, the laser polarized target group at Argonne has been using multi-mode fiber optic patch cords for a variety of applications. In this paper, we describe our design for transporting high power laser beams with optical fibers currently in use at IUCF.

One difficulty in utilizing laser polarized gas targets for nuclear and particle physics experiments is operating targets developed in "small science" laboratories in a "big science" accelerator environment. Usually, one has to choose between placing the lasers close to the target i.e. behind the radiation safety interlock and placing the lasers outside the accelerator vault which requires a long beam transport path. In most cases, placing the lasers close to the target has been selected because of its similarity to the laboratory setting and the large cost of beam transport optics. The use of multi-mode optical fibers for the beam transport dramatically lowers the cost and allows for greater flexibility in the location of the laser system.

The widespread use of optical fibers in the communications industry has made application of this technology to laser polarized targets easy and inexpensive. The fibers used in this work were purchased as "patch cords" in which the multi-mode silica fiber is packaged in protective tubing and terminated with industry standard connectors. This makes installing, interchanging, and replacing fibers as easy as electrical cables. For coupling light into and out of the fiber we use industry standard coupling/collimation packages which consist of an aspheric lens mounted in a body with mating connectors for the patch cords. The connector mating ensures that the fiber end is located and the focal length of the lens.

The main advantages of using fiber beam transport for laser polarized targets are low cost, reproducible allignment and high transmission efficiency. We have installed a 65 meter long transport system for the laser-driven H/D target installed at IUCF at a cost of less than $500. For the 5 Watts of 770 nm laser light we use, the transmission per meter is larger than 0.998 lead-

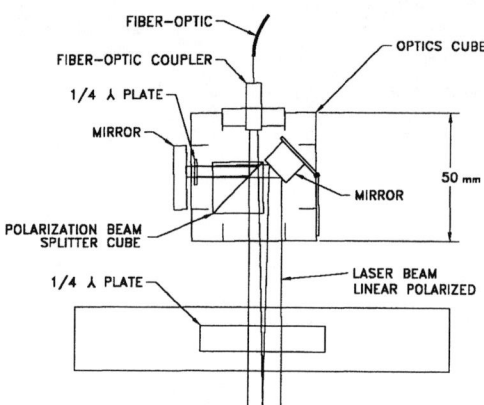

FIGURE 1. Optics used for re-polarizing laser light at IUCF.

ing to negligible transmission loss. Coupling the free space laser beams into the optical fiber is not difficult due to the large numerical aperture of the multi-mode fiber. This also allows one to put two free space beams into the same optical fiber. Input coupling efficiency can be as high as 80% and is more typically 75% even when two free space beams are accommodated. The output intensity profile is largely insensitive to fluctuations in the position of the input beam leading to more stability in the target operations.

The chief disadvantage of multi-mode fiber transport for use in laser polarized targets is depolarization of the laser beam. This can be overcome by "re-polarizing" the laser beam after the fiber. An example of how we did this at IUCF is shown in Figure 1. The output beam from the fiber is split into its two linear polarization components and one is rotated 90° by passing twice through a $\lambda/4$ plate. These two beams are pointed towards the target and converted to circular polarization using a second $\lambda/4$ plate. As can be seen in Figure 1, the output beam from the fiber coupler has a non-negligible divergence. This is due to the relatively large 400 μm core diameter of the fiber. We use this divergence to expand the beam to the size of the optical pumping region in our target. Applications which require a smaller beam divergence can be achieved using a fiber with a smaller core diameter.

In summary, the use of multi-mode fiber optic transport has allowed us to install a laser polarized target at IUCF while keeping our lasers outside the accelerator vault in an environmentally controlled room. As a consequence our lasers are stable and require very little maintenance. This work is supported by U.S. Department of Energy, Nuclear Physics Division, under contract No. W-31-109-ENG-38.

Polarisation and Compression of ^{3}He for Magnetic Resonance Imaging purposes

D.G. Geurts*, J.F.J. van den Brand*, H.J. Bulten*,
M. Ferro-Luzzi†, K. Nicolay#, H.R. Poolman*

*Vrije Universiteit, 1081 HV Amsterdam, the Netherlands
†NIKHEF, 1009 DB Amsterdam, the Netherlands
#Bijvoet Center for Biomolecular Research, Univ. Utrecht, the Netherlands

Magnetic Resonance Imaging is often used in medical science as a diagnostic tool for the human body. Conventional MRI uses the NMR signal from the protons of water molecules in tissue to image the interior of the patient's body. However, for certain areas such as the lungs and airways, the usage of a highly polarised gas yields better results [1]. We are currently constructing an apparatus that uses polarised ^{3}He gas to produce detailed images of those signal-deficient moyeties. We also plan to study possible uptake of polarised ^{3}He gas by the circulatory system to image other organs.

The use of polarised noble gases for MRI experiments was first proposed by Gatzke et al. in 1993 [2]. The first successful images were taken in 1994, using ^{129}Xe to image the lungs of a mouse [1]. There are, however, some advantages for using ^{3}He over ^{129}Xe. Firstly, the production of polarised ^{129}Xe requires the presence of Rubidium, a toxic substance whereas for the production of polarised ^{3}He high purity gas without any contaminants can be used. Furthermore, the magnetic moment of ^{3}He is larger than that of ^{129}Xe, resulting in a higher MRI signal.

We already constructed a polarised ^{3}He gas target for nuclear physics purposes [3]. Polarised ^{3}He is produced via direct optical pumping of the metastable state [4]. If a weak electric discharge is maintained in a low-pressure ^{3}He gas, then a small fraction of the atoms ($\approx 10^{-6}$) will be in the long-lived 2^3S_1 metastable state. Circularly polarised pumping light incident upon the sample along a weak applied magnetic field excites transitions between the 2^3S_1 and 2^3P_0 states. Angular momentum is thus transferred from the pumping light to the metastable atoms, and the metastable atoms become polarised. Transfer of polarisation to the ground-state atoms is achieved in an efficient manner through metastability exchange collisions. We obtained polarisation levels of up to 68 % with typical pump-up rates in the order of

1×10^{18} atoms/s (Fig. 1).

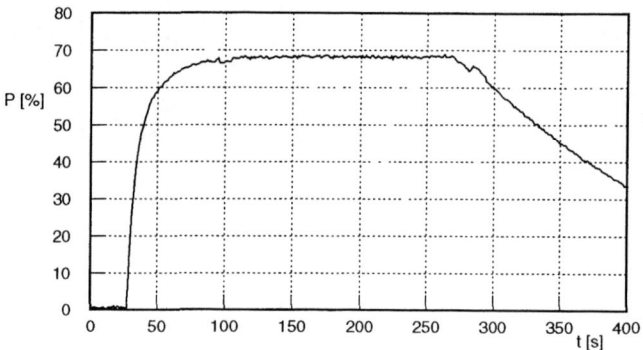

FIGURE 1. Polarisation of ^{3}He as a function of time for a sealed cell at about 1 mbar. At $t \simeq 270$ s the laser light is blocked.

Since this optical pumping technique requires ^{3}He gas at low pressure (typically 1 - 2 mbar), a compressor is needed to reach polarised ^{3}He gas pressures of about 1 bar. To prevent loss of polarisation the compressor needs to be non-magnetic. We have designed and are constructing a compression system that combines mechanical and cryogenic compression of polarised ^{3}He. This is accomplished by mechanically compressing the low-pressure ^{3}He gas with PTFE bellows into a storage cell that is cooled to liquid nitrogen temperature (≈ 80 K). After mechanical compression to about 300 mbar the storage cell is heated to room temperature, resulting in polarised ^{3}He at 1 bar.

Tests of the compression system for polarised ^{3}He gas are in progress. Subsequently we will perform MRI experiments with the gas on sealed cells and animals. In a later stage, efficient MRI pulse-schemes will be developed in order to allow us to image various parts in the human body with high resolution.

This research is sponsored by FOM/ NWO.

REFERENCES

1. M.S. Albert *et al.*, *Nature* **370**, 199 (1994)
2. M. Gatzke *et al.*, *Phys. Rev. Lett.* **70**, 690 (1993).
3. See the contribution of H.J. Bulten *et al.* in these proceedings.
4. F.D. Colegrove *et al.*, *Phys. Rev.* **132**, 2561 (1963).

The UNH Polarized ^{3}He Program

F. W. Hersman, R. H. Carrier, V. R. Pomeroy

Department of Physics, University of New Hampshire, Durham, NH 03824

Abstract. The UNH Nuclear Physics Group is developing polarized ^{3}He cells for use in approved experiments with electron beams and neutron beams. We also have begun to apply our expertise to magnetic resonance imaging experiments I list our projects and review several areas where we have been focusing our efforts.

I NUCLEAR PHYSICS

The UNH Nuclear Physics Group began developing polarized ^{3}He cells in 1992, primarily to support physics proposals to measure the neutron electric form factor and nuclear spin structure of ^{3}He. Those experiments were subsequently approved at the Saskatchewan Accelerator Laboratory and the MIT-Bates Linear Accelerator. While neither of those experiments has been completed (Saskatchewan discontinued its electron program due to funding cuts, and the continuous polarized beam is not yet available at Bates), our program has grown. We recently received approval of our proposal to study the spin structure of ^{3}He at the Jefferson Laboratory, and are collaborators on an inclusive measurement and a deuteron knockout measurement at MIT-Bates.

Last year we had an opportunity to contribute polarized ^{3}He cells for several experiments requiring a polarized neutron beam at LANSCE (see the contribution by V. R. Pomeroy in this volume). Several cells were prepared and tested at UNH for that purpose. The best of them yielded a maximum polarization of 64% after pumping for several days.

II MAGNETIC RESONANCE IMAGING

The UNH group has joined a collaboration including researchers form the Harvard-Smithsonian Astrophysical Observatory, the Brigham and Women's Hospital, MIT, Schlumberger-Doll Research, and North Carolina State University to perform hyperpolarized noble gas magnetic resonance imaging. The scientific motivation for these studies includes biomedical studies of normal and pathological tissues, geophysical studies of sandstones related to oil recovery, studies of foam rheology, studies of diffusion properties, and development of imaging techniques. The UNH group has provided roughly a dozen cells for various studies.

III NUMERICAL SIMULATIONS

To support our design efforts, we have developed a polarization simulation code (Figure 1). Our code calculates the attenuation of the laser light and the polarization of the rubidium throughout the cell volume, properly accounting for diffusion, spin destruction, and nuclear spin, based on the equations of Wagshul and Chupp. We have extended this calculation to allow arbitrary axially symmetric cell shapes, arbitrary laser wavelength distributions, diverging and converging laser optics, and in three fully independent dimensions. We plan to optimize cell geometries for each of our applications.

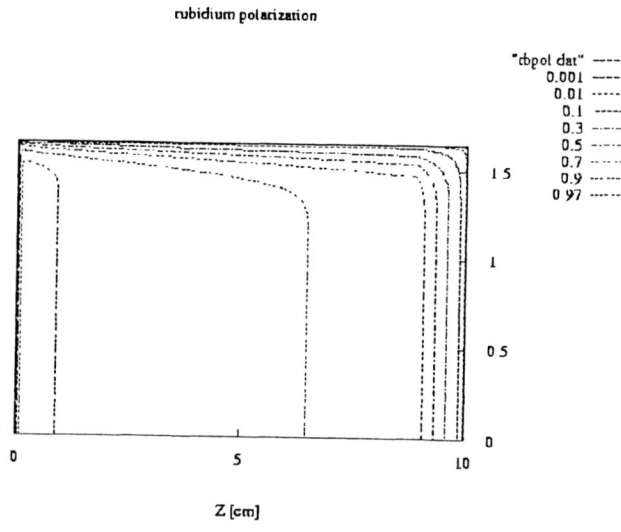

FIGURE 1. Rubidium polarization contours plotted against cell radius and length for a cell at 180°C and a 15 W laser. The circularly polarized light enters from the left where much light is lost due to the depolarized boundary layer. While propagating through the cell the light intensity lowers as does the rubidium polarization.

Polarized ^{3}He Spin Filters for Polarized Neutron Experiments at LANSCE

Vance R. Pomeroy, F.W. Hersman, M.B. Leuschner

University of New Hampshire

Abstract. At the University of New Hampshire we have a program of polarizing ^{3}He samples for use as targets in electron scattering experiments, neutron spin filters, and for magnetic resonance imaging. The objective of this poster was to give an introduction to polarized ^{3}He neutron spin filters and to outline the fabrication of spin-exchange polarized ^{3}He neutron spin filters at UNH for two recent measurements at LANSCE.

Polarized ^{3}He can be used as a neutron polarizer since there exists a broad ($\Gamma = 270 keV$) unbound resonance ($J^\pi = 0^+$) in the intermediate state ^{4}He*, for the reaction

$$n +^3 He \longrightarrow ^4 He^* \longrightarrow p + t.$$

There is a strong spin dependence in the cross section since the spin of the ^{3}He nucleus is carried mostly by the unpaired neutron and thus only neutrons with spin antiparallel to that of the ^{3}He nucleus will be absorbed. The ratio of the cross section of antiparallel neutron spin capture to that of the total cross section was measured to be 1.010 ± 0.032 [1], which indicates a small contribution from nonresonant scattering.

For a beam of unpolarized neutrons passing through a sample of polarized ^{3}He, the polarization and transmission are given by:

$$P_n = \frac{t^+ - t^-}{t^+ + t^-} = \tanh[nlP_{^3He}\sigma(v)]$$

$$T_n = \frac{t^+ + t^-}{2} = e^{-nl\sigma(v)} \cosh[nlP_{^3He}\sigma(v)]$$

where

$$t^\pm = e^{-nl\sigma(v)(1 \pm P_{^3He})},$$

and n is the ^{3}He number density, l is the length of the sample, $P_{^3He}$ is the ^{3}He polarization and $\sigma(v)$ is the absorption cross section for unpolarized neutrons with velocity v. For unpolarized 25.3 meV neutrons the cross section is

$$\frac{1}{2}\sigma_{\uparrow\downarrow}(v) = \sigma(v) \approx 5327b,$$

where $\sigma_{\uparrow\downarrow}(v)$ is the capture cross section for the neutron spin antiparallel to the ^{3}He spin. For a 3.3 atmosphere 10 cm long spinfilter at 43 % polarization, $P_n \simeq 73$ % at 0.1 eV.

We have produced neutron spin filter cells which were used in two recent experiments at the Los Alamos Neutron Science Center (LANSCE). The glassware was made by the Princeton Glassblowing Shop. The geometry of cells was cylindrical with a length of 10 cm and a diameter of approximately 3.3 cm. The walls of the cell were made from resized iron free Corning 1720 glass, and the windows were made from flat, boron free, iron free, Corning 1720 glass. The cells were connected to a Pyrex manifold which was then connected to the high vacuum system. A rubidium ampule is connected to this manifold and then a portion of the rubidium metal is distilled into a holding bulb. The ampule is then removed. The cell is then placed into an oven and baked at 450 °C for four days under high vacuum. After baking, some rubidium is chased into the cell using a torch. The cell is then left under vacuum for another day before filling.

In order to achieve high densities of ^{3}He gas, we designed a filling system using liquid helium. The surface temperature of each cell was monitored in several locations. Once the cell had reached liquid helium temperature, the cells were filled with nitrogen and then the ^{3}He gas. These gases were flown through individual getter pumps and cryo traps for purification. Once the gas has been introduced into the cell and the pressure has stabilized, the cell is sealed off from the vacuum system using a torch. The final pressure of nitrogen was 100 torr, and from 3.3 to 6 atmospheres for ^{3}He. Lifetimes ranged between 38 and 90 hours.

The first experiment measured the polarization of 25 meV to 10 eV neutrons passing through the spin filter to an accuracy approaching 0.1 %. Analysis of this data is ongoing at the University of Indiana. The second experiment measured the parity-violating neutron spin rotation in the n-^{139}La p-wave resonance. This experiment required two polarized ^{3}He samples, one as a spin filter and the other as a spin analyzer. The data is being analyzed at Kyoto University in Japan.

REFERENCES

1. L. Passell and R.I. Schermer,*Phys. Rev.* **150**, 146 **(1966)**.

Optimization of Spin-Exchange Cell Geometry for a Laser Driven Target

E. C. Schulte, R. J. Holt, B. R. Owen, E. L. Thorsland

Department of Physics, University of Illinois, Urbana, Illinois, 61801, USA

Abstract. A simple conductance and flow rate model is used to optimize the Illinois Laser Driven Target spin-exchange cell geometry. The diameter of the spin-exchange cell conductance limiter and the density of potassium vapor are varied to find the highest figure of merit.

Introduction

The purpose for optimizing the Illinois Laser Driven Target [1] spin-exchange cell is is to find a system which will yield the highest target figure of merit and to determine the effect if the nuclei in recombined molecules retain half their polarization. The simulation will then be used to design a system which will test for the presence of molecular nuclear polarization. In our geometry the diameter of an aluminum conductance limiter (the exit of the spin-exchange cell) and the potassium vapor density are allowed to vary. Maximum figures of merit are calculated in the cases where only atomic nuclei are polarized and where the nuclei of molecules retain 50% of the atomic nuclear polarization.

Method

This optimization used a simple model of conductance and rate equations. Measured atomic fractions and atomic polarizations helped set the cell conditions to yield the assumptions in Table 1. Other parameters are also given in Table 1. Also, the potassium polarization depended upon the density of hydrogen and the target and spin-exchange cells were kept at $227°C$.

The nuclear polarizations were calculated using equations (1) and (2).

$$p = P_K \frac{\gamma_{HSE}}{\gamma_{HSE} + \gamma_{wl}} \alpha \chi \Gamma \qquad (1)$$

$$p_{mol} = P_K [\frac{1}{2} \times \frac{\gamma_{HSE}}{\gamma_{HSE} + \gamma_{wl}} (1-\alpha)\chi\Gamma] \qquad (2)$$

TABLE 1. Assumptions from IUCF data with a 4 mm diameter conductance limiter.

Assumptions		Model Parameters		
		Fixed	Varied	
Atomic Fraction	≈ 0.4	K Density $\leq 5 \times 10^{11} cm^{-3}$	Hole Diam	0.01 to 6.7 mm
P_e	≈ 0.6	Flow Rate $1 \times 10^{18} atoms/s$	K Temp	$30°C$ to $200°C$

where α is the atomic fraction, χ is the dilution factor [2], and Γ is the spin temperature equilibrium [3] probability. Employing these equations and the calculated target thickness L, the target figures of merit could then be calculated using equations (3) and (4).

$$FoM = p^2 \times L \qquad (3)$$
$$FoM_{MNP} = (p + p_{mol})^2 \times L \qquad (4)$$

Using the parameters in Table 1 and equations (3) and (4), two cases were examined. First, equation (3) was maximized to give the highest figure of merit assuming only atomic nuclear polarization. Second, equation (4) was maximized to give the highest figure of merit assuming molecular nuclei retained 50% of the polarization. The optimization for hydrogen yielded two conductance limiter diameters, 6.7 mm and 6.5 mm. These resulted in figures of merit of $4.6 \times 10^{13} cm^{-2}$ and $4.5 \times 10^{13} cm^{-2}$, respectively, for atomic nuclear polarization, and $6.2 \times 10^{13} cm^{-2}$ for both cases assuming the presence of molecular nuclear polarization.

Conclusion

The optimization procedure produced two spin-exchange cell conductance limiter diameters. The larger gives the maximum figure of merit for atomic nuclear polarization only. The smaller gives the maximum figure of merit assuming molecular nuclear polarization. If nuclei in molecules retain half their polarization after recombination, we project that the figure of merit for the target should stay at approximately the same value as it is in the case where only atomic nuclei are polarized.

REFERENCES

1. Owen, B., et. al., *Proceedings of the 12th International Symposium on High-Energy Spin Physics (Spin 96)*, edited by C. W. de Jager, et. al., Singapore: World Scientific 1997, pp. 490-491.
2. Chupp, T. E., R. J. Holt, and R. G. Milner, *Annu. Rev. Nucl. Part. Sci.* **45** 373 (1994); and references therein.
3. Walker, T. G. and W. Happer *Reviews of Modern Physics.* **69**, No. 2, 629 (1997); and references therein.

Mechanical Filter for Alkali Atoms

Dmitrij Toporkov[⊗] and Bogdan Wojtsekhowski[#]

⊗ Budker Institute for Nuclear Physics, 630090 Novosibirsk, Russia
\# Thomas Jefferson National Accelerator Facility, Newport News, VA 23606

The recent past has shown a significant progress in the spin exchange optical pumping technology /1-3/. Spin-exchange optical pumping is a process whereby photon angular momentum is transferred to target nuclei through spin-exchange collisions with polarized alkali-metal atoms. In spite of their small numbers, alkali-metal atoms could significantly dilute the nucleon polarization in the target, because of there are a large number of unpolarized nucleons in the alkali nuclei. Also, if potassium atoms are used in the pumping cell with a density of one percent of the hydrogen density, the losses of the high energy electrons in the storage ring due to the presence of potassium is two time higher than the losses due to the hydrogen.

Design considerations

Here we present an idea how significantly reduce the amount of alkali-metal atoms in a target cell without too large a loss of the flux of polarized hydrogen atoms. The idea is based on the difference of the velocities of the hydrogen and alkali atoms. When an atom hits the surface, it spends some time on the surface and than goes out with a velocity corresponding to the temperature of the surface and some random direction in the surface reference frame.

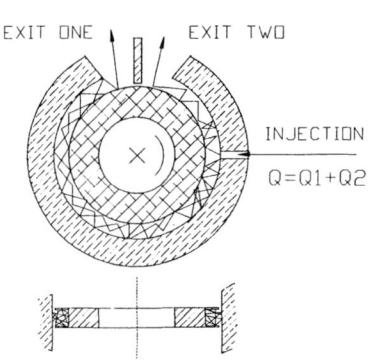

FIGURE 1. Mechanical Filter

Let us assume that the mixture of atoms flows into the filter shown in Fig.1 which consists of a fixed outer cylindrical wall and three other walls which could be rotated. There is a stopper, which is positioned between the two exit holes. The mixture of atoms is injected into the filter through the entrance hole. If the inner part of the device is not rotated the mixture of atoms will divide into two flows that come out through the exits with the intensities corresponding to the conductances and the concentration at the entrance of the filter. The rotation of the inner part affects mainly the alkali-metal atoms, which have much smaller velocities then the alkali-hydrogen atoms at the same temperature. The metal atoms start to move preferably in the direction of rotation while the hydrogen atoms are still not sensitive to the motion of the walls. Due to this effect it is possible significantly suppress the amount of alkali-metal atoms in the flow to exit two.

Table 1 presents the results of Monte-Carlo simulation of the gas flow. N_{coll} is the average number of the atom's collisions with the walls, N_{part} the number of the particles left the filter correspondingly through the first/second exit. The dimensions of the filter (1x1 cm^2, disk diameter d = 10 cm) and the rotation speed Ω were chosen to be similar to those of the turbo pump. The azimuthal extension of the exit holes is equal to 0.5 radians. For each situation 1008 atoms were injected into the filter at the position $\varphi = \pi/2$ radians counting clockwise from the stopper. The reduction of the potassium fraction by a factor of 15 for the given geometry of the filter could be achieved at the rotation velocity of 48.000 rpm and about 40% of the total flux of polarized hydrogen atoms from the source could be directed into the storage cell. The average number of wall collisions of the hydrogen atoms which left the filter through exit two is about 100. The results are averaged over the Maxwell distribution at T = 500° K.

TABLE 1. Intensity of flow through exits one and two

Ω, M	rpm exit	0.0 one/two	6.000	12.000	48.000	60.000	90.000
87	N_{coll}	341/ 140	237/ 97	162/ 63	52/ 14	42/ 4	32/ 2
	N_{part}	220/ 788	647/ 361	836/ 172	995/ 13	996/ 12	1003/ 5
39	N_{coll}	341/ 138	282/ 113	209/ 81	72/ 29	60/ 11	43/ 7
	N_{part}	222/ 786	533/ 475	736/ 272	981/ 27	996/ 12	1002/ 6
1	N_{coll}	344/ 141	333/ 133	321/ 131	252/ 99	230/ 92	180/ 66
	N_{part}	220/ 788	274/ 734	324/ 684	594/ 414	667/ 341	773/235

Conclusion

If used in LDS the filter could reduce by a factor of 10-15 the relative concentration of the alkali-metal atoms in the target and relax the existing limitation on the alkali density in the pumping cell. As a result the novel lasers (suitable for Rb) could be used in LDS and even higher flux and polarization of the hydrogen atoms could be achieved.

REFERENCES

1. Fedchak J. (ANL), " The Argonne Laser-Driven D Target", presented at this conference.
2. Miller M. (Urbana), " IUCF Test of an LDS Target", presented at this conference.
3. Stenger J. (Erlanger), " Nuclear Spin Polarized H and D by Means of Spin-Exchange Optical Pumping", presented at this conference.

Beam Saturation in the Munich Atomic-Beam-Source

R. Hertenberger, Y. Eisermann, A. Hofmann, A. Metz,
P. Schiemenz, S. Trieb, G. Graw

LMU, Sektion Physik Universität München, Am Coulombwall 1, D-85748 Garching [1]

Abstract. A Stern-Gerlach atomic beam ion source (ABS) with ECR ionizer and cesium gas-target as charge exchange units is under construction. The goal is to achieve intense negative beams of polarized H^- and D^- of high brightness for injection into the 14 MV MP tandem accelerator with beam intensity of about 2 μA on target. At the entrance of the ECR ionizer the observed intensity of the neutral beam of polarized atoms is $6.4 \pm 0.4 \times 10^{16}$ H/sec. Comparing beam transport calculations with beam intensity measurements under various conditions allows for a consistent description of the beam formation in the ABS with respect to degree of dissociation ($\simeq 80\%$), peaking ($\simeq 1.6$), beam absorption ($\simeq 40\%$) and pumping speeds.

I THE ATOMIC BEAM: ABSORPTION, PEAKING AND DEGREE OF DISSOCIATION

The atomic beam source works reliably. Its output was increased during the last 1.5 years from 5.2 to 6.4×10^{16} H/sec. In the following we try to disentangle the different contributions to the limitation of the ABS beam intensity. The atomic or molecular flux ϕ_{H1} and ϕ_{H2} into a compression tube detector installed at the entrance of the ECR ionizer is given by:

$$\phi_{H1} = \phi_0 \times P_{H1} \times \Omega_1 T \times Abs_{H1} \times 2 \times D \qquad (1)$$

$$\phi_{H2} = \phi_0 \times P_{H2} \times \Omega_2 \times Abs_{H2} \qquad (2)$$

The H_2 flux through the dissociator is denoted ϕ_0, $P_{H1,2}$ is the forward peaking relative to a $\cos(\theta)$ distribution, $\Omega_1 T$ the opening angle of the sextupole system times transmission through the sextupoles (Monte Carlo estimate), Ω_2 the

---

[1] Supported by the BMBF, the DFG and the Beschleuniger Laboratorium München

opening angle of the compression tube, $Abs_{H1,2}$ the beam absorptions, D the degree of dissociation.

Atomic and molecular beams saturate in the same way, fits to experimental data give the same exponential reductions of 60 % at working points of 2.2 mbar l/s of gas throughput.

The scattering of H_1 beam on residual gas has been measured by varying the density of the residual gas in the vacuum chambers of the ABS. The observed absorptions of $Abs_{H1,2}=0.6$ are reproduced using an effective cross section of $\sigma = 1\times 10^{-14}$ cm^2 and an exponential absorption law. A considerable part of the beam is absorbed in the magnets. Using molecular beams and comparing with and without sextupole magnets an absorption of 15 % was observed.

Comparing the pressure rises in chambers 1 and 2 as a function of throughput ϕ_0 with the expected rises from an atomic beam emerging the nozzle with a $\cos(\theta)$ intensity distribution results in a peaking factor of 1.6 for H_1. The saturation of the flux into the compression tube (measured without sextupoles in the ABS) can be simulated using the cross section value of $\sigma = 1\times 10^{-14}$ cm^2 for scattering on residual gas suggesting that intra beam scattering is not yet dominant.

Comparing measured beam intensities with intensities calculated according to formula (1) allows to estimate a degree of dissociation of D = 77 % at the working point. This is in reasonable agreement with a measurement at the IUCF ABS [1]. Investigating a similar dissociator they used a quadrupole mass spectrometer which was placed directly into the beam.

II CONCLUSION

The Munich ABS source produces beams of neutral polarized H atoms in two spinstates of 6.4×10^{16} H/s into a compression tube of 10 mm diameter and 100 mm length. Monte Carlo calculations of the beam intensity describe measurements using the experimentally determined values for beam absorption and peaking. The beam intensity is lost primarily due to the limited pumping speed in the first chamber (about 20 %) and due to insufficient conductances of the first two sextupole magnets (about 15 %). It is a challenge to optimize the geometry of the sextupoles, since focussing properties of the magnets lower immediately when opening their gap-width to provide for better pumping. For a more detailed presentation see refs. [2,3].

REFERENCES

1. T. Wise et al., NIM A336 (1993) 410-422
2. K. El Abiary, diploma thesis, LMU Munich, 1996
3. R. Hertenberger et al., Rev.Sci.Inst (1997) to be published

Optimized Atomic Beam Sources

S. Lemaitre*, R. Brüggemann[†], V. Nelyubin[‡] and
H. Paetz gen. Schieck[†]

*Department of Physics and Astronomy, University of North Carolina, NC 27599-3255
[†]Institut für Kernphysik, Universtät Köln, Zülpichers tr. 77, D-50937 Köln
[‡]St. Petersburg Nuclear Physics Institute, Gatchina, Russia

Abstract. Extensive numerical optimizations were performed for atomic beam sources (ABSs) operated at different beam velocities. An optimized magnet system was proposed for the polarized gas target for the ANKE spectrometer at COSY/Jülich [1]. For Hydrogen 24 % more flux into the storage cell is expected compared to the HERMES ABS design [2]. Numerical optimizations for the ABS of the polarized ion source at TUNL [3] which operates at a lower beam velocity/temperature promise an improvement of about 18 %.

INTRODUCTION

A computer code was developed for the simulation and optimization of atomic beam sources. Special emphasis was laid on the following features: 1. easy application to different sources/geometries, 2. easy application of different optimization algorithms, 3. easy extendable code to include simulations of other particle species (e.g. charged particles), and 4. simultanous simulation of different particle species. For an optimal fulfillment of these requirements the code was realized in an object oriented programming language (C++, [4]).

Different optimizers were implemented and applied to the above mentioned sources. Best results were obtained with a very simple optimizer (successive modification of parameters until optimization is reached). A more sophisticated optimization procedure (Levenberg-Marquardt) which was successfully applied to monochromatic ion beams [5] failed in the atomic beam case.

OPTIMIZED AB FLUX INTO A STORAGE CELL

The optimization showed that an increasing maximum flux into the storage cell is inevitably correlated with the need to decrease the nozzle temperature T_N (see straight line through maxima in fig. 1). This is true for all simulated

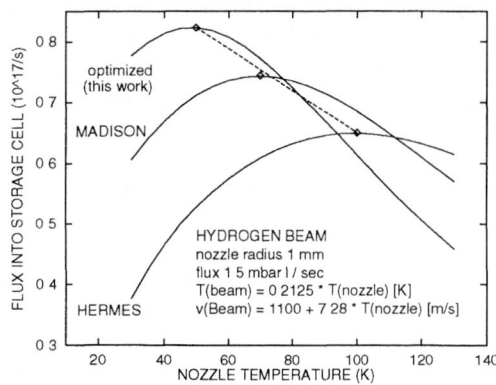

FIGURE 1. Simulated flux of H atoms (2 states) into a storage cell as function of the nozzle temperature using beam parameters from [6]. The optimized ABS (this work) is compared whith the simulated flux of the HERMES [2] and MADISON [7] ABSs. The flux is normalized to the maximum measured flux of the HERMES ABS, $0.65 \times 10^{17} \mathrm{s}^{-1}$ [2].

setups. With the parameters used [6] it was not possible to yield the high flux of the optimized setup at higher nozzle temperatures than $T_N = 50\text{--}60$ K.

OPTIMIZED AB DENSITY FOR AN ECR IONIZER

A hybrid variant was investigated combining the existing electromagnets [3] with new permanent magnets allowing for adaptation of the focal length to beam velocities from 850 to 1300 m/s. By this means an almost constant atomic beam acceptance can be achieved over this range. It is under discussion to operate the dissociator at a considerably higher gas throughput involving also slightly higher beam velocities.

REFERENCES

1. S. Lemaître et al., Annual Report 1996, Forschungszentrum Jülich, Germany
2. F. Stock, U. N. Funk, B. Lorentz, B. Povh, and E. Steffens, in [8].
3. T. B. Clegg et al., *Nucl. Instr. Methods*, **A357**, 200 (1995)
4. R. Brüggemann. Diploma thesis, Universität zu Köln, 1997.
5. S. Lemaître et al., *Rev. Sci. Instr.*, **67(3)**, 1145 (1996)
6. B. Lorentz. U. Funk. Diploma theses, Heidelberg, 1993 + 1995.
7. T. Wise et al., *Nucl. Instr. Methods*, **A336**, 410 (1993)
8. L. Sydow and H. Paetz gen. Schieck, editors. *Proc. of the Int. Workshop on Pol. Beams and Pol. Gas Targets, Köln, 1995.* Singapore, World Scientific, 1996.

Design Studies for the ABS of ANKE in COSY Jülich by 3-dimensional Magnetic Field Calculations

V.Nelyubin* and H.Seyfarth[†]

*St.Petersburg Nuclear Physics Institute, Gatchina 188350, Russia
[†]Institut für Kernphysik, Forschungszentrum Jülich, D-52425 Jülich, Germany

Abstract. Field distributions of the ANKE dipole magnets have been calculated numerically. Results for the stray field at the polarized atomic beam source and the storage cell gas target in COSY-Jülich are presented. The necessary soft-iron shields around the rf-transition units of the atomic beam source are discussed. Furthermore, a dependence of the focussing properties of adjacent permanent sextupole magnets on their relative azimuthal orientation and on the deviation from co-axiality has been investigated.

The polarized atomic beam source (ABS) and the storage cell gas target of the magnet spectrometer ANKE in the COSY storage ring in Jülich, Germany are under construction [1]. The center of the 40 cm long storage cell will be located about 50 cm from the yoke of the D2 spectrometer dipole magnet. At the maximum magnetic field strength of 1.6 T in the 20 cm high gap of D2 appreciable magnetic stray fields are expected at the storage cell and the ABS.

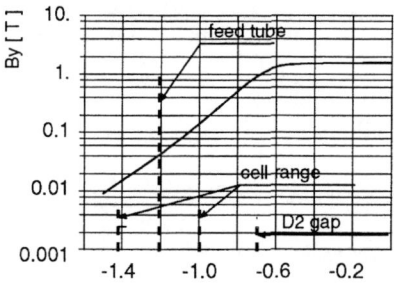 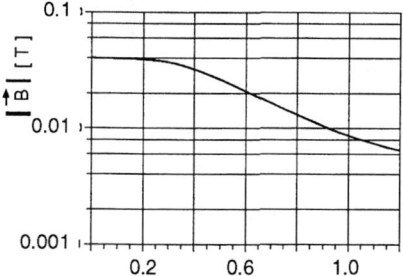

FIGURE 1. Calculated vertical stray field component B_y along the COSY-beam axis at $B_{gap}=1.6$ T (z=0 in the center of D2).

FIGURE 2. Magnitude of the calculated total magnetic field strength along the ABS-beam axis.

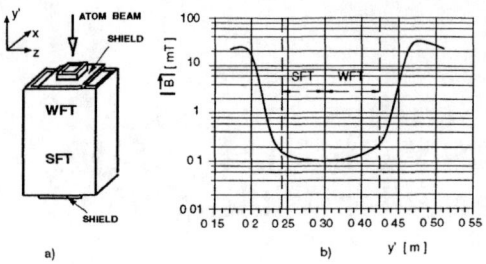

FIGURE 3. Holding-field magnet of the WF and SF transition units with the additional iron shield (a) and the calculated field along the ABS-beam axis (b)

Field distributions have been calculated using the computer code MAFIA [2]. Fig. 1 shows the vertical field component B_y along the COSY beam for B_{gap}=1.6 T. Fig. 2 shows the magnitude of the B-field along the ABS-beam (y') axis (for definition of the coordinate system see [1]). These data allow the following conclusions: (i) At the storage cell the stray field B_y is strong enough to be used as holding field for deuterium. Near the symmetry plane of the magnets ($y = 0$) $B_x, B_z \ll B_y$. (ii) The rf-transition units are operated at field strengths of 10^{-4} to 10^{-3} T and have to be shielded. (iii) At the turbomolecular pumps, positioned at about y=120 cm, the stray field is below the field limit of 0.013 T specified by the manufacturer [3].

The shielding for the rf-transition units that is necessary to reduce the stray field from D2 by two orders of magnitude has been designed with use of the MAFIA code. Even when the magnet yokes of the transition units are used as a shielding the ambient field in the transition region is still too high. The calculated field with additional 5 mm thick iron plates and additional short tubes. As shown in Fig. 3, is sufficiently low to operate the transition units.

The dependence of the total focussing properties of two adjacent sextupole magnets upon their relative azimutal and axial position has been studied as well. In the ANKE ABS the last two sextupole magnets have apertures of 30 mm diameter and the distance between them is 15 mm. The calculations show that relative azimutal orientation from parallel to antiparallel magnetization in corresponding segments of the two magnets reduces the flux into the storage cell by 3% only. For axial misalignment by less than 0.5 mm the reduction is smaller than 1%.

REFERENCES

1. H. Seyfarth, *contribution to the present proceedings*.
2. The MAFIA Collaboration. *The Manual for the Code Users (July 1994)*.
3. BALZERS/PFEIFFER, *Das Arbeiten mit Turbomolekularpumpen*.
4. Based on a code obtained from D.Toporkov, BINP Novosibirsk.

Tests of a Prototype Large-Bore, Low-Power $2 \leftrightarrow 4$ RF Transition Unit

R.S. Raymond, Randall Lab of Physics
University of Michigan, Ann Arbor, MI 48109
September 18, 1997

Abstract. The Michigan ultra-cold polarized hydrogen jet requires a $2 \leftrightarrow 4$ RF transition unit with a large bore, low power input for cryogenic operation, and a static magnetic field parallel to the atomic beam. The prototype unit has a cylindrical RF cavity with clear bore of 7 cm, loaded with a dielectric ring. Transition efficiency has been measured using a maser run in transient mode, by observing free induction decay. In tests on a room-temperature polarized beam, for static fields of under 100 G, we have measured efficiencies of 95% with less than 100 mW of RF power.
This work is supported by a grant from the U.S. Department of Energy.

The Michigan ultra-cold polarized hydrogen jet[1] requires a hydrogen $2 \leftrightarrow 4$ RF transition unit with a large-bore, low power input for cryogenic operation, and a static (and therefore RF) magnetic field parallel to the beam direction. To meet these novel requirements we have built and tested a room-temperature prototype cavity consisting of a metal cylinder with end plates, loaded with a dielectric ceramic ring. The ring is in the cylinder midplane. The atomic beam passes through large holes in the end plates and along the axis of the cylinder and ring. RF power is fed in via a small coupling loop on one side of the cavity, in the midplane and outside of the ceramic ring. The desired mode, TE_{015}, gives an axial RF magnetic field.

The prototype cavity uses a ceramic ring 11 cm O.D., 7.5 cm I.D. and 2.2 cm thick, with a dielectric constant of 13.9.[2] Metal parts of the prototype cavity are aluminum, with the cylinder 15.24 cm I.D., the holes in the end plates 7 cm I.D., and the inside length, for resonant frequencies of about 1425 MHz, about 5 cm. Teflon rings hold the ceramic ring centered in the cylinder. The Q value for the TE_{015} mode is about 7000.

For room-temperature beam tests of the prototype cavity we built a segmented solenoid allowing independent variation of the central field and the

---

[1] V.G. Luppov, Ultra-Cold Methods for Polarized Atomic Hydrogen, this conference
[2] The ceramic is SMAT-14, from Transtech, Inc.

axial field gradient. The assembled unit was then mounted between a small ground-state atomic beam source and a maser polarimeter, with a quartz tube through the transition unit to carry the room-temperature beam.

The maser consists of a Teflon-coated quartz bulb inside a loaded high-Q cavity tuned to 1420.4 MHz. Coils inside magnetic shields provide a uniform field of .1 mT over the bulb. For use as a polarimeter the maser is run in the transient mode, with the beam intensity below the threshold for self masing. We excite the cavity with a short pulse of RF power and record the resulting signal from free induction decay. The integral of this signal over a fixed time period is proportional to the difference in numbers of atoms in states 2 and 4. To measure the transition efficiency of the transition unit we measure the integrated signals, S, with RF power to the transition unit on and off, and a background, B. The transition efficiency, E, is

$$E = \frac{1}{2}\left(1 \pm \frac{S_{on} - B}{S_{off} - B}\right) \quad (1)$$

The + and - are for efficiencies greater and less than 50% respectively; there is a 180° phase change of the decay signal at 50% efficiency. S and B for these tests were typically 5 and .3 (arbitrary units), respectively. By moving the cavity and plates we could vary the resonant frequency, and so were able to measure the transition efficiency for static fields from 1.8 to 20 mT. Maximum efficiencies for each field are listed in Table 1.

TABLE 1.

RESULTS

STATIC FIELD (mT)	TRANSITION EFFICIENCY
20.0	75%
15.0	88%
10.5	92%
8.0	94%
5.6	95%
2.6	95%
1.8	95%

At low fields, less than 100 mW of RF power were required for efficiencies of over 95%. At higher fields, the efficiency was limited by the RF power available, about 300 mW.

We are continuing studies of the prototype and beginning the design of a cryogenic version. Tests indicate that there are no large changes in the characteristics of the ceramic at low temperatures, but support and cooling of the ring pose significant engineering challenges.

This work was supported by the U.S. D.O.E.

The Polarized Storage Cell Gas Target for the ANKE Spectrometer at COSY - Status and Design Considerations

H.Seyfarth* for the ANKE-ABS working group[†]

Institut für Kernphysik, Forschungszentrum Jülich, D-52425 Jülich, Germany

Abstract. An atomic beam source for polarized hydrogen and deuterium atoms is under construction. The source will be used to inject polarized atoms into a thin-walled storage-cell target at the magnet spectrometer ANKE in COSY-Jülich.

Unpolarized measurements on meson production [1] and deuteron break-up [2] will be performed at the spectrometer ANKE (**A**pparatus for the investigation of **N**uclear and **K**aon **E**jectiles) being installed in the COSY storage ring [3]. Later, these studies will make use of the polarized COSY-proton beam and the polarized hydrogen and deuterium storage cell target.

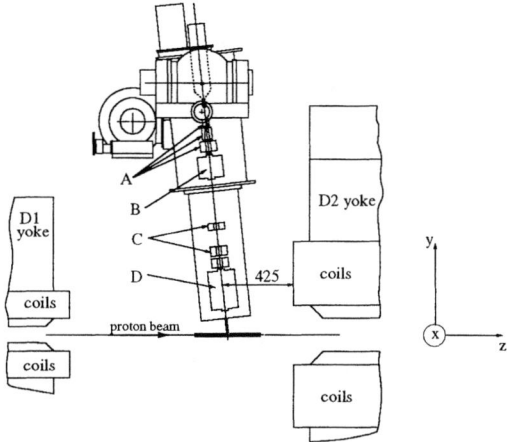

FIGURE 1. Sideview of the ANKE ABS (A and C: first and second group of separating permanent sextupole magnets, B: mean-field rf-transition unit, D: strong- and weak-field rf-transition units). D1 is a beam-bending, D2 the spectrometer dipole magnet.

Fig. 1 shows a side view of the polarized atomic beam source (ABS) and the 40 cm long storage-cell tube. The cell center is positioned 50 cm from the iron yoke of the D2 spectrometer magnet. Because of the limited space the ABS will be mounted vertically. A small inclination of 6 degrees shall prevent drizzling of quartz powder, possibly produced in the dissociator (e.g. [4]), into the storage cell. The upper vacuum vessel contains the nozzle and skimmer chambers (with two and one turbomolecular pump [5], respectively). A third chamber with a cryopump [6] contains the medium-field transition unit and the first group of sextupole magnets [7]. The three chambers are separated by adjustable baffles. The nozzle will be cooled by a cryocooler [8]. Studies have been started to replace the Cu heat bridge by a cryogenic heat pipe [9]. The upper vacuum vessel and the components are being manufactured, the design of the lower one and the storage-cell set-up is in progress. The magnetic stray field from D2, the necessary shielding, and the possibility to use it as holding field at the storage cell have been studied by numerical calculations [10].

† R.Brüggemann[1], N.Koch[2], V.Koptev[3], S.Kotov[3], S.Lemaitre[1], A.Lehrach[4], H.Loevenich[5], R.Maier[4], R.Nellen[4], V.Nelyubin[3], H.Pohl[5], D.Prasuhn[4], K.Rith[2], H.Paetz gen. Schieck[1], R.Schleichert[4], O.W.B.Schult[4], H.Seyfarth[4], A.Souslov[6], E.Steffens[2], H.J.Stein[4], A.Vassiliev[3], K.Zwoll[5]

[1] *Institut für Kernphysik, Universität zu Köln, D-50937 Köln, Germany*
[2] *Phys. Inst. II, Universität Erlangen-Nürnberg, D-91058 Erlangen, Germany*
[3] *High En.Phys.Dpt., St.Petersburg Nucl.Phys.Inst., 188350 Gatchina, Russia*
[4] *Inst. für Kernphysik, Forschungszentrum Jülich, D-52425 Jülich, Germany*
[5] *Zentrallabor für Elektronik, Forschgsz. Jülich, D-52425 Jülich, Germany*
[6] *on leave of abs. to [4] from Ural Techn. State Univ., Ekaterinenburg, Russia*

REFERENCES

1. COSY Prop.Exp.No.18 (K meson, spokesman K. Sistemich) and No.55 (a_0 meson, spokesman V. Tchernyshev), Letter of Intent Exp.No.35 (ϕ meson, spokesman M. Sapozhnikov)
2. COSY Prop.Exp.No.20 (spokesman V. Komarov)
3. O.W.B. Schult et al., Nuclear Physics **A583**, 629 (1995)
4. H. Okamura et al., AIP Conf. Proc. **293**, 84 (1994)
5. Pfeiffer, model TPH 2200
6. Leybold, model COOLVAC 3000
7. S. Lemaitre et al., contribution to these proceedings
8. Leybold, model RGS 120
9. A. Vassiliev et al., contribution to these proceedings
10. V. Nelyubin et al., contribution to these proceedings

Cryogenic Ne Heat Pipe Nozzle Cooling System for an Atomic Beam Source

A. Vassiliev*, V. Koptev*, S. Kotov* and H.Seyfarth[†]

*St.Petersburg Nuclear Physics Institute, Gatchina 188350, Russia
[†] Institut für Kernphysik, Forschungszentrum Jülich, D-52425 Jülich, Germany

Abstract. The main goal of this work is to construct and test an inexpensive and efficient ABS nozzle cooling system which allows to maintain a stable nozzle temperature and, if necessary, to achieve fast and well controlled temperature changes in a wide temperature range (30 K - 300 K).

The work presented here is part of the development of the polarized atomic beam source (ABS), which is under construction for the ANKE spectrometer [1] in the COSY storage ring in Jülich. The main goal is to construct and test an inexpensive and efficient ABS nozzle cooling system which allows to stabilize and quickly change the nozzle temperature in a wide temperature range (30 K - 300 K). The basic operating principle is to transfer the heat by vaporization of a suitable liquid on the nozzle and by liquefaction of the gas on the cold head of a cryo-cooler. This is known as a "heat pipe". The principle scheme of a heat pipe is shown in Fig. 1.

For the cryogenic heat pipe this process is described in ref. [2]. The nozzle temperature according to ref. [3] has to be in the range $60K \leq T_n \leq 120$ K. In this temperature range the estimated heat flow is 10 W$\leq W_n \leq 20$ W [3]. The gas suited to be used in the heat pipe has to be selected according to the desired temperature range. Only H_2, D_2, He and Ne have a triple-point temperature T_{tr} lower than 60 K.

The analysis of thermodynamic data shows that Ne is the best choice in this temperature range. With the use of a test set-up measurements have been performed to get experience with the operation of a cryogenic heat pipe and to understand the thermodynamics of vaporisation and liquefaction of the device [4]. Studies with a 400 mm long heat pipe of 30 mm outer diameter for the gas flow and 2mm inner diameter for the liquid flow filled with 20 g of Ne have shown the following results: 1) A heat power of up to 20 W can be

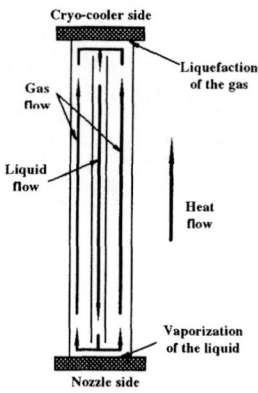

FIGURE 1. The principle set-up of the heat pipe.

transported with a temperature difference that does not exceed 6 K between the lower and upper end of the heat pipe. This corresponds to a pressure between 1 and 12 bar, or a gas (liquid) temperature, associated with the pressure on the vaporization line of 31 to 38 K; 2) No dependence of the heat pipe characteristics on the mass of liquid Ne has been observed. This allows to minimize the amount of Ne inside the heat pipe. Thereby the pressure resulting from total evaporation of the liquid in an emergency situation is minimized as well. Furthermore, the investigations showed that the heat pipe is capable to transfer heat power while the dissociator-nozzle is stabilized within one degree at temperatures between 60 and 120 K. Operation of the atomic beam sources at HERMES and IUCF have shown that the nozzle has to be warmed up approximately once per week to remove accumulated ice. A solid copper heat transfer line would result in a ten times longer downtime of the source. The results of the present investigations are encouraging. Design and construction of the heat-pipe cooling system for the dissociator nozzle of the polarized atomic beam source have been started.

REFERENCES

1. O.W.B. Schult et al., *Nuclear Physics A583, 629 (1995)* .
2. P.J. Brennan, L. Thienel, T. Swanson, M. Morgan, *28th AIAA Thermophysics Conference, July 6-9, 1993, AIAA paper No. 93-2735.*
3. F. Stock et al., *Nucl. Instr. And Meth. In Phys. Res. A 343 (1994) 334..*
4. A. Vassiliev, V. Koptev, S. Kotov, H. Seyfarth. *Preprint PNPI, NP-32-1997,* n. *2171..*

ION-EXTRACTION POLARIMETRY FOR A POLARIZED ^{1}H/^{2}H INTERNAL TARGET

Z.-L. Zhou[*,†], M. Ferro-Luzzi[‡,∥], J.F.J. van den Brand[‡,∥], H.J. Bulten[∥] and J. Lang[¶]

[*] Univ. of Wisc.-Mad., [†] MIT-LNS, [‡] NIKHEF, [∥] Vrije Univ.-Amsterdam, [¶] ETH-Zürich

Although the technique of polarized internal gas targets has many advantages, in that spin-dependent scattering from chemically and isotopically pure atomic species of high polarization can be realized, it provides a challenge for obtaining precise information on the target polarization. There exist many mechanisms ([1] and references therein) through which the target nuclei can be depolarized. Furthermore, for ^{1}H/^{2}H targets, a dilution of the target polarization by molecules becomes inevitable. Consequently, knowledge of the fraction of these molecules is required. Care should be taken that the polarization measurement samples the target polarization in the same manner as the nuclear or particle physics experiment under consideration. Methods, which only measure the polarization of certain parts of a target, or that have a different spatial sensitivity, may introduce systematic uncertainties.

We report on the development of a polarimeter [2] for measuring *in-situ* the tensor polarization of a deuterium internal target [1] in an electron ring. The method takes advantage of the ionization of atoms and molecules by the stored electron beam passing through the target cell. The number of ions produced along the storage cell is directly proportional to the product of target density and beam current. The total target polarization can be obtained independent of its spatial and temporal variations by uniformly extracting these ions from the cell, measuring their atomic and molecular fractions, and by directly determining their nuclear polarization.

We refer to Ref. [1] for the experimental setup. Fig. 1 (Left-Top) shows the deuterium mass spectra obtained by scanning the Wien filter magnetic field. The solid (dashed), dotted, and dash-dotted curves were obtained with the 1-4 transition turned on (off), the sextupoles turned off and the atomic beam shuttered in the ABS. The spectra reveal a significant dilution of the target polarization by molecules in the present setup which could be improved in the future. Detailed studies show that about 45% of molecules in the target were originated from undissociated molecular beam, 40% due to diffused flow from the ABS and to residual D_2 gas in the target chamber, and 15% to recombined molecules, which were potentially polarized [3].

Measurements of the $P_{zz}(D)$ were performed by selecting atomic ions (D^+) through the Wien filter and then accelerating them to bombard a tritiated titanium foil. The neutrons from the ^{3}H$(\vec{d},n)\alpha$ reaction were counted in two scintillators, located at 0° and 90° with respect to the polarization axis. Since the count rate is directly proportional to P_{zz}, one can extract P_{zz} by forming a super-ratio $R = \sqrt{(N^+(0°) \cdot N^-(90°))/(N^-(0°) \cdot N^+(90°))}$, which largely cancels variations in luminosity and differences in detection efficiency. Here, N^+ (N^-) represents the rates for the high (low) value of P_{zz}. Fig. 1 (Left-Bottom) gives the count rates in the 90° and 0° neutron counters measured as a function of time. The measurements were performed with a target holding field (B_{target}) of 22 mT and a cycle of four 16 second steps: background, no RF transition, the 1-4 and 3-5 transitions turned on, and the 1-4 and 2-6 transitions turned on.

For each hyperfine state, P_{zz} as a function of B_{target} can be exactly calculated. Fig. 1 (Right) shows the measured super-ratios as a function of B_{target}, and for various RF transition schemes. The curves show the super-ratios as expected from the P_{zz} diagram and from the population of the hyperfine states for the corresponding RF transition schemes. The data are well described by the calculated values over the whole measured range of magnetic field.

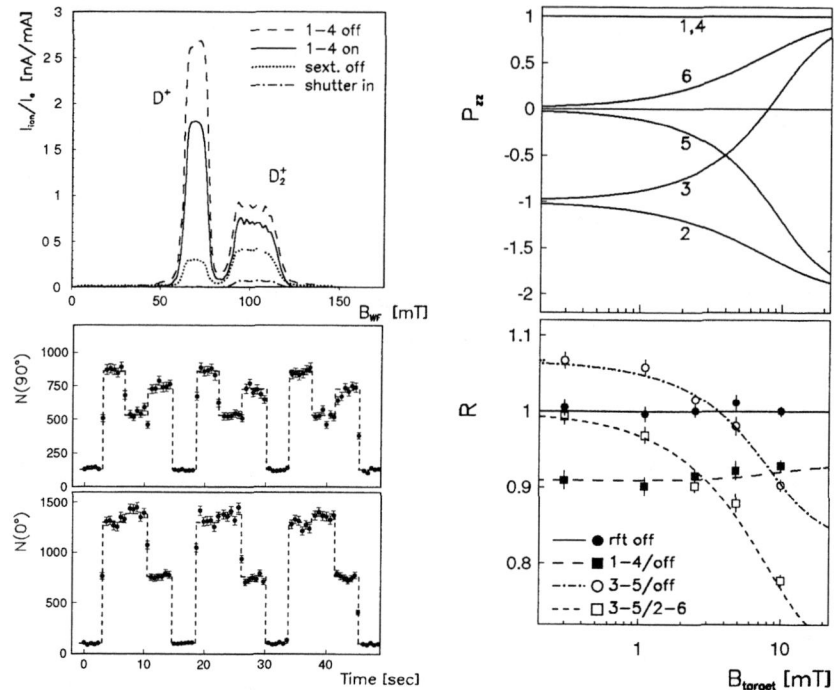

FIGURE 1. *Left-Top:* ion current as a function of the Wien filter magnetic field for various atomic beams produced in the ABS. *Left-Bottom:* Counts in the 90° and 0° neutron counters as a function of time with the target polarization sequentially changing. *Right:* Measured super-ratios (R) as a function of target holding field for various RF transition schemes in the ABS. Pzz for all hyperfine states are also shown.

We refer to Ref. [2] for a detailed discussion of the results and systematic uncertainties.

This work was supported in part by the US NSF (PHY-9316221), NATO (CRG920219), Swiss National Foundation, Dutch FOM and NWO.

[1] M. Ferro-Luzzi, contrib. to this conf.; Z.-L. Zhou et al., NIM **A378**, 40 (1996).
[2] Z.-L. Zhou et al., NIM **A379**, 211 (1996); submitted to NIM, (Sep. 1997).
[3] J.F.J. van den Brand et al., contrib. to this conf.; PRL **78**, 1235 (1997).

The Photocathode Gun of The Polarized Electron Source at NIKHEF

C.W. de Jager[1*], V.Ya. Korchagin[2], B.L. Militsyn[1],
V.N. Osipov[2], N.H. Papadakis[1&], S.G. Popov[2†],
M.J.J. van den Putte[1], Yu.M. Shatunov[2], Yu.F. Tokarev[2]

(1) NIKHEF, P.O. Box 41882, 1009 DB Amsterdam, The Netherlands;
(2) Budker Institute for Nuclear Physics, Novosibirsk, 630090, Russia;
(*) Present address; Thomas Jefferson Laboratory, Newport News, VA 23606, USA
(&) Present address; IASA, P.O. Box 17214, 10024 Athens, Greece.

Abstract. A description is given of the 100 kV photocathode gun of the 400 kV Polarized Electron Source (PES) used for internal target physics on polarized light nuclei at the Amsterdam Pulse Stretcher (AmPS) storage ring. The gun provides 2 μs long pulses of polarized electrons with a current up to 50 mA, at a repetition rate of 1 Hz, and a polarization degree up to 85%. Using the pulsed power supply, no deterioration of the vacuum during gun operation has been observed. An operational lifetime (1/e), using an InGaAsP photocathode, of 180 hours has been measured. The injection efficiency of the polarized beam from PES into the linear Medium Energy Accelerator is measured to be $\approx$ 30 %.

A drawing of the gun is given in Figure 1. The gun is realized with a double high voltage insulator and double vacuum chambers. The inner vacuum chamber (IVC) (4) is defined by the double insulator; the protection insulator (11), and the gun insulator (6). Its vacuum is better than 10^{-12} mbar. Both insulators have been manufactured from 11 porcelain rings, interlaced with metallic gradient electrodes. An external voltage divider (not shown) lowers the voltage on adjacent gradient electrodes by 11.1 kV.

The IVC contains the anode electrode (5), and the cathode electrode (7). The cathode electrode holds the photocathode, which is mounted in a dedicated holder. Photocathodes are transferred from the preparation set-up (PS) (where its surface is prepared to the Negative Electron Affinity state) by means of a magnetic manipulator through an UHV valve connected to the port on the left (12). The PS is permanently connected to the gun. The photocathode in the gun can be exchanged with a freshly activated photocathode in PS within a typical time of 15 minutes, without opening the IVC of the gun. The surface

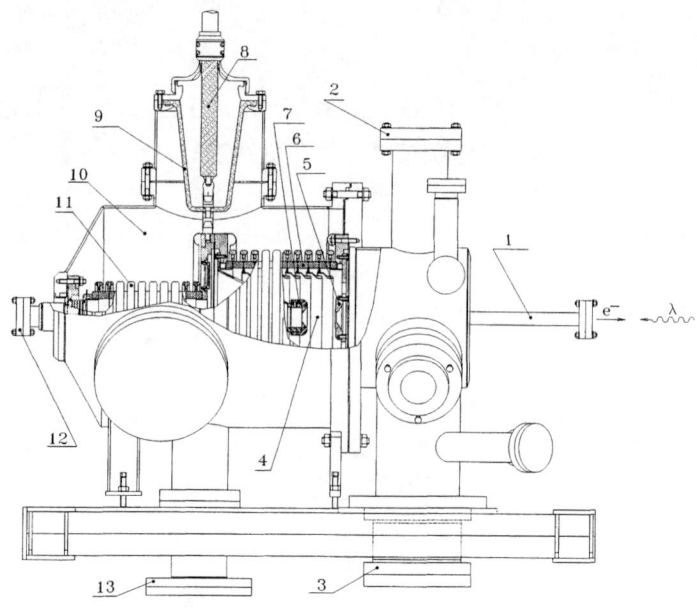

FIGURE 1. The photocathode gun.

of the photocathode placed in the cathode electrode is illuminated with laser light pulses entering through the beam pipe (1).

The IVC is placed in a guard vessel (GV) (10). Its vacuum is of order 10^{-7} mbar. On the top of the GV the high voltage supply cable (8) is connected. It is shielded from the other parts of the gun by means of a plastic insulator (9).

The vacuum of IVC is maintained by ion pumps connected to two ports. At the first port (3) a 250 l/s ion pump is connected. At the second port (2) a 100 l/s ion pump and a NEG pump are installed. Both these ion pumps are equipped with titanium getters. GV is pumped through a port (13) with an ion pump.

With 100 kV DC supplied to the gun, the vacuum in IVC deteriorated to $2 \cdot 10^{-11}$ mbar. Consequently the photocathode lifetime was limited to 4 hours. This deterioration was believed to originate from microsparkes. To eliminate this phenomenon a pulsed high voltage power supply was installed. It generates a gaussian pulse with an amplitude of -100 kV, and a total width of 600 μs.

Photocathode lifetime measurements have been performed on a strained layer InGaAsP photocathode. The photocathode current has been measured during a run of approximately one week. The exponential-function fitted to data points has a lifetime (1/e) of 180 hours.

Photocurrent saturation at GaAs(Cs,O)

A.S. Jaroshevich[1], M.A. Kirillov[2], D.A. Orlov[1], A.G. Paulish[1], H.E. Scheibler[1] and A.S. Terekhov[1,2]

[1] *Institute of Semiconductor Physics, 630090 Novosibirsk, Russia*
[2] *Novosibirsk State University, 630090 Novosibirsk, Russia*

Abstract. The photocurrent from p^+-GaAs photocathodes under the intensive pulse illumination is experimentally studied within the temperature range of 77-300 K. The line-shape of the photocurrent pulses is evidently influenced by surface photovoltage effect and is found to be a weak function of temperature. This observation proves that hole tunnelling is the main component of the restoring current.

Photocurrent saturation is a major problem for the use of NEA-photocathodes in the polarized electron sources. When a photocathode is illuminated with high light intensity, its quantum efficiency (QE) falls due to surface photovoltage [1]. The value of the surface photovoltage depends on the current density of photoelectrons arriving to the surface and restoring current density of holes. The holes can reach the surface in two ways: by thermionic diffusion over the barrier and by tunnelling. The intermediate case, which is thermotunnelling, can also take place. However, the actual kind of the hole transport is not experimentally examined up to now. To elucidate the nature of the restoring current, the photocurrent (PC) was measured under the intensive pulse illumination at 293 K and 77 K.

The measurements were carried out in the vacuum-sealed parallel plate photodiodes which included transmission mode GaAs-photocathode and a metal anode. The active 2 μm thick p-GaAs(100) layer was doped with Zn ($6 \cdot 10^{18}$ cm^{-3}) and was approximately 20 mm in diameter. For the experiments we selected several photodiodes. QE of the photocathodes in these diodes were in the range of 5÷30% at λ=790 nm. A photocathode was illuminated by laser diode light pulses (λ=790 nm) with the duration of 3 μs and peak power up to 2 mW. The light was focused to the spot with a diameter of about 170 μm. The line-shape of the laser pulses was near rectangular. The PC pulse line-shape was measured by a digital oscilloscope. For low-temperature measurements the diodes were directly immersed in liquid nitrogen.

Fig.1 shows two closely spaced PC pulses measured at 77 K (a) and 293 K (b). The measurements were performed at light intensity of 0.5 W/cm² for the photocathode with QE=8%. For photocathodes with higher values of QE (15-30%), the photocurrent saturation effect was small because of high values of NEA. The stationary PC of above 500 mA/cm² was reached for the photocathode with QE=28% at the maximum used light intensity (5 W/cm²). In this condition QE decreased only by a factor 0.6, and PC was still much smaller than the value of the saturated photocurrent.

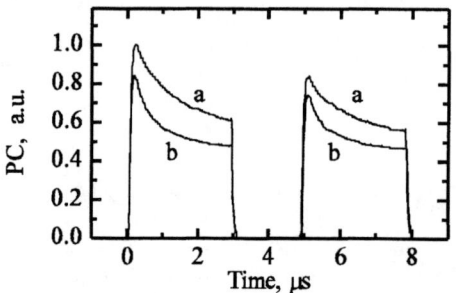

Fig.1 Two closely spaced photocurrent pulses at 77 K (a) and 293 K (b).

At the beginning of the light pulse, the photocurrent is initially high. The amplitude of the pulse at 77 K is higher than at 293 K (by about 20%) because of the increase of QE at low temperatures [2]. Under illumination, the surface charge decreases due to photoelectrons accumulated at surface states. As a result the band bending decreases, vacuum level rises, and PC falls. These processes form the decay of PC. When the current ot the photoelectrons arriving to the surface is compensated by the restoring current, the PC comes to a stationary value. Ratio between the stationary and the initial values of photocurrents is the same within 10% for both temperatures. This means that the restoring current and stationary photovoltage are weak functions of temperature.

The amplitude of the second closely spaced PC pulse is smaller than the amplitude of the first one (see Fig.1) because the surface charge is not completely restored during the delay time between the pulses. The time constant of the decay of the interpulse effect depends on the recombination rate of electrons accumulated at the surface, which, in its turn, is determined by the density of the restoring current. The measurements with different delay time showed that as the temperature increased from 77 K to 293 K, the decay time decreased by a factor not more than 1.5.

These results experimentally prove that the restoring current is mainly tunneling. Indeed, if the restoring current is thermionic or thermotunnelling, the value of the hole current to the surface over the surface barrier should decrease by many orders of magnitude at low temperature.

This work was partly supported by Russian Foundation for Basic Research (Grant 96-02-19063) and State Federal Program "Surface Atomic Structure" (Grant 95-4.2).

[1] A. Herrera-Gomez, G. Vergara and W.E. Spicer, *J.Appl.Phys.* **79**, 7318 (1996)
[2] A.S. Terekhov and D.A. Orlov, *Proc. SPIE* **2550**, 157 (1995)

Transverse Energy Measurements on NEA-GaAs Photocathodes

S. Pastuszka*, D. Kratzmann*, D. Schwalm*, A.S. Terekhov[†], and A. Wolf*

*Max-Planck-Institut für Kernphysik and Physikalisches Institut der Universität, 69029 Heidelberg, Germany,
and [†]Institute of Semiconductor Physics, 630090 Novosibirsk, Russia

For studies of electron-ion interactions at low relative energies at the Heidelberg Test Storage Ring TSR [1], intense electron beams with low energy spreads, especially in the transverse degrees of freedom, are required. Since photoemission from NEA-GaAs cathodes appears to be suitable for the production of such electron beams [2], an electron gun operating with a photocathode was developed and tested [3].

Using a new method based on the adiabatic transverse expansion of an electron beam in a spatially decreasing magnetic field, the mean transverse energy (MTE) of photoemitted electrons was measured systematically [3]. Adiabatic expansion of an electron beam causes a transfer of energy from the transverse degrees of freedom into the longitudinal one increasing the mean longitudinal energy by $(1 - B_f/B_0)\langle E_{\perp 0}\rangle$. Here, B_f and B_0 are the final and the initial field strengths and $\langle E_{\perp 0}\rangle$ denotes the initial mean transverse energy of the electrons before the expansion of the beam. Measuring the shift of the mean longitudinal energy by retarding field analysis for different magnetic expansion ratios B_f/B_0, the MTE can be determined with high accuracy, as was tested for thermionic emission [4]. Furthermore, the design of our electron gun allows us to 'block' low-energy electrons with an adjustable potential barrier in front of the photocathode, enabling the measurement of the MTE as a function of the mean *longitudinal* emission energy $\langle E_{\|0}\rangle$ of the electrons.

Our GaAs transmission-mode cathodes are prepared using a closed cycle technique and are activated with Cs and either O_2 or NF_3 in an XHV setup (10^{-12} mbar) [5]. Quantum efficiencies of 20-25 % (at 670 nm) and long life times could be achieved for both types of activation in a reproducible way. Figure 1 shows MTE-values measured on a cathode which was illuminated from the front with 800 nm light from a laser diode for different types of activation layers at room temperature ($k_B T = 25$ meV). Electrons with energies above the conduction band minimum are found to be thermalized with the

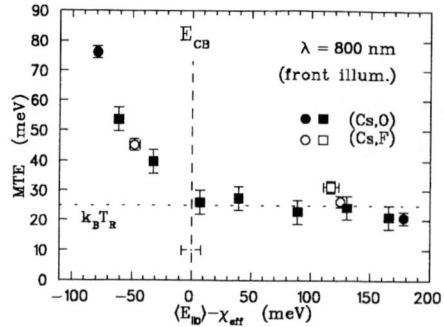

FIGURE 1. Mean transverse energy as a function of the mean longitudinal emission energy with respect to the conduction band minimum E_{CB}, namely $\langle E_{\parallel 0} \rangle - |\chi_{eff}|$. χ_{eff} denotes the value of NEA. Filled symbols represent activations with Cs and O_2, while open symbols represent activations of the same cathode with Cs and NF_3.

lattice temperature of the cathode, while electrons which have suffered energy losses prior to their emission into vacuum show a non-Maxwellian distribution with enhanced transverse energies. Thus, the mean transverse energy of the extracted electrons increases the more low energy electrons contribute to the beam. In fact, the MTE of the photoemitted electrons is mainly determined by their longitudinal emission energy and was shown to be independent of the composition of the activation layer on the same cathode. By cooling the cathode with liquid nitrogen ($k_B T \approx 7\,\text{meV}$) the MTE of the high-energy electrons could be reduced to $\approx 14\,\text{meV}$. Deviations from the expected value of 7 meV are probably due to surface roughness or slight inhomogeneities of the activation layer, which will be further studied. —— This work has been funded in part by the German Federal Minister for Education, Science, Research and Technology (BMBF) under Contract No. 06 HD 854 I(3). A.S.T. was supported in part by the Russian Foundation for Basic Research (Grant 96-02-19063).

REFERENCES

1. S. Pastuszka, U. Schramm, M. Grieser, C. Broude, R. Grimm, D. Habs, J. Kenntner, H.-J. Miesner, T. Schüßler, D. Schwalm, and A. Wolf, Nucl. Inst. Meth. A **369** (1996) 11
2. R.L. Bell, *Negative Electron Affinity Devices*, Oxford Clarendon Press (1973)
3. S. Pastuszka, Dissertation, MPI H-V14-1997, Heidelberg (1997)
4. S. Pastuszka, Diploma Thesis, MPI H-V3-1994, Heidelberg (1994), and S. Zwickler, D. Habs, P. Krause, S. Pastuszka, D. Schwalm, and A. Wolf, Proc. Workshop on Photocathodes for Polarized Electron Sources for Accelerators, Stanford, SLAC-432 (1994) 446
5. S. Pastuszka, A.S. Terekhov, and A. Wolf, Appl. Surf. Sci **99** (1996) 361-365

Polarization Anomalies in Polarized Electron Emission and Luminescence from Highly p-Doped Semiconductor Structures

A.V.Subashiev and E.P.German

*St.Petersburg State Technical University,
195251 St.Petersburg, Russia* [1]

Though strained semiconductor layers can give polarization potentially close to 100 %, polarization losses (PL) in PES at different stages of emission make the emitted electron polarization less than 90 %. For the theoretical evaluation of the PL the three step emission model (involving optical-excitation →transport-to-the-surface →emission-from-the-band-bending-region processes) gives an adequate approach. PL in two latter stages can be minimized by reducing the lifetime of the excited electrons in the structure. Therefore the PL in the optical excitation set an ultimate limitation to the maximum polarization from PES.

We show that PL the excitation can be experimentally separated and estimated on the base of the low-temperature studies of the polarization dependence on the optical excitation energy. At low temperatures the optical spin orientation of the electrons in highly p-doped semiconductor structures with splitted valence band is modified by the hole Fermi-distribution in the valence band. When the top of the heavy-hole band is empty, the absorption edge is shifted to higher energies. Then, near the edge the transitions from the light-hole band can predominate, while at higher energies large contribution of the heavy-hole band changes the sign of the average polarization to the opposite. Thus, abrupt changes in the electronic polarization and, consequently, in the polarized luminescence and photoemission are predicted [1]. When temperature is raised, the Fermi-level shifting and the smearing of the Fermi-distribution lead to the heavy-hole domination in the edge absorption, and the spin orientation becomes a smooth function of the excitation energy and does not change its sign.

We have considered theoretically the circularly polarized light absorption, polarized photoluminescence and polarized emission from p-doped GaAs struc-

---

[1] This work is supported by CRDF Grant No. RPI-351. The partial support by the RFBR (Grant No. 96-02-19187a) and by INTAS, project 94-1561, is also gratefully acknowledged.

tures with different doping level on a $Ga_{1-x}P_x$ substrate.

The results of the calculation of the low-temperature polarization of photoluminescence for several acceptor concentrations are presented in Fig. 1 a. with the experimental results for the low-temperature polarized luminescence [2]. The excitation-energy dependence of the electron polarization is found to be very sensitive to the Fermi-level position and to the light-hole band tailing, which smears the anomaly. The electron polarization at the excitation at T = 297 K as a function of excitation energy $E = \hbar\omega - E_g$ is shown in Fig. 1 b. The PL at the absorption edge are caused by the light-hole band tailing.

In GaAs strained layer cathodes with valence band splitting $\Delta = 40$ meV and doping $N_a = 3 \times 10^{18}$ cm^{-3} losses in excitation are ≈ 5 %. Overdoping of the emitting GaAs layer increases PL to 20 %, while larger valence band splitting could effectively block PL.

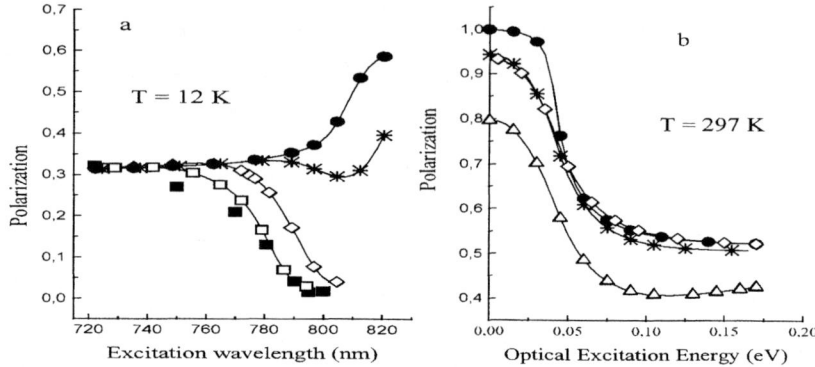

FIGURE 1. Polarization of photoluminescence (a) and photoemission (b) from GaAs strained layers; p= 5×10^{17} (•), 10^{18} (∗), 3×10^{18} (◊), 5×10^{18} (□), and 10^{19} cm^{-3} (△), (■) - experimental data of Ref.[2].

In conclusion, we have shown that in strained layer PES the polarization losses in excitation are sensitive to the doping level and band tailing caused by structural defects and can be as high as 20 %. At low temperatures the electron polarization can have anomalously strong dependence on the excitation energy, doping and temperature. Low temperature luminescence is shown to be effective in PES sructure characterization.

REFERENCES

1. Subashiev, A.V. and German, E.P., *Nanostructures:97, Proceedings of the International Symposium*, St.Petersburg, 1997, p. 130; German, E.P. and Subashiev, A.V., *JETP Lett.*, **65**, 909 (1997).
2. Mair, R., *SLAC-rep-488*, (1996).

Surface Potential Fluctuations in Negative Electron Affinity State Formation

B.D. Oskotskij, A.V.Subashiev and L.G.Gerchikov

*St.Petersburg State Technical University,
195251 St.Petersburg, Russia* [1]

The fluctuation potential in the Band Bending Region (BBR) of the surface activated to the negative electron affinity state was shown to be the main cause for the abrupt slowing down of the rate of spin relaxation at the surface, wide spread of the photoemitter parameters and also for the increase of photovoltage relaxation time [1]. In the report the results of the theoretical studies of the electron potential fluctuations by the surface potential modelling are presented.

We have studied the electron potential distribution at a NEA semiconductor surface, which is formed by Cs-originated donor-type centers in the activation layer at the surface and the acceptor-type centers in the GaAs layer. Random spatial distribution of acceptors and donors generates an inhomogeneous fluctuation potential in the surface plane. We consider the fluctuation potential generated by (1) randomly distributed in the BBR ionized acceptor; (2) random distribution of the ionized Cs-originated donor centers in the activation layer; (3) randomly distributed structural defects, which introduce short-range random potential.

The numerical stimulation included calculations of the surface potential distribution, correlation function of the surface potential and determination of the percolation level (PL) position. The evolution of the potential distribution with doping level was studied at different stages of the activation process modeled as the increase of the concentration of the surface deep donor centers. The surface pseudo ground state was modeled as a result of the electron subsequent redistributions caused by the Coulomb repulsion between the electrons bound to the neutral donors at the surface. As a result a minimum of the total surface electron energy was achieved. The surface chemical potential, one-particle electronic energies and the density of the surface states were calculated.

---

[1] This work is supported by CRDF Grant No. RPI-351, by the RFBR Grant No. 96-02-19187a) and by INTAS Grant No. 94-1561.

TABLE 1. Surface Electronic States Parameters

N_a (cm^{-3})	E_{max} (meV)	δE (meV)	E_{pl} (ev)	$\tilde{E}_{pl}$ (meV)	$\tilde{E}_{pl,corr}$ (meV)
10^{17}	290	60	294	170	286
10^{18}	269	89	280	156	277
3×10^{18}	249	113	268	144	270
5×10^{18}	251	122	262	138	266
10^{19}	225	134	257	113	260

We have found that the potential distribution transforms with the increasing of the donor concentration in the strongly correlated spatial distribution, while the PL shifts upward as a result of the potential asymmetry.

The parameters of the random potential distribution and the position of the PL of the optimally activated GaAs surface (band bending $V = 300$ meV) for several acceptor concentrations (N_a) are presented in Table 1. Here E_{max} is the energy of the maximum in the surface potential distribution, is the mean square width of the potential distribution, E_{pl} is the PL energy in the surface potential, $\tilde{E}_{pl}$ is the PL energy corrected for the energy shift via the states localization in the BBR quantum well, $E_{pl,corr}$ is the PL energy in the surface potential of highly correlated impurity distribution for high donor states concentration. The value $\tilde{E}_{pl}$ is in line with the experimentally found energy of the spin relaxation switching off [2], while calculated δE value gives the estimate for the width of the emitted electron energy distribution. The value of $E_{pl,corr}$ obtained by the random surface potential modelling is close to the value calculated analytically in Ref. [3] which verifies the reliability of the numerical calculations.

In conclusion, two-dimensional numerical simulation model of a semiconductor NEA surface is presented. We show that the calculated parameters of the electronic surface state distribution and the position of the surface states percolation level can account for the spin relaxation switching off in BBR, the electron energy distribution and the long-range tails in relaxation of the photovoltage.

REFERENCES

1. L.G. Gerchikov, B.D. Oskotskij, and A.V. Subashiev, *Proceedings of the 12th Intern. Symposium on High Energy Spin Physics*, (World Scientific, 1996), p.746.
2. Herman, C.,Drouhin, H.-J., Lampel, G., et al., *Spectroscopy of Nonequilibrium Electrons and Phonons*, edited by C. V. Shank and B. P. Zakharchenya, (Elseiver Science, B. V., 1992) p. 135.
3. Bello, M.S., Levin, E.I., Shklovskii, B.I. *JETP* **80** 1246 (1981).

Elucidation of activation layer model by means of measurements of photoelectron energy distribution curves.

S. Pastuszka[1], D. Kratzmann[1], D. Schwalm[1], A. Wolf[1],
D.A. Orlov[2], A.G. Paulish[2], H.E. Scheibler[2],
A.S. Terekhov[2,3], O.E. Tereshchenko[2,3]

[1] *Max-Plank-Institut für Kernphysik and Physikalisches Institut der Universität, 69029 Heidelberg, Germany*
[2] *Institute of Semiconductor Physics, 630090 Novosibirsk, Russia*
[3] *Novosibirsk State University, 630090 Novosibirsk, Russia*

Abstract. The comparative study of (Cs,NF_3) and (Cs,O_2) activation procedures and the measurements of the vacuum level position versus activation layer thickness show that the activation layer for optimally activated GaAs photocathodes with QE of about 20-30% should be treated as a dipole layer. For thick activation layers, the saturation of the vacuum level at the lowest position is observed. The saturation proves, that the activated surface should be treated as a heterojunction.

We did two experiments to clarify the nature of the activation layer. In the first experiment the (Cs,O) and (Cs,F) activation layers were compared. Having in mind the difference in chemistry of the bulk CsF and Cs_xO_y compounds, one can hope to observe some differences in the parameters of the (Cs,O) and (Cs,F) activation layers. Otherwise, one should conclude, that oxygen or fluorine atoms do not demonstrate any specific chemistry inside the activation layers and, therefore, that "chemical compound"-based models such as the heterojunction and cluster models should be rejected. In the second experiment we measured the position of the vacuum level versus the activation layer thickness. We hoped to see, that for thick activation layers, the position of the vacuum level would not depend on the activation layer thickness. This would be a direct evidence that a new material is grown on the GaAs surface, and the activated semiconductor surface should be treated as a heterojunction.

The Zn-doped GaAs(100) (p=$8 \cdot 10^{18}$ cm^{-3}) 2-μm thick epitaxial layers

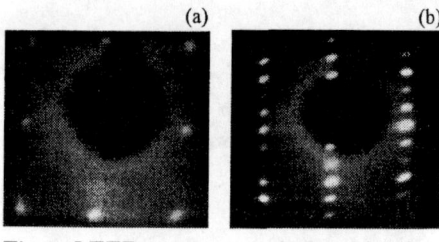

Fig.1 LEED patterns of GaAs(100).

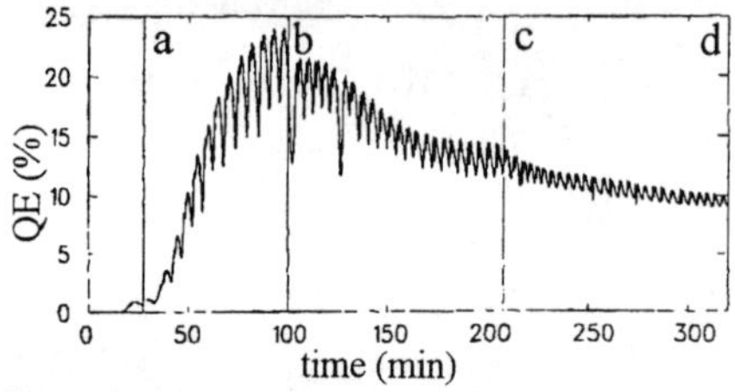

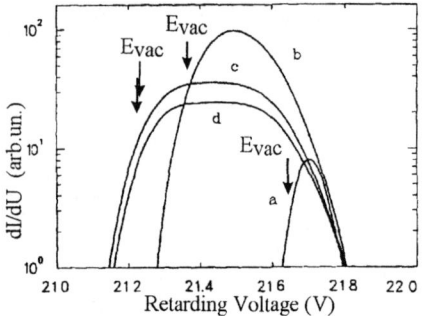

Fig.2 Activation curve of GaAs.

Fig.3 EDC of GaAs-photocathode.

were used. The layers were treated in HCl–isopropanol solution in the glove-box under N_2-atmosphere. Hermetic vessel was used to transfer the samples from the glove box to UHV without a contact with air. After heat cleaning at 350° C, Auger spectrometer did not detect surface contaminations. LEED pattern showed well-ordered As-rich (1x1) reconstruction (Fig1a). This surface was activated to NEA-state, but the QE was low. To reach QE$\geq$ 20 % at 670 nm, the annealing at 530-580° C was needed. After such annealing the LEED showed Ga-stabilized (4x1) or c(8x2) reconstructions (Fig1b).

The samples were first activated with (Cs,NF_3), heat cleaned, and then activated with (Cs,O_2). Activations started with exposing the GaAs to Cs vapor until photocurrent maximum was reached. Then the surface was exposed to both Cs and NF_3, or to Cs and O_2. The photocurrent evolution during (Cs,NF_3) and (Cs,O_2) activations and the final QE were practically the same.

The photocurrent evolution for (Cs,O_2) activation is shown in Fig.2. Vertical lines indicate the points (a,b,c,d) where electron distribution curves (EDC) are measured. The measured EDC are shown in Fig.3. It is seen, that at the beginning of the activation (points a,b) the position of the vacuum level (E_{vac}) depends on activation layer thickness. For thick activation layers (points c,d) the vacuum level does not depend on the activation layer thickness.

In conclusion, to reach the maximum escape probability, the GaAs(100) surface should be Ga-stabilized with (4x1) or c(8x2) reconstructions. From the start of the activation of such surface and up to the maximum of QE, the activation layer is thin and should be considered as the dipole layer. When such surface is activated to the lowest position of the vacuum level, the activated surface should be treated as a heterojunction.

This work was partly supported by Russian Foundation for Basic Research (Grant 96-02-19063) and State Federal Program "Surface Atomic Structure" (Grant 95-4.2).

Surface Charge Limit Observed in an NEA Photocathode of a 100 keV Polarized Electron Gun

K. Togawa [1], T. Nakanishi[1], S. Okumi[1], C. Takahashi[1],
C. Suzuki[1], F. Furuta[1], T. Ida[1], K. Wada[1], Y. Kurihara[2],
H. Matsumoto[2], T. Omori[2], Y. Takeuchi[2], M. Yoshioka[2],
T. Baba[3] and M. Mizuta[3]

(1) Dept. of Physics, Nagoya University, Chikusa-ku, Nagoya-464, Japan
(2) High Energy Accelerator Research Organization (KEK), Tsukuba, Ibaraki 305, Japan
(3) Fundamental Research Laboratories, NEC Corp., Tsukuba, Ibaraki 305, Japan

Our group has continued the development of polarized electron source (PES), since the PES is expected to do the essential part in the future electron-positron linear colliders, such as JLC (Japan Linear Collider). As shown in Fig.1, JLC requires a highly-polarized intense beam with a multi-bunch structure. 85 (72) micro bunches with a bunch separation time of 1.4 ns (2.8 ns) make a 120 ns (200 ns) macro pulse for X-band (C-band) scheme. The electron population of each bunch is $\sim 2\times 10^{10}$ electrons and the bunch width is ~ 700 ps at the source [1]. The total charge of the macro pulse is on the order of 10^{12} electrons.

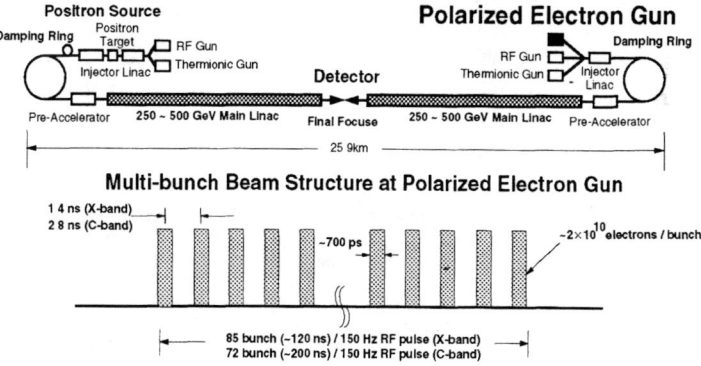

FIGURE 1. Schematic diagram of JLC

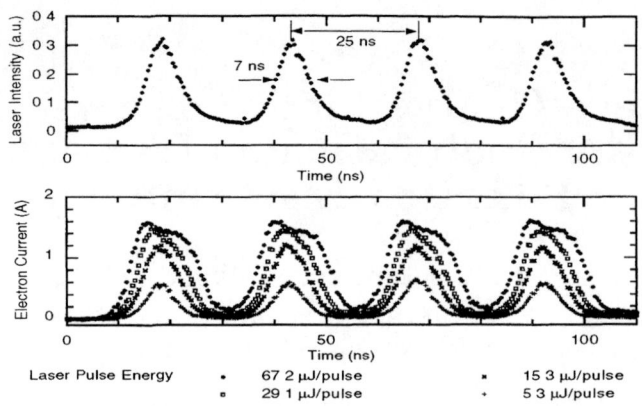

FIGURE 2. Time profiles of laser pulse and multi-bunch electron beam

The polarized electron beam is produced by using a GaAs-type semiconductor photocathode with a negative electron affinity (NEA) surface. In case of extracting intense beam from the NEA photocathode, the surface charge limit (SCL) problem must be overcome. The SCL effect, which was first observed at SLAC, is a photocurrent limitation by the surface charge of electrons which are trapped in the band bending region. In order to investigate the SCL effect, we have constructed a 100 keV polarized electron gun at Nagoya University. The SCL effect was clearly observed in a 100 nm-thick GaAs with a normal doping (5×10^{18}cm^{-3})[2]. Next, a 95 nm-thick AlGaAs-GaAs superlattice with a modulation doping (a high doping of 4×10^{19}cm^{-3} in only a 5 nm thick surface and a lower doping of 5×10^{17}cm^{-3} in the interior of the photocathode), has been tested using a quadruple-bunch Ti:sapphire laser (7 ns FWHM, 25 ns bunch separation time, 10 Hz repetition rate, 748 nm wavelength). The laser pulse shape and the 70 keV multi-bunch electron beam shapes for various laser energies are shown in Fig.2. In this case, the SCL effect was not observed in all bunches, while the characteristic features of space charge limit (the bunch shape becomes a flattop and the bunch width becomes wide with increasing the laser intensity) could be observed. The high charge of 92 nC (5.8×10^{11} electrons) in $\sim$100 ns macro pulse including four bunches could be extracted. This result demonstrates the modulation doped superlattice is a promising photocathode for generation of the multi-bunch electron beam for the linear colliders.

REFERENCES

1. JLC Design Study, KEK, April, 1997.
2. T. Nakanishi et al., in another report of this workshop.

Polarized Electrons in ELSA

(preliminary results)

S. Nakamura, W. von Drachenfels, D. Durek, F. Frommberger,
M. Hoffmann, D. Husmann, B. Kiel[†], F.J. Klein, D. Menze,
T. Michel[†], T. Nakanishi[‡], J. Naumann[†], T. Reichelt, C. Steier,
T. Toyama[*], S. Voigt, M. Westermann

Universität Bonn, Physikalisches Institut, Nußallee 12, D-53115 Bonn, Germany
[†]*Università Erlangen, Physikalisches Institut IV, D-91058 Erlangen, Germany*
[‡]*Nagoya University, Department of Physics, Nagoya-464, Japan*
[*]*KEK, Tsukuba-305, Japan*

Abstract. Polarized electrons have been accelerated in the electron stretcher accelerator ELSA for the first time. Up to 2.1 GeV the polarization of the electron beam supplied by the 120 keV polarized electron source has been measured with a Møller polarimeter. Preliminary results of polarization measurements at high energies and the performance of the source are presented.

EXPERIMENTAL SETUP

The polarized electron source [1] operates in 50 Hz pulsed mode with a pulse length of 1 μsec using a flashlamp pumped Ti:Sapphire laser. At $\lambda = 750$ nm pulses up to 100 mA have been extracted from a GaAs superlattice crystal, delivered by the Nagoya group [2]. A polarization of the electron beam at the source of about 66 % is measured by Mott scattering at 120 keV. The lifetime of the source (1/e-intensity) depends on the beam intensity and is up to 50 h at about 50 mA.

The polarized electrons from the source are injected into the 20 MeV linac, then accelerated in the booster synchrotron up to 1.2 GeV and transferred to ELSA. During further acceleration in ELSA to energies between 1.27 and 2.1 GeV the beam crosses several depolarizing resonances. The influence of these resonances on the beam polarization has been studied with a Møller polarimeter.

It uses a 40 μm Vacoflux foil with an inclination of 21° to the beam. The spin polarization of the foil is (8.27±0.26)% at saturation (100 Gauss). For the asymmetry measurements a 2-arms-coincidence mode with 4 detectors is used. Averaged over the acceptance of $\Theta^* = (90 \pm 15)°$ an asymmetry coefficient $a_{zz} = -0.76 \pm 0.02$ has been calculated (using Monte Carlo simulations). A Faraday cup and detectors sensitive to Bremsstrahlung events are used as intensity monitors.

RESULTS

Up to 1.27 GeV, the beam polarization can be transferred from the source to the experimental area with small depolarization (Fig. 1a).

The loss of polarization due to crossing of the 3rd imperfection resonance at 1.32 GeV could be avoided by a harmonic correction with a vertical closed orbit bump (Fig. 1b).

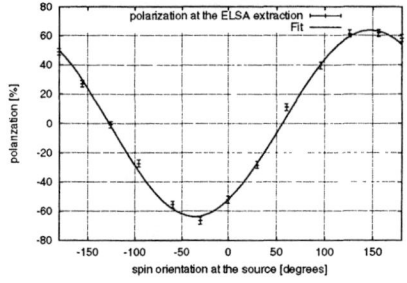
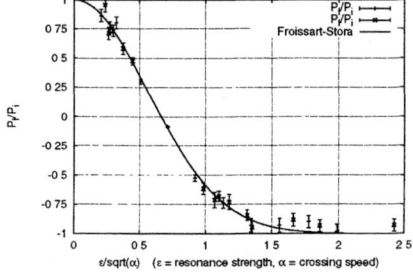

(a) Dependence of the final polarization on the spin orientation at the source.

(b) Polarization after crossing of the third imperfection resonance (extraction energy 1.37 GeV).

FIGURE 1. Polarization of the extracted beam.

With a ramping speed optimized for the resonances at 1.32 and 1.5 GeV (but without any correctors) a polarization of about 45 % could be conserved up to 1.9 GeV. Strong depolarization occurred at 2.0 GeV. Therefore above 2.0 GeV fast tunejump quadrupoles are essential to conserve polarization. The construction of two quadrupoles (with ferrite yoke) has been started [3].

REFERENCES

1. S. Nakamura, PhD thesis, in preparation
2. T. Nakanishi et al., proceedings of this workshop
3. C. Steier, D. Husmann, Correction of Depolarizing Resonances in ELSA, proceedings of the 1997 PAC, Vancouver

APOLLON at DESY: Spin-Dependent Photoproduction of Charm

C.A. Miller

TRIUMF, 4004 Wesbrook Mall, UBC Campus, Vancouver BC, Canada V6N 2J2

on behalf of the APOLLON Working Group

Abstract. APOLLON is a proposal to measure the polarization asymmetry of J/ψ photoproduction in a fixed target experiment at HERA in an effort to provide information on the gluon spin distribution in the nucleon. Inelastic production will be identified via the $\mu^+ \mu^-$ decay channel, resulting in a statistical precision of $\delta A = 0.05$ in a 12 month run. In the LO CSM, the corresponding precision in $\Delta G(x)/G(x)$ is 0.15 at $x \sim 0.4$.

There is now intense interest in the contribution of gluons to the nucleon spin. A process in which the gluons enter in leading order is photon-gluon-fusion (PGF). At fixed target energies, the only PGF channel that is distinguishable from ordinary quark scattering is charm production. The inelastic J/ψ channel combines a clear experimental signature with a theoretical interpretation of the spin asymmetries in the well-studied color-singlet model (CSM) [1]. Therefore this is the charm production channel that is presently proposed by APOLLON.

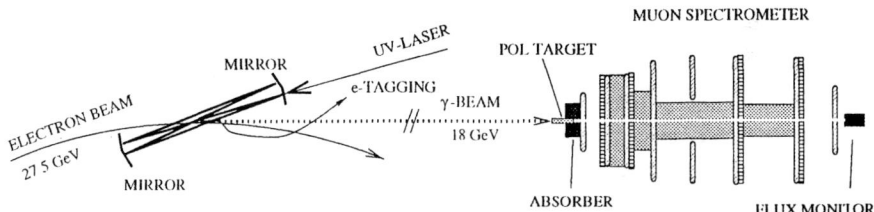

FIGURE 1. The basic layout of the APOLLON experiment.

Fig. 1 shows the basic layout of the APOLLON experiment. A highly polarized 18 GeV photon beam is incident on a solid ^{6}LiD polarized target. The photon beam is generated by Compton back-scattering (CBS) of an ultraviolet laser beam on the 27.5 GeV longitudinally polarized HERA electron beam. The electron polarization enhances the photon spectrum shape, but does not influence the nearly 100% photon polarization inherited from the uv laser beam. The intense 257 nm beam from a frequency doubled Ar ion laser is enhanced manyfold by an external coherent mirror cavity containing the Compton interaction point in the HERA beam. The energy of each useful photon is tagged through detection of the corresponding low energy recoil electrons. The target polarization is inverted once for each of the anticipated 200–300 fills per year. Once per month, the photon beam helicity is also flipped by simultaneously switching the electron and laser beam helicities to maintain the same photon spectrum shape.

A simple toroidal spectrometer accepts $\sim 75\%$ of the $J/\psi \to \mu^+\mu^-$ phase space. The invariant mass resolution is $\sim 20\%$, and the J/ψ energy resolution from only the muon opening angle is better than 1 GeV. The muons are distinguished from hadrons and electromagnetic showers by absorbers immediately following the target and in the magnetic field of the spectrometer. Finally, the signals from the gamma beam flux and position monitor could also provide an additional faster electron polarization measurement for tuning the HERA beam.

The table summarizes the projected performance based on a running time of 12 months. The systematic error is expected to be small compared to the statistical precision of $\delta A = 0.05$ in the J/ψ inelastic asymmetry, which can be interpreted in the leading order CSM to correspond to a precision of 0.15 in $\Delta G(x)/G(x)$ at $x \sim 0.4$ [2]. This is a larger x value than those probed by approved competing experiments. It is anticipated that a NLO CSM may become available in time to interpret the results.

Laser cavity multiplication factor	500
18 GeV Photon rate	4 MHz
Rate of J/ψ events	7.5/h
Rate of inelastic J/ψ events	1.25/h
Total number of J/ψ events	30000
Total number of inelastic ($z < 0.9$) J/ψ events	5094
Error on the total J/ψ asymmetry δA	0 043
Corrected for detector resolution $\delta \bar{A}$	0.05
Final error on $\Delta G/G$ (LO CSM)	**0.15**

REFERENCES

1. E.L. Berger, D. Jones, *Phys. Rev.* **D23**, 1521 (1981).
2. J.Ph. Guillet, *Z. Phys.* **C3975** (1988).

A P-even Test of Time-Reversal Invariance in $\vec{p}$–$\vec{d}$ Scattering at COSY-Jülich

P.D. Eversheim*, F. Hinterberger*, J. Bisplinghoff*, R. Jahn*,
J. Ernst*, H. Paetz gen. Schieck[†], W. Kretschmer[‡]
and H.E. Conzett[§]

*Institut für Strahlen- und Kernphysik[1] Universität Bonn, D-53115 Bonn, Germany
[†]Institut für Kernphysik, Universität zu Köln, D-50937, Germany
[‡]Physikalisches Institut, Universität Erlangen-Nürnberg, D-91058, Germany
[§]Lawrence Berkeley Laboratory, University of California, Berkeley, CA 94720, USA

Abstract. At the cooler synchrotron COSY at Juelich a novel (P-even, T-odd) true null test of time-reversal invariance (TRI) was proposed and accepted that allows to measure TRI to an accuracy of 10^{-6}. The observable of interest is the total cross-section asymmetry $A_{y,xz}$, which is measured in an transmission experiment of a circulating vector polarized proton beam through an internal tensor polarized atomic deuteron target. This experiment uses the COSY facility in three respects: As an accelerator, as an ideal forward spectrometer and as an detector.

THE TIME-REVERSAL INVARIANCE EXPERIMENT

Symmetry tests with nuclei imply the presence of the strong interaction. In contrast to parity-violation (PV) the origin of a time-reversal invariance (TRI) violating force is not identified yet. Though very precise TRI tests have been performed that violate parity and time-reversal simultaneously [1], less precise experimental limits exist for TRI tests that conserve parity.
For polarized particles many systematic error sources can be eliminated by flipping the spin and extending the spin-flipping scheme, so that the quantities of interest are extracted from double ratios [2]. Finally, a true null-test increases the experimental accuracy substantially. In this respect the term

---

[1]) This work was supported by the BMWF and the Forschungszentrum Jülich, Germany

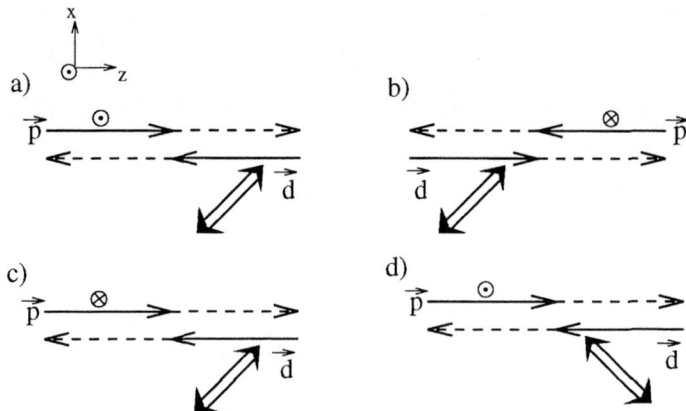

FIGURE 1. Pictorial demonstration that a time-reversed situation is prepared by either a proton or a deuteron spin-flip. a) The basic system is shown. b) The time reversal operation is applied (momenta and spins are reversed and the particles are exchanged). In order to have a direct comparison between situation a) and b), two rotations $R_y(\pi)$ or $R_x(\pi)$ by 180° around the y- or x- axis are applied, leading to the situations c) and d), respectively. This is allowed, since the scattering process is invariant under rotations. ⊙ is for proton spin up (y-direction), ⊗ is for proton spin down, and ⟺ is the deuteron tensor polarization.

"true" stresses the concept that the intended test has to be completely independent from dynamical assumptions. On this basis, Conzett [3] proposed to measure the total cross-section asymmetry $A_{y,xz}$ of vector polarized spin 1/2 ($\vec{p}$) particles interacting with tensor polarized spin 1 ($\vec{d}$) particles (cf. Fig. 1).

The quantity of interest $A_{y,xz}$ is proportional to the transmission asymmetry $T_{y,xz}$, which in turn is the relative difference of the transmission factors $T^{+/-}$. The transmission factors measure the decrease of the circulating y-polarized proton beam when it continuously passes through the xz-tensor polarized deuteron target. The superscripts + and − characterize mutually time-reversed situations that are prepared by the proton and deuteron spin alignment. Thus, the COSY ring not only serves as an accelerator, but also as an ideal forward spectrometer and detector. This experiment has the potential to lower the present experimental bounds of P-even, T-odd matrix elements by an other order of magnitude.

REFERENCES

1. I.S. Altarev et al., Phys. Lett. B276 (1992) 242
2. P.D. Eversheim et al., Phys. Lett. B234 (1990) 253
3. H.E. Conzett in Proc. High Energy Spin Physics, ed.: K.-H. Althoff and W. Meyer (Springer-Verlag, 1991) 589

The Polarized Ion Source at the Cooler Synchrotron COSY in Jülich

R. Gebel[1], P.D. Eversheim[2], M. Altmeier[2], O. Felden[2], M. Glende[1]

[1] *Institut für Kernphysik, Forschungszentrum Jülich, D-52425 Jülich, Germany*
[2] *Institut für Strahlen- u. Kernphysik, Universität Bonn, D-53625 Bonn, Germany*

Abstract. The polarized ion source at the cooler synchrotron COSY in Jülich has been designed for the cooler synchrotron COSY and the injector cyclotron JULIC. The source is based on the colliding beams concept [1] and delivers an pulsed H$^-$ current of over 20 μA within an emittance ϵ of less than 0.5 π mm mrad ($\beta\gamma$ normalized). The polarization P is in excess of 85 %. A substantial enhancement of the H$^-$ current results from the pulsing of the atomic beam part and the improvement of the neutral cesium intensity.

Introduction

The polarized ion source for COSY was built by a collaboration[1] of three groups from the universities Bonn, Erlangen and Köln [7] with the intention to charge the COSY ring during the stripping injection period of 20 ms to its space charge limit. Therefore a minimum polarized H$^-$ current of 1.5 μA is required in the injection beam line of COSY. With respect to the losses and the acceptance of the 45 MeV injector cyclotron JULIC, the polarized ion source for COSY has to provide at least 15 μA within an $\beta\gamma$-normalized emittance of 0.5 π mm mrad.

---

[1]) CBS collaboration: P.D. Eversheim[1], M. Altmeier[1], O. Felden[1], W. Kretschmer[2], R. Weidmann[2], A. Glombik[2], H. Paetz gen. Schieck[3], S. Lemaitre[3], M. Eggert[3], R. Gebel[4], M. Glende[4]
[1] Institut für Strahlen- und Kernphysik, Universität Bonn, Germany
[2] Physikalisches Institut, Universität Erlangen, Germany
[3] Institut für Kernphysik, Universität zu Köln, Germany
[4] Institut für Kernphysik, Forschungszentrum Jülich, Germany

The Polarized Colliding Beams Source

The polarized colliding beams source at COSY comprises three major components: the atomic beam source, the cesium beam source and the charge exchange and extraction region, shown schematically in figure 1. The nuclear polarized atomic $\vec{H}^0$ beam meets inside the charge exchange region the fast neutral Cs^0 beam and swaps charges according to the reaction $\vec{H}^0 + Cs^0 \rightarrow \vec{H}^- + Cs^+$. The negatively charged $\vec{H}^-$ ions are extracted from the charge exchange region by electric fields. The ions are transferred to the injector cyclotron, passing a Wien Filter to provide the proper spin alignment.

The ground state atomic beam source produces an intense polarized atomic hydrogen beam. Firstly, hydrogen gas molecules are dissociated in an inductively coupled 500 W rf discharge. The atoms are cooled to about 30 K by passing the aluminum nozzle of 20 mm length and 3 mm diameter. The atoms are considerably slowed down [2] with the consequences that shorter sextupole magnets can be used, the first tapered sextupole magnet accepts an increased solid angle and the dwell time of the atoms in the charge exchange region is increased in proportion to the decrease of the beam velocity. These beneficial effects are reduced in part by gas scattering in and outside the nozzle [4].

The fast neutral Cs^0 beam [5] for the charge exchange reaction in the solenoid is produced in two steps. First, Cs vapour is thermally ionized on a hot (1100 $°C$) porous tungsten surface at a beam potential of about 40-50 kV. It is difficult [3] to transport this perveance dominated Cs^+ beam further than about 450 mm, this beam had to be focused into the charge exchange solenoid by a magnetic quadrupole triplet. Usually the charged Cs^+ beam is deflected by means of the Cs^+ deflector in front of the solenoid into a Faraday cup. Only for the injection period in COSY, the neutralizer, placed between the quadrupoles and the solenoid, is filled with cesium vapour. The charged fast Cs^+ beam becomes neutralized, and enters the charge exchange region inside the solenoid through an

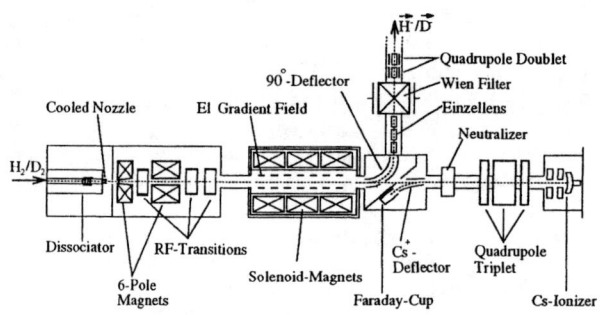

FIGURE 1. Scheme of the Polarized Colliding Beams Source at COSY.

orifice in the 90° deflector. The neutralizer comprises a cesium oven, a cell filled with Cs vapour, and a magnetically driven flapper valve between the oven and the cell. The neutralizer efficiency was measured to exceed 90%.

In the charge exchange solenoid various H^- beam properties can be adjusted. Since the neutral polarized atomic hydrogen beam is selectively ionized by the charge exchange reaction, only little unpolarized H^- background is produced that could reduce the nuclear beam polarization. By varying the solenoid field, the transversal emittance can be traded for polarization. The energy spread of the beam is tuned by the magnitude of the electrical drift field inside the solenoid.

At the position of the charge exchange region a hydrogen intensity with $I_{H^0} > 4 \cdot 10^{16}$ atoms/sec has been verified for continuous operation. The RF transitions have been tested in the Bonn polarized ion source. They showed efficiencies in excess of 90% [6].

Results of the Polarized CBS

The H^- beam intensity was increased by a consequent pulsing of the Cs^0 and H^0 beams. The improvement of the intensities is shown in figure 2. In continous operation of the source (open dots) a H^- output of 5 μA was reached. The filled dots with labels show the development of the pulsed beam. The development stages in pulsed operation are pulsing the dissociator (RF), the injection of hydrogen (H_2) and oxygen (O_2), the higher cesium intensity (Cs) and the pulsed

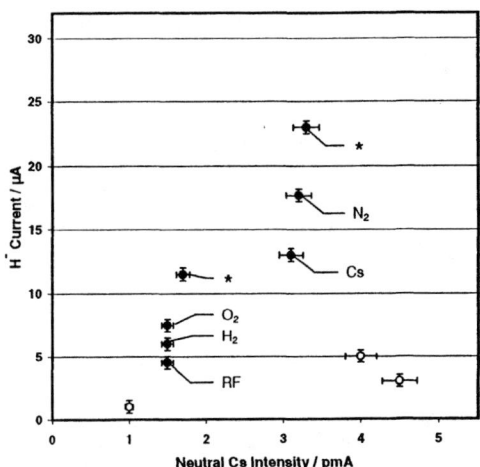

FIGURE 2. Development of the intensities at the CBS. Filled dots represent pulsed and open dots continous operation. A label indicates the major modification.

admixture of nitrogen at 30 K (N_2). The asterisk (*) indicates a slight optimization of the dissociator and the skimmer geometry.

Figure 3 shows the timing sequence for the H^0 beam and the H^- output. The H^- current output of the source depends sensitively on the relative fluxes of the H_2, O_2 and N_2 gases and on their timing with respect to the dissociator RF. The O_2 and N_2 admixtures ($\sim 10^{-3}$) reduce the recombination at the dissociator glass tube and nozzle. A 10 ms delay of the H^- output and the vacuum pressure rise in the dissociator recipient is observed with respect to the prompt response of the photodiode, detecting the discharge light from the dissociator. Peak currents of 23 μA H^- have been extracted from the CBS.

The source is used to study the depolarizing resonances in COSY. Inside COSY a beam polarization of $P > 80\%$ and behind the injector cyclotron JULIC a polarization of $P > 85\%$ has been measured. Polarization could be measured up to a beam momentum of 2 GeV/c.

REFERENCES

1. W. Haeberli *Nucl. Instrum. Methods*, **62**, 355 (1968).
2. H.G. Mathews et al. *Nucl. Instrum. Methods*, **213**, 155 (1983).
3. T. Wise and W. Haeberli *Nucl. Instr. Methods B*, **6**, 566 (1985).
4. W. Haeberli *Helv. Phys. Acta*, **59**, 513 (1986).
5. H. Paetz gen. Schieck et al. *AIP Conf. Proc.*, **293**, 97 (1993).
6. R. Weidmann et al. *Rev. Sci. Instrum.*, **67**, No. 3 part II, 1357 (1996).
7. P.D. Eversheim et. al., *Proceedings of the International Workshop on Polarized Beams and Polarized Gas Targets, Cologne June 6-9 1995*, World Scientific Publishing, 224 (1996).

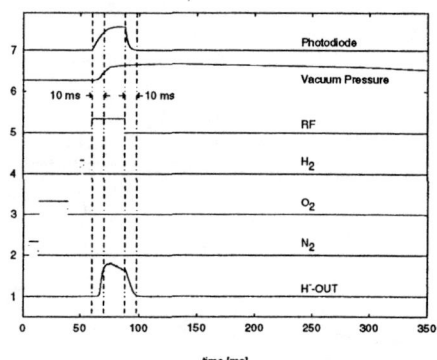

FIGURE 3. The timing sequence of the pulsed CBS beam. The vertical scale of the vacuum pressure is logarithmic, all other scales are linear.

Performance of the polarized ion source POLIS used at the AGOR accelerator facility.

H. R. Kremers, A. G. Drentje

KVI, zernikelaan 25,9747 AA Groningen, The Netherlands.

Abstract. The operation of the KVI polarized ion source POLIS presently running on proton beams is discussed. The best polarization values which have been measured, are 70 +/-0.5 % and -56 +/- 0.5 %. The asymmetry is due to 7 +/- 1 % offset in polarization, which originates from state mixing in the magnetic field of the ECR ionizer.

POLIS is an atomic beam type polarized ion source with a cryo cooled dissociator nozzle (30K). The source was constructed in a joint effort with the Forschungszentrum Karlsruhe, Cyclotron-Laboratory (where it served as a prototype)[1]. The ionizer section based on the ECR principle is modified in such a way that source extraction voltages up to 40 kV, needed for injection into AGOR, are possible. Following successful tests and deuteron polarization measurements at low energies [2], the source was connected to AGOR. Most parts are now adapted to the AGOR - KVI control system.

AGOR is a new super-conducting cyclotron, which came into operation in 1996. It started to deliver beams for experiments in November, in fact with a long lasting program of 190 MeV polarized protons. As a consequence, the first measurement of proton polarization could be done only during these experiments, and all attempts to improve the polarization degree had to take place also during these experiments.

Efforts to get the weak field (WF) transition in operation have given a present best value of the polarization of 70 +/- 0.5 % of the theoretical value. This value is using the well known analyzing power of the pp reaction at 190 MeV. The observed asymmetries could be measured with good statistics in less than a 5 seconds, using the SALAD set up at KVI. The best value of polarization which have been seen with the strong field (SF) unit was 56 +/- 0.5 %. The optimal polarization as function of frequency was measured as a function of the change in static magnetic field (fig1)

The difference in absolute values of the polarization degree between the WF and SF is due to the fact that the state 2 and 4 are not pure states in the magnetic field of the ECR ionizer. This effect has been measured in a comparison with a proton beam from a different ion source and is 7 +/- 1 %.

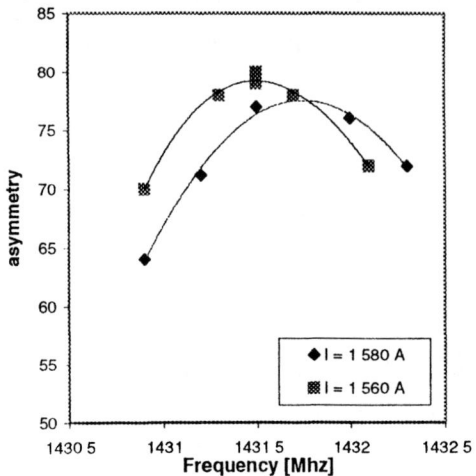

FIGURE 1. Shift of the optimal frequency due to the change in the static magnetic field.

The output of the source can be in the order of 40-70 μA. In the present experimental program we cut away 95 % of the intensity such that less than 1 μA is injected into the cyclotron. Running time of the source is usually 1 to 2 weeks before the dissociator and nozzle become contaminated. We do not use nitrogen to create a layer onto the nozzle (50K). In case nitrogen is being used (35K) to prevent recombination, the polarization degree tends to be slightly higher, but the running time is reduced to 3 to 5 days. The source has been in operation during more than 70 % of the 33 weeks scheduled for experiments with AGOR since November 1996. Our present activities concentrate on the improvement of the strong field transition unit, the dissociator performance and on reducing depolarization effects in the ECR ionizer.

REFERENCES

1. L. Friedrich, E . Huttel and P. A. Schmelzbach, Rev. Sci. Instrum. 63 (4), April 1992
2. L. Friedrich, E. Huttel, H. R. Kremers, A. G. Drentje, in Proc. Int. Workshop on Polarized Beams and Polarized Gas Targets, Cologne, 1995, H. Paetz gen. Schieck, L. Sydow, eds, World Scientific , p198..202

Polarized Negative Ion Source at the Kyoto University Tandem Accelerator

M. Nakamura, S. Kuwamoto, S. Takahashi, M. Hirose, K. Imai,
T. Murakami, M. Yosoi, M. Yoshimura[1], Y. Mori*[†], A. Takagi*,
K. Ikegami*, and M. Kinsho*

Department of Physics, Kyoto University, Kitashirakawa, Kyoto 606-01, Japan
* *High Energy Accelerator Research Organization(KEK), Tsukuba, Ibaraki 305, Japan*
† *High Energy Accelerator Research Organization(KEK)-Tanashi, Tokyo 188, Japan*

Abstract. A polarized H^- and D^- ion source is newly constructed. Atomic H or D beams from a dissociator with a cooled nozzle enter a system of permanent sextupole magnets and a radio-frequency transition(RFT) where they are focused and polarized. They enter a downstream electron-cyclotron-resonance(ECR) heated plasma ionizer from which positive ions are extracted. Negative ions are produced from the positive beam by charge exchange in rubidium vapor. Beam test results are briefly described.

Kyoto University tandem accelerator was upgraded to an 8UDH pelletron in 1990 in order to support new research programs in the field of nuclear and its related sciense. In 1991 construction began on a new atomic-beam-type polarized source in cooperation with the National Laboratory for High Energy Physics (KEK). We intend to measure the magnetic moment of ^{11}Be that is produced by the reaction $^{10}Be(d,p)^{11}Be$ using polarized deuterons.

Schematic view of the ion source is shown in Fig. 1. Atoms are obtained from rf dissociation of H_2 or D_2 inside a Pyrex tube placed on the main source axis [1]. The oscillator operates at 14.5 MHz. Dissociated atoms emerge through an aperture in a copper nozzle which is cooled to 77K by liquid nitrogen. The four permanent sextupole magnets are used [2]. The use of permanent magnets based on Neodym-iron-boron alloy allows the design of very compact sextupoles with the magnetic field at the surface as high as 1.4 T. The atoms focused by the sextupoles pass through an RFT unit where the electron polarization may be transferred to the nucleus. The unit consists of a helical wire coil producing an rf field at 8 MHz (for H) along the beam

---

[1] Present Address:Research Center for Nuclear Physics,Osaka University,Osaka 567,Japan

direction and a C-shaped magnet producing on axis a linearly varying static transverse magnetic field. The nuclear spin polarized atomic beam then enter the ECR ionizer [3,4]. The plasma is confined axially via a magnetic mirror field and radially by a sextupole field created by permanent magnets. Heating of plasma is accomplished via 100 W of rf power fed radially into the region between the mirror coils from a 2.45 GHz magnetron source. In this device the atoms pass though a dense plasm where they are stripped by collisions with fast electrons. The emerging polarized positive ion beam is then accelerated and passed through a canal which is heated to cause charge-exchange collisions with rubidium vapor. The emerging polarized negative ion beam is accelerated up to 80 kV and injected to the tandem accelerator.

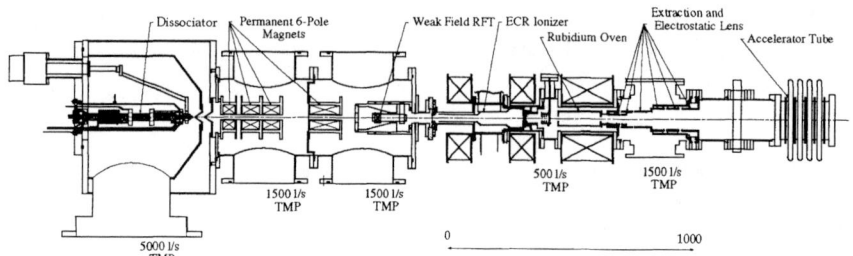

FIGURE 1. Schematic layout of the Kyoto University Polarized Ion Source.

Neutral beams and positive ion beams were produced successfully at KEK in 1992. Early in 1993 the ion source was moved from KEK to our tandem laboratory. Negative ion beams were produced stably in 1995 and H^- ions were injected to the tandem. However the obtained beam polarization turned out to be very low. We examined our atomic beam transport system by using a Monte-Carlo simulation code [5]. From the simulation we found an important source of unpolarized background.

It is clear that we should solve the problems which became evident duringthe beam test before strating the research program.

We are grateful to K. Hatanaka for giving us valuable suggestions.

REFERENCES

1. F.Stock et al., *Nucl. Instr. Meth.* **A343**, 334 (1994).
2. G.Graw et al., *Proc.of Int. Workshop on Polarized Ion Sources and Polarized Gas Jets*, Tsukuba, ed. Y.Mori, KEK Report **90-15**, 138 (1990).
3. L.Friedrich et al., *Nucl. Instr. Meth.* **A 272**, 906 (1988).
4. T.B.Clegg and M.B.Schneider, *Proc.of Int.Workshop on Polarized Source and Targets*, Montana, eds. S.Jacard and S.Mango, *Helv.Phys.Acta* **59**, 53 (1986).
5. A. Roberts, private communication.

Polarized Beams at the 10 GeV Machine of JINR (Dubna)

Yu. K. Pilipenko, P. A. Rukoyatkin, L. S. Zolin

Joint Inst. for Nuclear Research, Dubna 141980, Moscow region, Russia

The spin physics program of the Laboratory of High Energies (LHE), JINR was started early in the 80s when polarized deuterons were accelerated at the Dubna 10 GeV accelerator (Synchrophasotron). The source of polarized deuterons POLARIS provides deuterons in both vector and tensor polarization mode [1]. Typical values of the vector and the tensor deuteron polarizations are $p_z^{\pm} = \pm 0.52$ and $p_{zz}^{\pm} = \pm 0.70$. The slow extraction system allows one to provide experimental setups in the main experimental hall with beams of momenta up to 9 GeV/c at an intensity of up to $5 \cdot 10^9$ $\vec{d}$/spill [2]. The first spin experiments at the LHE have been carried out to study inclusive and binary reactions with vector and tensor polarized deuterons [3]. Some of these experimental data obtained with the magnetic spectrometers ALPHA and ANOMALON [4] are shown in Fig.1. After installing the Saclay-ANL proton polarized target (PPT) at Dubna in 1995, the LHE spin program is supplemented with spin correlation experiments [5].

To measure a polarization of the extracted deuteron beam, the two-arm magnetic spectrometer ALPHA is used as a high energy polarimeter [6]. The deuteron polarization components $(p_z^{\pm}, p_{zz}^{\pm})$ are measured by detecting dp elastic scattering on a hydrogen target. The vector A_y and the tensor A_{yy} analyzing power for this reaction at p_d=3 GeV/c are known to a high precision [7], so the polarization can be measured with a high absolute precision $(\Delta p_z(syst.) \simeq 2\%)$ at this fixed momentum.

However, it is necessary to readjust repeatedly the extraction system for the 3 GeV/c using this polarimetry method in long time runs at other beam momenta. For experiments with the tensor polarized deuteron beam a more convenient method to check periodically the beam polarization was tested. It is based on the measurement of the tensor analyzing power T_{20} in the reaction $\vec{d} + A \rightarrow p(0°) + X$. The T_{20} has an extreme value of $\simeq -1$ at the ratio of the proton to the deuteron momentum $p_p/p_d = 2/3$ (it corresponds to the internal momentum of proton in the deuteron k=0.3 GeV/c, Fig.1b). The value of $T_{20}(k = 0.3)$ does not show any energy dependence in the region

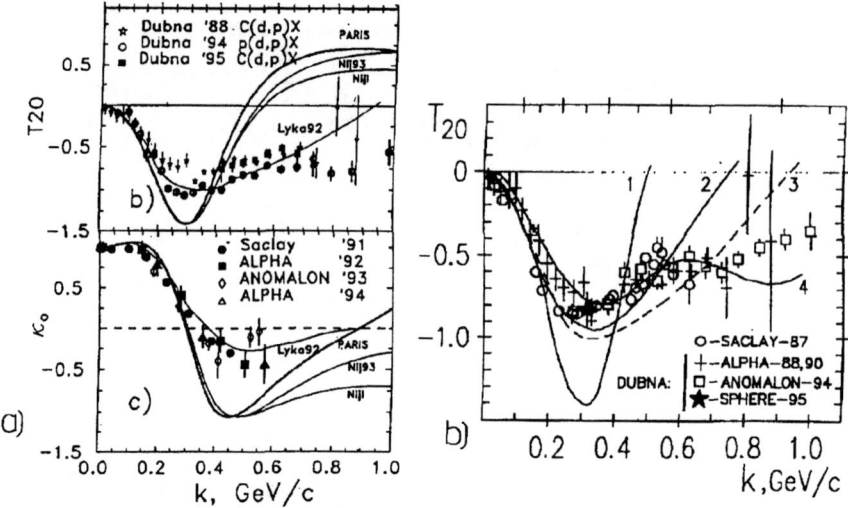

FIGURE 1. Tensor analyzing power (T_{20}) and polarization transfer (κ_o) for deuteron breakup, measured on polarized deuteron beams with momenta from 3.5 to 9 GeV/c (k is the internal momentum of the proton in the deuteron).

from 3 to 9 GeV/c and no theoretical reasons for its appearance at higher energies (i.e. the method has not limitation at high energies). For rather a high proton production rate at 0° ($\frac{d\sigma}{dpd\Omega} \simeq 100 mb GeV^{-1} sr^{-1} c$ at p_d=9, p_p=6 GeV/c), the breakup reaction can be recommended as an effective express method of tensor polarization monitoring of a high energy deuteron beam. As an example, the measurement result of $T_{20}(\vec{d} \to p)$ in a long time run for studying the polarized deuteron fragmentation into cumulative pions ($\vec{d} \to \pi$), is shown at Fig.1b (the point denoted by a star). Similar measurements were repeated 24 times during the five-day run and confirmed a long time stability of the POLARIS operation.

In 1995, a quasi-monochromatic polarized neutron beam was prepared to study the behaviour of $\vec{n}\vec{p}$ cross section differences over an energy range of 1.2-3.65 GeV using the PPT target [5]. The neutrons beams at the Dubna 10 GeV accelerator operate on the basis of a slowly extracted deuteron beam. Neutron beams, formed at 0° by means of vector polarized deuteron breakup, inherit the polarization of primary deuterons. The angular spread and yield of neutrons at a zero angle behave as $\sigma_\theta \sim 1/p_n$, $Y \sim p_n^2$. The PPT neutron beam line is equipped with the spin rotator magnet ($B \times l = 2.7 T \times m$) to turn neutron spins from the vertical direction to the longitudinal one (Fig.2). The next table shows neutron fluxes at different neutron momenta obtained with 20cm CH-target at the outlet of the 2.5m long iron collimator (ϕ 30mm) [8].

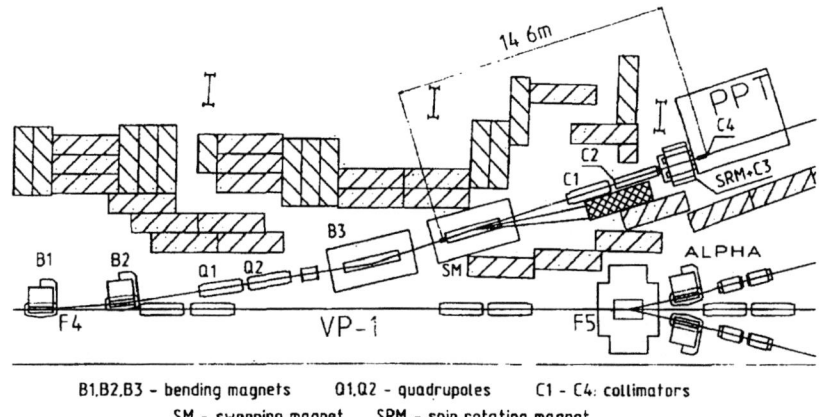

FIGURE 2. Neutron beam line for experiments with the polarized proton target [8].

P_n, GeV/c	1.13	1.5	1.77	2.25	4.5
$I_n/I_d(10^9)$	$7.6 \cdot 10^3/0.1$	$3.5 \cdot 10^4/0.3$	$7.7 \cdot 10^4/0.5$	$1.6 \cdot 10^5/0.8$	$1.0 \cdot 10^6/1.0$

At present, the main ring of the new 6 $GeV/nucleon$ superconducting accelerator (Nuclotron) is mounted and successfully tested. After new injection and extraction systems have been constructed, spin physics studies with polarized deuteron and neutron beams will be continued at this new machine.

The new injector of the Nuclotron will allow to provide polarized deuteron beams with intensity up to $I_d = 10^{11} \vec{d}/burst$. This will open a prospect to probe the deuteron spin structure at high internal momenta unavailable up to now. A small internal beam emittance of the Nuclotron gives an opportunity to use the polarized gas internal target with a storage cell. At luminosity $L \geq 1 \cdot 10^{29} cm^{-2} s^{-1}$ and taking into account a low density of the gas target, it makes possible studying, for instance, the Pomeron spin structure by means of the measurement of asymmetry A_{pp} in $p\vec{p}$ elastic scattering in the region of Coulomb-nuclear interference ($-t \simeq 10^{-3} GeV^2/c^2$) [9]. A_{pp} can be measured with an uncertainty δA_{pp} of 1% at $L \simeq 1 \cdot 10^{29} cm^{-2} s^{-1}$, which corresponds to the target density $\rho_t = 10^{13} atoms/cm^2$, the internal beam intensity $I = 10^{11} p$ and $T_N = 0.8 \mu s$ (the time of one turn in the Nuclotron). The internal gas target has well-known advantages of thin internal targets: a low density allowing the registration of low energy recoils, small beam losses and the possibility to be used (in the time sharing regime) simultaneously with running external beam experiments.

REFERENCES

1. V.P.Ershov *et al.*, *Intern. Workshop on Polarized Beams and Polarized Gas Targets*, Cologne, Germany June 6-9, 1995, p.193.
2. I.B.Issinsky *et al.*, Acta Physica Polonica B **25**, 673 (1994)
3. See: *Proc. Intern. Workshops:* "Dubna Deuteron-91", Dubna,1992; "Dubna Deuteron-93", Dubna, 1994; "Dubna Deuteron-95", Dubna, 1996.
4. A.A.Nomofilov *et al.*, Phys. Lett. B **325** 327 (1995); T.Aono *et al.*, Phys. Rev. Lett. **74** 4997 (1995); L.S.Azhgirey *et al.*, Phys. Lett. B **387** 37 (1996); Phys. Lett. B **391** 22 (1997);
5. Lehar F.,*et al.*, Nucl. Instrum. Methods Phys. Res., A **356**, 58 (1995).
6. V.G.Ableev *et al.*, Nucl. Instrum. Methods Phys. Res., A **306**, 73 (1991).
7. V.Ghazikkhanian *et al.*, Phys.Rev.C**43**, 1532 (1991).
8. A.Kirillov *et al.*, Preprint JINR E13-96-210, Dubna,1996.
9. N.Akchurin *et al.*, Phys. Rev. D **48** 3026 (1993).

APPENDICES

APPENDIX A

PROGRAM FOR THE SEVENTH INTERNATIONAL WORKSHOP ON POLARIZED GAS TARGETS AND POLARIZED BEAMS

SUNDAY, AUGUST 17, 1997
7-9 PM   **Registration and Reception**
                University Inn, Century Room - 21st Floor

MONDAY, AUGUST 18, 1997 - 141 Loomis
8-8:50 AM   **Registration**
8:50 AM   **Welcome** - D. Campbell (Illinois)

^{3}He Targets - Discussion Leader: C. Jones
9:00 AM   **New Results from Spin Exchange Optical Pumping**
                G. Cates (Princeton)
9:45 AM   **The HERMES Target** - D. De Schepper (Argonne)
                Coffee Break

^{3}He Targets - Discussion Leader: Z. E. Meziani
10:40 AM   **The AMPS ^{3}He Target** - H. Bulten (NIKHEF)
11:15 AM   **Running a Polarized ^{3}He Target in Doubly Polarized Electron Scattering at MAMI**
                J. Becker (Mainz)
11:35 AM   **Polarized High Pressure ^{3}HeTarget at MAMI**
                D. Rohe (Mainz)
                Lunch

^{3}He Polarimeters & HELION '97 - Discussion Leader: T. Chupp
1:30 PM   **The E154 Polarimeter** - M. Romalis (Princeton)
1:50 PM   **^{3}He Target Polarimetry at HERMES**
                A. Dvoredsky (Caltech)
2:10 PM   **Highlights from HELION '97** - A. Tanaka (RCNP)
                Coffee Break

H, D Targets - Discussion Leader: H. Paetz gen. Schieck
3:10 PM   **The HERMES H Target** - J. Stewart (Liverpool)
3:45 PM   **D Target Performance at AMPS**
                M. Ferro-Luzzi (NIKHEF)
4:20 PM   **The IUCF Target** - F. Rathmann (Wisconsin)

TUESDAY, AUGUST 19, 1997 - 141 Loomis
H, D Targets - Discussion Leader: E. Steffens
8:30 AM	**EDDA at COSY** - H. Rohdjess (Bonn)
9:05 AM	**Cryogenic Atomic Beam Source at VEPP-3**
	D. Toporkov (Novosibirsk)
9:40 AM	**Ultra-Cold Methods for Polarized Atomic Hydrogen**
	V. Luppov (Michigan)
	Coffee Break

Laser-Driven H/D Targets - Discussion Leader: W. Cummings
10:35 AM	**The Argonne Laser-Driven D Target: Recent Developments and Progress**
	J. Fedchak (Argonne)
11:10 AM	**Nuclear Spin Polarized H and D by Means of Spin-Exchange Optical Pumping**
	J. Stenger (Erlangen)
11:45 AM	**IUCF Tests of an LDS Target** - M. Miller (Illinois)
	Lunch

Target Polarimetry and Beam Interaction
  Discussion Leader: E. Steffens
1:45 PM	**The HERMES Polarimeter** - B. Braun (Erlangen)
2:10 PM	**Beam-Induced Depolarization at HERA**
	H. Kolster (Munich)
3:05 PM	**Krannert Art Museum Tour**

TUESDAY, AUGUST 19, 1997 - Contributed Talks - Evening Session
Session A: Atomic Beam and Ion Sources - 8PM - 144 Loomis
  Discussion Leader: E. Steffens
Polarization Technology at the High Energy Laboratory of the Joint Institute for Nuclear Research (Dubna)
Y.K. Pilipenko (JINR)
A Polarized Atomic Beam Target for COSY-Julich
D. Eversheim (Bonn)
Polarized Beam for the IUCF Cooler Injector Synchrotron
V.P. Derenchuk (IUCF)
A Microwave based Dissociator for the HERMES-ABS
N. Koch (Erlangen)

Session B: Optical Pumping - 8PM - 136 Loomis
   Discussion Leader: W. Cummings

$^3\vec{\text{He}}$ Neutron Spin Filter at ILL
H. Humblot (Langevin)

A Novel Approach for a Spin-Exchange High Density $^3\vec{\text{He}}$ Target for Accelerator Based Physics Programs
R.C. Welsh (UMich)

Development of Optical-Pumping Polarized Deuteron Target
T. Tamae (Tohoku)

Performance of the Laser Driven Polarized Hydrogen at IUCF
R.V. Cadman (Illinois)

Session C: Polarized Electron Sources - 8PM - 158 Loomis
   Discussion Leader: C. Sinclair

Status of the Helium Afterglow Injector for MAMI
Said Essabaa (IPN)

Photocathode Research at SLAC
G. Mulhollan (SLAC)

5 MeV Mott Polarimeter for Rapid Precise Electron Beam Polarization Measurements
S. Price (TJNAF)

Polarized Beam Transport at the TJNAF Accelerator: Simulations and Preliminary Measurements
J. Grames (Illinois)

WEDNESDAY, AUGUST 20, 1997 - 141 Loomis
Beam Polarimetry - Discussion Leader: A. D. Krisch

8:30 AM	**Novel Beam Polarization Measurements** at IUCF
	P. Pancella (W. Michigan)
9:05 AM	**Beam Polarimetry at HERA**
	W. Lorenzon (Michigan)
9:40 AM	**RF Polarimetry**
	Ya. S. Derbenev (Michigan)

Coffee Break

Polarized Electron Sources - Discussion Leader: G. Mulhollan

10:35 AM	**Jefferson Laboratory- Status and Plans**
	C. Sinclair (TJNAF)
11:10 AM	**The MAMI Source** - E. Reichert (Mainz)
11:45 AM	**Polarized Electrons at MIT-Bates**
	M. Farkhondeh (MIT)

Lunch
Polarized Electron Sources - Discussion Leader: G. Mulhollan
1:45 PM   **The SLAC Polarized Electron Source**
             J. Clendenin (SLAC)
2:20 PM   **The Polarized Electron Beam Source at NIKHEF**
             H. Bulten (NIKHEF)
2:55 PM   **Frequency Modulated Laser Diode Injection**
             M. Poelker (TJNAF)
             Coffee
3:15 PM   **Poster Session** - outside Rooms 136, 144, &
             158 Loomis

POSTER SESSION
1A Fiber Optic Applications in Laser Polarized Targets
   W. Cummings (ANL)
1B A p-even Test of Time-Reversal Invariance in p-d Scattering at COSY-Julich - D. Eversheim (Bonn)
1C The Polarized Ion Source at the Cooler Synchrotron COSY in Julich - R. Gebel (IKP-COSY)
1D Polarization and Compression of ^{3}He for Magnetic Resonance Imaging Purposes - D. Geurts (Free University)
2A The UNH Polarized ^{3}He Program - B. Hersman (UNH)
2B Beam Formation in the Munich Atomic Beam Source
   R. Hertenberger (Munchen)
2C Performance of the Polarized Ion Source POLIS used at the AGOR Accelerator Facility - H.R. Kremers (KVI)
2D Optimized Atomic Beam Optics - S. Lemaitre (Koln)
3A The Photocathode Gun of the Polarized Electron Source at NIKHEF - B. Militsyn (NIKHEF)
3B Polarized Negative Ion Source for the Kyoto University Tandem Accelerator - M. Nakamura (Kyoto)
3C Design Studies for the ABS of ANKE in COSY/Julich by 3-Dimensional Magnetic Field Calculations using the Computer Code MAFIA - V. Nelyubin (St. Petersburg)
3D Photocurrent Saturation and Surface Photovoltage at GaAs (Cs,O) - D. Orlov (Inst. of Semiconductor Physics)
4A Cold Electron Beams by Photoemission from NEA-GaAs Photocathodes - S. Pastuzka (Max-Planck)
4B Studies of Polarized ^{3}He Cell Wall Related Spin Relaxation
   V. Pomeroy (UNH)

4C Tests of a Prototype Large-Bore, Low-Power 2↔4 RF Transition Unit - R. Raymond (UMich)

4D Calculation of Spin Exchange Cell Geometry - E. Schulte (Illinois)

5A Status of the Polarized Gas-Storage Cell Target for the ANKE Spectrometer in COSY-Julich - H. Seyfarth (Julich)

5B Draft Proposal for an Experiment to Measure the Spin-Dependent J/psi-Production using a Real Photon Beam at HERA M. Spengos (DESY)

5C Polarization Anomalies in Polarized Electron Emission A. Subashiev (St. Petersburg)

5D Surface Potential Fluctuations in Negative Electron Affinity State Formation - A. Subashiev (St. Petersburg)

6A Elucidation of Activation Layer Model by Means of Measurements of Photoelectron Energy Distribution Curves A. Terekhov (BINP)

6B Surface Charge Limit Observed by an NEA Photocathode of a 100 keV Polarized Electron Gun - K. Togawa (Nagoya U.)

6C Mechanical Filter for Alkali Atoms D. Toporkov (BINP) and B. Wojtsekhowski (TJNAF)

6D Measurement of Polarization Observables for the ^{3}He(d,p) ^{4}He Reaction - T. Uesaka (RIKEN)

7A Heat-Pipe Cooling of the Dissociator Nozzle in the ABS of ANKE at COSY-Julich - A. Vassiliev (St. Petersburg)

7B Ion-Extraction Polarimetry for a Polarized Hydrogen and Deuterium Internal Targets - Z. Zhou (MIT)

7C Polarized Deuteron and Neutron Beams at 10 GeV Machine of **JINR (Dubna)** - L. Zolin (Dubna)

WEDNESDAY, AUGUST 20, 1997
Evening Session - 141 Loomis
NMR Tomography with Polarized Noble Gases
 Discussion Leader: G. Cates

8:00 PM **Polarized ^{3}He and ^{129}Xe NMR Imaging**
 T. Chupp (Michigan)

8:30 PM **Polarized ^{3}He Tomography at Mainz**
 T. Grossman (Mainz)

9:00 PM **Polarized Noble Gas MRI** - J. Brookeman (Virginia)

THURSDAY, AUGUST 21, 1997 - 141 Loomis
Polarized Electron Sources - Discussion Leader: C. Sinclair
9:00 AM **Strained Layer Semiconductors**
 C. W. Tu (UC San Diego)
9:35 AM **Spin Polarization of Photoelectrons from Ordered Semiconductor Alloys**
 S.-H. Wei (NREL)
 Coffee Break
Polarized Electron Sources - Discussion Leader: J. Clendenin
10:30 AM **Emission of Polarized ps-electron Bunches from III-V Semiconductor Cathodes** - P. Hartmann (Mainz)
11:05 AM **Highly Polarized Electrons from Super-Lattice Photocathodes** - T. Nakanishi (Nagoya Univ.)
11:40 AM **Laser Sources for MIT-Bates and IASA**
 H. Avramopoulos (Athens)
 Lunch
Electron Polarimeters - Discussion Leader: C. W. de Jager
1:30 PM **The Compton Polarimeter at NIKHEF**
 I. Passchier (NIKHEF)
1:55 PM **Moller Polarimetry at Jefferson Lab**
 D. Dale (Kentucky)
2:20 PM **Spin Dependence in Mott Scattering of 14 MeV Electrons from Heavy Nuclei**
 J. Sromicki (ETH- Zurich
 Coffee Break
Polarized Ion Sources - Discussion Leader: P. Schmelzbach
3:15 PM **Review of High Intensity Ion Sources**
 T. Clegg (Duke)
3:50 PM **Status of the D⁻ Source** - A. Takagi (KEK)
7:00 PM **Banquet - Miko Restaurant, 405 W. University Avenue, Urbana**

FRIDAY, AUGUST 22, 1997 - 141 Loomis
Polarized Ion Sources - Discussion Leader: P. Schmelzbach
8:30 AM **The RCNP Ion Source** - K. Hatanaka (Osaka)
9:05 AM **H⁻ from a Storage Cell** - A. Belov (INR, Moscow)
9:40 AM **Polarized Beams for Precision Experiments**
 A. Zelenski (TRIUMF, INR, Moscow)
 Coffee Break

Developments in Atomic Beams - Discussion Leader: D. K. Toporkov

10:40 AM	**A New Method to Produce Cold Atomic Hydrogen and Deuterium Beams** -V. Varentsov (St. Petersburg)
11:15 AM	**Nuclear Polarization in Recombined Atoms** J. van den Brand (NIKHEF)
11:50 AM	**Has the Atomic Beam Source Reached a Hard Limit?** - E. Steffens (Erlangen)
	Adjournment

APPENDIX B

LIST OF PARTICIPANTS

Yasushi Arimoto
Research Center for Nuclear Physics
Osaka University (Suita Campus)
10-1, Mihogaoka Ibaraki
Osaka 567
Japan
Phone: 81-6-879-8931
E-mail: arimoto@rcnp.osaka-u.ac.jp
Fax: 81-6-879-8899

Kurt Aulenbacher
Institut fur Kernphysik
Universitat Mainz
D-55099 Mainz
Germany
Phone: 49-6131-395804
E-mail: aulenbac@kph.uni-mainz.de
Fax: 49-6131-395802

Hercules Avramopoulos
Department of Electrical & Computer Engineering
National Technical University of Athens
9 Iroon Politechniou Str.
157 73 Zographou, Athens
Greece
Phone: 30-1-772-2076
E-mail: hav@cc.ece.ntua.gr
Fax: 30-1-772-2077

Douglas H. Beck
Loomis Laboratory of Physics
University of Illinois
1110 West Green Street
Urbana, IL 61801
USA
Phone: (217) 244-7994
E-mail: d-beck@uiuc.edu
Fax: (217) 333-1215

Juergen Becker
Institut fur Physik, EXAKT
Johannes-Gutenberg-Universitat Mainz
Johann-Joachim-Becher-Weg 45
D-55099 Mainz
GERMANY
Phone: 49-6131-395918
E-mail: becker@kph.uni-mainz.de
Fax: 49-6131-39-3428

Alexander S. Belov
Institute for Nuclear Research
Russian Academy of Sciences,
60th Oct. Anniv.
Prospect, 7a
117312 Moscow
Russia
Phone: 7-095-334-0962;
(812)-855-6971
E-mail: belov@al20.inr.troitsk.ru;
belov@iucf.indiana.edu
Fax: 7-095-135-2268; (812)-855-6645

Berndt Braun
DESY
Notkestrasse 85
D-22603 Hamburg
GERMANY
Phone: 49-40-8998-2026
E-mail: bbr@axher1.desy.de
Fax: 49-40-8998-4034

James Brookeman
Department of Radiology
University of Virginia
Health Sciences Center
Charlottesville, VA 22908
USA
Phone: (804) 982-3152
E-mail: jrb5m@virginia.edu
Fax: (804) 982-1618

Henk Bulten
NIKHEF-K
P.O. Box 41882
1009 DB Amsterdam
THE NETHERLANDS
E-mail:
henkjan@paramount.nikhefk.nikhef.nl
Fax: 31-20-592-2165

Robert V. Cadman
Imdiana University Cyclotron Facility
2401 Milo B. Sampson Lane
Bloomington, IN 47408
USA
Phone: (812) 855-9365
E-mail: cadman@uinpluxa.npl.uiuc.edu
Fax: (812) 855-6645

Gordon D. Cates
Department of Physics
Princeton University
Jadwin Hall, P.O. Box 708
Princeton, NJ 08544
USA
Phone: (609) 258-4414
E-mail: cates@pupgg.princeton.edu
Fax: (609) 258-2496

Timothy E. Chupp
1049 Randall Lab. of Physics
University of Michigan
500 East University Avenue
Ann Arbor, MI 48109
USA
Phone: (313) 747-2514
E-mail: chupp@umich.edu
Fax: (313) 763-9694

Thomas B. Clegg
Department of Physics & Astronomy
University of North Carolina
Phillips Hall, CB# 3255
Chapel Hill, NC 27599-3255
USA
Phone: (919) 962-2079
E-mail: clegg@tunl.duke.edu
Fax: (919) 962-0480

James E. Clendenin
Stanford University
Stanford Linear Accelerator Center
P.O. Box 4349, MS 66
Stanford, CA 94309
USA
Phone: (415) 926-2962
E-mail: clen@slac.stanford.edu
Fax: (415) 854-7268

Kevin Coulter
Physics Department
University of Michigan
2071 Randall Lab.
Ann Arbor, MI 48109-1120
USA
Phone: (313) 764-3284
E-mail: kpc@umich.edu
Fax: (313) 764-6843

William J. Cummings
Physics Division, Bldg. 203
Argonne National Laboratory
9700 South Cass Avenue
Argonne, IL 60439
USA
Phone: (630) 252-3431
E-mail: cummings@anl.gov
Fax: (630) 252-3903

Dan Dale
Department of Physics
University of Kentucky
Lexington, KY 40506
USA
Phone: (606) 257-2504
E-mail: dale@zeppo.pa.uky.edu
Fax: (606) 323-2846

Kees de Jager
TJNAF
12000 Jefferson Avenue
Newport News, VA
USA
Phone: (757) 269-5254
E-mail: kees@jlab.org
Fax: (757) 269-5235

Dirk De Schepper
Physics Division,
Bldg. 203-Rm. C-258
Argonne National Laboratory
9700 South Cass Avenue
Argonne, IL 60439
USA
Phone: (630) 252-4012
E-mail: dirk@anlmep.phy.anl.gov
Fax: (630) 252-3903

Yaroslav S. Derbenev
2071 Randall Lab. of Physics
University of Michigan
500 East University Avenue
Ann Arbor, MI 48109-1120
USA
Phone: (313) 936-1027
E-mail: derbenev@miphys.physics.lsa.umich.edu
Fax: (313) 936-0794

Vladimir P. Derenchuk
Indiana University Cyclotron Facility
2401 Milo B. Sampson Lane
Bloomington, IN 47408
USA
Phone: (812) 855-9130
E-mail: laddie@iucf.indiana.edu;
laddie@cyclops.iucf.indiana.edu
Fax: (812) 855-6645

Tim Drummer
Department of Physics
Florida State University
Tallahassee, FL 32301
USA
Phone: (904) 644-4189
E-mail: tim@nrb119c.physics.fsu.edu
Fax: (904) 644-9848

Andrea P. Dvoredsky
Kellog Radiation Lab
California Institute of Technology
106-38
Pasadena, CA 91125
USA
Phone: (818) 395-4836
E-mail: andread@krl.caltech.edu
Fax: (818) 564-8708

Said Essabaa
Institut de Physique Nuclear
Universite Paris-Sud
Bat. 106
91406 Orsay Cedex
FRANCE
Phone: 33169157778
E-mail: essabaa@ipncls.in2p3.fr
Fax: 33169157735

Dieter Eversheim
ISKP
Universitat Bonn
Nussallee 14-16
D-53115 Bonn
GERMANY
Phone: 228-73-5299 or 2201
E-mail: evershei@iskp.uni-bonn.de
Fax: 49-228-732-505

Manouchehr Farkhondeh
Bates Linear Accelerator Center
P.O. Box 846, 21 Manning Road
Middleton, MA 01949-2846
USA
Phone: (617) 253-9298
E-mail: manouch@bates.mit.edu
Fax: (617) 253-9599

James Fedchak
Physics Division, Bldg. 203
Argonne National Lab
9700 South Cass Ave.
Argonne, IL 60439-4843
USA
Phone: (630) 252-4037
E-mail: fedchak@anlmep.phy.anl.gov
Fax: (630) 252-3903

Massimiliano Ferro-Luzzi
NIKHEF
P. O. Box 41882
1009 DB Amsterdam
The Netherlands
Phone: 31-20-592-2018
E-mail: ferro@nikhefk.nikhef.nl
Fax: 31-20-592-5155

Ralf Gebel
IKP-COSY
Leo-Brandt-Str.
D-52425 Julich
GERMANY
Phone: 49-2461-61-3097
E-mail: r.gebel@fz-juelich.de
Fax: 49-2461-61-3930

Dennis Geurts
Dept. of Astronomy and Physics
Free University of Amsterdam
de Boelelaan 1081
1081 HV Amsterdam
The Netherlands
Phone: 3120-4447906
E-mail: dennisg@nikhef.nl
Fax: 3120-4447999

Martin Glende
IKP COSY
Leo-Brandt-Str.
D 52425 Julich
Germany
Phone: 49-24-6161-3097
E-mail: m.glende@fz-juelich.de
Fax: 49-24-6161-3930

Joe Grames
MS-12H
TJNAF
12000 Jefferson Avenue
Newport News, VA 23606
USA
Phone: (757) 269-7097
E-mail: grames@cebaf.gov
Fax: (757) 269-7363

Tino Grossmann
Institut fur Kernphysik
Johannes-Gutenberg-Universitat Mainz
Staudinger Weg 7
D-55099 Mainz
Germany
Phone: 49-6131-393693
E-mail: grossmann_t@dipmza.physik.uni-mainz.de
Fax: 49-6131-393428

Peter Hartmann
Institut fur Kernphysik
Universitat Mainz
Johann-Joachim-Becher-Weg 45
D-55128 Mainz
Germany
Phone: 49-6131-39-5808
E-mail: hartmann@kph.uni-mainz.de
Fax: 49-6131-39-2964

Kichiji Hatanaka
Research Center for Nuclear Physics
Osaka University
Mihogaoka 10-1, Ibaraki
Osaka 567
JAPAN
Phone: 81-6-879-8949
E-mail: hatanaka@rcnp.osaka-u.ac.jp
Fax: 81-6-879-8899

Bill Hersman
Department of Physics
University of New Hampshire
Durham NH 03824
USA
Phone: (603) 862-3512
E-mail: hersman@unh.edu
Fax: (603) 862-2998

Ralf Hertenberger
Sektion Physik
University of Munich
Am Coulombwall 1
D-85748 Garching
GERMANY
E-mail: ralf.hertenberger@physik.uni-muenchen.de
Fax: 49-89-3201-4146

Hubert Humblot
Institut Laue Langevin
Avenue des Martyrs, B.P. 156
38042 Grenoble
France
Phone: 33-04-7620-7470
E-mail: humblot@ill.fr
Fax: 33-04-7648-3906

Norbert Koch
Physikalisches Institut
Universitat Erlangen
Erwin-Rommel-Str. 1
91058 Erlangen
Germany
Phone: 49-40-8998-4682
E-mail: koch@dxhrb3.desy.de
Fax: 49-40-8998-4034

Roy J. Holt
Loomis Laboratory of Physics
University of Illinois
1110 West Green Street
Urbana, IL 61801-3080
USA
Phone: (217) 244-6039
E-mail: r-holt@uiuc.edu
Fax: (217) 333-1215

Cathleen E. Jones
Kellogg Radiation Lab.
California Institute of Technology
1201 California Blvd., 106-38
Pasadena, CA 91125
USA
Phone: (818) 395-4584
E-mail: cjones@krl.caltech.edu
Fax: (818) 564-8708

Hauke Kolster
HERMES-Munich
DESY
Notkestr. 85
D-22607 Hamburg
GERMANY
Phone: 49-40-8998-2026
E-mail: kolster@hermes.desy.de
Fax: 49-40-8998-4034

Robert Kowalczyk
Physics Division
Argonne National Laboratory
9700 South Cass Avenue
Argonne, IL 60439
USA
E-mail: bob@triton.phy.anl.gov

H. Rob Kremers
KVI Groningen
Zernikelaan 25
9747 AA Groningen
THE NETHERLANDS
Phone: 31-50-363-2571
E-mail: kremers@kvi.nl
Fax: 31-50-363-4003

Alan D. Krisch
Randall Lab. of Physics
University of Michigan
500 East University Avenue
Ann Arbor, MI 48109-1120
USA
Phone: (313) 936-1027
E-mail: krisch@umich.edu
Fax: (313) 936-0794

Sergio Lemaitre
Department of Physics and Astronomy
University of North Carolina
278 Phillips Hall
Chapel Hill, NC 27599-3255
USA
Phone: (91) 962-2079
E-mail: lemaitre@physics.unc.edu
Fax: (919) 962-0480

Bernd Lorentz
Department of Physics
University of Wisconsin
1150 University Avenue
Madison, WI 53706
USA
E-mail: lorentz@uwnuc0.physics.wisc.edu
Fax: (608) 262-3598

Wolfgang Lorenzon
Physics Department
University of Michigan
2071 Randall Lab, 1120
Ann Arbor, MI 48109-1120
USA
Phone: (313) 647-6825
E-mail: lorenzon@umich.edu
Fax: (313) 936-0794

Bill Lozowski
Indiana University Cyclotron Facility
2401 Milo B. Sampson Lane
Bloomington, IN 47408
USA
Phone: (812) 855-9365

Vladimir Luppov
Department of Physics
University of Michigan
Randall Lab.
Ann Arbor, MI 48109-1120
USA
Phone: (313) 764-5111
E-mail: luppov@miphys.physics.lsa.umich.edu
Fax: (313) 763-9027

Zein-Eddine Meziani
Dept. of Physics, Barton Hall
Temple University
1900 N. 13th Street
Philadelphia, PA 19122-0090
USA
Phone: (215) 204-5971
E-mail: meziani@vm.temple.edu
Fax: (215) 204-2569

Michael A. Miller
Loomis Laboratory of Physics
University of Illinois
1110 West Green Street
Urbana, IL 61801
USA
Phone: (217) 333-6544
E-mail: miller5@uiuc.edu
Fax: (217) 333-1215

Gregory Mulhollan
MS66
Stanford Linear Accelerator Center
P.O. Box 4349
Stanford, CA 94309
USA
Phone: (415) 926-4889
E-mail: mulholla@slac.stanford.edu
Fax: (415) 926-2407

Masanobu Nakamura
Deprtment of Physics
Kyoto University
Kitashirakawa
Kyoto 606
JAPAN
Phone: 81-075-753-3853
E-mail: nakamura@ne.scphys.kyoto-u.ac.jp
Fax: 81-075-753-3887

Shinsuke Nakamura
Physikalisches Institut
Universitat Bonn
Nussallee 12
D-53115 Bonn
GERMANY
Phone: 49-228-73-7764
E-mail: nakamura@elsa4.physik.uni-bonn.de
Fax: 49-228-73-3620

Tsutomu Nakanishi
Department of Physics,
Faculty of Science
Nagoya University
Chikusa-ku
Nagoya 464
JAPAN
Phone: 052-789-2898
E-mail: nakanisi@spin.phys.nagoya-u.ac.jp
Fax: 052-789-2897

Vladimir Nelyubin
Institut fur Kernphysik
St. Petersburg Nuclear Physics Institute
Gatchina
Leningrad District 188350
Russia
Phone: 007-81271-30087
E-mail: nelyubin@rec03.pnpi.spb.ru
Fax: 007-81271-37196

Dmitry Orlov
Inst. of Semiconductor Physics
Lavrent'eva 13
630090 Novosibirsk
Russia
Phone: 7-3832-357508
E-mail: orlov@thermo.isp.nsc.ru
Fax: 7-3832-351771

Brynnen R. Owen
Loomis Laboratory of Physics
University of Illinois
1110 West Green Street
Urbana, IL 61801
USA
Phone: (217) 244-7958
E-mail: owen@coel.cen.uiuc.edu
Fax: (217) 333-1215

Paul V. Pancella
Physics Department
Western Michigan University
Kalamazoo, MI 49008-5151
USA
Phone: (616) 387-4962
E-mail: paul.pancella@wmich.edu
Fax: (616) 387-4939

Igor Passchier
NIKHEF
P.O. Box 41882
1009 DB Amsterdam
The Netherlands
Phone: 31-20-592-2147
E-mail: igorp@nikhefk.nikhef.nl
Fax: 31-20-592-5155

Stefan Pastuszka
Physics Department
MPI fur Kernphysik
PF 103980
D-69029 Heidelberg
Germany
Phone: 49-6221-516504
E-mail: pas@mickey.mpi-hd.mpg.de
Fax: 49-6221-516602

Yuri K. Pilipenko
Laboratory of High Energies
Joint Institute for Nuclear Research
Joliot Curue 6
RU-141980, Dubna Moscow Region
RUSSIA
Phone: 7-096-21-65044
E-mail: pilyuk@lhe02.jinr.dubna.su
Fax: 7-096-21-65891 or
7-096-21-65599

Matt Poelker
MS-16
TJNAF
12000 Jefferson Avenue
Newport News, VA 23606
USA
Phone: (757) 269-7357
E-mail: poelker@cebaf.gov
Fax: (757) 269-5279

Vance R. Pomeroy
Physics Department
University of New Hampshire
DeMeritt Hall
Durham, NC 03824
USA
Phone: (603) 862-2645
E-mail: vance.pomeroy@unh.edu
Fax: (603) 862-2998

Hans R. Poolman
NIKHEF-K
P.O. Box 41882
1009 DB Amsterdam
THE NETHERLANDS
Phone: 31-20-592-2020
E-mail: hansrp@nikhef.nl
Fax: 31-20-592-2155

Scott Price
TJNAF
12000 Jefferson Avenue
Newport News, VA 23606
USA
Phone: (757) 269-7091
E-mail: jsprice@jlab.org
Fax: (757) 269-5279

Frank Rathmann
Nuclear Physics
University of Wisconsin-Madison
1150 University Avenue
Madison, WI 53706
USA
Phone: (608) 263-7089
E-mail: rathmann@uwnuc4.physics.wisc.edu
Fax: (608) 262-3598

Richard S. Raymond
Randall Lab. of Physics
University of Michigan
Ann Arbor, MI 48109
USA
Phone: (313) 764-5113
E-mail: rraymond@umich.edu
Fax: (313) 763-9027

Erwin Reichert
Institut fur Physik
Johannes-Gutenberg Universitat Mainz
Johann-Joachim-Becher-Weg 45
D-55099 Mainz
GERMANY
Phone: 49-6131-392729
E-mail: reichert@dipmza.physik.uni-mainz.de
Fax: 49-6131-39-2991

Heiko Rohdjess
ISKP
Universitat Bonn
Nussallee 14-16
D-53115 Bonn
GERMANY
Phone: 49-228-73-2506
E-mail: rohdjess@iskp.uni-bonn.de
Fax: 49-228-73-2505

Daniela Rohe
Institut fur Kernphysik, EXAKT
Universitat Mainz
Staudingerweg 7
D-55099 Mainz
GERMANY
Phone: 49-6131-39-3693
E-mail: rohe@kph.uni-mainz.de
Fax: 49-6131-39-3428

Mikhail V. Romalis
Physics Department
Princeton University
P.O. Box 708
Princeton, NJ 08544
USA
Phone: (609) 258-4647
E-mail: romalis@pupgg.princeton.edu
Fax: (609) 258-2496

Hans Paetz gen. Schieck
Institut fur Kernphysik
Universitat zu Koln
Zulpicher Strasse 77
D-50937 Koln
GERMANY
Phone: 49-221-470-3620
E-mail: schieck@ikp.uni-koeln.de
Fax: 49-221-470-5168

Pierre A. Schmelzbach
F1, Accelerator Division
Paul Scherrer Institute
CH-5232 Villigen-PSI
SWITZERLAND
Phone: 41 56 310 2111
E-mail: schmelzbach@psi.ch
Fax: 41 56 310 3383

Elaine Schulte
Loomis Laboratory of Physics
University of Illinois
1110 West Green Street
Urbana, IL 61801
USA
Phone: (217) 244-4732
E-mail: schulte@uinpluxa.npl. uiuc.edu
Fax: (217) 333-1215

Peter Schwandt
Indiana University Cyclotron Facility
2401 Milo B. Sampson Lane
Bloomington, IN 47408
USA
Phone: (812) 855-9365
E-mail: schwandt@iucf.indiana.edu
Fax: (812) 855-6645

Hellmut Seyfarth
Institut fur Kernphysik
Forschungszentrum Julich
D-52425 Julich
GERMANY
Phone: 49-2461-616083
E-mail: sey@ikpd01.dnet.kfa-juelich.de
Fax: 49-2461-61-3930

Charles K. Sinclair
TJNAF, MS-12A2
12000 Jefferson Avenue
Newport News, VA 23606
USA
Phone: (757) 269-7679
E-mail: sinclair@cebaf.gov
Fax: (757) 269-5024

James Sowinski
Indiana Univ. Cyclotron Facility
2401 Milo B. Sampson Lane
Bloomington, IN 47408
USA
Phone: (812) 855-9365
E-mail: sowinski@iucf.indiana.edu
Fax: (812) 855-6645

Jerzy Sromicki
ETH Zurich
CH-8093 Zurich
Switzerland
E-mail: sromicki@psi.ch

Erhard Steffens
Physikalisches Institut
Universitat Erlangen-Nurnberg
Erwin-Rommel-Str. 1
D-91058 Erlangen
GERMANY
Phone: 49-9131-857093
E-mail: steffens@physik.uni-erlangen.de
Fax: 49-9131-15249

Jorn Stenger
Room 26-256
Massachusetts Institute of Technology
77 Massachusetts Avenue
Cambridge, MA 02139
USA
Phone: (617) 253-5926
E-mail: stenger@physik.uni-erlangen.de
Fax: (617) 253-4876

James Stewart
DESY/HERMES
Notkestrasse 85
D-22607 Hamburg
GERMANY
Phone: 49-40-8998-3982
E-mail: stewart@dxhrb1.desy.de
Fax: 49-40-8998-4034

Arsen V. Subashiev
Experimental Physics Department
St. Petersburg State Technical University
Politechnicheskaya 29
195251 St. Petersburg
Russia
Phone: (812) 552-7574
E-mail: arsen@subashiev.hop.stu.neva.ru
Fax: (812) 247-20-88

Akira Takagi
Accelerator Lab.
KEK, High Energy Accelerator Research Org.
1-1 Oho, Tsukuba
Ibaraki 305
JAPAN
Phone: 81-298-64-5215
E-mail: takagi@kekvax.kek.jp
Fax: 81-298-64-3182

Masayoshi Tanaka
Kobe Tokiwa College
Ohtani-cho 2-6-2
Nagata, Kobe 653
JAPAN
Phone: 06-8798931
E-mail: tanaka@rcnpax.rcnp.osaka-u.ac.jp
Fax: 06-8798899

Kazuaki Togawa
Department of Physics
Nagoya University
Chikusa-ku
Nagoya 464-01
JAPAN
Phone: 81-52-789-2894
E-mail: togawa@spin.phys.nagoya-u.ac.jp
Fax: 81-52-789-2897

Tadaaki Tamae
Laboratory of Nuclear Science
Tohoku University
Mikamine,Taihaku-ku
Sendai 982
JAPAN
Phone: 81-22-743-3434
E-mail: tamae@thkln1.lns.tohoku.ac.jp
Fax: 81-22-743-3401

Alex Terekhov
Institute of Semiconductor Physics
Budker Institute for Nuclear Physics
Lavrenteva 13
630090 Novosibirsk
RUSSIA
Phone: 7-3832-356874
E-mail: terek@thermo.isp.nsc.ru
Fax: 7-3832-351771

Dmitri K. Toporkov
Budker Institute for Nuclear Physics
Lavrentjeva 11
630090-Novosibirsk 90
RUSSIA
Phone: (630) 252-3626
E-mail: toporkov@anlmep.phy.anl.gov;
tdm@inp.nsk.su
Fax: (630) 252-3903

Evgeni Tsentalovich
Massachusetts Institute of Technology
Bates Linear Accelerator Center
P.O. Box 846, 21 Manning Road
Middleton, MA 01949-2846
USA
Phone: (617) 253-9507
E-mail: tsentalovich@aesir.mit.edu
Fax: (617) 253-9599

Charles W. Tu
Dept. of Electrical & Computer
Engineering
University of California at San Diego
9500 Gilman Drive
La Jolla, CA 92093-0407
USA
Phone: (619) 534-4687
E-mail: ctu@ucsd.edu
Fax: (619) 457-3739

Jo van den Brand
NIKHEF-K
P.O. Box 41882
1009 DB Amsterdam
THE NETHERLANDS
E-mail:
jo@paramount.nikhefk.nikhef.nl

Victor Varentsov
St. Petersburg Institute
 for Informatics and Automation
39, 14th Line
St. Petersburg 199178
Russian
Phone: 7-812-2180625
E-mail: varen@vep.lnpi.spb.su
Fax: 7-81271-35175

Alexandre Vassiliev
Lab. of Cryogenic & Superconductivity
Technique
St. Petersburg Nuclear Physics Institute
188350 Gatchina, Leningrad District
Russia
Phone: 7812-71-46542
E-mail: vassilie@hep486.pnpi.spb.ru
Fax: 7812-71-37196

Wolther von Drachenfels
Physikalisches Institut
Universitat Bonn
Nussallee 12
D-53115 Bonn
GERMANY
Phone: 49-228-73-3219
E-mail:
drachen@elsa4.physik.uni-bonn.de
Fax: 49-228-73-3620

Su-Huai Wei
National Renewable Energy Lab
1617 Cole Blvd., MS-3213
Golden, CO 80401
USA
Phone: (303) 384-6666
E-mail: shw@nrel.gov
Fax: (303) 384-6531

Katsuya Yonehara
Department of Physics
Konan University
Phone: 81-6-879-8931
E-mail: yonehara@rcnp.osaka-u.ac.jp
Fax: 81-6-879-8899

Jonathan Zerger
1049 Randall Lab. of Physics
University of Michigan
500 East University Avenue
Ann Arbor, MI 48109
USA
E-mail: zerger@umich.edu
Fax: (313) 763-9694

Tom Wise
Department of Physics
University of Wisconsin
1150 University Avenue
Madison, WI 53706
USA
E-mail: wise@uwnuc0.physics.wisc.edu

Anatoli Zelenski
TRIUMF
4004 Wesbrook Mall
Vancouver, B.C.V6T 2A3
CANADA
Phone: (604) 222-7302
E-mail: zelenski@triumf.ca
Fax: (604) 222-1074

Zi-lu Zhou
Laboratory for Nuclear Science
Massachusetts Institute of Technology
77 Massachusetts Avenue, 26-452
Cambridge, MA 02139
USA
Phone: (617) 235-5875
E-mail: zzhou@mitlns.mit.edu
Fax: (617) 258-5440

Leonid Zolin
Joint Institute for Nuclear Research
Joliot Curue 6
RU-141980 Dubna, Moscow Region
Russia
Phone: 7-096-21-62031
E-mail: zolin@moonhe.jinr.dubna.su
Fax: 7-096-21-65891; 7-096-21-65599

APPENDIX C

AUTHOR INDEX

A
A-3 Collaboration at MAMI, 36
A1-Kollaboration, 41
Abe, K., 434
Agranoff, B.W., 200
Alarcon, R., 26, 79, 316, 389
Alley, R., 250
Altmeier, M., 419, 503
Andresen, H.G., 36, 296
ANKE-ABS group, 477
Annand, J., 36
APOLLON Group, 499
Arianer, J., 440
Arnold, J.D., 119
Assamagan, K.A., 446
Aulenbacher, K., 36, 229, 296, 440
Avramopoulos, H., 311

B
B2-Collab. at MAMI, 229
Baba, T., 300, 495
Bachert, P., 208
Bailey, K., 129, 148, 437
Barkhuff, David, 240
Bartsch, P., 41
Bauer, T., 26, 316
Baumann, D., 41
Beck, D.H., 451
Becker, J., 36, 41, 208
Belov, A.S., 362, 411, 422
Bermuth, J., 41, 208, 296
Beuchel, K., 36
Bisplinghoff, J., 501
Blinov, B.B., 119
Blume-Werry, J., 36
Bock, M., 208
Bodek, K., 326
Boersma, D., 26, 316
Bogorad, P.L., 46, 213

Böhm, R., 41
Botto, T., 26, 79, 316, 389
Bouwhuis, M., 79, 316, 389
Brack, J., 148, 437
Braun, B., 156
Brookeman, James R., 213
Brüggemann, R., 471
Bulten, H.J., 26, 79, 316, 389, 459, 481
Buttazoni, S., 41
Bychkov, M.A., 119

C
Cadman, R.V., 148, 437
Caprano, T., 41
Cardman, L.S., 446, 451
Carrier, R.H., 461
Cates, G.D., 3, 46, 213
Chen, J.P., 321
Chupp, T. E., 46, 200, 431
Ciarrocca, M., 311
Clawiter, N., 41
Clegg, Thomas B., 336
Clendenin, J., 250, 443
Cohen, S., 440
Comfort, J., 79
Conti, D., 326
Conzett, H.E., 501
Coulter, K.P., 46, 200, 431
Cummings, W.J., 129, 148, 437, 457

D
de Bever, L.J., 41
de Jager, C.W., 26, 79, 260, 316, 389, 483
de Lange, D.J., 79, 389
de Lange, Eduard E., 213
de Vries H., 26, 79, 316, 389
Daehnick, W., 89

Dale, D.S., 321
Daniel, Thomas M., 213
Davydenko, V.I., 372
De Schepper, D., 16
Deninger, A., 41, 208
Derbenev, Ya.S., 191
Derber, S., 41
Derenchuk, Vladimir P., 422
Ding, M., 41
Distler, M., 41
Dodson, George, 240
Doets, M., 79
Dolfini, S., 79
Dombo, Th., 36
Doskow, J., 89
Doyle, B., 321
Drentje, A.G., 507
Drescher, P., 36, 296
Driehuys, Bastiaan, 213
Durek, D., 497
Dutto, G., 372
Dvoredsky, A.P., 53
Dzemdzic, M., 89

E
Ebbes, A., 41
Ebert, M., 36, 41, 208
EDDA Collaboration, 99
Eisermann, Y., 469
Ent, R., 26, 79, 316, 389
Ernst, J., 501
Ershov, V.P., 411
Esin, S.K., 362
Essabaa, S., 440
Euteneuer, H., 229, 296
Eversheim, P.D., 419, 501, 503
Ewald, I., 41
Eyl, D., 36

F
Farkhondeh, M., 240
Fedchak, J., 129, 148, 437
Felden, O., 419, 503

Ferro-Luzzi, M., 26, 79, 316, 389, 459, 481
Fimushkin, V.V., 119, 411
Fischer, H., 36, 296
Flammang, R., 89
Fox, B., 148, 437
Frascaria, R., 440
Fraser, D., 311
Frey, A., 36
Friedrich, J. , 41
Friedrich, J.M., 41
Frisch, J., 250
Frommberger, F., 497
Furuta, F., 300, 495

G
Gai, G.I., 411
Gao, H., 129, 148, 437
Garwin, E., 443
Gasparian, A., 321
Gebel, R., 419, 503
Geiges, R., 41
Gerchikov, L.G., 491
German, E.P., 489
Geurts, D., 26, 79, 316, 459
Glamazdin, A., 321
Glende, M., 419, 503
Gorbenko, V., 321
Gorringe, T., 321
Grabmayr, P., 36
Grames, J., 446, 451
Graw, G., 469
Grimm, K., 296
Grosshauser, C., 139, 148, 437
Großmann, T., 36, 41, 208

H
Haeberli, W., 89, 326
Hall, S., 36
Hamian, A.A., 372
Hammel, Th., 296
Hansknecht, J., 270, 446
Happer, William, 213
Hardie, J.H., 89

Hartmann, P., 36, 229, 296
Harvey, M., 26, 316
Hatanaka, K., 352
Hatziefremidis, A., 311
Hauger, M., 41
Hehl, T., 36
Heil, W., 36, 41, 208, 430
Heimberg, P., 26, 79, 316, 389
HERA polarimeter group, 181
Herberg, C., 36
HERMES Collaboration, 16, 53, 69, 156, 162, 181
Hersman, F.W., 461, 463
Hertenberger, R., 469
Hiemer, A., 419
Highinbotham, D., 26, 79, 316, 389
Hinterberger, F., 501
Hirose, M., 509
Hoffmann, J., 36, 229, 296
Hoffmann, M., 497
Hofmann, A., 469
Hofmann, D., 41, 208, 430
Hofmann, H., 296
Holt, R.J., 148, 437, 465
Honegger, A., 41
Horinaka, H., 300
Houbavlis, T., 311
Hughes, E.W., 46
Humblot, H., 430
Husmann, D., 497

I

Ida, T., 300, 495
Igarashi, Z., 347
Ignatiev, Alexei A., 381
Ihloff, Ernie, 240
Ikegami, K., 347, 509
Imai, K., 509
Ireland, D., 36
Isaeva, L.G., 109

J

Jahn, R., 501
Jaroshevich, A.S., 485

Jennewein, P., 41, 229
Johnson, J.R., 46
Jones, C.E., 129, 148, 437
Jourdan, J., 41

K

Kabuß, E.-M., 296
Kahrau, M., 41
Kaiser, K.-H., 229, 296
Kauczor, H.U., 208
Kellie, J., 36
Kiel, B., 497
Kilian, Wolfgang, 139
Kinney, E., 148, 437
Kinsho, M., 347, 509
Kirby, R., 443
Kirillov, M.A., 485
Kirwin, W., 437
Kistryn, S., 326
Klein, A., 41
Klein, F., 36, 497
Klenov, V., 362, 372
Knight-Scott, Jack, 213
Knopp, M.W., 208
Köbis, S., 296
Koch, Norbert, 381, 427
Kolster, H., 162
Konno, Osamu, 434
Konstantinov, S.H., 260
Koptev, V., 479
Korchagin, V.Ya., 260, 483
Korn, M., 41
Korsch, W., 321
Kotov, S., 479
Kotseroglou, T., 250
Kowalczyk, R.S., 129, 148, 437, 457
Kratzmann, D., 487, 493
Kreidel, H.J., 229, 296
Kreitner, K.F., 208
Kremers, H.R., 507
Kretschmer, W., 501
Kroes, F.B., 260
Krygier, K.W., 41
Kubon, G., 41

Kulikov, M.V., 411
Kumar, K.S., 46
Kurihara, Y., 300, 495
Kutuzova, L.V., 411
Kuwamoto, S., 509

L

Lang, J., 79, 326, 389, 481
Lauer, L., 41, 208
Lazarenko, B.A., 109
Leberig, M., 229
Leduc, M., 36, 208
Lemaitre, S., 471
Leuschner, M.B., 463
Levy, C.D.P., 372
Liburg, M. Yu., 411
Liesenfeld, A., 41
Lin, A.M.T., 119
Lopes-Ginja, A., 296
Lorentz, B., 89
Lorenzon, W., 181
Lu, Z.-T., 148, 437
Luppov, V.G., 119

M

Maas, F.E., 296
Mack, D.J., 446
Magnes, J., 129
Maier, Therese, 213
Markou, V., 311
Maruyama, T., 443
Matsumoto, H., 495
Matsuyama, T., 300
Menze, D., 497
Merkel, H., 41
Merle, K., 41
Merle, P., 41
Metz, A., 469
Meyer, H.O., 89
Meyerhoff, M., 36
Michel, T., 497
Militsyn, B.L., 260, 483
Miller, C. A., 499
Miller, G., 36

Miller, M.A., 148, 437
Mishnev, S.I., 109
Mitchell, J.H., 451
Miyase, Haruhisa, 434
Mizuta, M., 300, 495
Mochalov, V.V., 119
Möller, H., 36
Mori, Y., 347, 509
Morozov, I.I., 372
Mugler III, John P., 213
Mühlbauer, M., 41
Mulhollan, G., 250, 443
Müller, U., 41
Murakami, T., 509

N

Nachtigall, Ch., 36, 296
Nagengast, W., 139, 148, 437
Nakagawa, Itaru, 434
Nakamura, M., 509
Nakamura, S., 497
Nakanishi, T., 300, 495, 497
Nanda, S., 321
Natter, A., 36
Naumann, J., 497
Navert, S., 326
Naviliat, O., 326
Nelyubin, V., 471, 473
Netchaeva, L.P., 362
Neuhausen, R., 41
Nicolay, K., 459
Nikolenko, D.M., 109
Nilgens, H., 208
Nishikawa, Itaru, 434
Norum, B., 26, 79, 316

O

Okumi, S., 300, 495
Omori, T., 300, 495
Orlov, D.A., 485, 493
Osipov, A.N., 109
Osipov, V.N., 483
Oskotskij, B.D., 491

Ostrick, M., 36
Otten, E.W., 36, 41, 208
Owen, B., 148, 437, 465
Owens, R.O., 36

P
Paetz gen. Schieck, H., 471, 501
Pancella, P.V., 89, 172
Papadakis, N., 260, 316, 483
Papakyriakopoulos, T., 311
Passchier, I., 26, 79, 316, 389
Pastuszka, S., 487, 493
Paulish, A.G., 485, 493
Petitjean, Th., 41
Pilipenko, Yu. K., 411
PINTEX collaboration, 172
Piot, P., 446
Pipes, M., 129
Plützer, S., 36, 296
Poelker, B.M., 270, 446, 451
Pollock, R.E., 89
Pomatsalyuk, R., 321
Pomeroy, V.R., 461, 463
Poolman, H.R., 26, 79, 316, 389, 459
Popov, S.G., 109, 260, 483
Pospischil, Th., 41
Prepost, R., 443
Price, J.S., 446, 451

Q
Quin, P., 89

R
Rachek, I.A., 109
Ranzenberger, Bernd, 139
Rathmann, F., 89
Raymond, R.S., 475
Reichelt, T., 497
Reichert, E., 36, 229, 296, 326
Rinckel, T., 89
Rith, K., 139, 148, 437
Roberts, T., 208
Rohdjess, H., 99
Rohe, D., 36, 41

Romalis, M.V., 46
Rosen, Matthew S., 200
Rosner, G., 41
Rukoyatkin, P.A., 511

S
Saha, A., 321
Schad, L.R., 208
Schäfer, M., 36
Scheibler, H.E., 485, 493
Schemies, M., 229, 296
Schiemenz, P., 469
Schilling, E., 296
Schmidt, F., 139, 148, 437
Schmieden, H., 36, 41
Schmor, P.W., 372
Schuler, J., 229, 296
Schulte, E., 148, 437, 465
Schultz, D., 250
Schwalm, D., 487, 493
Schwartz, B., 89
Semenov, P.A., 119
Serdobintsev, G.V., 260
Seyfarth, H., 473, 477, 479
Shatunov, Yu.M., 260, 483
Shestakov, Yu.V., 109
Sidorov, A.A., 109
Shevelev, S.V., 260
Sick, I., 41
Sinclair, C.K., 218, 446, 451
Six, E., 26, 79, 316
Sluijk, T.G.B.W., 260
Smith, T.B., 46, 431
Sowinski, J., 148, 437
Sperisen, F., 89, 148, 437
Sprengard, R., 36
Sromicki, J., 326
Steffens, Erhard, 381, 399
Steffens, K.-H., 36, 296
Steier, C., 497
Steigerwald, M., 229, 296, 326
Steijger, J.J.M., 26, 79, 316, 389
Stenger, J., 139, 148, 437
Stephan, E., 326

Stewart, J., 69
Stibunov, V.N., 109
Subashiev, A.V., 489, 491
Sugawara, Masumi, 434
Surkau, R., 36, 41, 208
Suzuki, C., 300, 495
Swanson, Scott D., 200
Szczerba, D., 26, 79, 316
Szeremeta, F., 440

T
Takagi, A., 347, 509
Takahashi, C., 300, 495
Takahashi, S., 509
Takahisa, K., 352
Takeuchi, Y., 495
Tamae, Tadaaki, 434
Tamura, H., 352
Tanaka, Eiji, 434
Tanaka, M., 58
Tang, H., 250, 443
Tasset, F., 430
Tawada, M., 300
Tedeschi, D., 89
Terekhov, A.S., 260, 485, 487, 493
Tereshchenko, O.E., 493
Thelen, M., 208
Thompson, A.K., 46
Thorsland, E., 148, 437, 465
Togawa, K., 300, 495
Tokarev, Yu.F., 260, 483
Toporkov, D.K., 109, 437, 467
Toyama, T., 497
Trautner, H., 296
Trieb, S., 469
Truwit, Jonathon D., 213
Tschalaer, Christopher, 240
Tsentalovich, Evgeni, 240
Tsubota, Hiroaki, 434
Tu, Charles W., 276
Turbabin, A.V., 362
Turner, J., 250

U
Ugorowski, P.B., 89
Unal, O., 79, 389

V
Valevich, A.I., 411
van Bakel, N., 79
van Buuren, L., 26, 79, 316
van den Brand, J.F.J., 26, 79, 316, 389, 459, 481
van den Putte, M., 26, 260, 316, 483
van Leeuwen, E.P., 260
van Oers, W.T.H., 372
Varentsov, Victor L., 381
Vasil'ev, G.A., 362
Vassiliev, A., 479
Vesnovsky, D.K., 109
Vodinas, N., 316
Voigt, S., 497
von Drachenfels, W., 497
von Harrach, D., 229, 296
von Przewoski, B., 89

W
Wada, K., 300, 495
Wagner, A., 41
Walcher, Th., 36, 41
Walker, M., 419
Warren, G., 41
Watson, R., 36
Wei, Su-Huai, 284
Weis, M., 41
Welsh, R., 46, 200, 431
Westermann, M., 497
Wight, G.W., 372
Williamson, S., 148
Wilms, E., 36
Wise, T., 89
Woehrle, H., 41
Wojtsekhowski, Bogdan, 467
Wolanski, M., 89
Wolf, A., 487, 493
Wolf, S., 41
Wurzinger, R., 440

Y
Yamaguchi, Nobuo, 434
Yamazaki, Hirohito, 434
Yang, Bin, 240
Yeremian, A.D., 250
Yokokawa, Tamio, 434
Yoshimura, M., 509
Yoshioka, M., 300, 495
Yosoi, M., 509

Z
Zalto, M., 229
Zeier, M., 41
Zejma, J., 326
Zelenski, A.N., 372
Zeps, V., 321
Zerger, John N., 431
Zerhouni, O., 440
Zevakov, S.A., 109
Zhao, J., 41
Zhou, Z.-L., 79, 389, 481
Zihlmann, B., 451
Zolin, Z., 511
Zwart, Townsend, 240